NUCLEAR REACTIONS FOR ASTROPHYSICS

Principles, Calculation and Applications of Low-Energy Reactions

Nuclear processes in stars produce the chemical elements for planets and life. This book shows how similar processes may be reproduced in laboratories using exotic beams, and how these results can be analyzed.

Beginning with one-channel scattering theory, the book builds up to complex reactions within a multi-channel framework. It describes both direct and compound reactions, making the connections to astrophysics. A variety of theories are covered in detail, including the adiabatic model and the CDCC method for breakup, eikonal models for stripping, R-matrix techniques, and the Hauser-Feshbach theory for compound nucleus reactions.

Practical applications are prominent in this book, confronting theory predictions with data throughout. The associated reaction program FRESCO is described, allowing readers to apply the methods to practical cases. Each chapter ends with exercises so that readers can test their understanding of the materials covered. Supplementary materials at www.cambridge.org/9780521856355 include the FRESCO program, input and output files for the examples given in the book, and hints and graphs related to the exercises.

IAN J. THOMPSON is a Nuclear Physicist in the Nuclear Theory and Modeling Group at the Lawrence Livermore National Laboratory, USA, having been Professor of Physics at the University of Surrey, UK, until 2006. His research deals with coupled-channels and few-body models for nuclear structure and reactions, especially concerning halo nuclei. He is a Fellow of the Institute of Physics.

FILOMENA M. NUNES is an Associate Professor in the Department of Physics and Astronomy, and at the National Superconducting Cyclotron Laboratory, of Michigan State University. Her research has focused mainly on direct nuclear reactions as a tool for nuclear astrophysics, with particular emphasis in breakup and transfer.

NUCLEAR REACTIONS
FOR ASTROPHYSICS

Principles, Calculation and Applications
of Low-Energy Reactions

IAN J. THOMPSON

*Lawrence Livermore National Laboratory, USA and
University of Surrey, UK*

and

FILOMENA M. NUNES

*National Superconducting Cyclotron Laboratory,
Michigan State University, USA*

CAMBRIDGE
UNIVERSITY PRESS

CAMBRIDGE UNIVERSITY PRESS
Cambridge, New York, Melbourne, Madrid, Cape Town, Singapore, São Paulo, Delhi

Cambridge University Press
The Edinburgh Building, Cambridge CB2 8RU, UK

Published in the United States of America by Cambridge University Press, New York

www.cambridge.org
Information on this title: www.cambridge.org/9780521856355

First published 2009

Printed in the United Kingdom at the University Press, Cambridge

A catalog record for this publication is available from the British Library

ISBN 978-0-521-85635-5 hardback

Contents

Preface

It was a rainy day in December and we were sitting in an office at the Nuclear Physics Center in Lisbon deeply involved in a heated discussion about the opening of this book. Should we follow the standard practice, or should we paint the big picture? True to our main motivation, after hours we finally agreed.

The human fascination for a clear starry sky is timeless. It has been around since the early days of mankind and includes the most diverse cultures. Only in the last century, nuclear physics has started to make a very important contribution to our understanding of these phenomena in the sky. And until the present day, many big questions connected to nuclear reactions remain to be answered. One of the prime examples listed amongst the eleven most important physics questions for our century is this: 'How and where are the heavy elements produced?'.

Why another book? For decades we have come across colleagues, including experimentalists, who would like to learn more about reactions. Some have become fluent in running reaction codes, but cannot find a book at the right level to learn the theory associated with the calculations they are performing. Probably the largest push toward embarking on the adventure of writing this book came after several years of teaching reaction theory to graduate students. The reference nuclear reaction books have been around for decades, and even though there have been some more recent efforts, nowhere could we find the appropriate level, detail, connection to the present experimental scene, the guiding motivation of astrophysics, and the content consistent with that motivation. So, five years ago, we convinced ourselves this was something worth doing.

Who is the book for? This book is primarily directed to physics graduate students with an interest in nuclear physics and astrophysics. It should serve as a practical guide to experimentalists that need a better understanding of the reaction theories available for the various processes. We hope it can also be a useful reference book for the experts in the matter.

What is different about this book? It contains the standard direct-reaction theory starting from the two-body scattering problem but, rather than expanding toward theories that have not been implemented, it focuses on those that are in use or are being developed. We have tried to present all derivations so that it is easier for the student to follow. We have also tried to make clear the limits of applicability of specific models, and to show examples that can be directly compared with data.

How is the book organized? The first two chapters were written at an introductory level, where the stage of nuclear astrophysics is set and the basic definitions are introduced. Next there are eighty pages of solid scattering theory, which is by far the biggest hurdle a student will have to overcome. This is the central theory component, together with the next two chapters on coupling potentials and structure models. We have provided a chapter on the most common approximations used in this field. More advanced chapters then cover specific types of reactions. And eventually we bring the reader back to astrophysics, introducing the reaction rates into reaction networks in stars and explosive environments.

Throughout the book, as the various reaction mechanisms are discussed, we provide specific examples of relevance to astrophysics and connect back to the astrophysical scenarios set in our first chapter.

In addition to the astrophysics motivation, we have kept in mind a strong connection to experiment. Here, calculations are important, so there is a chapter dedicated to numerical methods. Data is important, so there is a chapter on experimental details. And the comparison between theory and experiment is important, thus the chapter on fitting data.

Another essential component of this project is the assisted hands-on experience. The book comes with a reaction code (Fresco), and for many examples addressed in the book we provide the inputs to the reaction code so that the readers can perform the calculation by themselves. An appendix for 'Getting started with Fresco' is also provided.

What is left out? Although we expanded on the number of pages significantly, it is clear that this book does not cover everything that could be contained in such a title as 'Nuclear Reactions for Astrophysics.' From the start, our decision was to focus on direct reactions, and leave out the whole area on central collisions and the specific field of heavy-ion fusion. For each type of reaction included, we prefer to present in detail a small number of models that are implemented and in common use. In this sense the book is not extensive, and should not be used as a review of the field.

What background is needed? A background in quantum mechanics and angular momentum theory is required, although no previous knowledge in scattering theory or nuclear physics is necessary. We have tried to make this a self-contained book and, in particular, scattering theory is developed from scratch.

Where to stop? Writing a book can be a never-ending task. It certainly took us longer than we had originally intended, nor is it in the perfect shape we had first envisaged, specially at the graphical level. However, as with many things in life, one has to know when to stop, and our sense is that in its present form this book can already be very helpful to students and researchers. We hope you can learn with it the techniques and the many interesting aspects of studying nuclear reactions for astrophysics. We will surely come back to our heated discussions on how best to present the material. Be it seen from the Café a Brasileira in Lisbon, the Horticulture Gardens in East Lansing or the Golden Gate Bridge in San Francisco, the sky will present itself with the same fascination as always.

Sources of quotations

Chapter 1 London: *Guardian* (23 August 2001); Chapter 2 *Pierre Curie with Autobiographical Notes*, translated by Charlotte and Vernon Kellogg, New York: Macmillan (1923), p. 167; Chapter 3 Statement of 1963, as quoted in *Schrödinger: Life and Thought* by Walter J. Moore, Cambridge: Cambridge University Press (1992), p. 1; Chapter 4 *A Dictionary of Scientific Quotations* by Alan Lindsay Mackay, Bristol: Institute of Physics Publishing (1991), p. 35; Chapter 5 'How Nobel Prizewinners Get That Way' (December 1969) by Mitchell Wilson, Washington: *The Atlantic*; Chapter 6 *Nature* **403**, 345 (27 January 2000), said when shown the results of a large quantum mechanics calculation; Chapter 7 private communication, Michigan, August 2008; Chapter 8 *Brighter Than a Thousand Suns: A Personal History of the Atomic Scientists*, by Robert Jungk, translated by James Cleugh, New York: Harcourt Brace (1958), p. 22; Chapter 9 letter to her brother; Chapter 10 *Communications in Pure and Applied Mathematics* **13** (1959) 1; Chapter 11 *Lise Meitner: A Life in Physics*, by Ruth Lewin Syme, Berkeley and Los Angeles: University of California Press (1997), p. 375; Chapter 12 'A life in physics': Evening Lecture at the International Center for Theoretical Physics, Trieste, Italy, supplement of the *IAEA Bulletin* (1968), 24; Chapter 13 *Dictionary of Scientific Quotations* by Alan Lindsay Mackay, Bristol: Institute of Physics Publishing (1991); Chapter 14 *Nuclear Principles in Engineering* by Tatjana Jevremovic, New York: Springer (2005), p. 397; Chapter 15 'Physics and Philosophy: The Revolution in Modern Science,' Lectures delivered at University of St. Andrews, Scotland, Winter 1955–56.

Acknowledgements

Long is the list of people that, in one form or another, made this book possible.

We start with acknowledging all those who provided figures for the book: Thomas Baumann, Barry Davids, Erich Ormand, Marc Hausmann, Neil Summers, Jon Whiting. We also thank Ruth Syme for help with the quotations.

Preliminary versions of the book were distributed to a few experts in the summer of 2007. The comments we got back were important to correct and improve the presentation. We thank Goran Arbanas, Edward Brown, Raquel Crespo, Jutta Escher, Frank Dietrich, Christian Forssén, Alexandra Gade, David Howell, Ron Johnson, Antonio Moro, Petr Navrátil, Jorge Pereira, Sofia Quaglioni, Hendrik Schatz, Andreas Schiller, Andrew Steiner, Paul Stevenson, Michael Thoennessen, Jeff Tostevin, Remco Zegers, and Michael Zhukov. We will forever be in debt to Ivan Brida, who went through half of the chapters with a magnifying glass and whose comments helped very much in bringing the language to the right level.

This book was originally to be written by three authors. Although in the end, Ana Eiró could not be involved in the actual writing, we would like to thank her for all her enthusiasm, the many discussions, and shaping the content to be included.

1

Nuclei in the Cosmos

There is a coherent plan in the universe, though I don't know what it's a plan for.

Fred Hoyle

In order to understand about the composition of stars and how they produce energy, we need to know about nuclei, and about the reactions which they undergo. This chapter provides an introduction to the description of nuclei, and surveys the range of scenarios in which important reactions occur. We begin with the Big Bang, then discuss energy production cycles in stars, and finish with an outline of some of the processes by which we think that heavy elements are produced in supernovae and other stellar enviroments. The more detailed discussion of nuclear physics begins in Chapter 2, to which the more advanced student is directed.

1.1 Nuclei

1.1.1 Properties of nuclei

Each isotope (A, Z), characterized by *mass number A* and charge Z, has in its ground state a rest mass $m_{A,Z}$. This total mass is less than the sum of the masses of the constituent protons and neutrons due to the binding energy of the system. Energy is released when the bound state is formed. The binding energy may be calculated by

$$B(A, Z) = (Zm_p + Nm_n - m_{A,Z})c^2, \qquad (1.1.1)$$

and is the energy required to break up the nucleus into its A constituent nucleons. The number of neutrons is $N = A - Z$. A unit atomic mass (1 u) has rest energy $mc^2 = 931.494$ MeV.

The *binding energy per nucleon B/A* dictates whether energy must be supplied or will be released in the fusion of two nuclei to form their composite. The values of $B(A, Z)/A$ are shown in Fig. 1.1 for all the long-lived isotopes. The larger the energy one needs to supply to release a nucleon, the more stable is the nucleus. The most

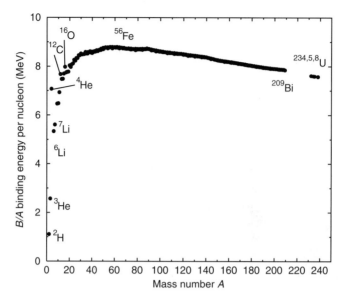

Fig. 1.1. Binding energies per nucleon, $B(A, Z)/A$, for all naturally occurring long-lived isotopes of A nucleons.

stable isotope is near ^{56}Fe, as seen from Fig. 1.1. If two nuclei A_1 and A_2 fuse to form $A = A_1 + A_2$, then the reaction is typically exothermic and energy is released if $A \lesssim 56$. If $A \gtrsim 56$ then fusion reactions are typically endothermic – energy is required – so we might expect the opposite process, *fission*, to be more likely. Fission occurs spontaneously for many nuclei $Z \gtrsim 90$, called the *actinides*.

The most stable nuclear isotopes for $Z \lesssim 20$ have $N \approx Z$, whereas heavier nuclei tend to have more neutrons, to compensate for the increased Coulomb repulsion. If we make a plot with N as the horizontal axis and Z as the vertical axis, we have the Segré chart of Fig. 1.2. Each row is a distinct chemical element, and the stable isotopes are the dark squares lying roughly along the diagonal. The naturally-occurring nuclei, with the longest lifetimes, are said to occupy the *valley of stability*. Neutron-rich nuclei are shown below, to the right of the valley of stability, out to the *neutron dripline*, the point beyond which one cannot form bound states, no matter how many neutrons are added to the system. There is a large gulf between observed isotopes and the predicted neutron dripline, especially for heavy elements.

Conversely, proton-rich nuclei, although they are not so numerous, can be seen above the central valley out to the dripline where proton emission (proton radioactivity) occurs. Most proton-rich nuclei for $A < 200$ have been observed. Nuclei between the driplines have ground states that are stable to nucleon emission, but may still slowly β-decay (see Section 2.2 for timescales) by the weak interaction (see for example the reactions 1.2.4), or radioactively decay (also slowly) by fission or α-particle emission.

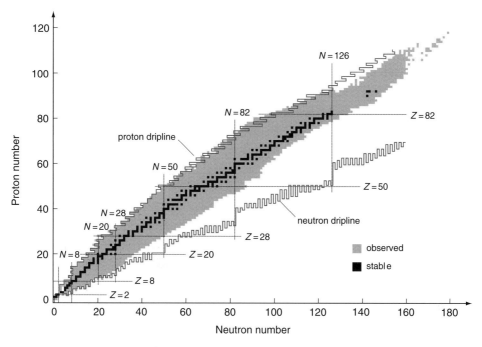

Fig. 1.2. Chart of stable and radioactive isotopes. Vertical and horizontal lines represent magic numbers. Figure courtesy of Marc Hausmann.

1.1.2 Nuclear reactions

If a nuclear reaction is performed in a laboratory, let the *projectile* be called A, the *target* called B, and the *residual nuclei* be C and D. The combination of A and B is called the *entrance channel*, and that of C and D is called an *exit channel* (more than one final channel may be possible). Then the reaction is labeled B(A,C)D, which is the common way of writing

$$A + B \rightarrow C + D. \qquad (1.1.2)$$

Given all the isotopic mass values, we may calculate the energy required or released. This energy, called the *Q*-value for the reaction, is

$$Q = (m_A + m_B - m_C - m_D)c^2. \qquad (1.1.3)$$

Exothermic reactions have $Q > 0$, whereas a $Q < 0$ reaction is endothermic.

1.1.3 Forces in nuclei

There are four forces in nature: the *strong, electromagnetic, weak* and *gravitational* forces. The strong or *nuclear* forces are dominant in binding nuclei, but the other forces still have important roles to play in nuclear astrophysics. The *electromagnetic* force is responsible for the Coulomb repulsion between protons in nuclei, and the decrease in binding for heavy nuclei seen in Fig. 1.1. The *weak* interaction plays a role whenever reactions involve neutrinos; we will see some examples of this later in this chapter (Eqs. (1.2.1) and (1.2.4)). The *gravitational* attraction is not significant inside nuclei, but is responsible for creating galaxies and stars in the first place, and then compressing them to the stage where nuclear reactions begin.

1.1.4 The Coulomb barrier

In order that a nuclear reaction takes place, the nuclei involved have to be close to each other, but this is hindered by the Coulomb repulsion between the protons, which acts at longer distances compared with the nuclear force of short range. The overall potential energy between two charged nuclei separated by a distance R therefore follows the pattern shown in Fig. 1.3. There is a repulsive *Coulomb barrier* of height V_B, and scattering at energies $E < V_B$ still exists because of quantum tunneling through the barrier.

The exponential reduction of reaction rates for charged particles reacting at low relative energies will be extremely important in all astrophysical scenarios,

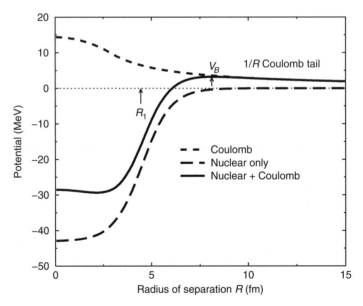

Fig. 1.3. The nuclear and Coulomb potential energies between a proton and ^{40}Ca as a function of the distance R between their centers, where R_1 is the radius of ^{40}Ca. The combined potential (solid line) has a maximum height of V_B, forming the Coulomb barrier.

and will very often be the limiting factor for nuclear reactions. We will see (Section 2.4) that reaction rates are defined by the quantity σ, called the *cross section*. Because cross sections $\sigma(E)$ drop rapidly with decreasing center-of-mass energy E, due to the Coulomb repulsion, we factorize out a simple energy dependence according to

$$\sigma(E) = \frac{1}{E} e^{-2\pi\eta} S(E) \qquad (1.1.4)$$

to define an *astrophysical S-factor* $S(E)$ which should vary less strongly with energy. The $1/E$ geometrical factor is associated with the wavelength of the incoming particle, and the exponential factor represents the penetrability through the Coulomb barrier. It depends on η, the *Sommerfeld parameter*, defined as $\eta = Z_1 Z_2 e^2/(\hbar v)$ (Eq. (3.1.71)) where $Z_1 Z_2 e^2$ is the product of charges and v the relative incident velocity. In Fig. 1.4 we show, in the upper panel, the cross section for the α capture on ^{3}He to synthesize ^{7}Be. The reaction cross section falls off rapidly as the energy decreases, whereas the S-factor, shown in the lower panel, is nearly constant.

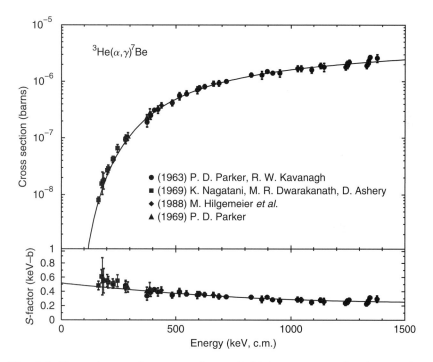

Fig. 1.4. Dependence of cross section and $S(E)$ on energy, for the reaction ^{3}He$(\alpha, \gamma)^7$Be. The solid curve is a calculation to be discussed in Appendix B.

1.2 Primordial nucleosynthesis

Having seen how nuclei and their reactions can be characterized, we now look at a range of nuclear reactions in astrophysics, starting at the beginning. A schematic illustration of the evolution of the Universe immediately after the Big Bang is presented in Fig. 1.5 and briefly described in this section.

Following the Big Bang, the Universe expanded and cooled with a temperature of $T_9 \approx 15/\sqrt{t}$ for time t in seconds and temperature T_9 in units of $GK = 10^9$ K. According to thermodynamics, this temperature corresponds to an energy of $E = k_B T$, where the Boltzmann constant is $k_B = 1.38 \times 10^{-23}$ J K^{-1} = 0.0861 MeV GK^{-1}. This means that the material in the expanding Universe had an average thermal energy of $E \approx 1.3/\sqrt{t}$ MeV.

For very early times, $t < 1$ s, the thermal energy E was greater than 1 MeV. In particular, E was greater than $(m_n - m_p)c^2 = 1.24$ MeV, the difference in the rest energies of the neutron and proton. At these times, therefore, there was enough radiative energy available to easily convert neutrons to protons, and back again, in

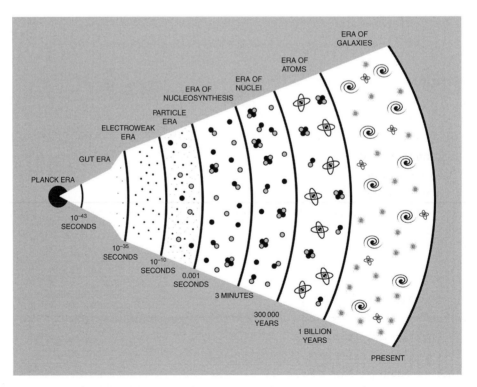

Fig. 1.5. History of the Universe, from the Big Bang to present times. Figure courtesy of Jon Whiting.

a statistical equilibrium by processes such as

$$n + e^+ \rightarrow p + \bar{\nu}_e$$
$$p + e^- \rightarrow n + \nu_e$$
$$e^- + e^+ \leftrightarrow \gamma + \gamma \tag{1.2.1}$$

where γ are photons, and ν_e and $\bar{\nu}_e$ are electron neutrinos and anti-neutrinos. The electrons e^- and positrons e^+ are commonly called *beta* (β) particles. The third reaction of (1.2.1) is the annihilation/production of electron-positron pairs in the high-temperature radiation environment.

Only after two seconds did the Universe cool down enough to enable the neutrons and protons to retain their identities ($E \lesssim 1\,\text{MeV}$); this was the beginning of the astrophysics of nuclei. At this point there begun a period of about 250 s in which a *primordial nucleosynthesis* took place, and neutrons and protons combined to form hydrogen and helium isotopes, and perhaps a few lithium nuclei. At the beginning of this period there were only neutrons and protons, with a relative number density determined by their mass difference according to the Saha equation

$$\frac{n_n}{n_p} \approx \exp\left[-\frac{(m_n - m_p)c^2}{k_B T}\right], \tag{1.2.2}$$

an equation that will be derived in Chapter 12. When $k_B T \gg (m_n - m_p)c^2$, we have $n_n \approx n_p$. As the temperature dropped, there was a *freeze-out* in which the small $m_n - m_p$ difference led to the residual neutron and proton ratio of $n_n/n_p \sim 1/8$. This ratio may be found from a calculation that balances the cooling rate with the actual transition rates of the reactions (1.2.1) above.

Two protons or two neutrons cannot form a bound state, but a neutron and a proton may collide and form a deuteron, abbreviated d or ^{2}H. This reaction releases energy ($Q = 2.226\,\text{MeV}$) in the form of a photon and the recoil energy of the deuteron:

$$n + p \rightarrow d + \gamma. \tag{1.2.3}$$

This is what we call a *capture reaction*. These deuterons react easily with other protons and neutrons, giving rise to a series of reactions, the dominant ones of which are illustrated in Fig. 1.6. Tritons ^{3}H (t) can be formed, the hydrogen isotope with one proton and two neutrons, as well as helium isotopes with two protons and either one neutron (^{3}He) or two neutrons (^{4}He). The binding energy of the deuteron is one of the most fortunate coincidences in the Universe. The Universe took about 7 minutes to cool down to 2.226 MeV, and the neutron lifetime is slightly above

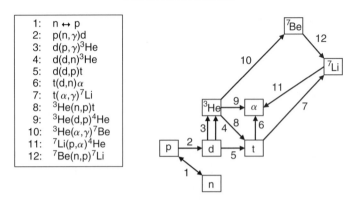

Fig. 1.6. The dominant reactions in primordial nucleosynthesis, after Kawano [1].

7 minutes. Had these two numbers not properly matched, there would have been no neutrons to initiate the whole primordial nucleosynthesis.

By time $t \approx 250$ s, the thermal energy E was 0.1 MeV, and all these primordial reactions came to a stop, except for the decays of neutrons, tritons and ^{7}Be. These last three nuclei were produced in primordial nucleosynthesis, but are not themselves stable, as they decay with lifetimes of 10.3 minutes, 12.3 years and 53 days, respectively, by *weak interactions* in what is called *β-decay*:

$$n \rightarrow p + e^- + \bar{\nu}_e,$$

$$t \rightarrow {}^3\text{He} + e^- + \bar{\nu}_e,$$

$$^7\text{Be} \rightarrow {}^7\text{Li} + e^+ + \nu_e. \tag{1.2.4}$$

Eventually, all the neutrons and radioactive nuclei transmuted into stable nuclei, such that only very small fractions of ^{7}Li, and practically no ^{6}Li remained. As a consequence, the initial composition of the Universe was almost entirely p, d, ^{3}He, ^{4}He, e$^-$, γ particles and neutrinos. This primeval ratio of abundances, listed in Table 1.1, can still be observed if we avoid regions where further reactions have taken place, such as in low-metal stars.

Very few nuclei heavier than helium are formed at this stage. One reason for this is that there are no stable nuclei with 5 nucleons, nor with 8 nucleons. The longest-lived isotopes with 5 nucleons are ^{5}He and ^{5}Li, but these emit neutrons and protons respectively. For element production there are thus bottlenecks at mass numbers of 5 and 8 that had to be later bridged by other means.

Very little further happened until the Universe reached time $t = 3.8 \times 10^5$ y, when the temperature and energies were low enough ($T \sim 4 \times 10^3$ K and $E \sim 0.4$ eV) for electrons to remain bound to nuclei in atoms. At that point, the atomic era started. After $t \sim 10^9$ y, stars and galaxies were formed, giving way to stellar

Table 1.1. *Isotopic abundances Y_i from primordial nucleosynthesis [2], defined by the fraction of nuclides i to the number of all nucleons. The nucleon number density is then $X_i = A_i Y_i$ of nucleons in that isotope of mass A_i. Normalization is $\sum_i X_i = 1$.*

Isotope	Nuclide fraction Y_i	Nucleon fraction X_i
$^1H = p$	0.75	0.75
$^2H = d$	2.44×10^{-5}	4.88×10^{-5}
^3He	1.0×10^{-5}	3.0×10^{-5}
^4He	0.062	0.2481
^6Li	1.1×10^{-14}	6.6×10^{-14}
^7Li	4.9×10^{-10}	34.3×10^{-14}

nucleosynthesis. Eventually some stars collapsed, heated up, and completely new cycles of nuclear reactions took place. This was how many heavier nuclei were produced. These processes continued to repeat themselves until the present day.

1.3 Reactions in light stars

After stars are formed by gravitational attraction, their continued contraction compresses the constituent gases and raises their temperature. If the star has a mass above a minimum of about 0.1 solar masses ($0.1\,M_\odot$), then the temperature rises to $T \sim 10-15 \times 10^6$ K and the density to $\rho \sim 10^2$ g cm^{-3}, and *nuclear hydrogen burning* can start. The release of energy in the resulting nuclear reactions is sufficient to stop further gravitational collapse, and the star remains in a phase of hydrostatic equilibrium. The compressive gravitational pressure is balanced by the expansive gas pressure of material heated by the nuclear reactions. Different initial stellar masses give rise, in this phase, to the range of main sequence stars represented in Fig. 1.7. The Hertzsprung-Russell diagram (H-R diagram) is a standard representation of stars in terms of their surface temperature and luminosity. Many stars are aligned roughly according to Stefan's law of $L \propto R^2 T^4$, and form what is called the main sequence. This corresponds to stars in their hydrogen-burning phase.

1.3.1 Proton-proton chains

The first series of nuclear reactions in a new star with mass $M < 1.5\,M_\odot$ is the *proton-proton chain*. This has the overall effect of converting 4 protons into one

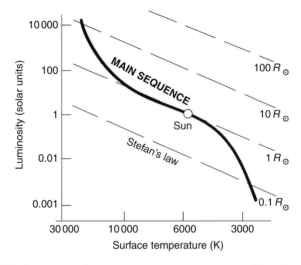

Fig. 1.7. H-R diagram: main sequence stars are represented by the thick curve, whereas the thin lines represent Stefan's law. Figure courtesy of Jon Whiting.

α particle (a ^{4}He nucleus), along with two positrons (2 e^+), two neutrinos (2 ν), and 26 MeV of released energy. It does not do this in one step, but via a chain that starts with

$$p + p \to d + e^+ + \nu_e \quad (Q = 1.44\,\mathrm{MeV}). \tag{1.3.1}$$

This reaction, involving neutrinos, proceeds by the *weak interaction*, and has a very low reaction rate. In fact, its rate is so low that it has never been measured directly. The slowness of this initial step is what is responsible for the long lifetime of stars in their hydrogen-burning phase.

Following the formation of the deuteron $d\,(^2\mathrm{H})$, a subsequent proton capture reaction

$$d + p \to {}^3\mathrm{He} + \gamma \quad (Q = 5.49\,\mathrm{MeV}) \tag{1.3.2}$$

may readily occur. The reaction $d + d \to {}^4\mathrm{He} + \gamma$ may also occur, but is less likely since protons are much more abundant than deuterons at this stage: about 1 deuteron for every 10^{18} protons.

A second proton capture on ^{3}He cannot succeed because ^{4}Li is unbound, but other possible reactions involving ^{3}He are (in Chain I):

$$^3\mathrm{He} + {}^3\mathrm{He} \to {}^4\mathrm{He} + 2p + \gamma \quad (Q = 12.86\,\mathrm{MeV}), \tag{1.3.3}$$

or

$$^3\mathrm{He} + {}^4\mathrm{He} \to {}^7\mathrm{Be} + \gamma \quad (Q = 1.59\,\mathrm{MeV}). \tag{1.3.4}$$

The ^{7}Be does not last long, but produces two α particles, using either Chain II

$$^7\text{Be} + \text{e}^- \rightarrow {}^7\text{Li} + \nu_e \tag{1.3.5}$$

$$^7\text{Li} + \text{p} \rightarrow {}^4\text{He} + {}^4\text{He}, \tag{1.3.6}$$

or Chain III

$$^7\text{Be} + \text{p} \rightarrow {}^8\text{B} + \gamma \tag{1.3.7}$$

$$^8\text{B} \rightarrow {}^8\text{Be} + \text{e}^+ + \nu \tag{1.3.8}$$

$$^8\text{Be} \rightarrow {}^4\text{He} + {}^4\text{He}, \tag{1.3.9}$$

where the ^{8}Be is a narrow resonance at 92 keV with width 6.8 eV. Resonances will be characterized in detail in Chapter 3: they are long-lived combinations of the reacting nuclei that are produced in specific circumstances. This particular ^{8}Be resonance will appear again later in this chapter when we discuss the triple-α process.

Figure 1.8 illustrates the three branches of the pp chain, in which the total energy released is the same (26.73 MeV). However, the energy carried by the neutrinos is lost to the star since neutrinos have a negligible probability for subsequent collisions. The energy retained in the star, Q_{ret}, is 26.20 MeV for Chain I, 25.66 MeV for Chain II, and 19.17 MeV for Chain III. This is the energy that is responsible for heating the star, and hence also its emission of light. It also, for a long while, prevents any further gravitational collapse.

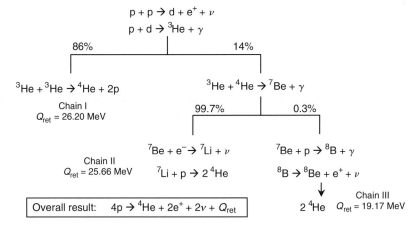

Fig. 1.8. The proton-proton chain has three branches, I, II, and III, which retain different quantities of energy Q_{ret} within the star.

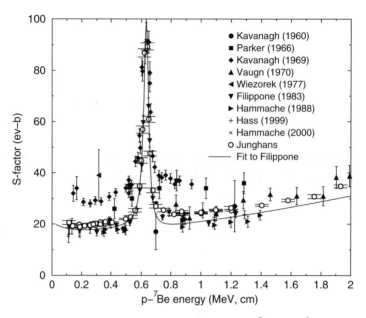

Fig. 1.9. Measured astrophysical S-factors $S(E)$ for the $^7\text{Be}(p,\gamma)^8\text{B}$ reaction, with the 1^+ resonance at 640 keV. The astrophysical rate is needed at ≈ 20 keV, requiring downward extrapolations from the experimental energies. The curve is an E1 direct capture model combined with an M1 resonance in a hybrid R-matrix treatment, fitted to the Filippone data.

Many of these reactions have been measured in the laboratory, all the way down to energies relevant for the stellar environment ($E \approx 0.1$ MeV). An example is shown in Fig. 1.4 for $^4\text{He}(^3\text{He},\gamma)^7\text{Be}$: cross sections decreasing exponentially at small energies are shown in the upper half and S-factors, nearly constant with energy, are shown in the lower half. As mentioned before, the cross section for fusions of charged particles is strongly dependent on energy because of the Coulomb barrier. Most of the energy dependence is given by the $e^{-2\pi\eta}/E$ factor in Eq. (1.1.4), which will be derived in Chapter 7.

For the $^7\text{Be}(p,\gamma)^8\text{B}$ reaction, the S-factor is only constant away from the 1^+ resonance at 640 keV that is prominent in Fig. 1.9. Reaction theory will be needed to describe how these resonances are superimposed upon the smoother non-resonant background cross sections. In addition, different data sets shown in Fig. 1.9 have different normalizations at low energies, where the measurement is hardest.

Both reactions, $^4\text{He}(^3\text{He},\gamma)^7\text{Be}$ and $^7\text{Be}(p,\gamma)^8\text{B}$, are not important for the energy production of a star. However they are connected to the amount of neutrinos emitted from the star and are very important contributions to the solar neutrino experiments [3].

1.3.2 Triple-α reaction

As mentioned earlier, there are no stable nuclei with $A = 5$ or 8 nucleons. We saw that the Big Bang production of ^{6,7}Li is very low, and so it is difficult to produce nuclei of mass $A \geq 9$ in stars. There are many α particles, but they do not form another binary bound state with themselves, with protons, or with neutrons. Fortunately, the narrow low-lying resonance in ^{8}Be $= \alpha + \alpha$ does trap them together in pairs with some small probability, and this probability is large enough for a third α particle to collide with the resonance pair to form a *triple-α* composite. This composite is an excited state in ^{12}C, and can either decay back to 3α, or, with a small branching ratio, decay to lower-energy bound states in ^{12}C by γ emission and e^+e^- production.

The non-resonant direct triple-α reaction does not produce enough carbon to explain the observed abundance. The key point is a narrow 0^+ resonance in the ^{8}Be $+ \alpha = {}^{12}$C* system that enhances the triple-α fusion reaction. This is shown in Fig. 1.10. This resonance is called the *Hoyle resonance*, after the person who predicted it in 1954 [4] on the basis that this was the only way to produce the measured quantities of ^{12}C in the Universe. The Hoyle resonance was subsequently found by experiments, at 287 keV above the 3α breakup threshold, with a narrow width of 8.3 eV [5]. The decay of the Hoyle resonance is via γ emission to the 2^+ excited state of ^{12}C at 4.44 MeV, or via direct e^+e^- decay to the 0^+ ground state. We will see in Chapter 3 that γ-decays cannot directly couple two 0^+ states.

The equilibrium population of the $2\alpha = {}^8$Be resonance can be determined by a Saha equation, like Eq. (1.2.2), where the numerator in the exponential is now the resonance energy of 92 keV. When the temperature $T_9 = 0.3$ and the density $\rho \sim 10^5$ g/cm^3, as typical for He burning, then the resonance population is $N(^8\text{Be})/N(\alpha) \sim 10^{-10}$. This then enables ^{8}Be $+ \alpha \rightarrow {}^{12}$C reactions, each of

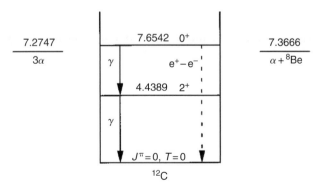

Fig. 1.10. Diagram showing the relevant states in ^{12}C for the triple-α reaction.

which releases an energy $Q = 7.27\,\mathrm{MeV}$. The net result is $3\alpha \rightarrow {}^{12}\mathrm{C}$, the triple-$\alpha$ reaction. Statistical equilibrium will be discussed in detail in Chapter 12.

Strongly connected to the triple-α reaction is the ${}^{12}\mathrm{C}(\alpha, \gamma){}^{16}\mathrm{O}$. The rate of this reaction, relative to the 3α capture, determines the post-He burning C/O ratio, which, in turn, affects the abundances of heavier elements produced in subsequent phases. At present, the C/O ratio is believed to play a crucial role in the last phases of a massive star. In particular, whether the final remnant following a supernova explosion is a neutron star or a black hole is affected by this ratio [6]. Because the 3α reaction is comparatively well known, the ${}^{12}\mathrm{C}(\alpha, \gamma){}^{16}\mathrm{O}$ reaction is still the most important source of uncertainty in the C/O ratio.

1.3.3 CNO cycles

In some stars there will be small fractions of carbon nuclei, either from the 3α reaction or from the remnants of earlier stars that have completed their evolutionary cycle. If these new stars are heavy enough ($M > 1.5\ M_\odot$), then the internal temperature is high enough ($T_9 > 0.03$) that there is another cycle that burns hydrogen into helium, but proceeds at a faster rate. This is the CNO cycle illustrated in Fig. 1.11, which uses the initial carbon as a catalyst: it is not consumed, but is regenerated at the end of the cycle, which proceeds as

$$^{12}\mathrm{C}\ (\mathrm{p},\gamma)\ {}^{13}\mathrm{N}\ (\mathrm{e}^+, \nu)\ {}^{13}\mathrm{C}\ (\mathrm{p},\gamma)\ {}^{14}\mathrm{N}\ (\mathrm{p},\gamma)\ {}^{15}\mathrm{O}\ (\mathrm{e}^+, \nu)\ {}^{15}\mathrm{N}\ (\mathrm{p},\alpha)\ {}^{12}\mathrm{C}.$$

The initial Coulomb barrier in the p+^{12}C reaction is higher than in the reactions of the pp chain but, if the temperature is high enough, this chain is faster because it does not involve the very slow p+p fusion reaction (1.3.1). The speed of the CNO cycle is still limited by weak interactions, namely a 10-minute lifetime for the β-decay of ^{13}N, and 2 minutes for ^{15}O. The total energy released in one CNO

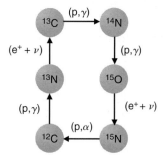

Fig. 1.11. The CNO cycle also converts four protons into one α particle, while the initial ^{12}C is a catalyst that is regenerated at the end of the cycle.

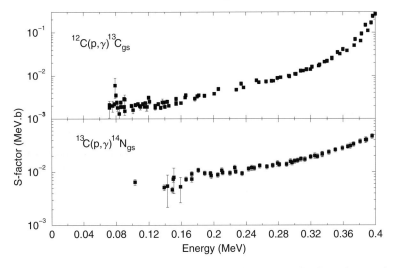

Fig. 1.12. S-factors for the cross sections of two important CNO cycle reactions.

cycle is the same (26.73 MeV) as in the pp chain, if the energy taken by escaping neutrinos is included.

The cross sections (or rather the S-factors) needed in the CNO cycle have to be extrapolated to low energies from those energies where measurements are possible. For the (p,γ) reactions on ^{12}C and ^{13}C, Fig. 1.12 shows that the S-factor extrapolation should be straightforward. For the ^{14}N$(p,\gamma)^{15}$O reaction in Fig. 1.13, however, there is a prominent resonance at 278 keV that makes the extrapolation non-trivial. In this case, better results should be expected from a model that is fitted to the resonance and the measured data above the resonance.

The basic CNO cycle returns to its starting point with the proton capture reaction via the production of a ^{16}O compound nucleus in an excited state (represented by ^{16}O*), which subsequently decays to ^{12}C $+ \alpha$:

$$^{15}\text{N} + \text{p} \rightarrow {}^{16}\text{O}^* \rightarrow {}^{12}\text{C}_{\text{gs}} + \alpha. \tag{1.3.10}$$

By this reaction, ^{12}C is regenerated at the end of the cycle. Occasionally, however, the excited state ^{16}O* will decay by γ emission. This is when ^{16}O, one of the most important nuclei for later organic life, is finally formed. This ^{16}O may capture another proton, leading to the additional bi-cycle loop shown in Fig. 1.14.

The ^{15}N$(p,\gamma)^{16}$O reaction mechanism is influenced even at low astrophysical energies by resonances, especially by the low-energy tail of the 1^- resonance at 338 keV shown in Fig. 1.13. We will see that theoretical models are essential here to determine how the several resonances combine with each other, and also with the non-resonant capture that forms a constant background.

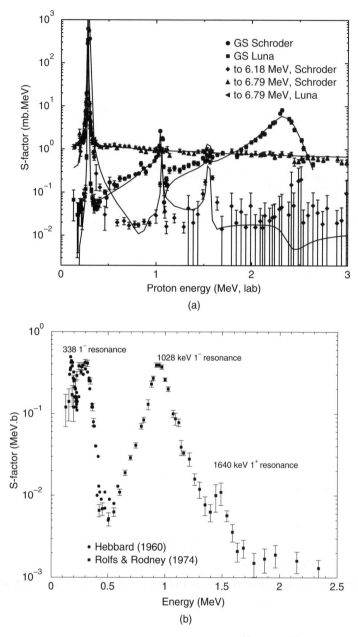

Fig. 1.13. (a) The S-factors for the cross sections of ^{14}N(p,γ)^{15}O reaction, to the ground state, and 6.18 MeV and 6.79 MeV excited states of ^{15}O. The curves are hybrid R-matrix fits discussed in Appendix B. (b) The S-factor for the ^{15}N(p,γ)^{16}O reaction.

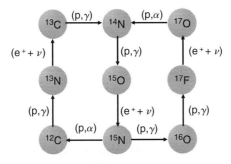

Fig. 1.14. The CNO bi-cycle is a breakout from the original CNO cycle, and produces ^{16,17}O and ^{17}F.

Additional cycles may also occur when a (p,γ) reaction on ^{17}O competes with the (p,α) reaction shown in Fig. 1.14. This leads to a new cycle involving ^{18}F and ^{18}O (not shown here), which after a (p,α) reaction, returns to ^{15}N in Fig. 1.11 and eventually regenerates ^{12}C for continued operation of the CNO cycles.

Eventually no new nuclear reactions are able to produce enough energy to stall further gravitational collapse. In light stars, less than about 8 solar masses, the star gradually contracts into a dwarf star. Outer layers of the star are shed off while the core continues to contract under gravity. If the remaining core mass is less than 1.4 solar masses, then it will compress to electron-degenerate matter, forming a white dwarf.

1.4 Heavy stars

In heavy stars (more than $8\,M_\odot$), all of the above pp, triple-α and CNO cycles occur in early evolutionary stages. These cycles produce residues of carbon and oxygen as before, but now there is sufficient gravitational pressure to compress and heat these residues so that further transmutation reactions may occur: the average thermal energy is sufficient to overcome the Coulomb barriers between the reacting nuclei. In this case, some reaction chains occur which do not eventually cycle back to ^{12}C, initiating the production of heavier elements.

In this section, we will discuss the main processes for heavy element production. In contrast to the production of light elements, here the number of reactions involved is very large and one can no longer enumerate specific reactions. Instead, we will mention the main mechanisms and the astrophysical sites in which these are most likely to occur. Sporadically, we will provide specific examples.

1.4.1 α-burning

Subsequent reactions with α particles, for example, produce heavier nuclei with N and Z both even, namely ^{16}O, ^{20}Ne, ^{24}Mg up to ^{28}Si, which is the dominant residue

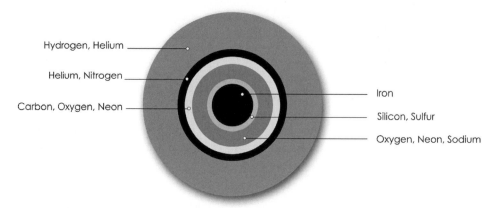

Hydrogen, Helium

Helium, Nitrogen

Carbon, Oxygen, Neon

Iron

Silicon, Sulfur

Oxygen, Neon, Sodium

Fig. 1.15. Layers in a red giant star before exploding in a supernova. Figure courtesy of Jon Whiting.

(or 'ash') of the process. These reactions only occur when the temperature is high enough; the Coulomb barriers involved are large because of the increased charge products Z_1Z_2. Some neutrons are also produced by (α,n) reactions, giving rise to nuclei with masses that are not multiples of four.

Stars go through a sequence of carbon, oxygen, neon burning and so on, as the temperature increases with progressively more gravitational contraction. We see from Fig. 1.1, however, that less and less nuclear energy is released in these successive stages, so there is diminishing return of energy in this advanced burning, and the stages pass progressively more quickly. Stars with these reactions, such as red giants in advanced stages of evolution, are therefore less luminous. A diagram of the composition of a red giant is given in Fig. 1.15. Eventually, nuclei near ^{56}Fe are produced (at the core of the massive star), and then no new nuclear reactions are able to produce enough energy to stall further gravitational collapse. This results in a rebound explosion where a new sequence of reactions can occur, to be discussed in Section 1.5.

1.4.2 s-process neutron reactions

Exothermic reactions ($Q > 0$) at temperatures just sufficient to surpass their entrance Coulomb barriers will never produce elements heavier than ^{56}Fe. Even at energies above all the Coulomb barriers, the general decrease of binding seen in Fig. 1.1 for large A implies, via the Saha equation, that the probability of producing, say, ^{208}Pb in equilibrium with ^{56}Fe, is extremely low. However, many heavier elements must be produced somewhere, as we see them in the Sun, according to the measured solar abundances of Fig. 1.16. There must therefore exist sites in

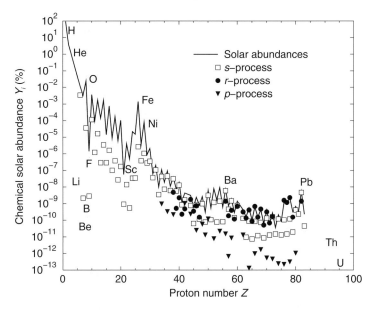

Fig. 1.16. Abundances of chemical elements in the Sun (percentages): observation and the contributions from three important nucleosynthesis processes.

the Cosmos which have sufficiently high temperature to enable the endothermic reactions producing heavy elements, and neutrons should primarily drive the process.

In contrast to proton captures, neutron captures are not hindered at very low energies. We will see later that neutron cross sections rise at low energies as $1/v$ for relative velocity v, and furthermore that the coefficient of v^{-1} *increases* for heavier nuclei. A sequence of neutron capture reactions may thus occur at moderate temperatures generating nuclei as heavy as uranium. This is what is referred to as the *s*-process, the '*s*' for slow.

In red giants (Asymptotic Giant Branch stars) during their α-burning stages, additional neutrons may be produced by (α,n) reactions, for example on ^{13}C or ^{22}Ne. These may be captured by seed nuclei in the iron group, and by progressively heavier nuclei, but at a rate much lower than their β-decay rates. The mechanism for the heavy element production in the *s*-process proceeds with (n,γ) on each stable nucleus (Z,N), producing neutron-rich isotopes $N+1, N+2, \ldots$ This gives nuclei that are a few nucleons away from stability, until a radioactive species β-decays by electron emission to the next element $(Z+1, N_\beta-1)$ in the Segré chart in an isotope with the same mass. This is called a *branching point*. Several branching-point nuclei are identified in Fig. 1.17. New (n,γ) captures can then be repeated on the isotopes with $Z+1$. Through this sequence of (n,γ) and β-decays, most stable

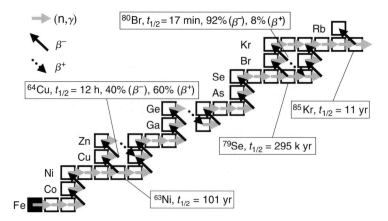

Fig. 1.17. Diagram illustrating the *s*-process with some key branching points.

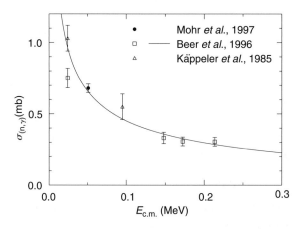

Fig. 1.18. Neutron capture reaction on ^{48}Ca, from [7]. Reprinted with permission from P. Mohr *et al.*, *Phys. Rev C* **56** (1997) 1154. Copyright (1997) by the American Physical Society.

species can be produced, with observable abundance ratios. A diagram illustrating a part of the *s*-process is shown in Fig.1.17. Because the rate of capture reactions is slower than the *β*-decay rates which bring nuclei back towards stability, the *s*-process proceeds close to the valley of stability in the Segré chart. In Fig. 1.16 we show the contribution of the *s*-process to the solar abundances. Observations of metal-poor stars show that the *s*-process is not universal and depends strongly on the metallicity of the star. This introduces some uncertainties on the final abundances due to the original composition of the star.

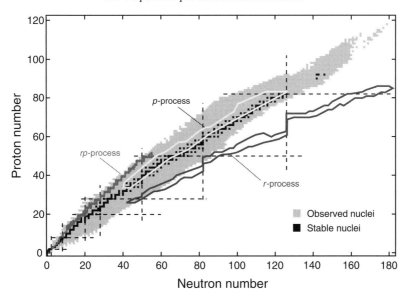

Fig. 1.19. Predicted paths of the *rp*- and *r*-processes, to the left and right, respectively, of the valley of stability. Figure courtesy of Marc Hausmann.

An example of an *s*-process reaction, ^{48}Ca(n,γ)^{49}Ca, has been measured by several researchers [7] and is shown in Fig. 1.18. We will revisit this capture rate when we discuss transfer reactions in Chapter 14 because transfer reactions offer an indirect method for extracting this type of astrophysical information.

1.5 Explosive production mechanisms

The *s*-process alone can only explain about half of the observed abundances of the heavy elements. This has led to a search for astrophysical sites in the Cosmos that are strongly time-varying, and where the equilibrium probabilities are not applicable, enabling an alternate path for the production of heavy elements. The most probable explosive scenario contributing to a large portion of the abundances of heavy elements is core-collapse supernovae. In this environment, high temperatures and a large abundance of neutrons enables a progression of capture reactions extending up to uranium. This is called the *r*-process for *rapid*, as opposed to the *s*-process, which is slow. The capture reactions, however, do not necessarily pass through the stable isotopes, but produce many neutron-rich isotopes that are radioactive, and eventually β-decay into other more stable species well after the explosion has finished. There are also some nuclei that cannot be produced by either the *s*- or the *r*-process mechanisms, suggesting the existence of an additional process which produces the *p*-nuclei. The isotopes involved in the *r*-process and the *p*-process are indicated in the Segré chart, Fig. 1.19.

Although not contributing significantly to the overall production of elements, other explosive environments where the proton density is high give rise to another mechanism, the *rp* (*rapid proton*) process. While the *s*-process proceeds along the valley of stability, the *rp*-process goes along proton-rich paths above and to the left of the valley on the Segré chart, as shown in Fig. 1.19.

1.5.1 *r-process neutron reactions*

It is widely believed that in supernovae, after core collapse, there are abundant neutrons produced, at least for a few seconds, leading to rapid capture sequences that extend the horizontal isotopic chains to the right, well beyond the first radioactive isotopes (the shaded region in Fig. 1.19). Neutrons are progressively captured by (n,γ) reactions until the production of extremely neutron-rich isotopes is limited by the increased probability of (γ,n) photo-disintegration reactions. This probability increases as the neutrons become less and less bound. The $n \leftrightarrow \gamma$ balance point between capture and disintegration defines the position of the *r*-process path on the Segré chart, and is thought to be between the lines shown in Fig. 1.19. Some of the nuclei that are produced will β-decay to heavier chemical elements (as with the *s*-process), giving a new seed nucleus for another series of neutron captures.

Closed-shell nuclei are usually very stable. These shell closures are believed to be at the places marked by the vertical lines in Fig. 1.19 at $N = 82$, 126, and 184, and here the *r*-process capture times become comparable to β-decay half lives. The rapid process slows down as β-decay wins over neutron capture, and the *r*-process path moves closer to the valley of stability. Nuclei where this happens are known as *waiting points* in the *r*-process. Waiting-point nuclei around $N = 126$ are depicted by the open squares in Fig. 1.20.

The nucleosynthesis produces heavier isotopes until eventually, in the actinide region, fission becomes more probable. It is possible that super-heavy elements $(Z > 114)$ may also be produced.

The neutron-rich nuclei will eventually experience a 'freeze-out' after the neutron flux has passed, and will no longer be replenished but will β-decay toward stability. This produces an isotopic population progressively drifting from the neutron-rich side toward the valley of stability. In the Segré chart, this corresponds to a drift diagonally up and to the left from the *r*-process path. In this sense, waiting-point nuclei are the progenitors for the production of stable nuclei of similar mass. The relative abundance of each stable isotope is approximately proportional to the lifetime of its originating nucleus on the *r*-process path. Peaks in the abundance of stable nuclei with $A = 80$, 130, and 192, arise due to the neutron closed shells at $N = 50$, 82, and 126. In Fig. 1.16 we show the contribution of the *r*-process to the solar abundances.

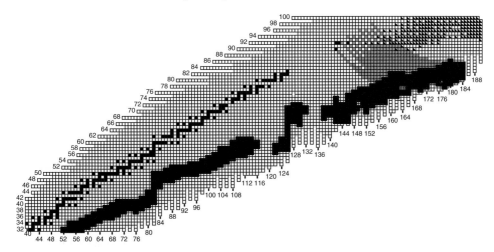

Fig. 1.20. Waiting-point nuclei (black open squares): nuclei produced in the *r*-process and that live long enough to be important signatures in the observed abundances of stable elements [9]. Reprinted with permission from H. Schatz and T. Beers, *Astro. J.* **579** (2002) 625.

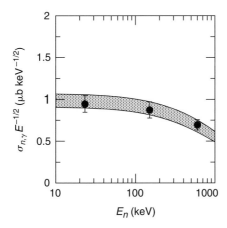

Fig. 1.21. Neutron capture reaction on ^{14}C [8]: comparing direct measurements (black circles) with the (n,γ) extracted from Coulomb dissociation (gray band); for more detail see Chapter 14. Figure courtesy of Neil Summers. Reprinted with permission from N. C. Summers and F. Nunes, *Phys. Rev. C.* **78** (2008) 069908. Copyright (2008) by the American Physical Society.

In connection to the *r*-process, we will study later the neutron capture on ^{14}C at very low energies (Appendix B.2.5). This reaction can have strong implications for the final abundances produced in the *r*-process in Type II supernovae. Data for ^{14}C(n,γ)^{15}C direct measurements and cross section determined indirectly from Coulomb dissociation[1] are compared in Fig. 1.21. As seen in Fig. 1.19, most nuclei

---

[1] A general theory for Coulomb dissociation will be introduced in Chapter 8, and applications discussed in Chapter 14.

involved in the *r*-process have not even been observed. Therefore, at present, the modeling of the *r*-process relies heavily on theory.

Numerous β-decay rates are important inputs to *r*-process network models. There are direct methods to measure this, but alternatively one can determine this information through charge-exchange reactions. In Fig. 1.22, data for ^{58}Ni(t,^{3}He) is shown as a function of the excitation energy of ^{58}Co [10, 11]. The mechanism for these reactions will be introduced in Chapter 4, and a discussion of how to extract the needed structure information from such data will be addressed in Chapter 14. This particular reaction is also very relevant to the mechanism of the explosions in supernovae.

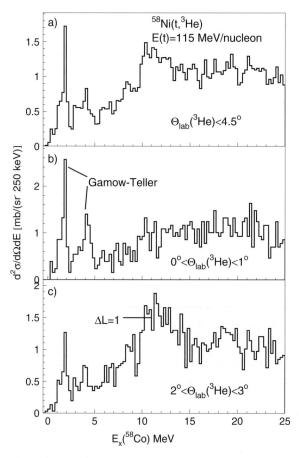

Fig. 1.22. Cross section for the charge exchange reaction ^{58}Ni(t,^{3}He) at 115 MeV/u as a function of the energy of the residual nucleus ^{58}Co [10]. Reprinted with permission from A. L. Cole *et al.*, *Phys. Rev. C* **74** (2006) 034333. Copyright (2006) by the American Physical Society.

1.5.2 The rp-process

The *rp*-process is not required to explain solar abundances but will occur whenever stellar material rich in hydrogen is suddenly heated to high temperatures. Main-sequence stars and red giants will have abundant H and He nuclei evolving according to the normal pp or CNO chains, but in some binary systems, transfers of material from the red giant to the smaller partner will lead to new kinds of thermonuclear explosive reactions.

The preferred sites for the *rp*-process are binary systems involving neutron stars. If there is H-rich massive transfer from the companion to the neutron star, then the temperature and density of that material will suddenly increase as it reaches the surface of the neutron star. There will be a rapid series of (p,γ) and (α,p) reactions that would normally be hindered by the Coulomb barriers, and which will produce a series of proton-rich nuclei up to the $A \sim 60$ mass region. At each step along this *rp*-process, the material may either capture another proton in a (p,γ) reaction, or wait for β-decay. The captures will lead to new isotopes until the proton dripline is reached, or until β-decays become fast enough to compete with the capture processes.

In addition there are intense X-ray fluxes being produced, which have been observed. For this reason, these binary systems are called *X-ray bursters*. With the highest temperatures, nuclei up to $A \sim 100$ may be generated. The *rp*-process becomes progressively hindered by the Coulomb barriers, and remains overall relatively unimportant for the production of nuclei and energy in stars.

A milder version of the *rp*-process can also occur in novae, in accreting material on white dwarfs. However, here, temperatures are not large enough for the process to reach the proton dripline.

1.5.3 The p-process

There are some neutron-deficient isotopes that cannot be made in the *r*- or *s*-processes, which justified naming an additional process. This process is represented by the white line in Fig. 1.19. It involves a sequence of (γ,n), (γ,p), and (γ,α) photo-disintegration on previous-generation stable nuclei. As it is triggered by photons, the process is sometimes called the gamma process.

One of the most likely sites for the *p*-process is in outer layers of core-collapse supernovae, while it heats due to the shock wave passing through it. In the *p*-process, all elements are produced, but in much-reduced quantities when compared to those resulting from the *s*- and the *r*-process. In Fig. 1.16 we show the contributions of the various processes to the abundances in the Sun. It is only for some proton-rich elements that the *p*-process becomes crucial. Also apparent from Fig. 1.16, light nuclei are not fully accounted for by the *s*-process.

1.6 Outlook

1.6.1 Implications for nuclear physics

The nuclides produced in the *s*-process are almost all long-lived enough to be targets in laboratory reaction measurements, and (n,γ) cross sections have been measured for many of these. The *rp*- and *r*-process nuclei, by contrast, have much shorter lifetimes, and are more difficult subjects for laboratory measurements. We will see that some of them have sufficient lifetimes to be produced in radioactive beams, and then used in subsequent secondary reactions to examine their properties. Those with even shorter lifetimes can still be produced as final states in secondary reactions, and some of their properties determined. Reaction theory will be needed to analyze the secondary reactions, and connect those measurements to the reaction rates relevant in astrophysical environments. In the next chapter, we examine nuclear reactions and see what properties of nuclei can be measured, and then in later chapters develop scattering theory so that we have a theoretical framework to describe these nuclear reactions. Eventually, in later chapters, we will close the circle by applying the various reaction theories to many of the astrophysical reactions we introduced in this chapter.

1.6.2 Nuclear astrophysics: an open field

Although the rest of the book focuses on the description of nuclear reactions, these contribute in multiple ways to many open questions in astrophysics. In all the standard processes producing the elements described in this chapter, there are abundance mismatches that need better constraints from nuclear physics, in particular, nuclear structure and nuclear reaction input. One of the greatest challenges has been blending nuclear physics input with astrophysics modeling in a way that meaningful constraints on the parameters can be made. In present-day modeling, there are specific conditions that need to be introduced artificially and are yet awaiting a better understanding. In some cases, there is uncertainty on the nucleosynthesis path (as in the *r*-process); in others, there is uncertainty on the endpoint (as in the weak *s*-process). Perhaps even more exciting are the new emerging ideas that have not been discussed here. There is the weak *s*-process introduced earlier to account for the light nuclei, but more recently the Light Element Primary Process (LEPP) was added to the list, to explain the early galactic adundances and, maybe related to it, the neutrino *p*-process, thought to occur in all core-collapse supernovae, and a possible site for the production of Sr and other elements beyond Fe in very early stages of galactic evolution. Studies aiming for a better understanding of the Universe will continue for decades to come. For more detailed information on astrophysics and the topics covered in this chapter we refer to textbooks such as Clayton [12], Rolfs and Rodney [13], Arnett [14], Pagel [15], Iliadis [16] and Bennett [17].

References

[1] L. Kawano, Fermi National Accelerator Laboratory, Report FERMILAB-Pub-92/04-A (1992).

[2] P. D. Serpico, S. Esposito, F. Iocco, G. Mangano, G. Miele and O. Pisanti, *J. of Cosm. and Astro. Phys.* **12** (2004) 010.

[3] R. G. H. Robertson, *Prog. Part. Nucl. Phys.* **57** (2006) 90.

[4] F. Hoyle, *Astro. J. Suppl.*, **1** (1954) 121.

[5] H. O. U. Fynbo *et al.*, *Nature*, **433** (2005) 136.

[6] T. A. Weaver and S. E. Woosley, *Phys. Rep.* **227** (1993) 65.

[7] P. Mohr *et al.*, *Phys. Rev. C* **56** (1997) 1154.

[8] N. C. Summers and F. M. Nunes, *Phys. Rev. C* **78** (2008) 011601R; *Phys. Rev. C* **78** (2008) 069908.

[9] H. Schatz and T. Beers, *Astro. J.* **579** (2002) 626.

[10] A. L. Cole *et al.*, *Phys. Rev. C* **74** (2006) 034333.

[11] R. Zegers, NSCL White Paper 2007.

[12] D. D. Clayton 1984, *Principles of Stellar Evolution and Nucleosynthesis*, Chicago: University of Chicago.

[13] C. E. Rolfs and W. S. Rodney 1988, *Cauldrons in the Cosmos*, Chicago: University of Chicago.

[14] D. Arnett 1996, *Supernovae and Nucleosynthesis*, Princeton: Princeton University Press.

[15] B. E. J. Pagel 1997, *Nucleosynthesis and Chemical Evolution of Galaxies*, Cambridge: Cambridge University Press.

[16] C. Iliadis 2007, *Nuclear Physics of Stars*, Weinheim: Wiley.

[17] J. Bennett, M. Donahue, N. Schneider, M. Voit, *The Essential Cosmic Perspective*, San Francisco; Toronto: Addison-Wesley.

2

Reactions of nuclei

I was taught that the way of progress was neither swift nor easy.
Marie Curie

In order to understand nuclear reactions, we have first in Section 2.1 to name the arrangement of nucleons in a nucleus in terms of the quantum-mechanical **state** of a nucleus, and then describe the different ways in which these nucleons may be rearranged during nuclear reactions. Reactions which proceed quickly, and thus called **direct reactions**, are distinguished in Section 2.2 from the comparatively slow reactions that also occur, which are called **compound nucleus reactions**. Almost all reactions involve the collision of two nuclei, and Section 2.3 shows how the conservations of mass, energy and momentum may be described in either non-relativistic or relativistic kinematics. Section 2.4 describes how the **rates** of nuclear reactions are measured in terms of **cross sections**, which have units of area. We show how these cross sections are different in the laboratory and center-of-mass coordinate frames of reference, then in the final subsection 2.4.4 how the cross sections may be determined from the wave functions that are solutions of a Schrödinger equation for the pair of reacting nuclei.

2.1 Kinds of states and reactions

2.1.1 States of nuclei

Nuclei are aggregations of Z protons and N neutrons in a particular configuration or state described by a wave function ϕ determined from quantum mechanics, given the strong and electromagnetic potentials V between the $A = N + Z$ constituent nucleons. A state is called *bound* if energy is needed to remove one or more nucleons to large distances. A bound state has thus a total binding energy $B(A, Z)$ which is positive, and therefore has a negative eigenenergy $E = -B(A, Z)$ for the Hamiltonian in the many-body Schrödinger equation. All the bound states are discrete, which means that they can be counted with integers b. Each bound state b has a specific spin I_b measured in units of $\hbar$, parity $\pi_b = \pm 1$, and eigenenergy $E_b < 0$, so the Schrödinger equation for an isolated nucleus is

$$[\hat{T} + V]\phi_b = E_b \phi_b, \qquad (2.1.1)$$

where $\hat{T}$ is the kinetic energy operator given in Chapter 5. The individual nucleons (neutrons and protons) all have spin $\frac{1}{2}$, which means that even-A nuclei have integer spins, and odd-A nuclei have half-integer spin values. The *parity* of a state is the factor ± 1 by which its wave function changes on the spatial reflection $\mathbf{r} \to -\mathbf{r}$. Nuclear and Coulomb forces do not change parities, so nuclear states may be labeled according to their parity.

The state of lowest energy is called the *ground state* E_0, and we measure *excitation energies* above the ground state as $\epsilon_b = E_b - E_0$, shown in Fig. 2.1 for ^{13}C. Nuclei, if left to themselves, will generally decay to their ground states by γ emission or electron capture: the exceptions are called *isomers*, which are nuclei in excited states with long lifetimes because for some individual reason the decays are hindered. All non-isomeric nuclei in laboratory conditions will be in their ground states, as they will also be in stars when the temperatures T satisfy $k_B T \ll \epsilon_b$.

Unbound states of nuclei include *resonances*, as well as the smooth background between resonances called the *non-resonant continuum*. Strictly speaking, the Schrödinger equation has eigensolutions at every continuous (unbound) positive energy, but traditionally only the resonances are listed as 'states' of a nucleus

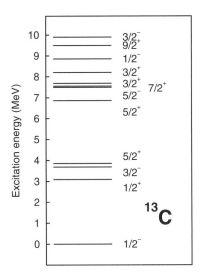

Fig. 2.1. Energy levels of the ^{13}C nucleus up to 10 MeV, drawn at their excitation energy ϵ_b and labeled by their spin and parity. They are bound up to 4.94 MeV, after which neutrons may have positive energy and escape.

in level diagrams. Each resonance has a *width* Γ, measured in MeV as (in most cases) the full width at half maximum of the resonance peak. Resonances can interfere with each other, and overlap each other at higher energies where the widths increase, and thus give more complicated patterns in experiments. A resonance can be regarded as a composite system that, because of its energy spread, lasts according to the uncertainty principle in proportion to $e^{-t/\tau}$ for a time duration of $\tau \sim \hbar/\Gamma$ called the lifetime. The half-life, after which half the nuclei will have decayed, is $t_{1/2} = (\ln 2)\tau$. When the corresponding decay channels are included in the picture, one can think of all radioactive nuclei as resonant states.

2.1.2 Kinds of reactions

In reactions of type B(A,C)D, the nuclei A and B usually start in their ground states. If they remain in their initial states, we have *elastic scattering*, written as B(A,A)B. The directions of motions of the two nuclei will have changed by the *scattering angle* θ shown in Fig. 2.2, while the relative kinetic energy of their motion E (defined on page 35) will remain unchanged.

If one or both of the incident nuclei A, B gets changed to an excited state during the reaction, this is called *inelastic* scattering. Excited states are often denoted by a * superscript, so B(A,A)B* is the reaction where B finishes in some excited state b, say. The relative kinetic energy after the reaction will be decreased by the amount ϵ_b of energy that has gone into exciting the nucleus B, and the Q-value will be $Q = -\epsilon_b$ for this endothermic reaction.

Perhaps during the reaction a proton or a neutron is moved from one nucleus to another. This is called a *transfer reaction*. If the nucleus A can be regarded as a

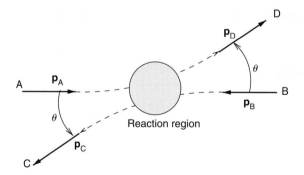

Fig. 2.2. Scattering angle in the B(A,C)D reaction showing the incoming and outgoing momenta. In the center-of-mass frame, the two θ angles are equal, and are called the *center-of-mass scattering angle*, θ_{cm}.

core C plus a neutron n, say, then a transfer reaction can be

$$(A = C + n) + B \rightarrow C + (D = B + n), \tag{2.1.2}$$

forming the new residual nucleus D that is composed of the original B together with the neutron bound to it. The final C and D may possibly be in excited states. If the (positive) separation energy of the neutron in A is S_A, and in D is S_D, then the Q-value satisfies $S_D = S_A + Q$, and the final kinetic energy will be increased by this Q, which may be positive or negative. The reaction d(d,n)^{3}He in Fig. 1.6, from primordial nucleosynthesis, is a transfer reaction where a proton is transferred from one deuteron to the other, the first being reduced to a neutron, and the second becoming d + p = ^{3}He with an energy release of $Q = 3.27$ MeV because the initial binding energy is 2.22 MeV and the final is 5.49 MeV.

Sometimes the two incident nuclei will *capture* each other, if the incident kinetic energy E is not too large. To form a composite state that lasts long enough to stop immediate escape back to the incident A + B configuration, either some energy has to be released by direct particle emission, or some resonance has to be excited with a long-enough lifetime, or both. The most frequent kind of direct emission is that of γ-rays, as in the primeval reactions d(p,γ)^{3}He or ^{3}He(α,γ)^{7}Be, both occurring without the assistance of resonances (see Chapter 1). Heavier nuclei than these have many more resonances, so captures of neutrons or protons on a larger nucleus is most likely to proceed first by exciting a narrow resonance, which may, after some average time τ, decay by emission of photons, neutrons, or other particles if there is enough energy. When direct γ processes dominate, the process is often called a *radiative capture* reaction A+B $\rightarrow$ C+γ, and the reverse process, C+γ $\rightarrow$ A+B, is called *photo-disintegration*. When resonance capture dominates, the process is often called a *fusion* reaction A + B $\rightarrow$ C*, as one kind of the *compound nucleus reactions* to be discussed in the next section.

Finally, there may be *breakup* reactions where, say, one participant B is broken into two or more fragments C and D that may be detected separately. We would write this reaction as A(B,C + D)A. The two parts C and D are sometimes together regarded as an excited state of B*, especially if that excited state can be counted as a particular resonance.

2.2 Time and energy scales

In the above description of kinds of reactions, we mentioned a variety of direct, resonant, and compound-nucleus reactions. These will be characterized also by the ranges of time and energy that are involved.

A *direct reaction* is one that proceeds the most quickly, and has rates (measured as cross sections) that vary most smoothly with incident kinetic energy. Transfer and breakup reactions are generally direct reactions of this kind. Sometimes reactions with *resonances* have peaks in the way the cross sections vary with energy. These are measured by the full widths at half maximum of these peaks Γ, and narrower resonances last for longer times in inverse proportion to their width. Finally, there are extremely narrow resonances from unbound *compound nucleus* states that, by the time they decay, will have lost practically all information about the direction of the incident nuclei, and will therefore decay isotropically.

2.2.1 Direct reactions

The fastest reactions only involve very few nucleons on the surface of the nucleus, or only the nucleus as a collective whole. These are called direct reactions, and are more likely to occur at high incident energies because then the reaction is typically finished more quickly and fewer internal collisions are possible. In these fast reactions, the directions of the final nuclei are much more influenced by the initial direction, and will typically have large cross sections at small θ_{cm}: large reaction rates in the direction of the incoming nuclei (see Fig. 14.3 for example).

Quantum mechanically, direct reactions are much more often modeled as a *one-step transition* between the initial and final scattering states. Most of the Big Bang reactions can be well described as one-step reactions, for which, we will see in Chapter 14, the distorted-wave Born approximation (DWBA) will prove very useful. Transfer processes such as A(d,p)B stripping reactions are usually modeled by the DWBA, assuming a direct-reaction mechanism. Of course, production of the A + d compound nucleus system is still possible at lower energies, but the decay of the compound nucleus gives isotropic angular distributions, which can be distinguished from the forward-peaked (d,p) cross sections, and subtracted if necessary.

One-step theories may be improved by including two and higher-order steps, as in a perturbation series. If some of the interaction potentials are strong, however, this series may not converge, and coupled-channels methods must be used, as discussed in Chapters 6 and 14.

2.2.2 Resonance reactions

We have seen many resonances in Chapter 1 as peaks in cross sections when plotted as functions of energy. From the theoretical point of view, resonances are longer-lived configurations of nucleons, and may be produced by many different mechanisms, hence having wide-ranging lifetimes. The simplest resonances are those that occur in elastic scattering, because of the nuclear attractive force combining with a Coulomb and/or centrifugal barrier to keep the colliding nuclei A and B together for some time before escaping again. For example, both neutrons and protons form resonances when interacting with an α particle, as will be examined in more detail in Chapters 3 and 10. These elastic resonances are called *shape resonances* because they arise from the shape of the effective elastic barrier. They do not last for very long, especially in light nuclei, so have relatively large widths Γ.

Slightly more complicated resonances may arise from multi-step processes, for which we see in Chapter 6 that coupled-channels models must be used. When only a few of the nucleons are involved, these are called *doorway resonances*, since they are near the entrance channel, but may lead to more complicated decay patterns. They will typically be longer lived than shape resonances, and will have smaller widths Γ.

At higher incident energies, some strongly collective resonances are found to occur. These are called *giant resonances*. In giant dipole resonances, for example, all the neutrons oscillate collectively against all the protons, and in giant monopole resonances there are 'breathing mode' oscillations, where the whole nucleus gets alternately larger and smaller in a spherically symmetric manner.

2.2.3 Compound nucleus reactions

The most complicated resonances are the *compound nucleus resonances*. These involve all the $A + B$ nucleons, live the longest, and so have the narrowest Γ values. No practical theory can predict all these resonances, but their widths may be determined experimentally from fitting to the results of experiments with high energy resolution.

One way of viewing compound nucleus reactions is to consider all the possible interactions between the nucleons of two nuclei when they are close together for a long time. There will be so many nucleon-nucleon scatterings in quick succession that the initial kinetic energy will be dispersed among all the nucleons. Eventually, all of that kinetic energy will be shared evenly between all nucleons of the fused, compound system. The energy will have been dissipated in a largely statistical manner, so that there is only a very small probability that any one nucleon will gain enough individual energy to escape from the compound nuclear system. This gives a long lifetime τ to the compound nucleus. Strictly speaking, it is still unbound and a resonance, because it is still possible for a nucleon to escape, but the rate of this, $1/τ$, will be very small, and hence its resonance width $Γ = \hbar/τ$ will also be very small.

The concept of a compound nucleus was first given by Niels Bohr, who formulated the *independence hypothesis* that the later decays are independent of the details of the initial channel. This means that the same compound nucleus may be formed in one of several ways, for example

$$\alpha + {}^{60}\text{Ni} \rightarrow {}^{64}\text{Ni}^*$$

$$\text{and} \quad \text{p} + {}^{63}\text{Cu} \rightarrow {}^{64}\text{Ni}^*, \tag{2.2.1}$$

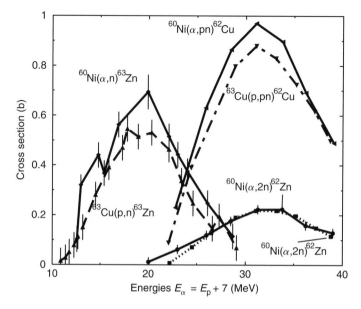

Fig. 2.3. Reaction products from two compound nucleus reactions showing that the production rates for the exit channel are independent of the entrance channel (cross-section data from Ghoshal [1]).

and the products from the two reactions should be similar. This is illustrated in Fig. 2.3 for several reactions producing ^{62}Cu, ^{62}Zn and ^{63}Zn, where the production rates are the same, whether generated by incident protons on ^{63}Cu, or by α particles on ^{60}Ni, provided the energy scales are shifted so that the compound nucleus excitation energies are the same.

2.3 Collisions

To model the collisions of two nuclei, even before we consider what happens while they are close, we need to accurately specify their initial and final energies and momenta, and how these are related by conservation laws. We consider first non-relativistic collisions, and later come back to the relativistic account.

2.3.1 Non-relativistic kinematics

Consider now the collision of two nuclei A, B of masses m_A and m_B respectively. We first use a coordinate system fixed in the laboratory (or the star) where they have individual velocities $\mathbf{v}_A$ and $\mathbf{v}_B$, from which we have their energies as $E_i = \frac{1}{2} m_i v_i^2$ ($i = $ A, B), and momenta $\mathbf{p}_i = m_i \mathbf{v}_i$.

Two-body kinematics

Suppose the two nuclei have instantaneous positions $\mathbf{r}_i$ ($i = A$ or B) in some fixed frame of reference. We can best describe the dynamics in terms of the two new coordinates: the center of mass position $\mathbf{S}$ and the relative vector $\mathbf{R}$, defined as

$$\mathbf{S} = (m_A \mathbf{r}_A + m_B \mathbf{r}_B)/m_{AB}$$
$$\mathbf{R} = \mathbf{r}_A - \mathbf{r}_B, \tag{2.3.1}$$

where $m_{AB} = m_A + m_B$. The total kinetic energy is

$$E_{\text{tot}} = \tfrac{1}{2} m_A v_A^2 + \tfrac{1}{2} m_B v_B^2 \tag{2.3.2}$$

and, using Eq. (2.3.1) solved for the $\mathbf{r}_i$ in terms of $\mathbf{S}$ and $\mathbf{R}$, this may be rewritten as

$$E_{\text{tot}} = \tfrac{1}{2} m_{AB} \dot{S}^2 + \tfrac{1}{2} \mu \dot{R}^2, \tag{2.3.3}$$

where the speeds $\dot{S} = |\dot{\mathbf{S}}|$ and $\dot{R} = |\dot{\mathbf{R}}|$ are the magnitudes of the $\dot{\mathbf{S}}$ and $\dot{\mathbf{R}}$ velocities respectively, and we have defined

$$\mu = m_A m_B / m_{AB}. \tag{2.3.4}$$

The total energy can therefore be seen as a sum of the energy of motion of the center of mass ($\tfrac{1}{2} m_{AB} \dot{S}^2$), and the *energy of relative motion*

$$E = \tfrac{1}{2} \mu \dot{R}^2. \tag{2.3.5}$$

The mass parameter μ associated with the relative motion is called the *reduced mass*. The 'relative energy' E will be used extensively in scattering theory.

An important case is that of laboratory experiments, when particle A is incident with energy E_A on a stationary nucleus B that is a target, illustrated in Fig. 2.4(a). In this case $\dot{S} = m_A / m_{AB} \mathbf{v}_A$, and

$$E = m_B / m_{AB} \, E_A = \tfrac{1}{2} \mu v_A^2, \tag{2.3.6}$$

and the remaining energy $E_{\text{tot}} - E = m_A / m_{AB} \, E_A$ is that from the speed $\dot{S}$ of the center of mass of the A + B system.

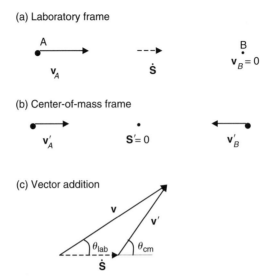

Fig. 2.4. Laboratory (a) and center-of-mass (b) velocities before the collision, when B is a stationary target. Part (c) shows the way the laboratory velocity $\mathbf{v}$ may be decomposed into a velocity in the c.m. frame $\mathbf{v}'$ and the velocity of that frame $\dot{\mathbf{S}}$ as $\mathbf{v} = \mathbf{v}' + \dot{\mathbf{S}}$. The azimuthal angles ϕ are out of the page, and are not changed by the changes of reference frame.

The center-of-mass coordinate system

If all the forces acting in the reaction are only *between* the nuclei and are not externally imposed, then it is most useful to change the reference frame to one in which the center of mass of A and B is at rest. This is called the *center-of-mass* (c.m.) frame. When the forces between A and B depend only on $\mathbf{R}$ and not at all on $\mathbf{S}$, then the velocity $\dot{\mathbf{S}}$ is constant, and the c.m. frame remains an inertial frame. We denote velocities in the center-of-mass frame by primes.

We may set the origin of the center-of-mass reference frame at the point $\mathbf{S}$, so the frame is defined by $\mathbf{S}' = \dot{\mathbf{S}}' = 0$ as shown in Fig. 2.4(b). The unprimed laboratory velocities and primed center-of-mass velocities are related in general by

$$\mathbf{v} = \mathbf{v}' + \dot{\mathbf{S}}$$

$$\text{or} \quad \mathbf{v}' = \mathbf{v} - \dot{\mathbf{S}}, \tag{2.3.7}$$

the vector addition illustrated in Fig. 2.4(c).

Conservation laws for collisions

In a reaction B(A,C)D that leads to final nuclei C and D, conservation laws limit the range of the energies and outgoing angles of nuclei C and D. Non-relativistically, we have separate *mass, energy* and *momentum* conservations:[1]

$$m_A + m_B = m_C + m_D, \tag{2.3.8}$$

$$Q + E_A + E_B = E_C + E_D, \tag{2.3.9}$$

$$\mathbf{p}_A + \mathbf{p}_B = \mathbf{p}_C + \mathbf{p}_D, \tag{2.3.10}$$

respectively, where Q (as before) is the internal energy released in the reaction. These laws apply in both the laboratory and (for primed quantities) in the center-of-mass coordinate frames.

Laboratory and center-of-mass scattering angles

Laboratory experiments may measure scattering angles, commonly when B is a stationary target. Most theories, however, predict cross sections as functions of center-of-mass angles θ_{cm}, which are different from the laboratory angles θ_{lab} because the center-of-mass frame is now moving in the laboratory with constant velocity

$$\dot{\mathbf{S}} = \frac{m_A}{m_{AB}} \mathbf{v}_A \tag{2.3.11}$$

when $\mathbf{r}_B = 0$ as in Fig. 2.4(a).

Consider some outgoing particle C. If it has laboratory velocity $\mathbf{v}_C$, then its velocity in the c.m. frame, by Eq. (2.3.7), is

$$\mathbf{v}'_C = \mathbf{v}_C - \dot{\mathbf{S}}. \tag{2.3.12}$$

Let us measure the angles θ of C from the incident beam direction: the direction of $\dot{\mathbf{S}}$. Then the lateral and parallel components of Eq. (2.3.12), according to the triangle of Fig. 2.4(c), give

$$v'_C \sin \theta_{cm} = v_C \sin \theta_{lab}$$

$$\dot{S} + v'_C \cos \theta_{cm} = v_C \cos \theta_{lab}$$

$$\phi_{cm} = \phi_{lab}, \tag{2.3.13}$$

from which we conclude that

$$\tan \theta_{lab} = \frac{v'_C \sin \theta_{cm}}{\dot{S} + v'_C \cos \theta_{cm}} = \frac{\sin \theta_{cm}}{\rho + \cos \theta_{cm}} \tag{2.3.14}$$

---

[1] There are also conservation laws for charge, angular momentum, baryon number, etc., but these are not needed here.

where we define $\rho = \dot{S}/v'_C$. To determine ρ we need to use the conservation laws. If E is the relative energy in the incident channel, then in the exit channel we have in the c.m. frame

$$Q + E = E_C + E_D = \tfrac{1}{2}m_C v'^2_C + \tfrac{1}{2}m_D v'^2_D$$
$$\mathbf{0} = m_C \mathbf{v}'_C + m_D \mathbf{v}'_D. \qquad (2.3.15)$$

Eliminating v'_D from these equations, we find $Q + E = \tfrac{1}{2}\frac{m_C}{m_D}(m_C + m_D)v'^2_C$. Combining this with Eqs. (2.3.6) and (2.3.11) we get

$$\rho = + \left[\frac{m_A m_C}{m_B m_D} \frac{E}{Q+E} \right]^{\frac{1}{2}}. \qquad (2.3.16)$$

In elastic scattering $A = C$, $B = D$ and $Q = 0$, so we have simply $\rho = m_A/m_B$. To find the c.m. angles in terms of the laboratory angles, Eq. (2.3.14) can be rearranged as a quadratic in $\sin \theta_{cm}$.

   In the remainder of this book we will drop the subscript cm and the primes, and use θ to directly refer to the scattering angles in the center-of-mass frame.[2]

2.3.2 Relative and center-of-mass wave functions

The above transformations from space fixed to center-of-mass coordinates have their counterpart in quantum mechanics. The Schrödinger equation for the motion of two particles A and B with total energy E_{tot} and some potential $V(\mathbf{r}_A - \mathbf{r}_B)$ that acts between them[3] is

$$\left[-\frac{\hbar^2}{2m_A}\nabla^2_{\mathbf{r}_A} - \frac{\hbar^2}{2m_B}\nabla^2_{\mathbf{r}_B} + V(\mathbf{r}_A - \mathbf{r}_B) - E_{tot} \right] \Psi(\mathbf{r}_A, \mathbf{r}_B) = 0. \qquad (2.3.17)$$

Using the same coordinate transformations that led to Eq. (2.3.3), but without assuming that B is at rest, the kinetic energy terms in this equation may be rewritten as a sum of operators using center-of-mass and relative coordinates. We may also use these coordinates for the wave function, yielding

$$\left[-\frac{\hbar^2}{2m_{AB}}\nabla^2_{\mathbf{S}} - \frac{\hbar^2}{2\mu}\nabla^2_{\mathbf{R}} + V(\mathbf{R}) - E_{tot} \right] \Psi(\mathbf{S}, \mathbf{R}) = 0. \qquad (2.3.18)$$

---

[2] Note also that they may sometimes be named 'c.m. scattering angles', just as the velocities $\mathbf{v}'$ are sometimes called 'c.m. velocities'. However, the angles, velocities and momenta *in* (not *of*) the center-of-mass frame are intended. Furthermore, $\mathbf{p}'_A$ and $\mathbf{p}'_B$ are sometimes called 'relative momenta', but this is misleading since they are not relative to the other particle, only to the center-of-mass frame. Since $\mathbf{p}'_A = -\mathbf{p}'_B$, they are equal and opposite momenta in the c.m. frame.

[3] We assume that $V(\mathbf{R}) \to 0$ when $R \to \infty$, so the total energy E_{tot} is also the total kinetic energy of the initial particles given by Eqs. (2.3.2) and (2.3.3).

We now look for separable solutions, of the form $\Psi(\mathbf{S}, \mathbf{R}) = \Phi(\mathbf{S})\psi(\mathbf{R})$. Substituting this in Eq. (2.3.18), one can show that it can be solved if we have solutions for the two separate equations

$$-\frac{\hbar^2}{2m_{AB}}\nabla_{\mathbf{S}}^2 \Phi(\mathbf{S}) = (E_{\text{tot}} - E)\, \Phi(\mathbf{S}) \qquad (2.3.19)$$

$$\text{and} \quad \left[-\frac{\hbar^2}{2\mu}\nabla_{\mathbf{R}}^2 + V(\mathbf{R})\right]\psi(\mathbf{R}) = E\,\psi(\mathbf{R}), \qquad (2.3.20)$$

for some separation constant E. We identify this constant by noting that Eq. (2.3.19) has plane wave solutions like $\Phi(\mathbf{S}) = A\exp(i\mathbf{K}\cdot\mathbf{S})$ for the free motion of the center of mass of the whole system, for any wave numbers $\mathbf{K}$ satisfying $E_{\text{tot}} - E = \hbar^2 K^2/2m_{AB}$. The E must therefore be the same relative energy defined in Eq. (2.3.5) earlier, so Eq. (2.3.20) must be the (non-trivial) Schrödinger equation that defines the real physics of the relative motion.

It is interesting and very useful to note that E is equal to the kinetic energy of a fictitious particle traveling with the relative velocity $\mathbf{v} = \mathbf{v}_A - \mathbf{v}_B$, and whose mass is the reduced mass μ. We can thus think of Eq. (2.3.20) as representing the collision of a particle of mass μ, initial velocity $\mathbf{v}$, momentum $\mathbf{p} = \mu\mathbf{v}$, and kinetic energy $E = \frac{1}{2}\mu v^2$, with a fixed scattering point at the center of the potential energy $V(\mathbf{R})$. Then $\mathbf{R}$ is the vector from the origin of the scattering potential to the position of the fictitious particle of mass μ. This usefully reduces a two-body scattering problem to the problem of scattering a single particle on a potential fixed around the origin.

When the potential is zero, Eq. (2.3.20) has plane-wave solutions like $\psi(\mathbf{R}) = A\exp(i\mathbf{k}\cdot\mathbf{R})$ for wave vector $\mathbf{k} = \mathbf{p}/\hbar$, whose magnitude is related to the energy by $E = \hbar^2 k^2/2\mu$. In terms of $\mathbf{k}$, the relative velocity is $\mathbf{v} = \hbar\mathbf{k}/\mu$. When the potential does not depend on the nuclear states, we have elastic scattering, where $|\mathbf{k}|$ is unchanged from the initial to the final state, changing only its direction. In general, we will see how the potential may be more complicated and change A to D and B to C, to produce the B(A,C)D reaction.

2.3.3 Relativistic kinematics

At medium and higher energies – above one or two hundred MeV per nucleon – we should take into account that, according to Special Relativity, the kinetic energy is not exactly $T = \frac{1}{2}mv^2$ because there are corrections that become significant as v approaches the speed of light c. In this subsection we just focus on establishing the correct relations between velocities, momenta and energy, but still keep a non-relativistic Schrödinger equation for determining the dynamics of the reactions. That is, we only use what is called *relativistic kinematics*.

We employ a 4-vector notation where the momentum 4-vector is

$$\mathbf{P} = m_0 \gamma(v)(c, \mathbf{v}) = (mc, \mathbf{p}) = (E/c, \mathbf{p}) \tag{2.3.21}$$

with invariant $P^2 = m^2 c^2 - p^2 = m_0^2 c^2$, for a particle of rest mass m_0 moving in a given frame with velocity $\mathbf{v}$, momentum $\mathbf{p}$, and total energy $E = m_0 c^2 + T$ for kinetic energy T.

When there are several particles, we cannot define a center-of-mass frame using Eq. (2.3.1) because the mass coefficients are frame dependent. What we can do is define a *center-of-momentum* (COM) frame[4] in which the summed 4-momentum of all the particles is purely time-like, with the spatial part being zero, $\sum \mathbf{p}_i = 0$ (Rindler [2, §30], Goldstein [3, §7.7]). In this frame, the summed 4-momentum for two particles is

$$\mathbf{P}_{\text{tot}} = ((m_A + m_B)c, \mathbf{p}_A + \mathbf{p}_B),$$
$$= ((m_A + m_B)c, \mathbf{0}) \tag{2.3.22}$$

in the COM frame, with invariant

$$P_{\text{tot}}^2 = (m_A + m_B)^2 c^2 \equiv M^2 c^2 \tag{2.3.23}$$

to define an invariant M, interpreted as the mass of the whole system in the COM frame. We evaluate M by considering the entrance $A + B$ channel in the laboratory frame, where we specify that A has kinetic energy T_A, B is at rest, and that they have rest masses m_{0A} and m_{0B}. Initially, therefore, $E_A = m_{0A} c^2 + T_A$ and $E_B = m_{0B} c^2$. The invariant is thus

$$M^2 c^2 = (m_A + m_{0B})^2 c^2 - p_A^2 = (m_{0A} + m_{0B})^2 c^2 + 2 m_{0B} T_A. \tag{2.3.24}$$

In the COM frame, $\mathbf{P}_{\text{tot}}$ has a first component of Mc, whereas in any other frame that component will be increased to $M\gamma c$, for $\gamma = 1/\sqrt{1 - (v/c)^2}$ corresponding to the relative motion of that frame with respect to the COM frame. Consider therefore the laboratory frame, so γ corresponds to the motion v_{com} of the COM frame. In the laboratory frame, the first component is $(m_A + m_{0B})c = (m_{0A} + m_{0B})c + T_A/c$, so

$$\gamma(v_{\text{com}}) = \frac{(m_{0A} + m_{0B})c + T_A/c}{Mc} = \frac{(m_{0A} + m_{0B})c^2 + T_A}{Mc^2}. \tag{2.3.25}$$

We may eliminate T_A using Eq. (2.3.24), giving

$$\gamma(v_{\text{com}}) = \frac{M^2 + m_{0B}^2 - m_{0A}^2}{2 m_{0B} M}. \tag{2.3.26}$$

---

[4] This reference frame is not the same as the center-of-mass (c.m.) frame.

We use this γ value to find the individual energies of A and B in the COM frame. Particle B at rest in the laboratory had a total energy of $E_B = m_{0B}c^2$, which now transforms to

$$E_B^{\mathrm{com}} = \gamma(v_{\mathrm{com}})E_B = \frac{M^2 + m_{0B}^2 - m_{0A}^2}{2M}c^2. \qquad (2.3.27)$$

In the COM frame the energies sum to Mc^2, so

$$E_A^{\mathrm{com}} = Mc^2 - E_B^{\mathrm{com}} = \frac{M^2 + m_{0A}^2 - m_{0B}^2}{2M}c^2. \qquad (2.3.28)$$

The 3-momenta of A and B in the COM frame must be equal and opposite. Their equal magnitude p may be found from $E_i^2 = p^2c^2 + m_{0i}^2c^4$ for either $i = $ A or B. The wave number for relative motion, which we will need for later wave equations, can be evaluated as

$$k = \frac{p}{\hbar} = \frac{\sqrt{(E_A^{\mathrm{com}})^2 - m_{0A}^2 c^4}}{\hbar c}. \qquad (2.3.29)$$

We will also need the Sommerfeld parameter η, which uses the relative velocity v_{rel} between A and B. This is most easily found via $\gamma \equiv \gamma(v_{\mathrm{rel}})$ as

$$\gamma(v_{\mathrm{rel}}) = \frac{T_A}{m_{0A}c^2} + 1, \qquad (2.3.30)$$

so

$$\beta \equiv \frac{v_{\mathrm{rel}}}{c} = \frac{\sqrt{\gamma^2 - 1}}{\gamma} \quad \text{and} \quad \eta = \frac{Z_A Z_B e^2}{\hbar v_{\mathrm{rel}}} = \frac{Z_A Z_B \alpha}{\beta} \qquad (2.3.31)$$

for fine structure constant $\alpha = e^2/\hbar c$.

In an exit channel, the variation of the reduced masses of particles C and D will automatically take into account the Q-value by Eq. (1.1.3). Their 4-momenta when summed will yield the same invariant Mc of Eq. (2.3.23). In the COM frame their 3-momenta will again sum to zero, so their energies may be found by equations similar to (2.3.27, 2.3.28). The wave number for the outgoing relative motion is given by an equation of the same form as Eq. (2.3.29). These all reduce to their non-relativistic limits for $T_A \ll m_{0A}c^2$, or $v_{\mathrm{rel}} \ll c$. Photons, with zero rest mass, must always be treated relativistically.

The above equations are sufficient to define the energies which enter into a quantum mechanical wave equation, and also the wave numbers which are needed to define its boundary conditions. This yields total cross sections, whereas we have not discussed the relativistic treatment of angles and differential cross sections: for these the reader is referred to Goldstein [3, §7.7]. Total cross sections, as ratios of fluxes, are not changed by Lorentz transformations.

This treatment of relativity is of course far short of field-theoretic treatments, for we rather simply use the above energies and wave numbers in the non-relativistic Schrödinger equation. In this way, the theoretical treatment may be at least consistent with the most accurate descriptions of experimental beams and final scattered particles.

2.4 Cross sections

The reaction rates of nuclear reactions are described as cross sections, which are defined as a ratio of outgoing flux to the incoming flux of the beam. There are differential cross sections for outgoing particles to specific angles, as well as integrated cross sections after integrating over all outgoing angles.

2.4.1 Differential cross sections

The angular distribution of particles scattered by some potential $V(\mathbf{R})$ is usually described by the *differential cross section* $\sigma(\theta, \phi)$ for polar angles θ measured from the beam direction, and azimuthal angle ϕ. The number of particles entering a given detector after a scattering reaction will depend on the solid-angular size of the detector $\Delta\Omega$, the number n of scattering centers in the target, and on the flux of the incident beam j_i.

The size of a detector is the solid angle measured in *steradians*, abbreviated sr. If the area of the detector is Σ and it is a distance R from the target, then it subtends a solid angle of $\Delta\Omega = \Sigma/R^2$. If we detect particles coming out in all directions, then $\Sigma = 4\pi R^2$ and $\Delta\Omega = 4\pi$.

The *flux* of incident particles j_i is measured as the number of particles per second per unit area. This is equivalent to the probability density of particles (per unit volume) multiplied by the velocity, so, for a wave function ψ and fixed particle velocity $\mathbf{v}$, the flux is the vector

$$\mathbf{j} = \mathbf{v}|\psi|^2. \tag{2.4.1}$$

The velocity, we saw, has magnitude $v = \hbar k/\mu$ for the relative motion.

We define the differential cross section $\sigma(\theta, \phi)$ as the number of particles N scattered per unit time, per unit scattering center and per unit incident flux, into a unit solid detection angle. When an incident flux of j_i scatters off n scattering centers into a solid angle $\Delta\Omega$, the cross section is therefore the coefficient σ in the equation

$$\frac{dN}{dt} = j_i \, n \, \Delta\Omega \, \sigma. \qquad (2.4.2)$$

It has units of area per solid angle. The area for nuclear-physics cross sections are sometimes given in units of $fm^2 = 10^{-30} \, m^2$, but more often with units of barns (b), where $1 \, b = 10^{-28} \, m^2 = 100 \, fm^2$, or millibarns (mb) where $10 \, mb = 1 \, fm^2$. Most commonly, the differential cross sections $\sigma(\theta, \phi)$ will be given in units of mb/sr.

If we consider just one scattering center $n = 1$, and measure the scattered *angular flux* in the final state as $\hat{j}_f(\theta, \phi)$ particles/second/steradian, then

$$\sigma(\theta, \phi) = \frac{\hat{j}_f(\theta, \phi)}{j_i}. \qquad (2.4.3)$$

The total number of elastically scattered particles is found by integrating $\sigma(\theta, \phi)$ over all the solid angles of detection

$$\sigma_{el} = \int_{4\pi} \sigma(\theta, \phi) \, d\Omega = \int_0^\pi \int_0^{2\pi} \sigma(\theta, \phi) \, d\phi \, \sin\theta d\theta. \qquad (2.4.4)$$

The unit of these *angle-integrated* cross sections is therefore an area. If $\sigma(\theta, \phi)$ were a constant σ_0 for all angles, then we would have $\sigma_{el} = 4\pi \sigma_0$.

The differential cross section is often written as $\frac{d\sigma}{d\Omega}$ to explicitly indicate the cross section per unit solid angle. This notation gives a more natural expression to Eq. (2.4.4) as, generically,

$$\sigma = \int_{4\pi} \frac{d\sigma}{d\Omega} \, d\Omega. \qquad (2.4.5)$$

In this book we will write most simply $\sigma(\theta, \phi)$, a function of angle, to denote the differential cross section $d\sigma/d\Omega$, and σ, without angular arguments, to denote angle-integrated cross sections.

2.4.2 Laboratory and center of mass measures

The total angle-integrated cross sections are the same in the laboratory and center-of-mass frames of reference, since they just measure the total number of incident particles that are deflected by that target. The differential cross sections, however, are different. If $\sigma(\theta, \phi)$ is the c.m. cross section, and $\sigma_{lab}(\theta_{lab}, \phi_{lab})$ the laboratory cross section as a function of laboratory angles, then integrals over a small solid angle must therefore be equal:

$$\sigma(\theta, \phi) \, d\phi \, \sin\theta d\theta = \sigma_{lab}(\theta_{lab}, \phi_{lab}) \, d\phi_{lab} \, \sin\theta_{lab} d\theta_{lab}. \qquad (2.4.6)$$

Using Eqs. (2.3.13) and (2.3.14), we may derive the relation

$$\sigma_{\text{lab}}(\theta_{\text{lab}}, \phi_{\text{lab}}) = \frac{(1 + \rho^2 + 2\rho \cos \theta)^{\frac{3}{2}}}{|1 + \rho \cos \theta|} \sigma(\theta, \phi) \qquad (2.4.7)$$

where we use ρ from Eq. (2.3.16).

2.4.3 Experimental and theoretical cross sections

The *cross section* is an effective meeting point between experiments and theory. Most often, experimentalists transform their measurements into the cross sections $\sigma(\theta, \phi)$ defined in the center-of-mass frame, as a function of c.m. angles.[5]

The task of the theorist is then to predict this c.m. cross section from some model. We have a prejudice for models based on good physical principles that have as few parameters as possible, so that they can be reliably fitted to known data, and so that extrapolations to new energies, as is often required in astrophysics, become more accurate. Microscopic theoretical models are those that start from a Hamiltonian with interaction potentials between the bodies. It would be best if these potentials were known in advance, second best if they were known from other experiments, but sometimes they may have to be fitted to whatever may be the current results so far.

This book focuses on calculating cross sections from defined Hamiltonians, but also considers phenomenological R-matrix models that fit data with a small number of parameters. It also shows how to take cross sections, which depend on energy, and average them over thermal energy distributions to find the *reaction rates* that are the ingredients to network calculations of nuclear reactions in stars.

2.4.4 Cross sections and scattering amplitudes

The scattering of one particle on another may be reduced, we saw earlier, to the dynamical problem of Eq. (2.3.20), involving only their relative coordinate **R**. If the energy of relative motion is E and the reduced mass is μ, then the corresponding wave number is related by $E = \hbar^2 k^2 / 2\mu$, or $k = \sqrt{2\mu E / \hbar^2}$. Remember that the relative velocity is $\mathbf{v} = \mathbf{p}/\mu = \hbar \mathbf{k}/\mu$.

A strict analysis of the scattering of a beam particle by a target nucleus should use wave packets which are localized in space. The projectile and target nuclei will be separately localized apart from each other, and then Schrödinger's time-dependent equation can be used to follow the subsequent evolution of these wave

---

[5] This transformation is unnecessary for angle-integrated cross sections, and must be reconsidered when there are more than two bodies in the final state.

packets under the influence of the potential between the two nuclei. We do not do this, but use *stationary-state scattering theory*, which determines the evolution of each separate Fourier component of a wave packet. In principle, if there are many energies in the wave packet, then a superposition of many stationary-state scattering waves for different E should be used. If, however, the projectile beam has only a narrow range of energies and is well collimated so that it points in nearly a single direction, then the same results can be obtained by using just *one* energy E. We may then set up the projectile beam in a single initial direction $\mathbf{k}_i$. The wave function for such a beam at position $\mathbf{R}$ will look like $\psi^{\text{beam}} = A \exp(i\mathbf{k}_i \cdot \mathbf{R})$.

We can always choose our center-of-mass coordinate system $\mathbf{R} = (x, y, z)$ so that the beam is a plane wave with momentum $\hbar k_i$ in the $+\hat{\mathbf{z}}$ direction, so more simply

$$\psi^{\text{beam}} = A e^{ik_i z}, \qquad (2.4.8)$$

for some amplitude A. In order to calculate the scattering cross sections, which are ratios of fluxes, we need to first determine the incident flux, the number of particles per unit area per unit time. Using Eq. (2.4.1), this is $j_i = v_i |A|^2$ with $v_i = \hbar k_i / \mu$ for the incident relative motion.

The scattered wave starts where the projectile and target are close together, at $R \approx 0$, and radiates outwards with increasing separation R. An outgoing radial wave will asymptotically be proportional to $e^{ik_f R}$ where k_f is the wave number for the outgoing relative motion. The coefficient of this scattered wave will vary with distance R and angles (θ, ϕ), but in such a way that, when integrated over all solid angles, the total flux will be a constant at every large R value. If the scattered wave can be written, for some $f(\theta, \phi)$, as

$$\psi^{\text{scat}} = A f(\theta, \phi) \frac{e^{ik_f R}}{R} \qquad (2.4.9)$$

with the same A in Eq. (2.4.8), then the final outgoing flux, from Eq. (2.4.1), will be

$$j_f = v_f |A|^2 |f(\theta, \phi)|^2 / R^2 \text{ particles/area/second.} \qquad (2.4.10)$$

The function $f(\theta, \phi)$ is called the *scattering amplitude*, has units of length, and in general is complex-valued. The reason for having the $1/R$ factor is to ensure that the flux follows an explicit inverse square law: on integrating j_f over the surface of a sphere of radius R (with $d\Sigma = R^2 d\Omega$),

$$\int j_f d\Sigma = \int v_f |A|^2 |f(\theta, \phi)|^2 R^{-2} R^2 d\Omega = v_f |A|^2 \int |f(\theta, \phi)|^2 d\Omega, \qquad (2.4.11)$$

we have a constant total flux that is independent of R.

The asymptotic form for the combined incident and scattering waves is thus

$$\psi^{\mathrm{asym}} = \psi^{\mathrm{beam}} + \psi^{\mathrm{scat}} = A \left[e^{ik_i z} + f(\theta, \phi) \frac{e^{ik_f R}}{R} \right]. \qquad (2.4.12)$$

The label 'asymptotic' here means that this is the form in the free space outside the range of the interaction potential between the particles.

The scattered angular flux per steradian is $\hat{j}_f = R^2 j_f$, namely

$$\hat{j}_f = v_f |A|^2 |f(\theta, \phi)|^2 \text{ particles/steradian/second.} \qquad (2.4.13)$$

The cross section has been defined as the ratio of scattered angular flux to the incident flux, and so is

$$\sigma(\theta, \phi) = \frac{v_f}{v_i} |f(\theta, \phi)|^2, \qquad (2.4.14)$$

which is independent of A and R, and has units of area/sr such as mb/sr. For elastic scattering $v_f = v_i$, and the velocity ratio (also called the 'flux factor') disappears. Often the flux factor will be absorbed into the definition of a new scattering amplitude

$$\tilde{f}(\theta, \phi) = \sqrt{\frac{v_f}{v_i}} f(\theta, \phi) \qquad (2.4.15)$$

so we have always

$$\sigma(\theta, \phi) = |\tilde{f}(\theta, \phi)|^2. \qquad (2.4.16)$$

We see that the cross section $\sigma(\theta, \phi)$ is independent of A, since A is an overall normalization that scales the incident and scattered waves equally, and the cross section is a ratio of outgoing to incident fluxes. We may therefore fix $A = 1$ without loss of generality.

The next chapter begins with solving the Schrödinger equation to find the scattering amplitudes $f(\theta, \phi)$ for spherical potentials with $A = 1$, so we have immediately the cross section for elastic scattering from Eq. (2.4.14) to compare to experiments. Later, these formulae will be generalized to particles with spin, to handle long-range Coulomb potentials, and then to more general B(A,C)D reactions, which require a multi-channel formulation because the nuclei have a variety of internal states that can be coupled together by the interaction potential.

Exercises

2.1 In the experiment of [4], the elastic scattering of ^{17}F on ^{14}N at 170 MeV was measured by positioning two position-sensitive silicon strip detectors (25 cm^2 in area) symmetrically around the beam axis covering $\theta_{lab} = 3-9$ degrees. A melamine ($C_3N_6H_6$) target was used with density $1.0\,mg/cm^2$ and the beam intensity was 2×10^6 pps.

(a) What was the angular coverage in the center-of-mass frame?
(b) Determine the solid angle corresponding to each detector.
(c) Estimate the order of magnitude of the total cross section obtained in that solid angle.
(d) Convert the differential cross section shown in Fig. 2 of [4] to the laboratory frame.

References

[1] S. N. Ghoshal, *Phys. Rev.* **80** (1950) 939.
[2] W. Rindler 1991, *Introduction to Special Relativity*, Oxford: Clarendon (2nd edn.).
[3] H. Goldstein 1980, *Classical Mechanics*, Reading: Addison-Wesley (2nd edn.).
[4] J. C. Blackmon *et al.*, *Phys. Rev. C* **72** (2005) 034606.

3

Scattering theory

I am now convinced that theoretical physics is actually philosophy.

Max Born

In order to find the cross sections for reactions in terms of the interactions between the reacting nuclei, we have to solve the Schrödinger equation for the quantum-mechanical wave function. Scattering theory tells us how to find these wave functions for the positive (scattering) energies that are needed. We start with the simplest case of finite spherical real potentials between two interacting nuclei in Section 3.1, and use a partial-wave analysis to derive expressions for the elastic scattering cross sections. We then progressively generalize the analysis to allow for long-ranged Coulomb potentials, and also complex-valued optical potentials. Section 3.2 presents the quantum mechanical methods to handle multiple kinds of reaction outcomes, each outcome being described by its own set of partial-wave *channels*, and Section 3.3 then describes how multi-channel methods may be reformulated using integral expressions instead of sets of coupled differential equations. We end the chapter by showing in Section 3.4 how the Pauli principle requires us to describe sets of identical particles, and by showing in Section 3.5 how Maxwell's equations for an electromagnetic field may, in the one-photon approximation, be combined with the Schrödinger equation for the nucleons. Then we can describe photo-nuclear reactions, such as photo-capture and disintegration in a uniform framework.

3.1 Elastic scattering from spherical potentials

When the potential between two interacting nuclei does not depend on the orientation of the vector between them, we say that the potential is *spherical*. In that case, the only reaction that can occur is elastic scattering, which we now proceed to calculate using the method of expansion in partial waves.

3.1.1 Partial-wave scattering from a finite spherical potential

We start our development of scattering theory by finding the elastic scattering from a potential $V(\mathbf{R})$ that is spherically symmetric and so can be written as $V(R)$. Finite potentials will be dealt with first: those for which $V(R) = 0$ for $R \geq R_n$, where R_n is the finite range of the potential. This excludes Coulomb potentials, which will be dealt with later.

We will examine the solutions at positive energy of the time-independent Schrödinger equation with this potential, and show how to find the scattering amplitude $f(\theta, \phi)$ and hence the differential cross section $\sigma(\theta, \phi) = |f(\theta, \phi)|^2$ for elastic scattering. We will use a decomposition in partial waves $L = 0, 1, \ldots$, and the spherical nature of the potential will mean that each partial wave function can be found separately.

The time-independent Schrödinger equation for the relative motion with c.m. energy E, from Eq. (2.3.20), is

$$[\hat{T} + V - E]\psi(R, \theta, \phi) = 0, \tag{3.1.1}$$

using spherical polar coordinates (θ, ϕ) such that $z = R\cos\theta, x = R\sin\theta\cos\phi$ and $y = R\sin\theta\sin\phi$. In this equation, the kinetic energy operator $\hat{T}$ uses the reduced mass μ, and is

$$\begin{aligned}
\hat{T} &= -\frac{\hbar^2}{2\mu}\nabla_R^2 \\
&= \frac{\hbar^2}{2\mu}\left[-\frac{1}{R^2}\frac{\partial}{\partial R}\left(R^2\frac{\partial}{\partial R}\right) + \frac{\hat{L}^2}{R^2}\right],
\end{aligned} \tag{3.1.2}$$

with $\hat{L}^2$ being the squared angular momentum operator

$$\hat{L}^2 = -\hbar^2\left[\frac{1}{\sin^2\theta}\frac{\partial^2}{\partial\phi^2} + \frac{1}{\sin\theta}\frac{\partial}{\partial\theta}\left(\sin\theta\frac{\partial}{\partial\theta}\right)\right]. \tag{3.1.3}$$

We will also use the z component of angular momentum, which is associated with the operator

$$\hat{L}_z = \frac{\hbar}{i}\frac{\partial}{\partial\phi}. \tag{3.1.4}$$

If we keep the z-axis as the beam direction as in Eq. (2.4.8), and illustrated in Fig. 3.1, the coordinates can be much simplified for the present case of spherical potentials. There is now no dependence on ϕ of the initial beam e^{ikz}, which implies that it is an eigensolution of $\hat{L}_z$ with eigenvalue $m = 0$. Furthermore spherical potentials $V(R)$ are independent of both θ and ϕ, which is equivalent to the

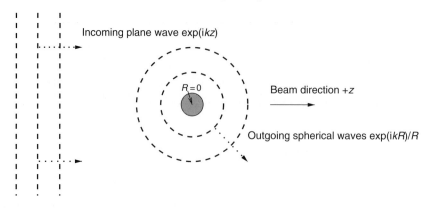

Fig. 3.1. A plane wave in the $+z$ direction incident on a spherical target, giving rise to spherically outgoing scattering waves.

Hamiltonian $\hat{T} + V$ commuting with all components of the angular momentum vector operator $\hat{\mathbf{L}}$, which we write as $[\hat{T}+V, \hat{\mathbf{L}}] = 0$.

These angular independences mean that, since the initial wave function is cylindrically symmetric and no potential breaks that symmetry, the final state must have a wave function that is cylindrically symmetric too, as well as its external scattering amplitude. Thus we need only consider wave functions $\psi(R, \theta)$ and amplitudes $f(\theta, \phi) = f(\theta)$ that are independent of ϕ. In quantum mechanical terms, the potential commutes with $\hat{L}_z$, so the $\hat{L}_z$ eigenvalue is conserved during the reaction. Its conserved value of $m = 0$ implies that the wave function and scattering amplitudes cannot vary with ϕ.

Our problem is therefore to solve

$$[\hat{T} + V - E]\psi(R, \theta) = 0. \tag{3.1.5}$$

The scattering wave functions that are solutions of this equation must, from Eq. (2.4.12), match smoothly at large distances onto the asymptotic form

$$\psi^{\text{asym}}(R, \theta) = e^{ikz} + f(\theta)\frac{e^{ikR}}{R}. \tag{3.1.6}$$

We will thus find a scattering amplitude $f(\theta)$ and hence the differential cross section $\sigma(\theta)$ for elastic scattering from a spherical potential.

Partial-wave expansions

The wave function $\psi(R, \theta)$ is now expanded using Legendre polynomials $P_L(\cos \theta)$, in what is called a *partial-wave expansion*. We choose these polynomials as they are eigenfunctions of both $\hat{L}^2$ and $\hat{L}_z$, with eigenvalues $L(L+1)$ and $m = 0$ respectively. We saw above that in the present case we only need solutions with $m = 0$, as these solutions are independent of ϕ. Furthermore, since the potential commutes with $\hat{L}$, we can find solutions with particular values of angular momentum L. These solutions for individual L are called *partial waves*.

Let us first write a single partial wave as the product of a Legendre polynomial $P_L(\cos\theta)$ for any $L = 0, 1, \ldots$, and a part that depends on radius, which we write as $\chi_L(R)/R$ for some function $\chi_L(R)$ that we have yet to determine. When operating on this partial-wave product, the ∇_R^2 in the kinetic energy operator gives

$$\nabla_R^2 P_L(\cos\theta)\frac{\chi_L(R)}{R} = \frac{1}{R}\left(\frac{d^2}{dR^2} - \frac{L(L+1)}{R^2}\right)\chi_L(R)\,P_L(\cos\theta). \qquad (3.1.7)$$

The $1/R$ factor of $\chi_L(R)$ was chosen so that the second derivative d^2/dR^2 appears in a simple form on the right-hand side of this equation.

Moreover, the $\{P_L(\cos\theta)\}$ together form an orthogonal and complete set over angles $0 \le \theta \le \pi$, satisfying the orthogonality and normalization condition

$$\int_0^\pi P_L(\cos\theta)P_{L'}(\cos\theta)\sin\theta\,d\theta = \frac{2}{2L+1}\delta_{LL'}. \qquad (3.1.8)$$

This means that any function of angle can be expanded as $\sum_L b_L P_L(\cos\theta)$ for some coefficients b_L. Thus any function of angle and radius can be expanded as $\sum_L b_L(R)P_L(\cos\theta)$ for the $b_L(R)$ now functions of R.

We can therefore expand the full wave function as

$$\psi(R,\theta) = \sum_{L=0}^{\infty}(2L+1)i^L P_L(\cos\theta)\frac{1}{kR}\chi_L(R) \qquad (3.1.9)$$

for functions $\chi_L(R)$ to be found. The explicit factors $(2L+1)i^L$ are built in for convenience, so that, as we will see, the $\chi_L(R)$ have simple forms in the limit of zero potential. The finiteness of $\psi(R,\theta)$ everywhere implies at least that $\chi_L(0) = 0$ always.[1]

We now substitute the partial wave sum of Eq. (3.1.9) in the Schrödinger equation Eq. (3.1.1) to find the conditions satisfied by the radial wave functions $\chi_L(R)$. For the kinetic energy applied to Eq. (3.1.9), we use the differential properties (3.1.7) of the Legendre polynomials. For the potential energy term, we use the fact that $V(R)$ is independent of θ. After multiplying on the left by $P_{L'}(\cos\theta)$, integrating over all angles, and using the orthogonality properties of Eq. (3.1.8), we conclude that for each L value there is a separate *partial-wave equation*

$$\left[-\frac{\hbar^2}{2\mu}\left(\frac{d^2}{dR^2} - \frac{L(L+1)}{R^2}\right) + V(R) - E\right]\chi_L(R) = 0. \qquad (3.1.10)$$

Coulomb functions

That $F_L(\eta, \rho)$ is regular means $F_L(\eta, \rho{=}0) = 0$, and irregularity means $G_L(\eta, \rho{=}0) \neq 0$. They are related by the Wronskian

$$G_L(\eta, \rho) \frac{\mathrm{d}F_L(\eta, \rho)}{\mathrm{d}\rho} - F_L(\eta, \rho)\frac{\mathrm{d}G_L(\eta, \rho)}{\mathrm{d}\rho} = 1$$

$$\text{or} \quad W(G, F) \equiv G\,F' - G'\,F = k. \tag{3.1.11}$$

Note that mathematics texts such as [1] usually define G' as $dG/d\rho$, but we denote this by $\dot{G}$. Since ρ is the dimensionless radius $\rho = kR$, we will use the prime for derivatives with respect to R, so $G' = k\dot{G}$, etc. The Wronskian is equivalently $G\dot{F} - \dot{G}F = 1$.

The Coulomb Hankel functions are combinations of F and G,

$$H_L^{\pm}(\eta, \rho) = G_L(\eta, \rho) \pm \mathrm{i}F_L(\eta, \rho). \tag{3.1.12}$$

Coulomb functions for $\eta = 0$

The $\eta = 0$ functions are more directly known in terms of Bessel functions:

$$F_L(0, \rho) = \rho\, j_L(\rho) = (\pi\rho/2)^{1/2} J_{L+1/2}(\rho)$$

$$G_L(0, \rho) = -\rho\, y_L(\rho) = -(\pi\rho/2)^{1/2} Y_{L+1/2}(\rho), \tag{3.1.13}$$

where the irregular spherical Bessel function $y_L(\rho)$ is sometimes written as $n_L(\rho)$ (the Neumann function). The J_ν and Y_ν are the cylindrical Bessel functions. The $\eta = 0$ Coulomb functions for the first few L values are

$$F_0(0, \rho) = \sin \rho,$$

$$G_0(0, \rho) = \cos \rho; \tag{3.1.14}$$

$$F_1(0, \rho) = (\sin \rho - \rho \cos \rho)/\rho,$$

$$G_1(0, \rho) = (\cos \rho + \rho \sin \rho)/\rho; \tag{3.1.15}$$

$$F_2(0, \rho) = ((3{-}\rho^2) \sin \rho - 3\rho \cos \rho)/\rho^2,$$

$$G_2(0, \rho) = ((3{-}\rho^2) \cos \rho + 3\rho \sin \rho)/\rho^2. \tag{3.1.16}$$

Their behaviour near the origin, for $\rho \ll L$, is

$$F_L(0, \rho) \sim \frac{1}{(2L{+}1)(2L{-}1)\cdots 3.1}\rho^{L+1} \tag{3.1.17}$$

$$G_L(0, \rho) \sim (2L{-}1)\cdots 3.1\, \rho^{-L}, \tag{3.1.18}$$

Box 3.1 (*Continued*)
and their asymptotic behaviour when $\rho \gg L$ is

$$F_L(0, \rho) \sim \sin(\rho - L\pi/2)$$

$$G_L(0, \rho) \sim \cos(\rho - L\pi/2) \tag{3.1.19}$$

$$H_L^{\pm}(0, \rho) \sim e^{\pm i(\rho - L\pi/2)} = i^{\mp L} e^{\pm i\rho}. \tag{3.1.20}$$

So H_L^{+} describes an outgoing wave $e^{i\rho}$, and H_L^{-} an incoming wave $e^{-i\rho}$.
Coulomb functions for $\eta \neq 0$ are described on page 62, and Whittaker functions on page 135.

Box 3.1 **Coulomb functions**

The spherical nature of the potential is crucial in allowing us to solve for each partial-wave function separately; this corresponds to angular momentum being conserved when potentials are spherical.

Equation (3.1.10) is a second-order equation, and so needs two boundary conditions specified in order to fix a solution. One boundary condition already known is that $\chi_L(0) = 0$. The other is fixed by the large R behavior, so that it reproduces the external form of Eq. (3.1.6). Since $f(\theta)$ is not yet known, the role of Eq. (3.1.6) is to fix the overall normalization of the $\chi_L(R)$. We show below how to accomplish these things.

As usual in quantum mechanical matching, both the functions and their derivatives must agree continuously. We therefore match interior and exterior functions and their derivatives at some *matching radius* $R = a$ chosen outside the finite range R_n of the nuclear potential.

Radial solutions for zero potential

For $R \geq a$ we have $V(R) = 0$, so at and outside the matching radius the radial wave functions must attain their external forms, which we name $\chi_L^{\text{ext}}(R)$. The free-field partial-wave equation may be simplified from Eq. (3.1.10), and rewritten using a change of variable from R to the dimensionless radius

$$\rho \equiv kR, \tag{3.1.21}$$

so the $\chi_L^{\text{ext}}(R)$ satisfy

$$\left[\frac{d^2}{d\rho^2} - \frac{L(L+1)}{\rho^2} + 1 \right] \chi_L^{\text{ext}}(\rho/k) = 0. \tag{3.1.22}$$

---

[1] It implies more precisely that $\chi_L(R) = O(R)$ as $R \to 0$.

This equation for the external form χ_L^{ext} is a special case for $\eta = 0$ of the more general *Coulomb wave equation*

$$\left[\frac{d^2}{d\rho^2} - \frac{L(L+1)}{\rho^2} - \frac{2\eta}{\rho} + 1\right] X_L(\eta, \rho) = 0, \qquad (3.1.23)$$

which has solutions defined in Abramowitz and Stegun [1, ch. 14] and described in more detail in Box 3.1.[2] This second-order equation has two well-known solutions that are linearly independent: the regular function $F_L(\eta, \rho)$ and the irregular function $G_L(\eta, \rho)$. A regular function is so named because it is zero at $\rho = 0$, and an irregular function because it is non-zero at $\rho = 0$. Any solution X_L can be written as $X_L = bF_L + cG_L$ for some constants b and c chosen according to the boundary conditions.

We may construct $H_L^\pm = G_L \pm iF_L$, which are also two linearly independent solutions of Eq. (3.1.23). By Eq. (3.1.20), these functions are asymptotically proportional to $e^{\pm ikR}$. Since the radial momentum operator is $\hat{p} = \frac{\hbar}{i}\frac{\partial}{\partial R}$, the linear combinations $H_L^\pm$ asymptotically have radial momentum eigenvalues $\pm \hbar k$, and this means that H_L^+ describes a radially *outgoing wave* and H_L^- an *incoming* wave.

The partial-wave expansion of Eq. (3.1.9) can be found for any function $\psi(R, \theta)$. In particular, it can be proved [2] that the partial-wave expansion for the incident plane wave is

$$e^{ikz} = \sum_{L=0}^{\infty}(2L+1)i^L P_L(\cos\theta)\frac{1}{kR}F_L(0, kR), \qquad (3.1.24)$$

using just the regular Coulomb functions $F_L(0, kR)$. Their appearance in this equation is the reason that $(2L+1)i^L$ were defined in Eq. (3.1.9), as we now have $\chi_L = F_L$ in the plane-wave limit when the potential is zero.

In terms of the $H_L^\pm = G_L \pm iF_L$, the plane-wave expansion is equivalently

$$e^{ikz} = \sum_{L=0}^{\infty}(2L+1)i^L P_L(\cos\theta)\frac{1}{kR}\frac{i}{2}[H_L^-(0, kR) - H_L^+(0, kR)]. \qquad (3.1.25)$$

From this equation we see that the incident beam has equal and opposite amplitudes of the radially ingoing wave H_L^- and the radially outgoing wave H_L^+. This describes a plane wave approaching the target, and leaving it again unchanged.

Radial solutions with a potential

Outside the potential we know the external form of Eq. (3.1.6), but not the scattering amplitudes $f(\theta)$. At the origin we know that $\chi_L(0) = 0$, but not the derivatives $\chi_L'(0)$.

---

[2] In this section we need just the special case of $\eta = 0$, but we have begun here with a definition of the complete Coulomb functions, as the $2\eta/\rho$ term will reappear in the next section when Coulomb potentials are introduced.

Because the boundary conditions are thus distributed at different radii, we have to use trial solutions integrated from one radial limit, and determine the unknown parameters by using the boundary condition at the other limit.

We may therefore integrate a trial solution $u_L(R)$ of Eq. (3.1.10) outwards, starting with $u_L(0) = 0$ and some finite $u'_L(0) \neq 0$ chosen arbitrarily, using

$$u_L''(R) = \left[\frac{L(L+1)}{R^2} + \frac{2\mu}{\hbar^2}(V(R) - E) \right] u_L(R) \tag{3.1.26}$$

and some numerical integration scheme for second-order ordinary differential equations. The true solution will be some multiple of this: $\chi_L(R) = Bu_L(R)$. In the external region outside the potential, $u_L(R)$ will be found to be a linear combination of two linearly independent solutions of Eq. (3.1.22), such as $H_L^+(0, kR)$ and $H_L^-(0, kR)$:

$$Bu_L(R) = \chi_L(R) \overset{R > R_n}{\longrightarrow} \chi_L^{\text{ext}}(R) = A_L \left[H_L^-(0, kR) - \mathbf{S}_L H_L^+(0, kR) \right], \tag{3.1.27}$$

for some complex constants B, A_L and $\mathbf{S}_L$. The $\mathbf{S}_L$ is called the *partial-wave S-matrix element*. (It will be unity for zero potential $V(R)$, as the solution must then be proportional to $F_L(0, kR)$ only, by Eq. (3.1.24).) The $\mathbf{S}_L$ is determined in general from $u_L(R)$ by matching the first and last terms of Eq. (3.1.27), and their derivatives, at the radius $R = a$, which is outside the nuclear range R_n. This is most conveniently done by constructing the inverse logarithmic derivative, which is called the R-matrix element

$$\mathbf{R}_L = \frac{1}{a} \frac{\chi_L(a)}{\chi_L'(a)} = \frac{1}{a} \frac{u_L(a)}{u_L'(a)}, \tag{3.1.28}$$

where the a^{-1} factor is used traditionally to keep the R matrix dimensionless. Given a trial solution $u_L(R)$ in the interior region, although its absolute normalization B is not yet known, its logarithmic derivative is the same as that of $\chi_L(R)$, and thus unambiguously determines $\mathbf{R}_L$. The $\mathbf{R}_L$ then determines the S-matrix element uniquely by matching with the inverse logarithmic derivative of $\chi_L^{\text{ext}}(R)$:

$$\mathbf{R}_L = \frac{1}{a} \frac{H_L^- - \mathbf{S}_L H_L^+}{H_L'^- - \mathbf{S}_L H_L'^+}, \tag{3.1.29}$$

implying

$$\mathbf{S}_L = \frac{H_L^- - a\mathbf{R}_L H_L'^-}{H_L^+ - a\mathbf{R}_L H_L'^+}. \tag{3.1.30}$$

The matrix elements $\mathbf{S}_L$ are thereby uniquely determined by the potential. Next we use them to find the scattering amplitude $f(\theta)$.

Equation (3.1.27) implies that the full scattering wave function has the external form of the partial-wave sum

$$\psi(R,\theta) \overset{R>R_n}{\to} \frac{1}{kR}\sum_{L=0}^{\infty}(2L+1)i^L P_L(\cos\theta)A_L[H_L^-(0,kR)-\mathbf{S}_L H_L^+(0,kR)], \quad (3.1.31)$$

which we now have to match with Eq. (3.1.6) in order to determine $f(\theta)$ in terms of the $\mathbf{S}_L$.

Substituting Eq. (3.1.25) in Eq. (3.1.6), equating to the right-hand side of Eq. (3.1.31), multiplying by kR and using the asymptotic forms of Eq. (3.1.20) for the $H_L^\pm$ functions, we have

$$\sum_{L=0}^{\infty}(2L+1)i^L P_L(\cos\theta)A_L\left[i^L e^{-ikR}-\mathbf{S}_L i^{-L}e^{ikR}\right]$$

$$=\sum_{L=0}^{\infty}(2L+1)i^L P_L(\cos\theta)\frac{i}{2}(i^L e^{-ikR}-i^{-L}e^{ikR})+kf(\theta)e^{ikR} \quad (3.1.32)$$

when both $R>R_n$ and $kR\gg L$. Collecting together the separate terms with e^{ikR} and e^{-ikR} factors, we find

$$e^{ikR}\left[\sum_{L=0}^{\infty}(2L+1)i^L P_L(\cos\theta)\left\{A_L\mathbf{S}_L i^{-L}-\frac{i}{2}i^{-L}\right\}+kf(\theta)\right]$$

$$=e^{-ikR}\left[\sum_{L=0}^{\infty}(2L+1)i^L P_L(\cos\theta)\left\{A_L i^L-\frac{i}{2}i^L\right\}\right]. \quad (3.1.33)$$

Since the $e^{\pm ikR}$ are linearly independent, and the two [...] expressions in this equation are independent of R, they must each be identically zero. Furthermore, using the orthogonality (3.1.8) of the Legendre polynomials, the second $\{\ldots\}$ expression must also be zero, which implies $A_L=i/2$. Substituting this result into the first zero [...] expression, we derive

$$f(\theta)=\frac{1}{2ik}\sum_{L=0}^{\infty}(2L+1)P_L(\cos\theta)(\mathbf{S}_L-1). \quad (3.1.34)$$

This important equation (3.1.34) constructs the scattering amplitude in terms of the S-matrix elements. The elastic differential cross section is thus

$$\sigma(\theta)\equiv\frac{d\sigma}{d\Omega}=\left|\frac{1}{2ik}\sum_{L=0}^{\infty}(2L+1)P_L(\cos\theta)(\mathbf{S}_L-1)\right|^2. \quad (3.1.35)$$

The resulting full scattering wave function is

$$\psi(R, \theta) = \sum_{L=0}^{\infty} (2L+1) i^L P_L(\cos\theta) \frac{1}{kR} \chi_L(R), \tag{3.1.36}$$

where the radial functions have external form for $R > R_n$ in detail as

$$\chi_L^{\text{ext}}(R) = \frac{i}{2} \left[H_L^-(0, kR) - \mathbf{S}_L H_L^+(0, kR) \right]. \tag{3.1.37}$$

Phase shifts

Each matrix element $\mathbf{S}_L$ is equivalently described by a *phase shift* δ_L for each partial wave by

$$\mathbf{S}_L = e^{2i\delta_L} \tag{3.1.38}$$

by taking complex logarithms as $\delta_L = \frac{1}{2i} \ln \mathbf{S}_L$. Phase shifts are thus defined up to additive multiples of π. We often add suitable integer multiples $n(E)$ of π,

$$\delta_L(E) = \frac{1}{2i} \ln \mathbf{S}_L + n(E)\pi \tag{3.1.39}$$

to make $\delta_L(E)$ continuous functions of energy E for each separate partial wave L, but no cross section should depend on this addition as $e^{2i\pi n} \equiv 1$.

In terms of the phase shift δ_L, the scattering amplitude can be written as

$$f(\theta) = \frac{1}{k} \sum_{L=0}^{\infty} (2L+1) P_L(\cos\theta) e^{i\delta_L} \sin\delta_L, \tag{3.1.40}$$

and the external form of Eq. (3.1.37) of the wave function is equivalently

$$\chi_L^{\text{ext}}(R) = e^{i\delta_L} \left[\cos\delta_L \, F_L(0, kR) + \sin\delta_L \, G_L(0, kR) \right]. \tag{3.1.41}$$

In this form we can see the reason for the name 'phase shift.' In the asymptotic region where both $R > R_n$ and $kR \gg L$, we may use Eq. (3.1.19) to write Eq. (3.1.41) as

$$\chi_L^{\text{ext}}(R) \to e^{i\delta_L} [\cos\delta_L \sin(kR - L\pi/2) + \sin\delta_L \cos(kR - L\pi/2)]$$

$$= e^{i\delta_L} \sin(kR + \delta_L - L\pi/2). \tag{3.1.42}$$

The oscillations are therefore shifted to smaller R when δ_L is positive, which occurs for attractive potentials (at least when they are weak). The oscillatory patterns shift to larger R when $\delta_L < 0$ for repulsive potentials. Physically, attractive potentials pull the oscillations into its range, and repulsive potentials tend to expel the scattering oscillations.

The external solution can be also be written as

$$\chi_L^{\text{ext}}(R) = F_L(0, kR) + \mathbf{T}_L H_L^+(0, kR), \tag{3.1.43}$$

where

$$\mathbf{T}_L = e^{i\delta_L} \sin \delta_L \tag{3.1.44}$$

is called the *partial wave T-matrix element*. By Eq. (3.1.43), $\mathbf{T}_L$ is the coefficient of $H_L^+(0, kR)$, an outgoing wave, and is related to the S-matrix element by

$$\mathbf{S}_L = 1 + 2i\mathbf{T}_L. \tag{3.1.45}$$

The scattering amplitude in terms of the $\mathbf{T}_L$ is

$$f(\theta) = \frac{1}{k} \sum_{L=0}^{\infty} (2L+1) P_L(\cos \theta) \mathbf{T}_L. \tag{3.1.46}$$

For zero potential, $\delta_L = \mathbf{T}_L = 0$, and $\chi_L^{\text{ext}}(R) = F_L(0, kR)$ only, the regular partial-wave component of the incident plane wave.

A third form of the external wave function is

$$\chi_L^{\text{ext}}(R) = e^{i\delta_L} \cos \delta_L \left[F_L(0, kR) + \mathbf{K}_L G_L(0, kR) \right], \tag{3.1.47}$$

with $\mathbf{K}_L = \tan \delta_L$, called the *partial-wave K-matrix element*. The S-matrix element in terms of this is

$$\mathbf{S}_L = \frac{1 + i\mathbf{K}_L}{1 - i\mathbf{K}_L}. \tag{3.1.48}$$

For zero potential, $\mathbf{K}_L = 0$. The K-matrix element may be directly found from the R-matrix element $\mathbf{R}_L$ of the interior solution at the matching radius a as

$$\mathbf{K}_L = -\frac{F_L - a\mathbf{R}_L F_L'}{G_L - a\mathbf{R}_L G_L'}, \tag{3.1.49}$$

where the arguments of the F_L and G_L are omitted for clarity. All of the above scattering phase shifts δ_L and partial-wave elements $\mathbf{T}_L, \mathbf{S}_L$, and $\mathbf{K}_L$ are independent of a, provided that it is larger than the range R_n of the potential.

From Eq. (3.1.49) we can see the consequences of $V(R)$ being real. In this case, the trial function $u(R)$ may be real, and hence also $\mathbf{R}_L$ by Eq. (3.1.28), $\mathbf{K}_L$ by Eq. (3.1.49), and hence δ_L will be real since $\tan \delta_L = \mathbf{K}_L$. It is for these reasons that scattering from a real potential is most often described by a (real) phase shift. This corresponds to the matrix element $\mathbf{S}_L = e^{2i\delta_L}$ having unit modulus, $|\mathbf{S}_L| = 1$.

Table 3.1. *Relations between the wave functions and the phase shifts, **K**-, **T**- and **S**-matrix elements, for an arbitrary partial wave. Partial-wave indices and the arguments of the Coulomb functions are omitted for clarity. The last two lines list the consequences of zero and real potentials.*

Using:	δ	**K**	**T**	**S**
$\chi(R) =$	$e^{i\delta}[F\cos\delta + G\sin\delta]$	$\dfrac{1}{1-i\mathbf{K}}[F + \mathbf{K}G]$	$F + \mathbf{T}H^+$	$\dfrac{i}{2}[H^- - \mathbf{S}H^+]$
$\delta =$	δ	$\arctan\mathbf{K}$	$\arctan\dfrac{\mathbf{T}}{1+i\mathbf{T}}$	$\dfrac{1}{2i}\ln\mathbf{S}$
$\mathbf{K} =$	$\tan\delta$	$\mathbf{K}$	$\dfrac{\mathbf{T}}{1+i\mathbf{T}}$	$i\dfrac{1-\mathbf{S}}{1+\mathbf{S}}$
$\mathbf{T} =$	$e^{i\delta}\sin\delta$	$\dfrac{\mathbf{K}}{1-i\mathbf{K}}$	$\mathbf{T}$	$\dfrac{i}{2}(1-\mathbf{S})$
$\mathbf{S} =$	$e^{2i\delta}$	$\dfrac{1+i\mathbf{K}}{1-i\mathbf{K}}$	$1+2i\mathbf{T}$	$\mathbf{S}$
$V = 0$	$\delta = 0$	$\mathbf{K} = 0$	$\mathbf{T} = 0$	$\mathbf{S} = 1$
V real	δ real	$\mathbf{K}$ real	$|1 + 2i\mathbf{T}| = 1$	$|\mathbf{S}| = 1$

The relations between the phase shifts and the K-, T- and S-matrix elements are summarized in Table 3.1.

Angle-integrated cross sections

From the cross section $\sigma(\theta) = |f(\theta)|^2$ given by Eq. (3.1.35), we may integrate over the entire sphere to find the angle-integrated cross section[3]

$$\sigma_{\text{el}} = \int_0^{2\pi} d\phi \int_0^{\pi} d\theta \sin\theta \, \sigma(\theta)$$

$$= 2\pi \int_0^{\pi} d\theta \sin\theta |f(\theta)|^2$$

$$= \frac{\pi}{k^2} \sum_{L=0}^{\infty} (2L+1)|1 - \mathbf{S}_L|^2$$

$$= \frac{4\pi}{k^2} \sum_{L=0}^{\infty} (2L+1)\sin^2\delta_L, \qquad (3.1.50)$$

using the orthogonality and normalization Eq. (3.1.8) of the Legendre polynomials.

---

[3] Note that this integrated cross section is sometimes called the *total* cross section because it is the total after integration over all angles. However, we reserve the term 'total' to include all non-elastic final states too, as will be used in subsection 3.2.1.

There exists an *optical theorem* which relates the angle-integrated cross section σ_{el} to the zero-angle scattering amplitude. Using $P_L(1) = 1$, we have

$$f(0) = \frac{1}{2ik} \sum_{L=0}^{\infty} (2L+1)(e^{2i\delta_L} - 1),$$ (3.1.51)

so

$$\text{Im} f(0) = \frac{1}{k} \sum_{L=0}^{\infty} (2L+1) \sin^2 \delta_L$$

$$= \frac{k}{4\pi} \sigma_{el}.$$ (3.1.52)

This relation exists because the incident and scattered waves at zero scattering angle have a fixed relative phase, and interfere in a manner that portrays the total loss of flux from the incident wave to the scattered waves.

Scattering using rotated coordinate systems

To find the scattering from an incident beam in direction $\mathbf{k}$ (not necessarily in the $+\hat{\mathbf{z}}$ direction) to direction $\mathbf{k}'$, the Legendre polynomial $P_L(\cos\theta) = P_L(\hat{\mathbf{k}} \cdot \hat{\mathbf{k}}')$ in Eq. (3.1.34) may be simply replaced using the addition theorem for spherical harmonics [2],

$$P_L(\cos\theta) = \frac{4\pi}{2L+1} \sum_{M=-L}^{L} Y_L^M(\hat{\mathbf{k}}) Y_L^M(\hat{\mathbf{k}}')^*.$$ (3.1.53)

where $\hat{\mathbf{k}}$ is the notation for a unit vector in the $\mathbf{k}$ direction. This, from Eq. (3.1.24), gives the partial-wave expansion of a plane wave in direction $\hat{\mathbf{k}}$ as

$$e^{i\mathbf{k}\cdot\mathbf{R}} = 4\pi \sum_{LM} i^L Y_L^M(\hat{\mathbf{R}}) Y_L^M(\hat{\mathbf{k}})^* \frac{1}{kR} F_L(0, kR).$$ (3.1.54)

Thus the amplitude $f(\mathbf{k}'; \mathbf{k})$ for scattering from direction $\mathbf{k}$ to $\mathbf{k}'$ is

$$f(\mathbf{k}'; \mathbf{k}) = \frac{2\pi}{ik} \sum_{LM} Y_L^M(\hat{\mathbf{k}}') Y_L^M(\hat{\mathbf{k}})^* (e^{2i\delta_L} - 1)$$ (3.1.55)

$$= \frac{4\pi}{k} \sum_{LM} Y_L^M(\hat{\mathbf{k}}') Y_L^M(\hat{\mathbf{k}})^* \mathbf{T}_L,$$ (3.1.56)

and the full scattering wave function depends on the incident momentum as

$$\psi(\mathbf{R}; \mathbf{k}) = 4\pi \sum_{LM} i^L Y_L^M(\hat{\mathbf{R}}) Y_L^M(\hat{\mathbf{k}})^* \frac{1}{kR} \chi_L(R).$$ (3.1.57)

In both cases, the vector **k** after the semicolon indicates the incident momentum. Note that a spherical harmonic, for its unit vector argument in the $+z$ direction, is $Y_L^M(\hat{\mathbf{z}}) = \delta_{M0}\sqrt{\frac{2L+1}{4\pi}}$, so another form of Eq. (3.1.24) is

$$\mathrm{e}^{\mathrm{i}kz} = \sqrt{4\pi} \sum_{L=0}^{\infty} \mathrm{i}^L \sqrt{2L+1}\; Y_L^0(\hat{\mathbf{R}}) \frac{1}{kR} F_L(0, kR) \tag{3.1.58}$$

for the plane wave in the $+\hat{\mathbf{z}}$ direction.

Coulomb functions for $\eta \neq 0$

The functions $F_L(\eta, \rho)$, $G_L(\eta, \rho)$ and $H_L^{\pm}(\eta, \rho)$ are the solutions of Eq. (3.1.23) for $\eta \neq 0$. In terms of a 'Coulomb constant'

$$C_L(\eta) = \frac{2^L \mathrm{e}^{-\pi\eta/2}|\Gamma(1+L+\mathrm{i}\eta)|}{(2L+1)!} \tag{3.1.59}$$

and the confluent hypergeometric function $_1F_1(a; b; z) \equiv M(a, b, z)$, the regular function defined in Abramowitz and Stegun [1, ch. 13] as

$$F_L(\eta, \rho) = C_L(\eta)\rho^{L+1}\mathrm{e}^{\mp\mathrm{i}\rho}\; _1F_1(L+1 \mp \mathrm{i}\eta; 2L+2; \pm 2\mathrm{i}\rho), \tag{3.1.60}$$

where either the upper or lower signs may be taken for the same result. The $_1F_1(a; b; z)$ is defined by the series expansion

$$_1F_1(a; b; z) = 1 + \frac{a}{b}\frac{z}{1!} + \frac{a(a+1)}{b(b+1)}\frac{z^2}{2!} + \frac{a(a+1)(a+2)}{b(b+1)(b+2)}\frac{z^3}{3!} + \cdots. \tag{3.1.61}$$

The two irregular functions have the corresponding definitions

$$H_L^{\pm}(\eta, \rho) = G_L(\eta, \rho) \pm \mathrm{i}F_L(\eta, \rho) \tag{3.1.62}$$

$$= \mathrm{e}^{\pm\mathrm{i}\Theta}(\mp 2\mathrm{i}\rho)^{1+L\pm\mathrm{i}\eta} U(1+L \pm \mathrm{i}\eta, 2L+2, \mp 2\mathrm{i}\rho) \tag{3.1.63}$$

where $U(a, b, z)$ is the corresponding irregular confluent hypergeometric function defined in [1, ch. 13]. The $\Theta \equiv \rho - L\pi/2 + \sigma_L(\eta) - \eta\ln(2\rho)$, and

$$\sigma_L(\eta) = \arg\Gamma(1+L+\mathrm{i}\eta) \tag{3.1.64}$$

is called the *Coulomb phase shift*. The functions may easily be calculated numerically [1, 3], also for complex arguments [4].

Their behavior near the origin is thus

$$F_L(\eta, \rho) \sim C_L(\eta)\rho^{L+1}, \quad G_L(\eta, \rho) \sim \left[(2L+1)C_L(\eta)\,\rho^L\right]^{-1}, \tag{3.1.65}$$

Box 3.2 (*Continued*)
noting that

$$C_0(\eta) = \sqrt{\frac{2\pi\eta}{e^{2\pi\eta} - 1}} \quad \text{and} \quad C_L(\eta) = \frac{\sqrt{L^2 + \eta^2}}{L(2L+1)} C_{L-1}(\eta). \tag{3.1.66}$$

A transition from small-ρ power law behavior to large-ρ oscillatory behavior occurs outside the classical turning point. This point is where $1 = 2\eta/\rho + L(L+1)/\rho^2$, namely

$$\rho_{\text{tp}} = \eta \pm \sqrt{\eta^2 + L(L+1)}. \tag{3.1.67}$$

In nuclear reactions η is usually positive, so with purely Coulomb and centrifugal potentials there is only one turning point. Classically, the turning point is at the distance of closest approach, R_{near} of Eq. (3.1.79), and these quantities are related by $\rho_{\text{tp}} = kR_{\text{near}}$ if the classical impact parameter b is related to the quantum mechanical partial wave L according to

$$k\,b = \sqrt{L(L+1)} \approx L + \tfrac{1}{2}. \tag{3.1.68}$$

The asymptotic behaviour of the Coulomb functions outside the turning point ($\rho \gg \rho_{\text{tp}}$) is

$$F_L(\eta, \rho) \sim \sin\Theta, \quad G_L(\eta, \rho) \sim \cos\Theta, \quad \text{and} \quad H_L^{\pm}(\eta, \rho) \sim e^{\pm i\Theta}. \tag{3.1.69}$$

Box 3.2 **Coulomb functions for $\eta \neq 0$**

3.1.2 Coulomb and nuclear potentials

In general, we saw in Chapter 1, nuclei have between them both a short-range attractive nuclear potential and a long-range Coulomb repulsion. This Coulomb component has the $1/R$ form shown in Fig. 1.3, and invalidates the theory presented above for finite-range potentials. We develop below a theory for a pure $1/R$ component, and then see how to add to it the effects of an additional finite-range correction. We still assume both the Coulomb and nuclear potentials to be spherical.

Pure point-Coulomb scattering

If we consider *only* the point-Coulomb potential between two particles with charges Z_1 and Z_2 times the unit charge e, we have

$$V_c(R) = Z_1 Z_2 e^2 / R. \tag{3.1.70}$$

For scattering with relative velocity v, we define a dimensionless *Sommerfeld parameter* η, as mentioned before, by

$$\eta = \frac{Z_1 Z_2 e^2}{\hbar v} = \frac{Z_1 Z_2 e^2 \mu}{\hbar^2 k} = \frac{Z_1 Z_2 e^2}{\hbar} \left(\frac{\mu}{2E}\right)^{\frac{1}{2}}. \qquad (3.1.71)$$

where the energy of relative motion is $E = \hbar^2 k^2/2\mu$. A beam in the direction $\mathbf{k}$ is no longer $e^{i\mathbf{k}\cdot\mathbf{R}}$ when $\eta \neq 0$. Fortunately, the Schrödinger equation with the Coulomb potential can be solved exactly, and the solution in terms of hypergeometric functions is

$$\psi_c(\mathbf{k}, \mathbf{R}) = e^{i\mathbf{k}\cdot\mathbf{R}} e^{-\pi\eta/2} \, \Gamma(1+i\eta) \, {}_1F_1(-i\eta; 1; i(kR - \mathbf{k}\cdot\mathbf{R})), \qquad (3.1.72)$$

defined using the gamma function $\Gamma(z)$ and the confluent hypergeometric function ${}_1F_1(a; b; z)$ of Eq. (3.1.61).

In partial-wave form, the generalization of the standard $+\hat{\mathbf{z}}$ plane-wave expansion Eq. (3.1.24) is

$$\psi_c(k\hat{\mathbf{z}}, \mathbf{R}) = \sum_{L=0}^{\infty} (2L+1)i^L P_L(\cos\theta) \frac{1}{kR} F_L(\eta, kR). \qquad (3.1.73)$$

We are now using the regular Coulomb function $F_L(\eta, kR)$ with $\eta \neq 0$. Details of F_L and G_L for general η, and their asymptotic forms for small and large ρ, are given in Box 3.2. In particular, the asymptotic form of $F_L(\eta, kR)$ is

$$F_L(\eta, kR) \sim \sin(kR - L\pi/2 + \sigma_L(\eta) - \eta\ln(2kR)), \qquad (3.1.74)$$

with the *Coulomb phase shift* $\sigma_L(\eta)$ given by Eq. (3.1.64). The logarithmic term in this expression is needed to accommodate the $1/R$ Coulomb potential. Thus the phase shift caused by the Coulomb potential is $\sigma_L(\eta)$, once $kR \gg \eta\ln(2kR)$.

The outgoing part in $\psi_c(k\mathbf{z}, \mathbf{R})$ is found by matching at large values of $R-z$ to

$$\psi_c(k\hat{\mathbf{z}}, \mathbf{R}) \overset{R-z\to\infty}{\longrightarrow} e^{i[kz+\eta\ln k(R-z)]} + f_c(\theta)\frac{e^{i[kR-\eta\ln 2kR]}}{R}. \qquad (3.1.75)$$

The $\psi_c(k\hat{\mathbf{z}}, \mathbf{R})$ has a scattering amplitude $f_c(\theta)$ which, using the phase shift in Eqs. (3.1.34, 3.1.38), is *formally* the partial-wave sum

$$f_c(\theta) = \frac{1}{2ik} \sum_{L=0}^{\infty} (2L+1)P_L(\cos\theta)(e^{2i\sigma_L(\eta)} - 1), \qquad (3.1.76)$$

in terms of the Coulomb phase shift. However, this series expression does not converge, because the Coulomb potential does not go to zero fast enough for large

The classical scattering orbit in a Coulomb potential is a hyperbola, in which the distance of the projectile in polar coordinates (R, α), starting at $\alpha = \pi$, is

$$\frac{1}{R(\alpha)} = \frac{1}{b} \sin \alpha - \frac{D_c}{2b^2}(1 + \cos \alpha), \qquad (3.1.77)$$

where $D_c = Z_1 Z_2 e^2 / E = 2\eta/k$ is the distance of closest approach in a head-on collision, and b is the impact parameter for a general collision: the distance between the target and a straight line continuing the initial trajectory. As $R \to \infty$ after the collision, α becomes the scattering angle θ, which from Eq. (3.1.77) is

$$\tan\frac{\theta}{2} = \frac{\eta}{bk}. \qquad (3.1.78)$$

The distance of closest approach for arbitrary scattering angle θ is

$$R_{\text{near}}(\theta) = \frac{\eta}{k}\left[1 + \text{cosec}\frac{\theta}{2}\right], \qquad (3.1.79)$$

and the classical differential cross section, using Eq. (3.1.78), is

$$\sigma(\theta) \equiv \frac{b(\theta)}{\sin\theta}\frac{db}{d\theta} = \frac{\eta^2}{4k^2 \sin^4(\theta/2)}. \qquad (3.1.80)$$

Box 3.3 **classical Coulomb scattering**

$R = L/k$, and the phase shifts $\sigma_L(\eta)$ never go to zero for large L. The series only has meaning if a screened Coulomb potential is used, and then the screening radius let tend to infinity: see Taylor [5, §14-a].

If, however, Eq. (3.1.72) is directly matched to Eq. (3.1.75) for large $(R-z)$ values, the asymptotic amplitude is found to be

$$f_c(\theta) = -\frac{\eta}{2k \sin^2(\theta/2)} \exp\left[-i\eta \ln(\sin^2(\theta/2)) + 2i\sigma_0(\eta)\right], \qquad (3.1.81)$$

which is called the *point-Coulomb scattering amplitude*. The point-Coulomb cross section is therefore

$$\sigma_{\text{Ruth}}(\theta) = |f_c(\theta)|^2 = \frac{\eta^2}{4k^2 \sin^4(\theta/2)}. \qquad (3.1.82)$$

and is called the *Rutherford cross section*, because it happens to be the same as in classical scattering theory (see Box 3.3).

Coulomb+nuclear scattering

With the nuclear potential included as well as the deviation at short distances of the Coulomb potential from the pure $1/R$ form, the scattering potential may be written as $V(R) = V_c(R) + V_n(R)$ for some finite-range potential $V_n(R)$ in addition to the point-Coulomb potential of Eq. (3.1.70). We assume that $V_n(R)$ is spherical.

The phase shift from $V(R)$ will be $\delta_L \neq \sigma_L(\eta)$, so we define an *additional nuclear phase shift* δ_L^n by

$$\delta_L = \sigma_L(\eta) + \delta_L^n. \tag{3.1.83}$$

To find the Coulomb-distorted nuclear phase shift δ_L^n, or equivalently the nuclear S-matrix element $\mathbf{S}_L^n = e^{2i\delta_L^n}$, we match to a generalization of the asymptotic form of Eq. (3.1.37), namely to

$$\chi_L^{\text{ext}}(R) = \frac{i}{2}[H_L^-(\eta, kR) - \mathbf{S}_L^n H_L^+(\eta, kR)]. \tag{3.1.84}$$

This equation simply uses Coulomb functions with their first argument η now not zero, but given by Eq. (3.1.71). Similarly, in terms of the nuclear phase shift δ_L^n, we generalize Eq. (3.1.41) to obtain

$$\chi_L^{\text{ext}}(R) = e^{i\delta_L^n} \left[\cos \delta_L^n \, F_L(\eta, kR) + \sin \delta_L^n \, G_L(\eta, kR) \right]. \tag{3.1.85}$$

Using Eq. (3.1.34) the scattering amplitude $f_{nc}(\theta)$ from the combined Coulomb and nuclear potential $V(R)$ will have a factor

$$e^{2i\delta_L} - 1 = (e^{2i\sigma_L(\eta)} - 1) + e^{2i\sigma_L(\eta)}(e^{2i\delta_L^n} - 1), \tag{3.1.86}$$

so that the partial-wave sums will be a combination of the point-Coulomb amplitude of Eq. (3.1.76) and an additional 'Coulomb-distorted nuclear amplitude' $f_n(\theta)$:

$$f_{nc}(\theta) = f_c(\theta) + f_n(\theta). \tag{3.1.87}$$

The new scattering amplitude for the nuclear potential in addition to the point-Coulomb scattering is found by using the second term in Eq. (3.1.86):

$$f_n(\theta) = \frac{1}{2ik} \sum_{L=0}^{\infty} (2L+1) P_L(\cos \theta) e^{2i\sigma_L(\eta)} (\mathbf{S}_L^n - 1). \tag{3.1.88}$$

This is therefore *not* just the amplitude due to the short-range forces alone.

Note that we can optionally multiply both f_c of Eq. (3.1.81) and f_n of Eq. (3.1.88) by a phase factor such as $\exp(-2i\sigma_0(\eta))$, as only relative phase makes any difference to the *Coulomb+nuclear cross section*

$$\sigma_{nc}(\theta) = |f_c(\theta) + f_n(\theta)|^2 \equiv |f_{nc}(\theta)|^2. \qquad (3.1.89)$$

Since this cross section diverges to infinity at small angles (see Eq. (3.1.82)), very often elastic scattering cross sections are presented numerically in terms of their ratio to Rutherford, written as[4]

$$\sigma/\sigma_{\text{Ruth}} \equiv \sigma_{nc}(\theta)/\sigma_{\text{Ruth}}(\theta), \qquad (3.1.90)$$

which becomes unity at small angles.

3.1.3 Resonances and virtual states

When cross sections are plotted as functions of energy, $\sigma(E)$, either angle-integrated or for a specific angle, these plots are called excitation functions. They often show a range of narrower and wider peaks, as discussed in subsection 2.2.2. Many of these peaks are caused by resonances, when the interacting particles are trapped together inside a potential barrier for some period of time τ. This gives rise to variations in the excitation function with energy widths $\Gamma \sim \hbar/\tau$. We will see that resonance variations can be peaks, but also sometimes interference dips, or a peak next to a dip; all these patterns may result from a single resonance interfering with other scattering mechanisms.

A resonance is described by its total spin J_{tot} and parity $\pi = \pm 1$ (which we combine in the notation J_{tot}^{π}),[5] along with its energy E_r and width Γ in units of energy. A resonance will show characteristically as a rapid rise of the scattering phase shift $\delta(E)$ as seen in Fig. 3.2 for the $3/2^-$ channel, and this corresponds to a time delay for the scattering of a wave packet of the order $\Delta t \sim \hbar d\delta(E)/dE$ [6].[6]

The form of the increase is often found to be like

$$\delta(E) = \delta_{\text{bg}}(E) + \delta_{\text{res}}(E) \qquad (3.1.91)$$

where

$$\delta_{\text{res}}(E) = \arctan\left(\frac{\Gamma/2}{E_r - E}\right) + n(E)\pi, \qquad (3.1.92)$$

---

[4] Written this way for simplicity, omitting the θ-dependencies on the left side.
[5] With the one-channel scattering considered so far, $J_{\text{tot}} = L$ and $\pi = (-1)^L$.
[6] This paper [6] explains why phase shifts which *decrease* are not candidates for resonances because this corresponds to a time *advance*, which is limited by causality to the size of target divided by velocity of incident relative motion.

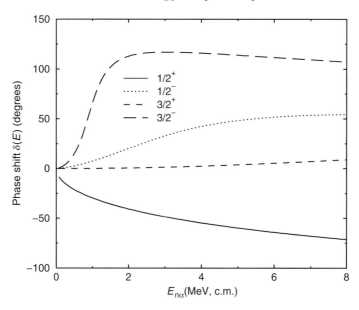

Fig. 3.2. Examples of resonant phase shifts for the $J^\pi = 3/2^-$ channel in low-energy n–α scattering, with a pole at $E = 0.96 - i0.92/2$ MeV. There is only a hint of a resonance in the phase shifts for the $J^\pi = 1/2^-$ channel, but it does have a wide resonant pole at $1.9 - i6.1/2$ MeV.

for $\Gamma > 0$, and for some *background phase shift* $\delta_{bg}(E)$ that varies only slowly around the resonance energy. The $n(E)$ is the integer depending on energy that may be optionally added to make $\delta_{res}(E)$ a continuous function of energy. If the background phase shift $\delta_{bg}(E) \approx 0$, then by Eq. (3.1.50) the resonance produces a contribution to the angle-integrated cross section of

$$\sigma_{el}^{res}(E) \simeq \frac{4\pi}{k^2}(2L+1)\sin^2\delta_{res}(E)$$

$$= \frac{4\pi}{k^2}(2L+1)\frac{\Gamma^2/4}{(E-E_r)^2 + \Gamma^2/4}, \qquad (3.1.93)$$

which shows a clear peak at $E \sim E_r$ with a full width at half maximum (fwhm) of Γ.

Because the propagation of a wave packet in the presence of a resonance experiences a time delay of $\tau \sim \hbar \, d\delta_{res}(E)/dE$, the form of Eq. (3.1.91) implies that $\tau \sim \hbar/\Gamma$ as earlier expected.

A resonance with the form of Eq. (3.1.93) with $\delta_{bg}(E) = 0$ is called a *pure Breit-Wigner resonance*. If $\delta_{bg}(E) \neq 0$, then Fig. 3.3 shows some of the other Breit-Wigner patterns that may be produced. If, moreover, $\delta_{bg}(E)$ varies with energy, a

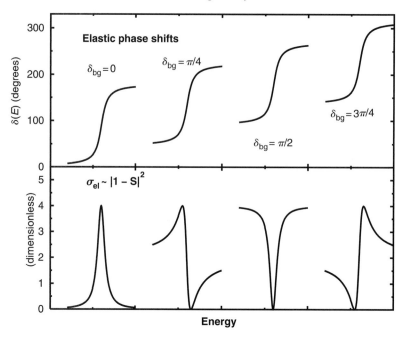

Fig. 3.3. Possible Breit–Wigner resonances. The upper panel shows resonant phase shifts with several background phase shifts $\delta_{bg} = 0$, $\pi/4$, $\pi/2$ and $3\pi/4$ in the same partial wave. The lower panel gives the corresponding contributions to the total elastic scattering cross section from that partial wave.

resonance may still exist even though the phase shift $\delta(E)$ does not pass $\pi/2$, such as with the $1/2^-$ scattering in Fig. 3.2.

The interference and cancellation effects shown in Fig. 3.3 for the angle-integrated elastic cross sections occur when there is a background phase shift in the *same* partial wave as the resonance. If there are contributing amplitudes from *different* (non-resonant) partial waves, then there can be no cancellation effects for the angle-integrated cross sections of Eq. (3.1.50) since these are *incoherent* in the partial-wave sum. A resonance can only give coherent interference with another partial wave if the *angular* cross sections are measured, since Eq. (3.1.34) is coherent in its partial-wave sum.

If the S matrix is calculated for a Breit–Wigner resonance using the parametrization of Eq. (3.1.91), then, using $\tan\delta_{\mathrm{res}}(E) = \frac{1}{2}\Gamma/(E_r - E)$ and Eq. (3.1.48),

$$\mathbf{S}(E) = \mathrm{e}^{2\mathrm{i}\delta_{bg}(E)}\frac{E - E_r - \mathrm{i}\Gamma/2}{E - E_r + \mathrm{i}\Gamma/2}. \tag{3.1.94}$$

This expression shows that if the function $\mathbf{S}(E)$ were continued to complex energies E, it would have a pole, where the denominator is zero, at $E_p = E_r - \mathrm{i}\Gamma/2$. These

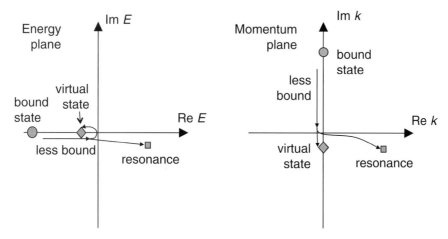

Fig. 3.4. The correspondences between the energy (left) and momentum (right) complex planes. The arrows show the trajectory of a bound state caused by a progressively weaker potential: it becomes a resonance for $L > 0$ or when there is a Coulomb barrier, otherwise it becomes a virtual state. Because $E \propto k^2$, bound states on the positive imaginary k axis and virtual states on the negative imaginary axis both map onto the negative energy axis.

are poles in the fourth quadrant of the complex energy plane, near but below the real energy axis. The existence of such a complex pole leads most theorists to *define* resonance by its pole position,[7] by the parameters J_{tot}^{π} and complex E_p.[8] In the example of low-energy n–α scattering in Fig. 3.2, calculations of scattering at complex energies show that there is indeed a pole at $E = 0.96 - i0.92/2$ MeV in the $3/2^-$ channel. There is only a hint of a resonance in the phase shifts for the $J^{\pi} = 1/2^-$ channel, but we find it does have a resonant pole at $1.9 - i6.1/2$ MeV. A wide resonance (one with large Γ) such as this one has a broad effect on scattering over a large energy range.

Mathematically, therefore, a resonance is a pole of the S matrix in the fourth complex energy quadrant, with $\text{Re} E_p > 0$ and $\text{Im} E_p < 0$. Resonances may also be characterized as *unbound states*: states that would be bound if the nuclear potential were stronger. In the complex plane, bound states are also poles, on the negative real energy axis and the positive imaginary k axis, as shown by the circles in Fig. 3.4. When the potential for a bound state becomes weaker, the pole moves towards zero energy according to the arrows, and then becomes a resonance (the squares in Fig. 3.4) if there is a potential barrier between large distances and the interior

---

[7] We should note, however, that there yet exist cross section peaks that are not associated with poles, as noted for example in [5].

[8] In the more general multi-channel theory to be presented in subsection 3.2.1, we will see more how many partial-wave channels may couple to the total J_{tot}^{π}. The resonance peaks will then occur in all the partial waves, and the definition in terms of an S-matrix pole is applicable to multi-channel as well as one-channel scattering.

nuclear attraction. This trapping barrier could be a repulsive Coulomb barrier for proton–nucleus scattering such as seen in Fig. 1.3, or a centrifugal barrier for either neutrons or protons in angular momentum $L > 0$ states. The width of the resonance is extremely sensitive to the height of these barriers. Very wide resonances, or poles a long way from the real axis, will not have a pronounced effect on scattering at real energies, especially if there are several of them. They may thus be considered less important physically.

The case of neutral scattering in $L = 0$ partial waves deserves special attention, since here there is no repulsive barrier to trap, for example, an s-wave neutron. There is no Breit-Wigner form now, and mathematically the S-matrix pole S_p is found to be on the negative imaginary k axis: the diamonds in Fig. 3.4. This corresponds to a negative real pole energy, but this is *not* a bound state, for which the poles are always on the positive imaginary k axis. The neutral unbound poles are called *virtual states*, to be distinguished from both bound states and resonances. The dependence on the sign of $k_p = \pm\sqrt{2\mu E_p/\hbar^2}$ means we should write the S matrix as a function of k not E. A pure virtual state has pole at $k_p = i/a_0$ on the negative imaginary axis, described by a negative value of a_0 called the *scattering length*. This corresponds to the analytic form

$$\mathbf{S}(k) = -\frac{k + i/a_0}{k - i/a_0}, \qquad (3.1.95)$$

giving $\delta(k) = -\arctan a_0 k$, or $k \cot \delta(k) = -1/a_0$. These formulae describe the phase shift behavior close to the pole, in the case for low momenta where k is not too much larger than $1/|a_0|$. For more discussion see, for example, Taylor [5, §13-b].

3.1.4 Nuclear currents or flux

When a charged particle is scattered from another particle, its acceleration during the reaction will in general lead to an exchange of energy with the electromagnetic field according to Maxwell's equations. The electromagnetic field depends more specifically on the *nuclear current* or *flux* of the charged particles, and therefore we need to find this current in terms of the scattering wave functions.

One definition of the nuclear probability current or flux was that given in Eq. (2.4.1), namely $\mathbf{j} = \mathbf{v}|\psi|^2$, but this definition is only preliminary as it uses for $\mathbf{v}$ the average beam velocity, and not a local property of the wave functions in the presence of potentials. A proper definition of the nuclear current $\mathbf{j}(\mathbf{r}, t)$ should at every position $\mathbf{r}$ satisfy the *continuity equation*

$$\nabla \cdot \mathbf{j}(\mathbf{r}, t) + \frac{\partial \rho(\mathbf{r}, t)}{\partial t} = 0 \qquad (3.1.96)$$

with probability density $\rho(\mathbf{r}, t) = |\psi(\mathbf{r}, t)|^2$. A current satisfying this continuity equation is called a *conserved current*.

Using first the *free-field* Schrödinger equation $\hat{T}\psi = i\hbar\frac{\partial \psi}{\partial t}$ with the kinetic energy operator $\hat{T} = -\frac{\hbar^2}{2\mu}\nabla^2$, we may evaluate the rate of density change for a free particle as

$$\frac{\partial \psi^*\psi}{\partial t} = \frac{\partial \psi^*}{\partial t}\psi + \psi^*\frac{\partial \psi}{\partial t}$$

$$= \frac{\hbar}{2i\mu}\left[\left(\nabla^2\psi\right)^*\psi - \psi^*\nabla^2\psi\right]$$

$$= \frac{\hbar}{2i\mu}\nabla \cdot \left[(\nabla\psi)^*\psi - \psi^*\nabla\psi\right]. \tag{3.1.97}$$

This yields a local conserved current or flux for the free-field case of

$$\mathbf{j}_{\text{free}} = \frac{\hbar}{2i\mu}(\psi^*\nabla\psi - \psi\nabla\psi^*), \tag{3.1.98}$$

and leads us to define the *free current operator* either as

$$\hat{\mathbf{j}}_{\text{free}} = \frac{\hbar}{2i\mu}(\overrightarrow{\nabla} - \overleftarrow{\nabla}), \tag{3.1.99}$$

where the arrows indicate the target of the differentiation, or better as a position-dependent operator

$$\hat{\mathbf{j}}_{\text{free}}(\mathbf{r}) = \frac{\hbar}{2i\mu}\left\{\delta(\mathbf{r}-\mathbf{r}_i)\,\nabla_i - \nabla_i\,\delta(\mathbf{r}-\mathbf{r}_i)\right\}, \tag{3.1.100}$$

where the $\mathbf{r}_i$ are the variables in terms of which the wave function is written as $\psi(\mathbf{r}_i)$, and the gradients ∇_i are with respect to the $\mathbf{r}_i$. The current operator of Eq. (3.1.100) can be used in normal matrix elements such as $\langle\phi|\hat{\mathbf{j}}_{\text{free}}(\mathbf{r})|\psi\rangle$ which integrate over the internal $\mathbf{r}_i$ coordinates, and yield the current as a function of $\mathbf{r}$.

We define the *electric* current j_q by means of the *electric current operator* $\hat{\mathbf{j}}_q$ that is proportional to the charge times the nuclear current of that charged particle. That is,

$$\hat{\mathbf{j}}_q = q\hat{\mathbf{j}} \quad \text{and} \quad \hat{\mathbf{j}}_{\text{free}}^q = q\hat{\mathbf{j}}_{\text{free}} \tag{3.1.101}$$

for a particle of charge q, to be used for example in subsection 3.5.2 for photo-nuclear reactions.

In the case of *interacting* particles because of a potential between them, ψ satisfies $[\hat{T} + \hat{V}]\psi = i\hbar\frac{\partial \psi}{\partial t}$, where we now allow the potential $\hat{V}$ to be a general operator that operates on functions of the $\mathbf{r}_i$. It may have derivatives (from a spin-orbit force) or may be parity- or partial-wave-dependent. These are all particular kinds of *non-localities*, where a non-local potential kernel such as $V(\mathbf{r}'_i, \mathbf{r}_i)$ removes flux at one position $\mathbf{r}_i$ and has it reappear at another place $\mathbf{r}'_i \neq \mathbf{r}_i$. In this case it should not be surprising that we have to adjust our definitions of the nuclear current, which is supposed to follow continuity.

Evaluating the rate of density change gives

$$\frac{\partial \psi^* \psi}{\partial t} = \left(\frac{1}{i\hbar} [\hat{T} + \hat{V}] \psi \right)^* \psi + \psi^* \frac{1}{i\hbar} [\hat{T} + \hat{V}] \psi$$

$$= \frac{\hbar}{2i\mu} \nabla \cdot \left[(\nabla \psi)^* \psi - \psi^* \nabla \psi \right] + \frac{1}{i\hbar} [\psi^* \hat{V} \psi - (\hat{V} \psi)^* \psi].$$

Thus $\dfrac{\partial \rho(\mathbf{r}, t)}{\partial t} = -\nabla \cdot \mathbf{j}_{\text{free}} - \dfrac{i}{\hbar} \langle \psi | \left[\delta(\mathbf{r}-\mathbf{r}_i) \hat{V} - \hat{V}^\dagger \delta(\mathbf{r}-\mathbf{r}_i) \right] | \psi \rangle.$

$$(3.1.102)$$

If the potential $\hat{V}$ were a local real function of position $\mathbf{r}_i$ only, then this last term disappears, and the free current operator of Eq. (3.1.99) gives a conserved current. In general, however, this is not the case, and it is a difficult task to find a second contribution $\hat{\mathbf{j}}_2$ to the current operator so that the sum

$$\hat{\mathbf{j}} = \hat{\mathbf{j}}_{\text{free}} + \hat{\mathbf{j}}_2 \qquad (3.1.103)$$

is a conserved current when the interaction potentials are present. If this additional term comes from the two-body interaction terms, it is called the *two-body current contribution*. It must satisfy

$$\nabla \cdot \hat{\mathbf{j}}_2 = \frac{i}{\hbar} [\delta(\mathbf{r}-\mathbf{r}_i) \, \hat{V} - \hat{V}^\dagger \, \delta(\mathbf{r}-\mathbf{r}_i)], \qquad (3.1.104)$$

so that the full current operator satisfies

$$\nabla \cdot \hat{\mathbf{j}} = \frac{i}{\hbar} [\delta(\mathbf{r}-\mathbf{r}_i) \, H - H^\dagger \, \delta(\mathbf{r}-\mathbf{r}_i)], \qquad (3.1.105)$$

which is equivalent to the original continuity equation (3.1.96).[9]

Later, in Section 4.7, we will see that Eq. (3.1.105) is by itself sufficient to solve our problems for one particular important application. For electric transitions at long distances and low momentum, there exists what will be called Siegert's theorem, which allows us to include two-body currents, because in that case their only relevant property is the one given by Eq. (3.1.104), that is, by the continuity

---

[9] The continuity equation is not sufficient by itself to define the complete current. It only defines the 'longitudinal' component of $\hat{\mathbf{j}}_2$, not its 'transverse' component, because adding $\nabla \times \mathbf{X}(\mathbf{r})$ to $\hat{\mathbf{j}}_2$ does not change $\nabla \cdot \hat{\mathbf{j}}_2$ whatever vector field $\mathbf{X}(\mathbf{r})$ might be chosen.

equation itself. However, we cannot from a purely nucleonic model calculate the effect of two-body currents $\hat{\mathbf{j}}_2$ in *magnetic* transitions, and in electric transitions with higher momentum transfers,[10] so in their calculations we will have to omit the terms depending on $\hat{\mathbf{j}}_2$ in this book.

3.1.5 Complex potentials

Often, the effective interactions between two nuclei give the best fit to experimental cross sections if they are allowed to have *negative imaginary* as well as real components, even though the microscopic interaction potentials between individual nucleons may be entirely real valued. The effective imaginary components arise from a variety of reasons, but principally because in nature there are more reactions occurring than can be described by the spherical potentials dealt with so far. These further reactions remove flux from elastic scattering, and we will see that this removal can be equivalently described by complex potentials.

In the next Section 3.2 we will see how to define multiple channels for the model wave functions, and later (subsection 11.5.2) it will be shown how eliminating such channels from our model induces imaginary potentials in the remaining components. Imaginary potentials are also present in a potential that reproduces amplitudes that have been averaged over a range of scattering energies (Section 11.5), especially if we average over many compound-nucleus resonances (subsection 11.5.3). The potentials that fit elastic scattering are therefore generally complex, of the kind outlined in Box 3.4 and again in Section 4.1.

If the potential is $V(R) + iW(R)$, then it is no longer Hermitian, and the S matrix is no longer unitary. If the imaginary part $W < 0$, then we have *absorptive potentials*, and a loss of flux. This can be made to approximate the flux leaving in the exit channels that are not explicitly in our model. Potentials with both real and imaginary parts are called *optical potentials*, since they describe both the refraction and absorption in the same way as a light wave passing through a cloudy refractive medium.

If the Schrödinger equation $[\hat{T} + V + iW]\psi = i\hbar\partial\psi/\partial t$ now has an imaginary potential, then we can calculate the rate of loss of flux as

$$\frac{\partial \psi^*\psi}{\partial t} = -\nabla \cdot \mathbf{j} + \frac{2}{\hbar}\, W \psi^*\psi. \tag{3.1.106}$$

---

[10] That is, only if we have a microscopic explanation of the potential non-localities in terms of a local theory (for example of meson exchanges) can we have a unique specification of local nuclear currents. This would be the case where we start from a fully gauge-invariant Lagrangian. In other cases, it may still be possible to construct a current which is conserved. For example, Sachs [7] finds a current for isospin-exchange forces, Riska [8] finds a current for spin-orbit potentials, and recently Marcucci *et al.* [9] find conserved currents for modern fully-fitted two- and three-nucleon potentials, but these derivations, however, do not give unique results

The interaction potential between a nucleon and a nucleus is usually well described by an attractive nuclear well of the form

$$V(R) = -\frac{V_r}{1 + \exp\left(\frac{R-R_r}{a_r}\right)}, \qquad (3.1.107)$$

which is called a 'Woods-Saxon' (or 'Saxon-Woods' or 'Fermi') shape. The central depth V_r is typically between 40 and 50 MeV, and the diffuseness a_r about 0.6 fm. The radius R_r is proportional to the size of the nucleus, and is commonly around $R_r = r_r A^{1/3}$ for a nucleus of A nucleons, with $r_r \approx 1.2$ fm. Similar potentials can be used for the interaction between two nuclei with mass numbers A_1 and A_2, if the radii are scaled instead as $R_r = r_r(A_1^{1/3} + A_2^{1/3})$, since this is proportional to the sum of the individual radii.

This potential is usually combined with an imaginary and a spin-orbit part. The imaginary part, which is present at higher scattering energies, as discussed in subsection 3.1.5, is also often given by a Woods-Saxon form

$$W(R) = -\frac{V_i}{1 + \exp\left(\frac{R-R_i}{a_i}\right)} \qquad (3.1.108)$$

for a similar geometry $R_i \approx R_r$ and $a_i \approx a_r$, and a depth V_i fitted to experiments giving $V_i \sim 10 - 20$ MeV depending on energy. Sometimes a surface-peaked imaginary contribution is also included, with a shape like the derivative of Eq. (3.1.108).

The spin-orbit potentials will be described in subsection 4.3.2.

Box 3.4 **Typical nuclear scattering potentials**

As $W < 0$, the imaginary potential acts as a *sink* of particles. The imaginary potential causes particles to be removed from the incident beam with an additional rate of

$$\left.\frac{\partial \rho}{\partial t}\right|_W = \frac{2}{\hbar} W \rho, \qquad (3.1.109)$$

for the density $\rho = |\psi|^2$. Thus, if the transport from the kinetic terms could be neglected, the probability density would decay as

$$\rho(t) \propto e^{2Wt/\hbar} = e^{-2|W|t/\hbar}. \qquad (3.1.110)$$

for the conserved current: that can only be obtained in definite meson-exchange models. The effect of these ambiguities is that we do not really know the details of the current at short nucleon-nucleon distances.

When complex potentials are introduced into scattering theory, all the previous scattering theory remains valid, but now the phase shifts δ_L become complex, and the moduli $|\mathbf{S}_L| \neq 1$. For absorptive potentials we have $|\mathbf{S}_L| < 1$.

There is now an *absorptive cross section* σ_A of flux disappearing from the elastic channel, which can be calculated as an integral of $W(R)$ multiplied by the probability density

$$\sigma_A = \frac{2}{\hbar v} \int [-W(R)] \, |\psi(\mathbf{R})|^2 \, d^3\mathbf{R}, \qquad (3.1.111)$$

where v is the velocity of the incident beam so $2/\hbar v = k/E$. In terms of the partial wave radial functions $\chi_L(R)$, the absorption cross section is

$$\sigma_A = \frac{2}{\hbar v} \frac{4\pi}{k^2} \sum_L (2L+1) \int_0^\infty [-W(R)] \, |\chi_L(R)|^2 \, dR. \qquad (3.1.112)$$

Since, in the L'th partial wave, the difference between the square moduli of the coefficients of the incoming and outgoing waves in Eq. (3.1.37) is $1 - |\mathbf{S}_L|^2$, we define a *reaction cross section* as the sum over all partial waves:

$$\sigma_R = \frac{\pi}{k^2} \sum_L (2L+1)(1 - |\mathbf{S}_L|^2). \qquad (3.1.113)$$

We can use the Schrödinger equations for χ_L and its complex conjugate for χ_L^* to integrate twice by parts and prove that

$$-\int_0^a \chi_L(R)^* \, W(R) \, \chi_L(R) \, dR = \frac{\hbar^2 k}{8\mu}(1 - |\mathbf{S}_L|^2), \qquad (3.1.114)$$

and hence, in this spherical potential case, that $\sigma_A = \sigma_R$. The next section will discuss more general Hamiltonians, and as these couple also to non-elastic exit channels, we will see that, in general, σ_R will be larger than σ_A by precisely the summed cross sections to these non-elastic reactions.

3.2 Multi-channel scattering

3.2.1 Multiple channels

In the reaction between two nuclei such as ^{12}C + d, a variety of mass rearrangements may be possible, such as ^{13}C + p, ^{14}N + γ, etc. We define each of these as a different *mass partition*, and label them according to variable x. We deal in this chapter only with two-body partitions, leaving three-body breakup (for example to ^{12}C + n + p) until Chapter 8, but in all cases the sum of the particle masses in each partition $\sum_i m_{xi}$ is almost the same for each x. The total is only *almost* the same because of relativistic effects, and hence the different Q-values describing the energy differences between a pair of partitions according to Eq. (1.1.3).

Each partition x will therefore consist of two bodies, one we call like the projectile, and the other we call target-like. Let the vector $\mathbf{R}_x$, extending from the target-like nucleus to the projectile-like one, describe their relative position. Let p and t label the state (the energy level) of the projectile- and target-like nuclei respectively, to distinguish their ground state from any excited states. Each state will have a definite spin and parity, so we label their spins by I_p and I_t. We denote the nucleus internal coordinates by ξ_p and ξ_t, which would be the spin states for nucleons, and sets of spin and radial coordinates for clusters of nucleons. We will use $\phi^{xp}_{I_p\mu_p}(\xi_p)$ and $\phi^{xt}_{I_t\mu_t}(\xi_t)$ to refer to the whole quantum states of the projectile and the target, respectively.

Let $\mathbf{L}$ be their relative angular momentum, as in the previous section. Now, however, we have to couple together $\mathbf{L}$, $\mathbf{I}_p$, and $\mathbf{I}_t$ to make some *total* angular momentum $\mathbf{J}_{\text{tot}}$. This can be done in two ways, either by coupling $\mathbf{I}_p$ to $\mathbf{I}_t$ first, or $\mathbf{L}$ to $\mathbf{I}_p$. Thereby we introduce two intermediate angular momenta, $\mathbf{S}$ or $\mathbf{J}_p$ respectively, so these two schemes are named as 'S basis' or 'J basis':[11]

'S basis'	Channel spin S	$\mathbf{I}_p + \mathbf{I}_t = \mathbf{S}$	$\mathbf{L} + \mathbf{S} = \mathbf{J}_{\text{tot}}$
'J basis'	Projectile J	$\mathbf{L} + \mathbf{I}_p = \mathbf{J}_p$	$\mathbf{J}_p + \mathbf{I}_t = \mathbf{J}_{\text{tot}}$

These are two complete and orthonormal basis schemes. The S basis has the advantage that compound-nucleus resonances typically have little mixing of channel-spin quantum number, whereas the J basis has the advantage that projectile spin-orbit forces are diagonal in this basis. Most often we will use the J basis for both bound and scattering states. Later we give the conversion formulae between the two basis schemes.

The set of all quantum numbers for a given total J_{tot}, in the J basis $\{xpt, LI_pJ_pI_t\}$, will be abbreviated by α. Each x denotes a mass partition, each $\{p, t\}$ denotes an *excited state pair*, and each α denotes a *partial-wave channel*. The unqualified noun 'channel' will refer to one of these according to context.

The relative coordinate $\mathbf{R}_x$ depends on the partition, so the radial wave function of relative motion will be written as $\psi_\alpha(R_x)$, in place of $\chi_L(R)$ for the one-channel case. The total system wave function is now written as $\Psi(\mathbf{R}_x, \xi_p, \xi_t)$, in place of $\psi(R, \theta)$ in the one-channel case, and will contain a sum over the partitions, where each partition is represented by a product of the internal states $\phi^{xp}_{I_p\mu_p}(\xi_p)$ and $\phi^{xt}_{I_t\mu_t}(\xi_t)$, as well as a wave function for their relative orbital motion. We assume for now that the projectile and target nuclei are distinguishable as far as the Pauli principle is

---

[11] Sometimes they are called 'LS' and 'jj' respectively, but those names are more appropriate for coupling *four* rather than three angular momenta.

concerned: the antisymmetrization of the wave function for identical particles will be discussed in Section 3.4.

In each partition x, we define first the state of two nuclei in relative motion with total angular momentum J_{tot} and projection M_{tot}. When we include the Clebsch-Gordon coefficients for coupling the angular momenta together, in the J basis the basis set of wave functions for a given partition x is

$$\Psi^{M_{\text{tot}}}_{xJ_{\text{tot}}}(\mathbf{R}_x, \xi_p, \xi_t)$$

$$= \sum_{LI_pJ_pI_tM\mu_pM_a\mu_t} \phi^{xp}_{I_p\mu_p}(\xi_p)\phi^{xt}_{I_t\mu_t}(\xi_t) i^L Y^M_L(\hat{\mathbf{R}}_x) \frac{1}{R_x}\psi^{J_{\text{tot}}}_\alpha(R_x)$$

$$\langle LM, I_p\mu_p|J_pM_a\rangle\langle J_pM_a, I_t\mu_t|J_{\text{tot}}M_{\text{tot}}\rangle$$

$$\equiv \sum_\alpha \left[\left[i^L Y_L(\hat{\mathbf{R}}_x) \otimes \phi^{xp}_{I_p}(\xi_p)\right]_{J_p} \otimes \phi^{xt}_{I_t}(\xi_t)\right]_{J_{\text{tot}}M_{\text{tot}}} \frac{1}{R_x}\psi^{J_{\text{tot}}}_\alpha(R_x)$$

$$\equiv \sum_\alpha |xpt:(LI_p)J_p, I_t; J_{\text{tot}}M_{\text{tot}}\rangle\, \psi^{J_{\text{tot}}}_\alpha(R_x)/R_x$$

$$\equiv \sum_\alpha |\alpha; J_{\text{tot}}M_{\text{tot}}\rangle\, \psi^{J_{\text{tot}}}_\alpha(R_x)/R_x, \tag{3.2.1}$$

where x on the left side selects those α on the right of the same partition. The symbol $\langle L_1M_1, L_2M_2|LM\rangle$ is the Clebsch-Gordon coefficient for coupling two angular momentum states L_1M_1 and L_2M_2 to a total of LM. The i^L coefficients are included for the same reason that they are in Eq. (3.1.9): to make the wave functions $\psi^{J_{\text{tot}}}_\alpha(R_x)$ revert to the standard Coulomb functions $F_L(\eta, k_\alpha R_x)$ in the absence of a nuclear potential.

In the S (channel spin) basis, the wave functions for given $J_{\text{tot}}M_{\text{tot}}$ are analogously

$$\Psi^{M_{\text{tot}}}_{xJ_{\text{tot}}}(\mathbf{R}_x, \xi_p, \xi_t) = \sum_{LI_pSI_t} \left[i^L Y_L(\hat{\mathbf{R}}_x) \otimes \left[\phi^{xp}_{I_p}(\xi_p) \otimes \phi^{xt}_{I_t}(\xi_t)\right]_S\right]_{J_{\text{tot}}M_{\text{tot}}}$$

$$\times \psi^{J_{\text{tot}}}_\beta(R_x)/R_x$$

$$\equiv \sum_\beta |xpt:L(I_p, I_t)S; J_{\text{tot}}M_{\text{tot}}\rangle\, \psi^{J_{\text{tot}}}_\beta(R_x)/R_x$$

$$\equiv \sum_\beta |\beta; J_{\text{tot}}M_{\text{tot}}\rangle\, \psi^{J_{\text{tot}}}_\beta(R_x)/R_x, \tag{3.2.2}$$

where β is the set of quantum numbers $\{xpt, LI_pI_tS\}$, and the sum over β on the right side is restricted to those with the same x value as on the left. Partial waves in the channel-spin basis are often labeled by $^{2S+1}L_{J_{\text{tot}}}$, for example $^{3}P_{2}$ for $S = 1$, $L = 1$ and $J_{\text{tot}} = 2$.

There is a unitary transformation between the radial wave functions in the S and J bases for a given partition x:

$$\psi_{\beta}^{J_{\text{tot}}}(R_x) = \sum_{\alpha} \langle \beta | \alpha \rangle \, \psi_{\alpha}^{J_{\text{tot}}}(R_x) \text{ and } \psi_{\alpha}^{J_{\text{tot}}}(R_x) = \sum_{\beta} \langle \alpha | \beta \rangle \, \psi_{\beta}^{J_{\text{tot}}}(R_x) \quad (3.2.3)$$

where the transformation matrix element, the same for all R_x, is

$$\langle \alpha | \beta \rangle = \sqrt{(2S+1)(2J_p+1)} \; W(LI_pJ_{\text{tot}}I_t; J_pS). \quad (3.2.4)$$

The symbol $W(abcd; ef)$ is the Racah coefficient of angular momentum recoupling theory.

The system wave function for a given $J_{\text{tot}}M_{\text{tot}}$ is a superposition of all partitions, as

$$\overline{\Psi}_{J_{\text{tot}}}^{M_{\text{tot}}} = \sum_{x} \Psi_{xJ_{\text{tot}}}^{M_{\text{tot}}} = \sum_{\alpha} |\alpha; J_{\text{tot}}M_{\text{tot}}\rangle \, \psi_{\alpha}^{J_{\text{tot}}}(R_x)/R_x, \quad (3.2.5)$$

where now the α sum is unrestricted. This wave function $\overline{\Psi}_{J_{\text{tot}}}^{M_{\text{tot}}}$ can be written as a function of the coordinates $(\mathbf{R}_x, \xi_p, \xi_t)$ of any single partition, because the coordinates of each individual partition enable a complete set of basis states to be defined. Note that by writing Eq. (3.2.5) we are *not* assuming the basis functions in the separate partitions are orthogonal. In the subsequent development, we will be careful to keep any terms arising from channel non-orthogonalities.

Limiting case of pure Coulomb monopole potentials

Since the one-channel wave function $\psi(R, \theta)$ reduces to a plane wave if the nuclear potentials are zero (as then $\chi_L = F_L$), so should the total wave function Ψ when summed over all total spins and projections $J_{\text{tot}}M_{\text{tot}}$. In the multi-channel case, the plane wave as an initial state must be supplemented by the m-quantum numbers of the approaching nuclei. A total system wave function must in the free-field limit reduce to

$$\Psi_{xpt}^{\mu_p\mu_t}(\mathbf{R}_x, \xi_p, \xi_t; \mathbf{k}_i) \overset{V=0}{\to} e^{i\mathbf{k}_i \cdot \mathbf{R}_x} \phi_{I_p\mu_p}^{xp}(\xi_p)\phi_{I_t\mu_t}^{xt}(\xi_t) \quad (3.2.6)$$

for a projectile in state p and target in state t in partition x, m-states μ_p and μ_t respectively, and initial momenta in the c.m. frame of $\pm\mathbf{k}_i$.

Summing over $J_{\text{tot}}M_{\text{tot}}$, with nuclear potentials

When scattering potentials are present, the total system wave function $\Psi_{x_ip_it_i}^{\mu_{p_i}\mu_{t_i}}$ will contain radial wave functions $\psi_{\alpha}^{J_{\text{tot}}}$ different from the Coulomb F_L functions, and for initial m-state projections μ_{p_i} and μ_{t_i} for the incoming nuclear states $x_i p_i t_i$, it will be a sum over all $J_{\text{tot}} M_{\text{tot}}$ that generalizes Eq. (3.1.57):

$$
\Psi^{\mu_{p_i}\mu_{t_i}}_{x_i p_i t_i}(\mathbf{R}_x, \xi_p, \xi_t; \mathbf{k}_i) = \sum_{J_{\text{tot}} M_{\text{tot}} x} \Psi^{M_{\text{tot}}}_{x J_{\text{tot}}}(\mathbf{R}_x, \xi_p, \xi_t)
$$

$$
\times \frac{4\pi}{k_i} \sum_{L_i M_i} Y^{M_i}_{L_i}(\mathbf{k}_i)^* \sum_{J_{p_i} m_i} \langle L_i M_i, I_{p_i}\mu_{p_i}|J_{p_i}m_i\rangle \langle J_{p_i}m_i, I_{t_i}\mu_{t_i}|J_{\text{tot}}M_{\text{tot}}\rangle \quad (3.2.7)
$$

in the J basis, for an incoming plane wave in direction $\mathbf{k}_i$.

We now combine the indices $x_i p_i t_i$ with the L_i and J_{p_i} of Eq. (3.2.7) to form a new value for the multi-index, $\alpha_i = \{x_i p_i t_i, L_i I_{p_i} J_{p_i} I_{t_i}\}$. This identifies the partial-wave channel in which there is an incoming plane wave, so the multi-channel radial functions will henceforth be labeled also with α_i, as $\psi^{J_{\text{tot}}}_{\alpha\alpha_i}(R_x)$. The α_i subscript describes the incoming channel, and α the outgoing channel.

The $\Psi^{M_{\text{tot}}}_{x J_{\text{tot}}}$ therefore depends on α_i, called the *incoming channel*, so to make this dependence explicit we rewrite Eq. (3.2.7) as

$$
\Psi^{\mu_{p_i}\mu_{t_i}}_{x_i p_i t_i}(\mathbf{R}_x, \xi_p, \xi_t; \mathbf{k}_i) = \sum_{J_{\text{tot}} M_{\text{tot}}} \sum_{\alpha\alpha_i} |\alpha; J_{\text{tot}}M_{\text{tot}}\rangle \frac{\psi^{J_{\text{tot}}}_{\alpha\alpha_i}(R_x)}{R_x} A^{J_{\text{tot}}M_{\text{tot}}}_{\mu_{p_i}\mu_{t_i}}(\alpha_i; \mathbf{k}_i) \quad (3.2.8)
$$

where we define an 'incoming coefficient'

$$
A^{J_{\text{tot}}M_{\text{tot}}}_{\mu_{p_i}\mu_{t_i}}(\alpha_i; \mathbf{k}_i)
$$

$$
\equiv \frac{4\pi}{k_i} \sum_{M_i m_i} Y^{M_i}_{L_i}(\mathbf{k}_i)^* \langle L_i M_i, I_{p_i}\mu_{p_i}|J_{p_i}m_i\rangle \langle J_{p_i}m_i, I_{t_i}\mu_{t_i}|J_{\text{tot}}M_{\text{tot}}\rangle. \quad (3.2.9)
$$

Here and elsewhere, the sum over α_i is not over all possible channels, but over just the partial waves that include the incoming state indices $x_i p_i t_i$ defined by the properties of the beam projectile and target. Similar expressions can be derived for the S basis.

It is by means of this expansion with the incoming channel index α_i that we determine the boundary conditions and scattering amplitudes below. In all channels $\alpha \neq \alpha_i$ there should only be outgoing flux.[12]

---

[12] To confirm the meaning of the multi-channel formalism, we briefly outline how it reduces to the previous one-channel theory of Section 3.1 in the case of structureless spin-zero particles interacting only by spherical elastic potentials. The only non-zero $\psi^{J_{\text{tot}}}_{\alpha\alpha_i}(R)$ for partial wave L will then be

$$
\psi^L_{\alpha\alpha}(R) = \chi_L(R)
$$

for $\alpha = \{111, L0L0\}$. The only non-zero $\Psi^{M_{\text{tot}}}_{x J_{\text{tot}}}(\mathbf{R}, \xi_p, \xi_t)$ are

$$
\Psi^M_{1L}(\mathbf{R}, \xi_p, \xi_t) = i^L Y^M_L(\hat{\mathbf{R}}) \chi_L(R)/R,
$$

Parity

Nuclear and Coulomb interactions do not change parity, so that each projectile and target state has a specific parity π_{xp} and π_{xt}. The parity for a partial wave L is $(-1)^L$, so the total parity of a partial-wave channel is $\pi = (-1)^L \pi_{xp} \pi_{xt}$. This must be the same for all partial waves, since parities are not mixed by Coulomb or nuclear couplings.

This also means that coupled-channels sets for a given J_{tot} can be subdivided into positive and negative parity subsets, and each subset calculated separately. We will therefore label coupled-channels sets not just by J_{tot} but by J_{tot}^π, and sums over J_{tot} are rewritten as over J_{tot}^π. The parities will therefore be recombined in sums like that of Eq. (3.2.8), where the first summation is now over $\{J_{\text{tot}}, \pi, M_{\text{tot}}\}$.

Multi-channel S matrix

The S-matrix element of subsection 3.1.1 is now generalized to a full *matrix* $\mathbf{S}_{\alpha\alpha_i}^{J_{\text{tot}}\pi}$ for each total angular momentum and parity J_{tot}^π, where α_i is the partial-wave channel with the incoming plane wave, and α is an outgoing channel.

This means that Eqs. (3.1.6, 3.1.84) are generalized to depend on the entrance channel α_i, and from Eqs. (3.2.6) and (3.2.8) we obtain for $R_x > R_n$:

$$\psi_{\alpha\alpha_i}^{J_{\text{tot}}\pi}(R_x) = \frac{\text{i}}{2}\left[H_{L_i}^-(\eta_\alpha, k_\alpha R_x)\,\delta_{\alpha\alpha_i} - H_L^+(\eta_\alpha, k_\alpha R_x)\,\mathbf{S}_{\alpha\alpha_i}^{J_{\text{tot}}\pi}\right]. \qquad (3.2.10)$$

The S matrix $\mathbf{S}_{\alpha\alpha_i}^{J_{\text{tot}}\pi}$ gives the amplitude of an outgoing wave in channel α that arises from an incoming plane wave in channel α_i, in addition to the scattering from a diagonal point-Coulomb potential. For all the non-elastic channels $\alpha \neq \alpha_i$ we have

$$\psi_{\alpha\alpha_i}^{J_{\text{tot}}\pi}(R_x) \overset{\alpha \neq \alpha_i}{=} H_L^+(\eta_\alpha, k_\alpha R_x)\,\frac{1}{2\text{i}}\mathbf{S}_{\alpha\alpha_i}^{J_{\text{tot}}\pi}, \qquad (3.2.11)$$

which is to be proportional to a purely outgoing wave. When $\alpha = \alpha_i$, Eq. (3.2.10) leads to a matching equation similar to Eq. (3.1.37) for the elastic channel.

The cross sections, we saw in subsection 2.4.4, depend on the channel *velocity*[13] multiplying the square modulus of an amplitude. It is therefore convenient to

---

and the only non-zero $\Psi_{xpt}^{\mu_p \mu_t}(\mathbf{R}_x, \xi_p, \xi_t; \mathbf{k})$ are

$$\Psi_{111}^{00}(\mathbf{R}, \xi_p, \xi_t; \mathbf{k}) = \psi(\mathbf{R}; \mathbf{k}) = \frac{4\pi}{k}\sum_{LM}Y_L^M(\mathbf{k})^*\,\Psi_{1L}^M(\mathbf{R}, \xi_p, \xi_t),$$

where $\psi(\mathbf{R}; \mathbf{k})$ is that of Eq. (3.1.57).

[13] Strictly a *speed*, but this is the most common terminology.

combine these velocity factors with the S matrix, by defining (for each $J_{\text{tot}}\pi$)

$$\tilde{\mathbf{S}}_{\alpha\alpha_i} = \sqrt{\frac{v_\alpha}{v_{\alpha_i}}} \mathbf{S}_{\alpha\alpha_i}, \tag{3.2.12}$$

where the velocities satisfy $\mu_\alpha v_\alpha = \hbar k_\alpha$. The combination S matrix $\tilde{\mathbf{S}}_{\alpha\alpha_i}$ may now be used to find the multi-channel cross sections, and its matrix elements may be more directly found from the boundary conditions of Eq. (3.2.10) expressed as

$$\psi_{\alpha\alpha_i}^{J_{\text{tot}}\pi}(R_x) = \frac{i}{2}\left[H_{L_i}^-(\eta_\alpha, k_\alpha R_x)\, \delta_{\alpha\alpha_i} - H_L^+(\eta_\alpha, k_\alpha R_x) \sqrt{\frac{v_{\alpha_i}}{v_\alpha}}\, \tilde{\mathbf{S}}_{\alpha\alpha_i}^{J_{\text{tot}}\pi} \right]. \tag{3.2.13}$$

Both $\mathbf{S}_{\alpha\alpha_i}$ and $\tilde{\mathbf{S}}_{\alpha\alpha_i}$ can be regarded as complex numbers in *matrices* $\mathbf{S}$ and $\tilde{\mathbf{S}}$. The second (column) index in these matrices refers to the incoming channel, and the first (row) index names the exit channel.

We can also define a partial-wave T matrix by $\mathbf{S} = \mathbf{I} + 2i\mathbf{T}$ where $\mathbf{I}$ is the identity matrix,[14] or

$$\mathbf{S}_{\alpha\alpha_i} = \delta_{\alpha\alpha_i} + 2i\mathbf{T}_{\alpha\alpha_i} \tag{3.2.14}$$

$$\tilde{\mathbf{S}}_{\alpha\alpha_i} = \delta_{\alpha\alpha_i} + 2i\tilde{\mathbf{T}}_{\alpha\alpha_i}, \tag{3.2.15}$$

noting that the velocity ratios in Eq. (3.2.12) are unity for the diagonal matrix elements, so the diagonal terms $\delta_{\alpha\alpha_i}$ are not affected.

In terms of these T-matrix elements, the scattering boundary conditions of Eq. (3.2.10) are simply

$$\psi_{\alpha\alpha_i}^{J_{\text{tot}}\pi}(R_x) = F_{L_i}(\eta_\alpha, k_\alpha R_x)\, \delta_{\alpha\alpha_i} + H_L^+(\eta_\alpha, k_\alpha R_x)\, \mathbf{T}_{\alpha\alpha_i}^{J_{\text{tot}}\pi}. \tag{3.2.16}$$

Multi-channel cross section

The scattering amplitude from an incoming elastic channel $(x_i p_i t_i)$ to (xpt) depends on the m-substates of the initial nuclei μ_{p_i}, μ_{t_i} and the final nuclei μ_p, μ_t, as well as on the scattering angle θ, as $\tilde{f}_{\mu_p\mu_t,\mu_{p_i}\mu_{t_i}}^{xpt}(\theta)$. We use the scattering amplitude $\tilde{f}$ calculated from the $\tilde{\mathbf{S}}$-matrix elements, so that, in contrast to Eq. (2.4.14), there are no further velocity factors for the flux ratio. The cross section for an unpolarized beam, where all $\mu_{p_i}\mu_{t_i}$ values are equally likely, is thus found by using Eq. (2.4.16), then summing over final m-states and averaging over the initial states:

$$\sigma_{xpt}(\theta) = \frac{1}{(2I_{p_i}+1)(2I_{t_i}+1)} \sum_{\mu_p\mu_t,\mu_{p_i}\mu_{t_i}} \left| \tilde{f}_{\mu_p\mu_t,\mu_{p_i}\mu_{t_i}}^{xpt}(\theta) \right|^2. \tag{3.2.17}$$

[14] We will often write $\mathbf{S} = 1 + 2i\mathbf{T}$ for simplicity.

By Eq. (3.2.16), in addition to the incoming wave in channel α_i, the $\psi_{\alpha\alpha_i}^{J_{\text{tot}}\pi}(R_x)$ have outgoing waves, which in the external region become

$$\psi_{\alpha\alpha_i}^{J_{\text{tot}}\pi}(R_x) \overset{R>R_n}{=} H_{L_\alpha}^+(\eta_\alpha, k_\alpha R_x)\mathbf{T}_{\alpha\alpha_i}^{J_{\text{tot}}\pi} \overset{R\to\infty}{\to} i^{-L_\alpha}e^{ik_\alpha R_x}\mathbf{T}_{\alpha\alpha_i}^{J_{\text{tot}}\pi}, \qquad (3.2.18)$$

and the $\Psi(\mathbf{R}_x, \xi_p, \xi_t; \mathbf{k}_i)$ have outgoing waves proportional to the scattering amplitude as

$$\langle \phi_{I_p\mu_p}^{xp}(\xi_p)\phi_{I_t\mu_t}^{xt}(\xi_t)|\Psi_{x_ip_it_i}^{\mu_{p_i}\mu_{t_i}}(\mathbf{R}_x, \xi_p, \xi_t; \mathbf{k}_i)\rangle \overset{R_x\geq R_n}{=} f_{\mu_p\mu_t,\mu_{p_i}\mu_{t_i}}^{xpt}(\theta)e^{ik_\alpha R_x}/R_x. \qquad (3.2.19)$$

Using Eq. (3.2.8), we may determine the scattering amplitudes $f(\theta)$ in terms of the $\mathbf{T}_{\alpha\alpha_i}^{J_{\text{tot}}\pi}$, for scattering from a beam in the $\mathbf{k}_i$ direction to the asymptotic $\hat{\mathbf{k}} = \hat{\mathbf{R}}$ direction differing by an angle θ:

$$f_{\mu_p\mu_t,\mu_{p_i}\mu_{t_i}}^{xpt}(\theta) = \sum_{J_{\text{tot}}\pi M_{\text{tot}}}\sum_{\alpha\alpha_i}i^{-L_\alpha}\langle\phi_{I_p\mu_p}^{xp}\phi_{I_t\mu_t}^{xt}|\alpha; J_{\text{tot}}M_{\text{tot}}\rangle$$
$$\times A_{\mu_{p_i}\mu_{t_i}}^{J_{\text{tot}}M_{\text{tot}}}(\alpha_i; \mathbf{k}_i)\,\mathbf{T}_{\alpha\alpha_i}^{J_{\text{tot}}\pi}, \qquad (3.2.20)$$

where the α sum is over partial waves consistent with the outgoing xpt, and the α_i must be consistent with the incoming beam specification $x_ip_it_i$.

We include the flux factor of Eq. (2.4.14) by using $\tilde{\mathbf{T}}$ instead of $\mathbf{T}$, to now give $\tilde{f}$, and expand the $|\alpha\rangle$ state and the $A(\alpha_i; \mathbf{k}_i)$ coefficients, cancelling the i^L factor. After also inserting appropriate Coulomb phases as in Eq. (3.1.88) and adding the diagonal pure Coulomb amplitude $f_c(\theta)$, we have a general scattering amplitude for Coulomb + nuclear reactions

$$\tilde{f}_{\mu_p\mu_t,\mu_{p_i}\mu_{t_i}}^{xpt}(\theta) = \delta_{\mu_p\mu_{p_i}}\delta_{\mu_t\mu_{t_i}}\delta_{xpt,x_ip_it_i}f_c(\theta)$$
$$+ \frac{4\pi}{k_i}\sum_{L_iLJ_{p_i}J_pm_imM_iJ_{\text{tot}}}\langle L_iM_i, I_{p_i}\mu_{p_i}|J_{p_i}m_i\rangle$$
$$\langle J_{p_i}m_i, I_{t_i}\mu_{t_i}|J_{\text{tot}}M_{\text{tot}}\rangle\langle LM, I_{p}\mu_p|J_pm\rangle\langle J_pm, I_t\mu_t|J_{\text{tot}}M_{\text{tot}}\rangle$$
$$Y_L^M(\mathbf{k})Y_{L_i}^{M_i}(\mathbf{k}_i)^* \tilde{\mathbf{T}}_{\alpha\alpha_i}^{J_{\text{tot}}\pi} e^{i[\sigma_L(\eta_\alpha)+\sigma_{L_i}(\eta_{\alpha_i})]}, \qquad (3.2.21)$$

and in the S (channel-spin) basis

$$\tilde{f}^{xpt}_{\mu_p \mu_t, \mu_{p_i} \mu_{t_i}}(\theta) = \delta_{\mu_p \mu_{p_i}} \delta_{\mu_t \mu_{t_i}} \delta_{xpt, x_i p_i t_i} f_c(\theta)$$

$$+ \frac{4\pi}{k_i} \sum_{L_i L S S_i S s s_i M_i J_{tot}} \langle I_{p_i} \mu_{p_i}, I_{t_i} \mu_{t_i} | S_i s_i \rangle$$

$$\langle L_i M_i, S_i s_i | J_{tot} M_{tot} \rangle \langle I_p \mu_p, I_t \mu_t | S s \rangle \langle L M, S s | J_{tot} M_{tot} \rangle$$

$$Y_L^M(\mathbf{k}) Y_{L_i}^{M_i}(\mathbf{k}_i)^* \, \tilde{\mathbf{T}}^{J_{tot} \pi}_{\beta \beta_i} \, e^{i[\sigma_L(\eta_\beta) + \sigma_{L_i}(\eta_{\beta_i})]}. \tag{3.2.22}$$

Note that $\tilde{\mathbf{T}}^{J_{tot}\pi}_{\alpha \alpha_i} = \frac{i}{2}[\delta_{\alpha\alpha_i} - \tilde{\mathbf{S}}^{J_{tot}\pi}_{\alpha\alpha_i}]$, and similarly for the S basis.

Polarized beams

In general, cross sections depend on the polarization of the beam, namely on any non-uniform distribution over initial m-states μ_{p_i} for the projectile of spin I_p. The properties of any beam are defined by its density matrix operator $\hat{\rho}$, whereby the expectation value of any operator $\hat{O}$ is given by the trace

$$\langle \hat{O} \rangle_{\hat{\rho}} = \sum_{\mu_{p_i} \mu'_{p_i} = -I_p}^{I_p} \langle I_p \mu_{p_i} | \hat{\rho} | I_p \mu'_{p_i} \rangle \langle I_p \mu'_{p_i} | \hat{O} | I_p \mu_{p_i} \rangle \tag{3.2.23}$$

$$\equiv Tr(\hat{\rho} \hat{O}). \tag{3.2.24}$$

The polarization properties of the density operator for the beam are usually represented by complex numbers t_{Qq} for $Q = 1, 2, \ldots, 2I_p$ and $0 \leq q \leq Q$, according to the construction

$$\hat{\rho} = \frac{1}{2I_p + 1} \sum_{Qq} t^*_{Qq} \hat{\tau}_{Qq}, \tag{3.2.25}$$

where the spherical tensor $\hat{\tau}_{Qq}$ is the operator with matrix elements

$$(\hat{\tau}_{Qq})_{\mu\mu'} = \sqrt{2Q+1} \, \langle I_p \mu, Qq | I_p \mu' \rangle. \tag{3.2.26}$$

The degree to which the beam polarization is reflected in the observed cross section $\sigma^{pol}_{xpt}(\theta)$ for any reaction channel xpt is given by the *tensor analyzing powers* T^{xpt}_{Qq} for this reaction, according to

$$\sigma^{pol}_{xpt}(\theta) = \sigma_{xpt}(\theta) \sum_{Qq} t^*_{Qq} T^{xpt}_{Qq}, \tag{3.2.27}$$

where $\sigma_{xpt}(\theta)$ is the cross section of Eq. (3.2.17) for an unpolarized beam.

These tensor analyzing powers can be calculated from the scattering amplitudes $f^{xpt}_{\mu_p \mu_t : \mu_{p_i} \mu_{t_i}}(\theta)$ as the ratio of traces

$$T^{xpt}_{Qq}(\theta) = \frac{Tr(\mathbf{f}\hat{\tau}_{Qq}\mathbf{f}^+)}{Tr(\mathbf{f}\mathbf{f}^+)} \tag{3.2.28}$$

$$= \sqrt{2Q+1} \frac{\sum_{\mu_p \mu_t \mu_{p_i} \mu_{t_i}} f^{xpt\,*}_{\mu_p \mu_t : \mu_{p_i} \mu_{t_i}}(\theta) \langle I_p \mu_{p_i}, Qq | I_p \mu'_{p_i} \rangle f^{xpt}_{\mu_p \mu_t, \mu'_{p_i} \mu_{t_i}}(\theta)}{\sum_{\mu_p \mu_t \mu_{p_i} \mu_{t_i}} |f^{xpt}_{\mu_p \mu_t : \mu_{p_i} \mu_{t_i}}(\theta)|^2}.$$

For a more complete description of polarization, see Gómez-Camacho and Johnson [10].

Integrated cross sections

The angle-integrated outgoing cross section to a non-elastic excited state pair *xpt* is

$$\sigma_{xpt} = 2\pi \int_0^\pi d\theta \sin\theta \; \sigma_{xpt}(\theta)$$

$$= \frac{\pi}{k_i^2} \frac{1}{(2I_{p_i}+1)(2I_{t_i}+1)} \sum_{J_{tot}\pi LJ\alpha_i} (2J_{tot}+1)|\tilde{\mathbf{S}}^{J_{tot}\pi}_{\alpha\alpha_i}|^2 \tag{3.2.29}$$

$$= \frac{\pi}{k_i^2} \sum_{J_{tot}\pi LJ\alpha_i} g_{J_{tot}} |\tilde{\mathbf{S}}^{J_{tot}\pi}_{\alpha\alpha_i}|^2, \tag{3.2.30}$$

where we have abbreviated a *spin weighting factor* to

$$g_{J_{tot}} \equiv \frac{2J_{tot}+1}{(2I_{p_i}+1)(2I_{t_i}+1)}. \tag{3.2.31}$$

The *reaction cross section* σ_R is defined as the flux leaving the elastic channel, and depends only on the elastic S-matrix element $\mathbf{S}_{\alpha_i\alpha_i}$ as

$$\sigma_R = \frac{\pi}{k_i^2} \frac{1}{(2I_{p_i}+1)(2I_{t_i}+1)} \sum_{J_{tot}\pi\alpha_i} (2J_{tot}+1)(1-|\mathbf{S}^{J_{tot}\pi}_{\alpha_i\alpha_i}|^2) \tag{3.2.32}$$

$$= \frac{\pi}{k_i^2} \sum_{J_{tot}\pi\alpha_i} g_{J_{tot}}(1-|\mathbf{S}^{J_{tot}\pi}_{\alpha_i\alpha_i}|^2), \text{ similarly.}$$

Remember that the sum over α_i is over just the partial waves that include the incoming state indices $x_i p_i t_i$, and that for elastic channels $\mathbf{S}_{\alpha_i\alpha_i} = \tilde{\mathbf{S}}_{\alpha_i\alpha_i}$.

The integrated *elastic cross section* is defined only for neutral scattering ($\eta = 0$), and is a generalization of Eq. (3.1.50):

$$\sigma_{\text{el}} = \frac{\pi}{k_i^2} \frac{1}{(2I_{p_i}+1)(2I_{t_i}+1)} \sum_{J_{\text{tot}}\pi\alpha_i} (2J_{\text{tot}}+1)|1 - \mathbf{S}_{\alpha_i\alpha_i}^{J_{\text{tot}}\pi}|^2. \tag{3.2.33}$$

It measures all the flux being elastically scattered by any non-zero angle. The *total cross section* is the sum of the reaction and elastic cross sections:[15]

$$\begin{aligned}
\sigma_{\text{tot}} &= \sigma_R + \sigma_{\text{el}} \\
&= \frac{2\pi}{k_i^2} \frac{1}{(2I_{p_i}+1)(2I_{t_i}+1)} \sum_{J_{\text{tot}}\pi\alpha_i} (2J_{\text{tot}}+1)[1 - \text{Re}\mathbf{S}_{\alpha_i\alpha_i}^{J_{\text{tot}}\pi}],
\end{aligned} \tag{3.2.34}$$

and is the sum of the fluxes leaving the incident direction for any reason: elastic scattering to another angle, or a reaction leading to any non-elastic channel.

The *absorption cross section* is the loss of flux caused by any imaginary potentials in the Hamiltonian. In multi-channel theory, we calculate it as the reaction cross section minus all the non-elastic outgoing cross sections:

$$\sigma_A = \sigma_R - \sum_{xpt \neq x_i p_i t_i} \sigma_{xpt}. \tag{3.2.35}$$

This reduces to the result of subsection 3.1.5 (namely $\sigma_A = \sigma_R$) when there are no non-elastic channels. With multiple channels, if the coupled equations are solved precisely in the presence of absorptive complex potentials, then σ_A will be positive, and will as before be an integral of $W(R) < 0$, now

$$\sigma_A = \frac{2}{\hbar v_i} \frac{4\pi}{k_i^2} \sum_{J_{\text{tot}}\pi\alpha_i\alpha} \int_0^\infty [-W_\alpha(R_x)] \, |\psi_{\alpha\alpha_i}^{J_{\text{tot}}\pi}(R_x)|^2 \, dR_x. \tag{3.2.36}$$

3.2.2 Coupled equations

The multi-channel wave functions of Eqs. (3.2.1) and (3.2.2) contain the channel wave functions $\psi_\alpha(R_x)$ or $\psi_\beta(R_x)$. In order to find these wave functions, we have to write down the coupled equations that they satisfy, starting from the Schrödinger equation for the whole system. Chapter 6 will show how to solve the coupled equations that we derive in this section.

For total energy E and Hamiltonian operator H, we have to solve

$$[H - E]\Psi_{J_{\text{tot}}\pi}^{M_{\text{tot}}} = 0. \tag{3.2.37}$$

---

[15] Note that the term 'total cross section' has several meanings in the literature. Sometimes it refers to the angle-integrated cross section of Eq. (3.1.50), but the term is best kept to refer to that of Eq. (3.2.34).

The results will be independent of M_{tot} when H is rotationally invariant: when H does not depend on any particular direction in space.

The total Hamiltonian H can be written equivalently for each partition x, in terms of that partition's internal Hamiltonians, kinetic energies, and interaction potentials:

$$H = H_{xp}(\xi_p) + H_{xt}(\xi_t) + \hat{T}_x(R_x) + \mathcal{V}_x(R_x, \xi_p, \xi_t), \qquad (3.2.38)$$

where $\mathcal{V}_x(R_x, \xi_p, \xi_t) \to 0$ as $R_x \to \infty$.

The internal nuclear states $\phi_{I_p}^{xp}(\xi_p)$ and $\phi_{I_t}^{xt}(\xi_t)$ satisfy eigen-equations for their Hamiltonians

$$H_{xp}(\xi_p)\phi_{I_p}^{xp}(\xi_p) = \epsilon_{xp}\phi_{I_p}^{xp}(\xi_p),$$

$$H_{xt}(\xi_t)\phi_{I_t}^{xt}(\xi_t) = \epsilon_{xt}\phi_{I_t}^{xt}(\xi_t), \qquad (3.2.39)$$

for eigenenergies ϵ_{xp} and ϵ_{xt} respectively.[16] The kinetic energy operator depends on the masses m_{xp} and m_{xt} via the partition's reduced mass $\mu_x = m_{xp}m_{xt}/(m_{xp}m_{xt})$ as

$$\hat{T}_x(R_x) = -\frac{\hbar^2}{2\mu_x}\nabla_{R_x}^2. \qquad (3.2.40)$$

The interaction potential $\mathcal{V}_x(R_x, \xi_p, \xi_t)$ is the residual interaction between nuclei p and t. For nucleons i in the projectile and j in the target, it may be written as the sum

$$\mathcal{V}_x(R_x, \xi_p, \xi_t) = \sum_{i\in p, j\in t} V_{ij}(\mathbf{r}_i - \mathbf{r}_j) \qquad (3.2.41)$$

of individual nucleon-nucleon forces V_{ij}. When the distance R_x between p and t becomes large asymptotically, the residual interaction goes to zero.

For convenience, we define a *joint* projectile and target Hamiltonian as $H_x \equiv H_{xp} + H_{xt}$, a joint eigenstate as $\phi^{xpt} \equiv \phi_{I_p}^{xp}\phi_{I_t}^{xt}$, and a joint eigenenergy as $\epsilon_{xpt} \equiv \epsilon_{xp}+\epsilon_{xp}$, so we have simply $H_x\phi^{xpt} = \epsilon_{xpt}\phi^{xpt}$ for the product of internal structures. The total Hamiltonian is then $H = \hat{T}_x + H_x + \mathcal{V}_x$ for any chosen x.

The coupled equations are now found by expanding the total wave function $\Psi_{J_{tot}\pi}^{M_{tot}}$ in either the S or J partial-wave basis. In the J basis, for example, $\Psi_{J_{tot}\pi}^{M_{tot}} = \sum_\alpha |\alpha; J_{tot}\pi\rangle \psi_\alpha(R_x)/R_x$, so from Eq. (3.2.37) for a given $J_{tot}\pi$ value, we have[17]

$$\sum_\alpha [H - E]|\alpha; J_{tot}\pi\rangle \psi_\alpha(R_x)/R_x = 0. \qquad (3.2.42)$$

---

[16] We omit the m-state quantum numbers from this equation since the energy does not depend on these.

[17] The M_{tot} has been omitted since the results should be the same for all M_{tot} values, and now we will often omit also the $J_{tot}\pi$ labels, and the incoming channel label α_i, when discussing a given coupled-channels set.

Projecting onto one of the basis states, by operating on the left by $R_{x'} \langle \alpha' |$,

$$\sum_\alpha R_{x'} \langle \alpha' | H - E | \alpha \rangle R_x^{-1} \, \psi_\alpha(R_x) = 0, \tag{3.2.43}$$

abbreviated as

$$\sum_\alpha (H-E)_{\alpha'\alpha} \, \psi_\alpha(R_x) = 0, \tag{3.2.44}$$

which gives a separate equation for each α' combination of quantum numbers. The set of all the equations for various α' is called the *set of coupled-channels equations*.

The Hamiltonian and energy matrix element was abbreviated by $\langle \alpha' | H - E | \alpha \rangle = (H-E)_{\alpha'\alpha}$. To evaluate all these, we note that

$$
\begin{aligned}
[H - E] | \alpha \rangle &= [H - E] \, | xpt : (LI_p) J_p, I_t; J_{\text{tot}} \pi \rangle \\
&= [\hat{T}_x + H_x + V_x - E] \, | xpt : (LI_p) J_p, I_t; J_{\text{tot}} \pi \rangle \\
&= [\hat{T}_x + \epsilon_{xp} + \epsilon_{xp} + V_x - E] \, | xpt : (LI_p) J_p, I_t; J_{\text{tot}} \pi \rangle \\
&= [\hat{T}_x + V_x - E_{xpt}] \, | xpt : (LI_p) J_p, I_t; J_{\text{tot}} \pi \rangle, \tag{3.2.45}
\end{aligned}
$$

where $E_{xpt} = E - \epsilon_{xp} - \epsilon_{xp}$ is the external kinetic energy for a given excited-state pair xpt.

This means that the matrix elements $\langle \alpha' | H - E | \alpha \rangle$ may be written in two ways, one by replacing H either by $\hat{T}_x + H_x + V_x$ for acting on the right-hand side, and the other by $\hat{T}_{x'} + H_{x'} + V_{x'}$ for acting on the left side. The first option is called the *prior* form of the matrix element, and the second the *post* form. Ideally, if all terms of the coupled equations are included and the equations are solved accurately, both choices will give the same results.[18] The *prior* form of the matrix element is thus

$$
\begin{aligned}
(H-E)_{\alpha'\alpha} &= R_{x'} \langle \alpha' | \hat{T}_x + V_x - E_{xpt} | \alpha \rangle R_x^{-1} \\
&= R_{x'} \langle \alpha' | \alpha \rangle R_x^{-1} [\hat{T}_{xL} - E_{xpt}] + R_{x'} \langle \alpha' | V_x | \alpha \rangle R_x^{-1} \\
&\equiv \hat{N}_{\alpha'\alpha} [\hat{T}_{xL}(R_x) - E_{xpt}] + \hat{V}_{\alpha'\alpha}^{\text{prior}}, \tag{3.2.46}
\end{aligned}
$$

where the partial-wave kinetic energy operator, the same as the one-channel operator of Eq. (3.1.10), is

$$\hat{T}_{xL}(R_x) = -\frac{\hbar^2}{2\mu_x} \left[\frac{d^2}{dR_x^2} - \frac{L_x(L_x+1)}{R_x^2} \right], \tag{3.2.47}$$

---

[18] On page 104 we will see that there is also a simpler first-order result, whereby that post and prior forms necessarily give the same first-order transition amplitudes.

where L_x is the orbital angular momentum in channel α. The coupling interactions *between* channels are either the *prior* or *post* matrix elements, defined as

$$\hat{V}^{\text{prior}}_{\alpha'\alpha} = R_{x'} \langle \alpha' | \mathcal{V}_x | \alpha \rangle R_x^{-1} \tag{3.2.48}$$

$$\hat{V}^{\text{post}}_{\alpha'\alpha} = R_{x'} \langle \alpha' | \mathcal{V}_{x'} | \alpha \rangle R_x^{-1} \tag{3.2.49}$$

respectively, and the norm overlap operators between the partial-wave basis states are

$$\hat{N}_{\alpha'\alpha} = R_{x'} \langle \alpha' | \alpha \rangle R_x^{-1}. \tag{3.2.50}$$

Within the same partition, $x' = x$, the norm overlaps are diagonal: $\hat{N}_{\alpha'\alpha} = \delta_{\alpha'\alpha}$. This suggests treating the matrix elements of $\hat{T}_{xL} - E_{\text{xpt}}$ *within* a partition separately from those *between* partitions.

With these definitions, we have from Eq. (3.2.43) one version of the coupled-channels equation set:

$$[\hat{T}_{x'L}(R'_x) - E_{x'p't'}]\psi_{\alpha'}(R'_x) + \sum_{\alpha} \hat{V}^{\text{prior}}_{\alpha'\alpha} \psi_{\alpha}(R_x)$$

$$+ \sum_{\alpha, x \neq x'} \hat{N}_{\alpha'\alpha}[\hat{T}_{xL} - E_{\text{xpt}}]\psi_{\alpha}(R_x) = 0,$$

which, on interchanging primes and the unprimed, gives a perhaps more natural

$$[\hat{T}_{xL}(R_x) - E_{\text{xpt}}]\psi_{\alpha}(R_x) + \sum_{\alpha'} \hat{V}^{\text{prior}}_{\alpha\alpha'} \psi_{\alpha'}(R_{x'})$$

$$+ \sum_{\alpha', x' \neq x} \hat{N}_{\alpha\alpha'}[\hat{T}_{x'L'} - E_{x'p't'}]\psi_{\alpha'}(R_{x'}) = 0. \tag{3.2.51}$$

The third terms in these equations are called *non-orthogonality terms* because they involve the overlap of basis functions $\langle \alpha' | \alpha \rangle$ between different mass partitions, and arise particularly in transfer reactions. We will see in Chapter 6 that they may be neglected in some circumstances, which would allow the coupled equations to be written in the more familiar form

$$[\hat{T}_{xL}(R_x) - E_{\text{xpt}}]\psi_{\alpha}(R_x) + \sum_{\alpha'} \hat{V}^{\text{prior}}_{\alpha\alpha'} \psi_{\alpha'}(R_{x'}) = 0. \tag{3.2.52}$$

The equations (3.2.51) used the *prior* form. This may be discerned from the fact that the interaction potential $\mathcal{V}_{x'}$ in the coupling matrix element $\hat{V}^{\text{prior}}_{\alpha\alpha'} = \langle \alpha | \mathcal{V}_{x'} | \alpha' \rangle$ refers to the *initial* channel α' rather than the final channel α. The same solutions should also result in the converse *post*-form matrix element, obtained when Eq. (3.2.46) is replaced by

$$(H-E)_{\alpha'\alpha} = R_{x'}\langle\alpha'|\hat{T}_{x'} + \mathcal{V}_{x'} - E_{x'p't'}|\alpha\rangle R_x^{-1}$$
$$= [\hat{T}_{x'L'} - E_{x'p't'}]\hat{N}_{\alpha'\alpha} + \hat{V}_{\alpha'\alpha}^{\text{post}}. \qquad (3.2.53)$$

The detailed construction of all the coupling potentials (whether post or prior) is the subject of Chapter 4.

3.2.3 Unitarity of the multi-channel S matrix

The multi-channel S matrices $\mathbf{S}_{\alpha\alpha_i}$ and $\tilde{\mathbf{S}}_{\alpha\alpha_i}$, for each total angular momentum and parity J_{tot}^{π}, have certain symmetry properties when the initial Hamiltonians have specific features. These concern physical properties such as Hermiticity, time invariance, and what is called reciprocity.

It may be that the couplings $\hat{V}_{\alpha\alpha'}$ are Hermitian, that is $\hat{V}_{\alpha\alpha'} = \hat{V}_{\alpha'\alpha}^*$ (whether post or prior). This is true if the coupling matrix is real and symmetric, but also hold more generally for self-adjoint or Hermitian couplings. Hermitian operators should be familiar in quantum mechanics since they have real eigenvalues and their eigenvectors form an orthogonal set.

For scattering, the consequence of Hermiticity is that the matrix $\tilde{\mathbf{S}}_{\alpha\alpha_i}$ of Eq. (3.2.12) is *unitary*:

$$\tilde{\mathbf{S}}^{-1} = \tilde{\mathbf{S}}^{\dagger} \equiv (\tilde{\mathbf{S}}^*)^{\mathrm{T}}, \qquad (3.2.54)$$

so $\tilde{\mathbf{S}}^{\dagger}\tilde{\mathbf{S}} = 1$. For α_i, α_i' as two incoming channels,

$$\sum_{\alpha} \tilde{\mathbf{S}}_{\alpha\alpha_i}^* \tilde{\mathbf{S}}_{\alpha\alpha_i'} = \delta_{\alpha_i\alpha_i'}, \qquad (3.2.55)$$

and, in particular

$$\sum_{\alpha} |\tilde{\mathbf{S}}_{\alpha\alpha_i}|^2 = 1. \qquad (3.2.56)$$

Each row of the $\tilde{\mathbf{S}}$ matrix is therefore a vector with unit norm. From $\tilde{\mathbf{S}}\tilde{\mathbf{S}}^{\dagger} = 1$ we can similarly prove that each column is a unit vector.

Unitarity implies that the absorption cross section of Eq. (3.2.35) is $\sigma_A = 0$, so that the reaction cross section (the flux leaving the entrance channel) is precisely equal to the sum of all the outgoing cross sections.

3.2.4 Detailed balance

The condition of *detailed balance* in a general statistical or stochastic system in classical physics is said to hold when the forward and reverse *transition probabilities* are equal for each transition. The transition probabilities in the present context are the square moduli of the S-matrix elements, which properly sum up to unity in the unitarity limit of Eq. (3.2.56). We therefore define as the *detailed balance relation* the equation

$$|\tilde{\mathbf{S}}_{\alpha\alpha_i}|^2 = |\tilde{\mathbf{S}}_{\alpha_i\alpha}|^2, \tag{3.2.57}$$

which makes all the forward and reverse transition probabilities equal.

This condition holds for the solutions of coupled Schrödinger equations if the forward and reverse couplings are identical, with the same magnitude and phase, namely that coupling matrix $\hat{V}_{\alpha\alpha'} = \hat{V}_{\alpha'\alpha}$ is symmetric. In this case, we have a somewhat stronger *reciprocity* condition where the $\tilde{\mathbf{S}}$ matrix itself is also *symmetric*:

$$\tilde{\mathbf{S}} = \tilde{\mathbf{S}}^{\mathrm{T}}. \tag{3.2.58}$$

This will be proved not here, but in subsection 10.3.2, where we give an explicit construction for $\tilde{\mathbf{S}}$. (Note that it is $\tilde{\mathbf{S}}$ of Eq. (3.2.12), *with* the velocity factors, which is symmetric, not the original $\mathbf{S}$ matrix.) This symmetry is sufficient for the detailed balance relation to hold.

We may also use the coupling matrices defined using the multi-channel wave functions of subsection 3.2.1 which are *not* symmetric, because of the i^L factors in Eq. (3.2.1). If these factors are used, the coupling matrices will satisfy rather $\hat{V}_{\alpha\alpha'} = (-1)^{L-L'}\hat{V}_{\alpha'\alpha}$, as also do the $\tilde{\mathbf{S}}$ matrices, but this is still sufficient for detailed balance, Eq. (3.2.57), to hold.

The symmetry condition (3.2.58) is distinct from unitarity, but real symmetric coupling matrices lead to both unitarity and symmetry of the $\tilde{\mathbf{S}}$ matrix. Hamiltonians as commonly used with complex potentials cannot be unitary, but can almost always be made to have symmetric matrix elements. For more discussion about unitarity, time reversal and reciprocity, see Taylor [5, §6-e, §17-e] and Satchler [11, §4.4, §9.5-6].

The above reciprocity results can also be proved for potentials that are *time-reversal invariant*. Complex potentials, however, never satisfy this condition, but still lead to reciprocity, which is hence true more generally than just for invariance under time reversal.

From the detailed balance of the $\tilde{\mathbf{S}}$ matrix of Eq. (3.2.57), we may derive a direct connection between the *total cross sections* for the forward and reverse reactions. From Eq. (3.2.29), the cross section from entrance channel $x_i p_i t_i$ to a distinct exit channel *xpt* is

$$\sigma_{xpt:x_ip_it_i} = \frac{\pi}{k_i^2} \frac{1}{(2I_{p_i}+1)(2I_{t_i}+1)} \sum_{J_{tot}\pi\alpha\alpha_i} (2J_{tot}+1)|\tilde{\mathbf{S}}_{\alpha\alpha_i}^{J_{tot}\pi}|^2. \tag{3.2.59}$$

The detailed balance $|\tilde{\mathbf{S}}_{\alpha\alpha_i}|^2 = |\tilde{\mathbf{S}}_{\alpha_i\alpha}|^2$ implies that the equivalent expression for the reverse reaction $\sigma_{x_ip_it_i:xpt}$, from entrance channel *xpt* to exit channel $x_ip_it_i$, satisfies

$$k_i^2(2I_{p_i}+1)(2I_{t_i}+1)\sigma_{xpt:x_ip_it_i} = k^2(2I_p+1)(2I_t+1)\sigma_{x_ip_it_i:xpt}, \tag{3.2.60}$$

so

$$\sigma_{x_ip_it_i:xpt} = \frac{k_i^2(2I_{p_i}+1)(2I_{t_i}+1)}{k^2(2I_p+1)(2I_t+1)}\sigma_{xpt:x_ip_it_i}. \tag{3.2.61}$$

This equation is therefore called the *principle of detailed balance*. The Hermiticity of the Hamiltonian leads to unitarity S matrices, but that by itself is only sufficient for detailed balance between the cross sections if the couplings are also real or can be made real by a unitary transformation.

A slightly different expression holds for photon channels. Although we will see in subsection 3.5.1 that they can be considered as spin 1 objects, the gauge condition implies that there are only two independent polarization projections. The number of photon m-states is hence not $(2s_\gamma+1)$, so this factor should be replaced by the value of 2.

Another slightly different expression holds for identical particles. If the reacting nuclei in either the entrance or exit channels are identical, then we will see in Section 3.4 that requirements of the Pauli principle dictate either symmetry or antisymmetry of the overall system wave function. Compared with Eq. (3.2.59), this induces an additional factor in the way the cross section depends on the S-matrix elements. This factor may be different in the forward and reverse channels, and affect the detailed balance equation (3.2.61). These factors will be determined in subsection 3.4.3, after isospin and (anti)symmetries have been defined.

3.3 Integral forms

Instead of defining cross sections in terms of S- or T-matrix elements contained in the boundary conditions for differential equations, it is also possible to give expressions for these matrix elements that are integrals over the wave functions with some part of the Hamiltonian. These *integral forms* for the S- or T-matrix elements should in principle yield identical results, but are useful since they may suggest a new range of approximations that may still be sufficiently accurate in the relevant physical respects.

3.3.1 Green's function methods

Up to now we have solved only homogeneous Schrödinger equations like $[E - H]\Psi = 0$. Sometimes we may need to solve inhomogeneous equations like $[E - H]\Psi = \Omega$ with outgoing boundary conditions, for some radial functions $\Omega(R)$ called *source terms*, as such equations arise as part of a coupled-channels set. The inhomogenous equation may be solved by differential methods as discussed in Chapter 6, but often it is useful to give an integral expression for its solution, and it is especially useful that there exist simple integrals giving directly the asymptotic outgoing amplitude of the solution, namely its T-matrix element. This section shows how to use Green's function methods to solve the inhomogeneous differential equation.

Integral solutions of inhomogeneous equations

Consider the general problem of solving the coupled equations similar to those of Eq. (3.2.52):

$$[T_{xL}(R) + V_c(R) - E_{xpt}]\psi_\alpha(R) + \sum_{\alpha'}\langle\alpha|V|\alpha'\rangle\psi_{\alpha'}(R') = 0. \tag{3.3.1}$$

for some given total angular momentum and parity J_{tot}^π and incoming channel α_i that we assume are all fixed, and not always written among the indices. Here we have separated out the point-Coulomb potential $V_c(R) = Z_{xp}Z_{xt}e^2/R$ (if present), and put all the other couplings, local or non-local, into the matrix elements of V.

The solutions must satisfy the standard outgoing boundary conditions of Eq. (3.2.13) for the given α_i. Suppose that all the $\psi_{\alpha'}(R')$ are known for which $\langle\alpha|V|\alpha'\rangle \neq 0$, in which case we may solve the inhomogeneous equation (3.3.1) for the wave function $\psi_\alpha(R)$ using the known source term

$$\Omega_\alpha(R) = \sum_{\alpha'}\langle\alpha|V|\alpha'\rangle\psi_{\alpha'}(R'). \tag{3.3.2}$$

This is to solve the inhomogeneous equation

$$[E_{xpt} - T_{xL}(R) - V_c(R)]\psi_\alpha(R) = \Omega_\alpha(R). \tag{3.3.3}$$

The outgoing-wave boundary conditions from Eq. (3.2.13) may be written in the T-matrix form of Eq. (3.2.16):

$$\psi_{\alpha\alpha_i}(R) = F_\alpha(R)\delta_{\alpha\alpha_i} + H_\alpha^+(R)\mathbf{T}_{\alpha\alpha_i}, \tag{3.3.4}$$

where we have reinserted α_i as the given incoming elastic channel. The $\mathbf{T}_{\alpha\alpha_i}$ is the nuclear T-matrix element for scattering in addition to the point-Coulomb potential. We confirm that in the limit of $V = 0$, when $\Omega_\alpha(R) = 0$, only the the elastic channel wave function is non-zero with value $\psi_{\alpha_i\alpha_i}(R) = F_{\alpha_i}(R)$, and $\mathbf{T}_{\alpha\alpha_i} = 0$.

Definition of $G^+(R, R')$

Let us use Green's function methods to find the outgoing solution of the linear equation

$$\left[\frac{d^2}{dR^2} - \breve{U}(R) + k_\alpha^2 \right] \psi_\alpha(R) = \frac{2\mu_x}{\hbar^2} \Omega_\alpha(R), \tag{3.3.5}$$

where $\breve{U}(R) \equiv 2\eta k_\alpha/R + L_\alpha(L_\alpha+1)/R^2$ is the sum of the Coulomb and centrifugal terms, and $k_\alpha^2 = 2\mu_x E_{xpt}/\hbar^2$.

Now the general source term $\Omega_\alpha(R)$ can always be formally written as a superposition of solutions for δ-function sources $\delta(R - R')$ at R', since

$$\Omega_\alpha(R) = \int \delta(R - R')\Omega_\alpha(R')dR'. \tag{3.3.6}$$

Thus all we need is the solution, a function of R, of Eq. (3.3.5) for a δ-function source at R'. We denote this solution by $G^+(R, R')$, since it depends on both R and R', and it is precisely the Green's function $G^+(R, R')$ satisfying

$$\left[\frac{d^2}{dR^2} - \breve{U}(R) + k_\alpha^2 \right] G^+(R, R') = \delta(R - R'). \tag{3.3.7}$$

The desired solution is therefore a superposition of all the $G^+(R, R')$ with amplitudes corresponding to the magnitude of the source term at R', namely $\frac{2\mu_x}{\hbar^2}\Omega_\alpha(R')$. This will give the wave function in terms of the integral expression

$$\psi_\alpha(R) = \delta_{\alpha\alpha_i}F_\alpha(R) + \frac{2\mu_x}{\hbar^2} \int G^+(R, R')\Omega_\alpha(R')dR', \tag{3.3.8}$$

since the homogeneous solution $F_\alpha(R)$ is present only in the elastic channel. This equation is often written more compactly in operator notation as

$$\psi_\alpha = F_\alpha + \hat{G}^+\Omega_\alpha, \tag{3.3.9}$$

where $\hat{G}^+$ is defined as the Green's integral operator that has the kernel function $2\mu_x/\hbar^2 \, G^+(R, R')$. Furthermore, because $\hat{G}^+\Omega_\alpha$ is the solution ψ of the differential equation $[E - \hat{T} - U_c]\psi = \Omega_\alpha$ with $\hat{T}$ the kinetic energy operator, the Green's operator with the Coulomb potential U_c can be written as

$$\hat{G}^+ = [E - \hat{T} - U_c]^{-1} \tag{3.3.10}$$

with the specified outgoing boundary conditions. Eq. (3.3.9) can therefore be written as

$$\psi_\alpha = F_\alpha + [E - \hat{T} - U_c]^{-1}\Omega_\alpha. \tag{3.3.11}$$

To find $G^+(R, R')$

For fixed R', when $R \neq R'$ we have from Eq. (3.3.7) that[19]

$$\left[\frac{d^2}{dR^2} - \breve{U}(R) + k^2 \right] G^+(R, R') = 0. \tag{3.3.12}$$

Since this is a second-order linear differential equation, any solution must be a linear combination of two fixed linearly independent solutions. We choose for these the regular Coulomb function $F(R)$ and the irregular function $H^+(R)$ for this partial wave L. With this choice, the unknown Green's function must satisfy

$$G^+(R, R') = f(R')F(R) + h(R')H^+(R), \tag{3.3.13}$$

where the coefficients f, h are as yet unknown functions of R'. Since these coefficients are different for different R' values, we now apply Eq. (3.3.12) separately for $R < R'$ and $R > R'$, avoiding $R = R'$.

To determine the $R < R'$ case, consider $R = 0$. Any solution of Eq. (3.3.12) must be zero at the origin in R, so $G^+(0, R') = 0$ for $R' > 0$. Since $H^+(0) \neq 0$ and $F(0) = 0$, we conclude from Eq. (3.3.13) that $h(R') = 0$ when $R < R'$.

To determine the $R > R'$ case, note that the $R \to \infty$ boundary condition of Eq. (3.3.4) implies that $G^+(R, R') \propto H_L^+(R)$. We conclude from Eq. (3.3.13) that $f(R') = 0$ when $R > R'$.

Summarizing the two results,

$$G^+(R, R') = \begin{cases} f(R')F(R) & \text{for } R < R' \\ h(R')H^+(R) & \text{for } R > R'. \end{cases} \tag{3.3.14}$$

To fix $f(R')$, $h(R')$ we use the differential equation (3.3.7). Integrating this with respect to R over the range from just below R' to just above, we find

$$\frac{d}{dR} G^+(R, R') \bigg|_{R \to R'+} - \frac{d}{dR} G^+(R, R') \bigg|_{R \to R'-} = 1, \tag{3.3.15}$$

so from Eq. (3.3.14) we have

$$h(R)H^{+\prime}(R) - f(R)F'(R) = 1. \tag{3.3.16}$$

Now $G(R, R')$ is itself continuous over $R \sim R'$:

$$h(R)H^+(R) - f(R)F(R) = 0. \tag{3.3.17}$$

---

[19] We are dealing with just one channel α here, so up to Eq. (3.3.21) we may omit that subscript for simplicity.

So, solving Eqs. (3.3.16, 3.3.17) simultaneously, we have

$$h(R) = \frac{F(R)}{W(F, H^+)}, \quad f(R) = \frac{H^+(R)}{W(F, H^+)}, \tag{3.3.18}$$

where $W(f, g) = fg' - f'g$ is the Wronskian for two functions $f(R)$ and $g(R)$. For our F, H^+, the Wronskian $W(F, H^+) = W(F, G) = -k$ from Eq. (3.1.11).

The full Green's function is therefore

$$G^+(R, R') = -\frac{1}{k} \begin{cases} H^+(R')F(R) & \text{for } R < R' \\ F(R')H^+(R) & \text{for } R > R' \end{cases} \tag{3.3.19}$$

$$= -\frac{1}{k} F(R_<)H^+(R_>) \tag{3.3.20}$$

where $R_< = \min(R, R')$ and $R_> = \max(R, R')$.

We wanted the solution of the original inhomogeneous equation (3.3.3). Now restoring the channel indices α and incoming index α_i, that solution is

$$\psi_\alpha(R) = \delta_{\alpha\alpha_i} F_\alpha(R) - \frac{2\mu_x}{\hbar^2 k_\alpha} \int F_\alpha(R_<) H_\alpha^+(R_>) \Omega_\alpha(R') dR'. \tag{3.3.21}$$

At large distances $R > \max(R', R_n)$, we have from Eq. (3.3.4)

$$\psi_\alpha(R) \rightarrow \delta_{\alpha\alpha_i} F_\alpha(R) + \mathbf{T}_{\alpha\alpha_i} H^+(R),$$

so

$$\mathbf{T}_{\alpha\alpha_i} H_\alpha^+(R) = -\frac{2\mu_x}{\hbar^2 k_\alpha} H_\alpha^+(R) \int F_\alpha(R') \Omega_\alpha(R') dR', \tag{3.3.22}$$

and we arrive at a very useful integral expression for the partial-wave T-matrix element:

$$\mathbf{T}_{\alpha\alpha_i} = -\frac{2\mu_x}{\hbar^2 k_\alpha} \int F_\alpha(R') \Omega_\alpha(R') dR'. \tag{3.3.23}$$

This may be rewritten in Dirac bra-ket notation as

$$\mathbf{T}_{\alpha\alpha_i} = -\frac{2\mu_x}{\hbar^2 k_\alpha} \langle F_\alpha^* | \Omega_\alpha \rangle \tag{3.3.24}$$

$$= -\frac{2\mu_x}{\hbar^2 k_\alpha} \langle F_\alpha^{(-)} | \Omega_\alpha \rangle. \tag{3.3.25}$$

The complex conjugation in Eq. (3.3.24) is necessary to cancel the conjugation implicit in the matrix elements. A $^{(-)}$ superscript in Eq. (3.3.25) is used for the same purpose, for reasons to be explained on page 98.

In the operator notation, Eq. (3.3.11) may be rewritten as

$$\psi = \phi + \hat{G}^+ \Omega$$
$$= \phi + \hat{G}^+ V \psi, \tag{3.3.26}$$

using ϕ to refer to the homogeneous solution present only in the elastic channel, and with the $+$ sign indicating outgoing boundary conditions of Eq. (3.3.4). An equation like (3.3.26) is called a partial-wave *Lippmann-Schwinger equation*, and in this notation the T matrix (3.3.25) is the integral

$$\mathbf{T} = -\frac{2\mu}{\hbar^2 k} \langle \phi^{(-)} | V | \psi \rangle \equiv -\frac{2\mu}{\hbar^2 k} \int \phi(R) V(R) \psi(R) dR. \tag{3.3.27}$$

In the multi-channel formulation, ψ and ϕ are interpreted as vectors (ϕ being only non-zero in the elastic channel), V as a matrix, and $\hat{G}^+$ is a matrix of integral operators.

3.3.2 Vector-form T matrix for plane waves

If the above analysis is repeated for every partial wave L in the spherical potential case, then the resulting $\mathbf{T}_L$ may be substituted in Eq. (3.1.56) to yield the angle-dependent scattering amplitude $f(\theta)$. If $V_c(R) = 0$, for example, the ϕ are the partial-wave components $F_L(0, kR)$ of a plane wave and the ψ are the partial waves χ_L for scattering with the potential. Summing over L gives

$$f(\theta) = -\frac{8\pi\mu}{\hbar^2 k_i} \sum_{LM} \int_0^\infty Y_L^M(\hat{\mathbf{k}}')^* F_{L'}(0, k'R) \, V(R) \, Y_L^M(\hat{\mathbf{k}}_i) \chi_L(R) \, dR. \tag{3.3.28}$$

Then, expanding $V(R)\delta_{LL'}\delta_{MM'} = \int_{4\pi} d\hat{\mathbf{R}} \, Y_L^M(\mathbf{R}) V(R) Y_{L'}^{M'}(\mathbf{R})$ for a spherical potential, we have

$$f(\theta) = -\frac{\mu}{2\pi\hbar^2} \int d\mathbf{R} \, e^{-i\mathbf{k}'\cdot\mathbf{R}} \, V(\mathbf{R}) \, \Psi(\mathbf{R}; \mathbf{k}_i) \tag{3.3.29}$$

$$= -\frac{\mu}{2\pi\hbar^2} \langle e^{i\mathbf{k}'\cdot\mathbf{R}} | V | \Psi(\mathbf{R}; \mathbf{k}_i) \rangle, \tag{3.3.30}$$

where θ is the scattering angle from the initial momentum $\mathbf{k}_i$ to the final momentum $\mathbf{k}'$ (with $|\mathbf{k}'| = |\mathbf{k}_i|$), and the bra-ket is a three-dimensional integral. This integral form is also called a 'T matrix', but is one that depends on initial and final $\mathbf{k}$ values, not on partial waves L or α. We call this *vector form* of the **T** matrix (as distinct from the partial-wave form **T**), with the notation $\mathbf{T}(\mathbf{k}', \mathbf{k})$.

The vector form **T** matrix is defined in relation to the two-body scattering amplitude as

$$f(\mathbf{k}'; \mathbf{k}) = -\frac{\mu}{2\pi\hbar^2}\mathbf{T}(\mathbf{k}', \mathbf{k}), \tag{3.3.31}$$

so we have just shown that for a plane-wave final state,

$$\mathbf{T}(\mathbf{k}', \mathbf{k}) = \langle e^{i\mathbf{k}' \cdot \mathbf{R}} | V | \Psi(\mathbf{R}; \mathbf{k}) \rangle. \tag{3.3.32}$$

3.3.3 Two-potential formula

If a channel potential can be composed of two parts $V(R) = U_1(R) + U_2(R)$, then it is possible to treat U_1 as the distorting potential and U_2 as the remaining interaction. Then, using the T-matrix integral expression for the scattering from their combined potential V, we can derive an exact *two-potential formula* involving the T-matrix *difference* between U_1 scattering and V scattering. This difference will be proportional to U_2.

For each partial wave, let us define solutions ϕ for the free field, χ for U_1 only, and ψ for the full case, and use Eq. (3.3.26) to write down the corresponding Lippmann-Schwinger equations:

Free:	$[E-T]\phi = 0$	$\hat{G}_0^+ = [E - T]^{-1}$	$\phi = F$
Distorted:	$[E-T-U_1]\chi = 0$	$\chi = \phi + \hat{G}_0^+ U_1\chi$	$\chi \to \phi + \mathbf{T}^{(1)}H^+$
Full:	$[E-T-U_1-U_2]\psi = 0$	$\psi = \phi + \hat{G}_0^+(U_1+U_2)\psi$	$\psi \to \phi + \mathbf{T}^{(1+2)}H^+$

From Eq. (3.3.27), the **T**-matrix integral $\mathbf{T}^{(1)} = -\frac{2\mu}{\hbar^2 k}\langle\phi^{(-)}|U_1|\chi\rangle$ describes the scattering from U_1 only. Similarly the T matrix $\mathbf{T}^{(1+2)}$ for the combined potentials $U_1 + U_2$ satisfies

$$-\frac{\hbar^2 k}{2\mu}\mathbf{T}^{(1+2)} = \int \phi(U_1 + U_2)\psi \, dR$$

$$= \int (\chi - \hat{G}_0^+ U_1\chi)(U_1 + U_2)\psi \, dR$$

$$= \int \left[\chi(U_1+U_2)\psi - (\hat{G}_0^+ U_1\chi)(U_1+U_2)\psi\right] dR. \tag{3.3.33}$$

Because the kernel function of Eq. (3.3.19) for the operator $\hat{G}_0^+$ is symmetric under $R \leftrightarrow R'$ interchange,

$$-\frac{\hbar^2 k}{2\mu} \mathbf{T}^{(1+2)} = \int [\chi (U_1 + U_2)\psi - \chi U_1 \hat{G}_0^+ (U_1 + U_2)\psi]\, dR$$

$$= \int [\chi (U_1 + U_2)\psi - \chi U_1 (\psi - \phi)]\, dR$$

$$= \int [\phi U_1 \chi + \chi U_2 \psi]\, dR \tag{3.3.34}$$

$$= \langle \phi^{(-)}|U_1|\chi \rangle + \langle \chi^{(-)}|U_2|\psi \rangle. \tag{3.3.35}$$

Note that in Eq. (3.3.34) the two terms in the integrand are products of three complex functions, and the product order is unimportant. In Eq. (3.3.35), by contrast, the functions are no longer interchangeable.

Thus $\mathbf{T}^{(1+2)} = \mathbf{T}^{(1)} + \mathbf{T}^{2(1)}$, defining an additional term

$$\mathbf{T}^{2(1)} = -\frac{2\mu}{\hbar^2 k} \int \chi U_2 \psi\, dR \tag{3.3.36}$$

as the scattering T-matrix contribution from coupling U_2, with U_1 appearing as a distorting potential in χ. The previous equation (3.3.35) is called the *two-potential formula*, and is an exact equation for both real and complex potentials U_1, U_2.

We use the $^{(-)}$ superscript to indicate complex conjugation for the left-hand wave function, as in Eq. (3.3.25). The reason for this notation can now be explained. The χ^* satisfies a boundary condition with an *incoming* boundary condition for all the asymptotic parts in addition to the plane wave. Since externally

$$\chi = F + \mathbf{T}^{(1)} H^+ \quad \to \quad \sin(kR) + \mathbf{T}^{(1)} e^{ikR}, \tag{3.3.37}$$

$$\text{we have} \quad \chi^* = F^* + \mathbf{T}^{(1)*} H^{+*} \quad \to \quad \sin(kR) + \mathbf{T}^{(1)*} e^{-ikR}, \tag{3.3.38}$$

where $\mathbf{T}^{(1)}$ is the scattering from the potential $U_1(R)$ that defines the homogeneous functions $\chi(R)$. Thus χ^* is asymptotically a plane wave plus some coefficient multiplying an *incoming* spherical wave e^{-ikR}, and hence is frequently written as $\chi^{(-)}$.

The exact wave function ψ is the solution of the implicit equation

$$\psi = \chi + \hat{G}_1^+ U_2 \psi, \tag{3.3.39}$$

using $\hat{G}_1^+ = [E - T - U_1]^{-1}$ with outgoing-wave boundary conditions. The first term χ represents the contribution present if $U_2 = 0$. By methods analogous to that of subsection 3.3.1, the $\hat{G}_1^+$ operator has the integral kernel

$$\hat{G}_1^+(R, R') = -\frac{1}{k}\chi(R_<)\chi^+(R_>),\qquad(3.3.40)$$

where $\chi^+(R)$ is the new irregular solution of $[E-T-U_1]\chi^+(R) = 0$ that equals $H^+(R)$ everywhere outside the range of $U_1(R)$.

3.3.4 Vector-form T matrix for distorted waves

When the distorting potential $U_1(R) \neq 0$ is spherical, let $\mathbf{X}(\mathbf{R}; \mathbf{k})$ be the system wave function like Eq. (3.2.1) obtained by solving the Schrödinger equation with that potential, and with incident plane wave $e^{i\mathbf{k}\cdot\mathbf{R}}$. The two-potential formula of Eq. (3.3.35) will now give the vector-form T matrix for the additional scattering by potential U_2, namely

$$\mathbf{T}(\mathbf{k}', \mathbf{k}) = \langle \mathbf{X}^{(-)}(\mathbf{R}; \mathbf{k}')|U_2|\Psi(\mathbf{R}; \mathbf{k})\rangle,\qquad(3.3.41)$$

where we can show that $\mathbf{X}^{(-)}(\mathbf{R}; \mathbf{k}') = \mathbf{X}(\mathbf{R}; -\mathbf{k}')^*$. The $\mathbf{X}^{(-)}$ has thus *incoming* spherical waves in its asymptotic boundary conditions, in addition to the elastic plane wave. The scattering amplitude f is still related to $\mathbf{T}$ by Eq. (3.3.31).

In the multi-channel case, we sum over all coupled-channels sets J_{tot}^π, and following Eq. (3.2.7) construct the wave functions $\Psi_{x_ip_it_i}^{\mu_{p_i}\mu_{t_i}}(\mathbf{R}_x; \mathbf{k}_i)$ for the potential U_1+U_2 with coupled radial functions $\psi_{\alpha\alpha_i}(R_x)$, and $\mathbf{X}_{x'p't'}^{\mu_{p'}\mu_{t'}}(\mathbf{R}_{x'}; \mathbf{k}')$ with uncoupled radial functions $\chi_{\alpha'}(R_{x'})$ for the spherical potential U_1 only. Both depend on the m-state projections of the nuclei in the relevant partition. The vector-form T matrix, in this general case, is now

$$\mathbf{T}_{x'p't':x_ip_it_i}^{\mu_{p'}\mu_{t'}:\mu_{p_i}\mu_{t_i}}(\mathbf{k}', \mathbf{k}_i)$$

$$= \langle \mathbf{X}_{x'p't'}^{\mu_{p'}\mu_{t'}(-)}(\mathbf{R}_{x'}; \mathbf{k}')|U_2|\Psi_{x_ip_it_i}^{\mu_{p_i}\mu_{t_i}}(\mathbf{R}_x; \mathbf{k}_i)\rangle\qquad(3.3.42)$$

$$= \sum_{J_{\text{tot}}M_{\text{tot}}\,\alpha'\alpha_i}\sum A_{\mu_{p'}\mu_{t'}}^{J_{\text{tot}}M_{\text{tot}}}(\alpha'; \mathbf{k}')^* A_{\mu_{p_i}\mu_{t_i}}^{J_{\text{tot}}M_{\text{tot}}}(\alpha_i; \mathbf{k}_i) \sum_\alpha \langle \chi_{\alpha'}|U_2|\psi_{\alpha\alpha_i}\rangle\qquad(3.3.43)$$

$$= -\frac{\hbar^2 k_{\alpha'}}{2\mu_{\alpha'}}\sum_{J_{\text{tot}}M_{\text{tot}}\,\alpha'\alpha_i}\sum A_{\mu_{p'}\mu_{t'}}^{J_{\text{tot}}M_{\text{tot}}}(\alpha'; \mathbf{k}')^* A_{\mu_{p_i}\mu_{t_i}}^{J_{\text{tot}}M_{\text{tot}}}(\alpha_i; \mathbf{k}_i)\, \mathbf{T}_{\alpha'\alpha_i}^{J_{\text{tot}}M_{\text{tot}}}\qquad(3.3.44)$$

where the $A_{\mu_p\mu_t}^{J_{\text{tot}}M_{\text{tot}}}(\alpha; \mathbf{k})$ are given by Eq. (3.2.9). This last equation gives the general multi-channel relationship between the *partial-wave* matrix elements $\mathbf{T}_{\alpha'\alpha_i}$ and the *vector-form* matrix elements $\mathbf{T}(\mathbf{k}', \mathbf{k}_i)$. Note that the wave number $k_{\alpha'}$ and reduced mass $\mu_{\alpha'}$ in the final channel are entirely determined by $x'p't'$. The final equation holds generally; also when U_1 is non-spherical and the $\chi_{\alpha''\alpha'}(R_{x''})$ are thus solutions of a set of coupled equations.

Equation (3.3.44) also yields another expression for the nuclear-only scattering amplitude in terms of the partial-wave T-matrix elements:

$$
f^{xpt}_{\mu_p\mu_t,\mu_{p_i}\mu_{t_i}}(\mathbf{k};\mathbf{k}_i) = -\frac{\mu_\alpha}{2\pi\hbar^2} \mathbf{T}^{\mu_p\mu_t:\mu_{p_i}\mu_{t_i}}_{xpt:x_ip_it_i}(\mathbf{k},\mathbf{k}_i)
$$

$$
= \frac{k_\alpha}{4\pi}\sum_{J_{tot}M_{tot}}\sum_{\alpha\alpha_i}A^{J_{tot}M_{tot}}_{\mu_p\mu_t}(\alpha;\mathbf{k})^* A^{J_{tot}M_{tot}}_{\mu_{p_i}\mu_{t_i}}(\alpha_i;\mathbf{k}_i)\mathbf{T}^{J_{tot}M_{tot}}_{\alpha\alpha_i}. \quad (3.3.45)
$$

3.3.5 Born series and approximations
One-potential scattering

For a fixed potential $U(R)$, solving the Lippmann-Schwinger equation $\chi = \phi + \hat{G}_0^+ U\chi$ should provide an exact solution for the wave function χ with potential U. This however is an implicit equation, as χ appears on both the left and right sides. To find it explicitly, we could perhaps sum the iterated *Born series*:

$$
\chi = \phi + \hat{G}_0^+ U[\phi + \hat{G}_0^+ U[\phi + \hat{G}_0^+ U[\cdots]]]
$$

$$
= \phi + \hat{G}_0^+ U\phi + \hat{G}_0^+ U\hat{G}_0^+ U\phi + \hat{G}_0^+ U\hat{G}_0^+ U\phi\hat{G}_0^+ U\phi + \cdots, \quad (3.3.46)
$$

from which the outgoing amplitude $\mathbf{T} = -\frac{2\mu}{\hbar^2 k}\langle\phi^{(-)}|U|\chi\rangle$ is

$$
\mathbf{T} = -\frac{2\mu}{\hbar^2 k}\left[\langle\phi^{(-)}|U|\phi\rangle + \langle\phi^{(-)}|U\hat{G}_0^+ U|\phi\rangle + \cdots\right]. \quad (3.3.47)
$$

The equation (3.3.46) may be illustrated by Fig. 3.5, where each node of the graph is an action of the potential U and each line a propagation by $\hat{G}_0^+$.

If the potential $U(R)$ is *weak*, in the sense that we could treat it as a perturbation, then we might truncate these series and still achieve sufficient precision. The first term in Eq. (3.3.47) is called the *plane-wave Born approximation* (PWBA):

$$
\mathbf{T}^{PWBA} = -\frac{2\mu}{\hbar^2 k}\langle\phi^{(-)}|U|\phi\rangle. \quad (3.3.48)
$$

This PWBA, when written explicitly with partial-wave radial functions, is

$$
\mathbf{T}^{PWBA}_L = -\frac{2\mu}{\hbar^2 k}\int_0^\infty F_L(0,kR)\, U(R)\, F_L(0,kR)\, dR. \quad (3.3.49)
$$

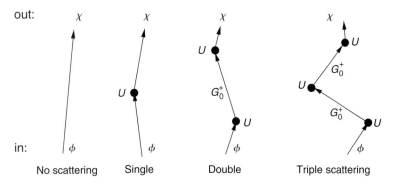

Fig. 3.5. Illustrating the Born series (3.3.46) for the wave function χ, as a sum of a homogeneous term with single, double, and higher-order rescattering contributions to the outgoing wave χ.

Substituting again these T-matrix elements into Eq. (3.1.46), or by a special case of Eq. (3.3.29), the three-dimensional form for the PWBA scattering amplitude is

$$f^{\text{PWBA}}(\theta) = -\frac{\mu}{2\pi\hbar^2} \int d\mathbf{R} \, e^{-i\mathbf{q}\cdot\mathbf{R}} \, U(\mathbf{R}), \qquad (3.3.50)$$

where the momentum transfer $\mathbf{q} = \mathbf{k} - \mathbf{k}_i$, so $q = 2k\sin\theta/2$. The PWBA amplitude is thus simply proportional to the Fourier transform of the potential. The PWBA is expected to be more accurate at very high energies when potentials are weak, such as in electron-nucleus scattering.

Two-potential scattering

From Eq. (3.3.35), the two-potential formula, the exact T-matrix expression, is again

$$\mathbf{T}^{(1+2)} = \mathbf{T}^{(1)} - \frac{2\mu}{\hbar^2 k}\langle\chi^{(-)}|U_2|\psi\rangle \qquad (3.3.51)$$

where, from Eq. (3.3.39), the exact wave function is the solution of the implicit equation $\psi = \chi + \hat{G}_1 U_2\psi$. We may again by iteration, therefore, form a Born series

$$\mathbf{T}^{(1+2)} = \mathbf{T}^{(1)} - \frac{2\mu}{\hbar^2 k}\left[\langle\chi^{(-)}|U_2|\chi\rangle + \langle\chi^{(-)}|U_2\hat{G}_1 U_2|\chi\rangle + \cdots\right]. \qquad (3.3.52)$$

We might expect this series to converge if U_2 is weak in a suitable sense. We do *not* here require that U_1 is weak.

Post and prior T-matrix integrals

The exact expression (3.3.51) is often called the *post T-matrix integral* because the solution χ for the first potential U_1 is in the *post* or final channel. A mirror *prior T-matrix integral* can also be derived where the χ is in the *prior* or entrance channel.

We may rewrite Eq. (3.3.52) as

$$\mathbf{T}^{(1+2)} = \mathbf{T}^{(1)} - \frac{2\mu}{\hbar^2 k}\left[\langle\chi^{(-)}| + \langle\chi^{(-)}|U_2\hat{G}_1 + \cdots\right]U_2|\chi\rangle, \qquad (3.3.53)$$

and define the expression in the square brackets as $\langle\psi^{(-)}|$ where

$$\psi^{(-)} = \chi^{(-)} + \hat{G}_1^- U_2^* \chi^{(-)} + \cdots$$
$$= \chi^{(-)} + \hat{G}_1^- U_2^* \psi^{(-)}. \qquad (3.3.54)$$

The Green's function $\hat{G}_1^-$ is the complex conjugate of $\hat{G}_1^+$, and thus describes incoming boundary conditions like Eq. (3.3.38). The wave function $\psi^{(-)}$ is thus a *full* solution satisfying $[E - T - U_1 - U_2]\psi^{(-)} = 0$ but with incoming boundary conditions. This wave function appears now in the *prior T-matrix integral*

$$\mathbf{T}^{(1+2)} = \mathbf{T}^{(1)} - \frac{2\mu}{\hbar^2 k}\langle\psi^{(-)}|U_2|\chi\rangle. \qquad (3.3.55)$$

The wave functions on the kets of Eqs. (3.3.51) and (3.3.55) are often written with a $(+)$ as a reminder that they are calculated with normal *outgoing* boundary conditions:

$$\mathbf{T}^{(1+2)} = \mathbf{T}^{(1)} - \frac{2\mu}{\hbar^2 k}\langle\chi^{(-)}|U_2|\psi^{(+)}\rangle = \mathbf{T}^{(1)} - \frac{2\mu}{\hbar^2 k}\langle\psi^{(-)}|U_2|\chi^{(+)}\rangle. \qquad (3.3.56)$$

If now in multi-channel theory we label the wave functions by the channels in which there is a boundary condition with a plane wave, the post and prior $\mathbf{T}$-matrix integrals for the reaction from entrance channel α_i to exit channel α are

$$\mathbf{T}^{(1+2)}_{\alpha\alpha_i} = \mathbf{T}^{(1)}_{\alpha\alpha_i} - \frac{2\mu_\alpha}{\hbar^2 k_\alpha}\langle\chi_\alpha^{(-)}|U_2|\psi_{\alpha_i}^{(+)}\rangle \quad \text{[post]}, \qquad (3.3.57a)$$

$$= \mathbf{T}^{(1)}_{\alpha\alpha_i} - \frac{2\mu_\alpha}{\hbar^2 k_\alpha}\langle\psi_\alpha^{(-)}|U_2|\chi_{\alpha_i}^{(+)}\rangle \quad \text{[prior]}. \qquad (3.3.57b)$$

Remember here that the $\chi^\pm$ are the wave functions with U_1 only, and that the $\psi^\pm$ are the full coupled wave functions with $V = U_1 + U_2$, so the potential in the matrix element is $U_2 = V - U_1$. The superscript $(+)$ refers to the normal boundary conditions with outgoing waves except in the elastic channel, while the minus $(-)$ superscript refers to the unusual *incoming* boundary condition in all waves (in addition to the plane-wave component in the elastic channel).

If the second terms of Eq. (3.3.57) are inserted in Eq. (3.3.44) and summed over all partial waves, there are analogous three-dimensional integrals for the vector-form T matrices $\mathbf{T}_{xpt:x_i p_i t_i}(\mathbf{k}', \mathbf{k}_i)$ from Eq. (3.3.42). If $\mathbf{X}$ is the system scattering function for potential $U(R)$ and Ψ with $V(R)$, then the scattering from V that is in addition to the U scattering has vector-form T-matrix expressions in two versions:

$$\mathbf{T}^{2(1):post}_{xpt:x_i p_i t_i}(\mathbf{k}', \mathbf{k}_i) = \langle \mathbf{X}^{(-)}_{xpt}(\mathbf{R}, \mathbf{k}')| \ V - U \ |\Psi_{x_i p_i t_i}(\mathbf{R}; \mathbf{k}_i)\rangle \quad [\text{post}], \qquad (3.3.58a)$$

$$\mathbf{T}^{2(1):prior}_{xpt:x_i p_i t_i}(\mathbf{k}', \mathbf{k}_i) = \langle \Psi^{(-)}_{xpt}(\mathbf{R}; \mathbf{k}')| \ V - U \ |\mathbf{X}_{x_i p_i t_i}(\mathbf{R}, \mathbf{k}_i)\rangle \quad [\text{prior}] \qquad (3.3.58b)$$

that should be equal. Both of these give scattering amplitudes $f = -\frac{\mu}{2\pi\hbar^2}\mathbf{T}$ according to Eq. (3.3.31), and then cross sections $\sigma = |f|^2$ as a function of angle. Strictly speaking, the Ψ and $\mathbf{X}$ both carry m-state quantum numbers: these carry over together to the $\mathbf{T}$ as in Eq. (3.3.42), and hence to the scattering amplitudes f and cross sections σ.

Distorted-wave Born approximation (DWBA)

If the series (3.3.52) is truncated after the first term, linear in U_2, then

$$\mathbf{T}^{DWBA} = \mathbf{T}^{(1)} - \frac{2\mu}{\hbar^2 k}\langle \chi^{(-)}|U_2|\chi\rangle \qquad (3.3.59)$$

is called the *distorted-wave Born approximation* (DWBA), because it is a matrix element using wave functions $\chi(R)$ which include U_1 as a distorting potential. It is a *first-order* DWBA because U_2 appears only linearly. It is particularly useful for exit channels where U_1 might be, say, a central optical potential that cannot by itself cause the transition. In this case $\mathbf{T}^{(1)} = 0$, and we have the convenient DWBA expression for the T matrix from incoming channel α_i to exit channel α:

$$\mathbf{T}^{DWBA}_{\alpha\alpha_i} = -\frac{2\mu_\alpha}{\hbar^2 k_\alpha}\langle \chi^{(-)}_\alpha|U_2|\chi_{\alpha_i}\rangle. \qquad (3.3.60)$$

There is similarly a second-order DWBA expression

$$\mathbf{T}^{2nd-DWBA}_{\alpha\alpha_i} = -\frac{2\mu_\alpha}{\hbar^2 k_\alpha}\left[\langle \chi^{(-)}_\alpha|U_2|\chi_{\alpha_i}\rangle + \langle \chi^{(-)}_\alpha|U_2\hat{G}^+_1 U_2|\chi_{\alpha_i}\rangle\right]. \qquad (3.3.61)$$

Figure 3.6 illustrates first- and second-order couplings, in contrast to including all couplings in a full coupled-channels solution.

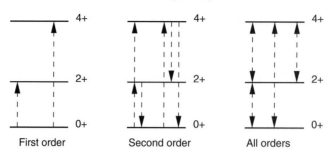

Fig. 3.6. First, second and all-order couplings within a set of 0^+, 2^+ and 4^+ nuclear levels, starting from the ground state.

First-order DWBA for transfer reactions

If we write the DWBA T-matrix element of Eq. (3.3.60) using the formalism for multiple mass partitions developed in subsection 3.2.2, then we have a DWBA expression that can be used for simple transfer reactions that couple one mass partition to another. In that formalism, we have

$$\mathbf{T}^{\text{DWBA}}_{\alpha\alpha_i} = -\frac{2\mu_\alpha}{\hbar^2 k_\alpha} \left\langle \chi^{(-)}_\alpha \left| R_x \langle \alpha | H - E | \alpha_i \rangle R^{-1}_{x_i} \right| \chi_{\alpha_i} \right\rangle. \tag{3.3.62}$$

where the inner matrix element $\langle \alpha | H - E | \alpha_i \rangle$ integrates over the internal nuclear coordinates ξ, and yields a non-local function of R_x and R_{x_i} that is then integrated in a matrix element with $\chi^{(-)}_\alpha(R_x)$ and $\chi_{\alpha_i}(R_{x_i})$. As discussed in subsection 3.2.2, the inner Hamiltonian H may be written in either *prior* or *post* forms. If we take the prior form of Eq. (3.2.46), and keep a diagonal optical potential U_α with each kinetic energy term, we have now

$$\mathbf{T}^{\text{dw-prior}}_{\alpha\alpha_i} = -\frac{2\mu_\alpha}{\hbar^2 k_\alpha} \left\langle \chi^{(-)}_\alpha \left| \hat{N}_{\alpha\alpha_i}[T_{x_i L_i} + U_{\alpha_i} - E_{x_i p_i t_i}] + \hat{V}^{x_i}_{\alpha\alpha_i} \right| \chi_{\alpha_i} \right\rangle$$

$$= -\frac{2\mu_\alpha}{\hbar^2 k_\alpha} \left[\langle \chi^{(-)}_\alpha | \hat{V}^{x_i}_{\alpha\alpha_i} | \chi_{\alpha_i} \rangle + \langle \chi^{(-)}_\alpha | \hat{N}_{\alpha\alpha_i}[T_{x_i L_i} + U_{\alpha_i} - E_{x_i p_i t_i}] | \chi_{\alpha_i} \rangle \right]$$

$$= -\frac{2\mu_\alpha}{\hbar^2 k_\alpha} \langle \chi^{(-)}_\alpha | \hat{V}^{x_i}_{\alpha\alpha_i} | \chi_{\alpha_i} \rangle \tag{3.3.63}$$

since the one-channel functions χ_{α_i} are found by $[T_{x_i L_i} + U_{\alpha_i} - E_{x_i p_i t_i}]\chi_{\alpha_i} = 0$. A similar calculation of the *post* DWBA matrix element yields

$$\mathbf{T}^{\text{dw-post}}_{\alpha\alpha_i} = -\frac{2\mu_\alpha}{\hbar^2 k_\alpha} \left\langle \chi^{(-)}_\alpha \left| [T_{xL} + U_\alpha - E_{xpt}]\hat{N}_{\alpha\alpha_i} + \hat{V}^x_{\alpha\alpha_i} \right| \chi_{\alpha_i} \right\rangle$$

$$= -\frac{2\mu_\alpha}{\hbar^2 k_\alpha} \langle \chi^{(-)}_\alpha | \hat{V}^x_{\alpha\alpha_i} | \chi_{\alpha_i} \rangle. \tag{3.3.64}$$

We therefore see that in these first-order DWBA expressions for transfer reactions, the non-orthogonality terms $\hat{N}_{\alpha\alpha_i}$ disappear. Moreover, if both Eqs. (3.3.63) and (3.3.64) are derived consistently from Eq. (3.3.62), then the above derivations show that they must yield the same T-matrix element.[20] Neither of these conclusions are true for second- and higher-order DWBA calculations with transfer reactions.

Approximate coupled-channels solutions

As discussed on page 92, the result of these integral expressions may be equally obtained by solving inhomogeneous differential equations. This means that the above PWBA and DWBA approximations may be also reached by modifications to the coupled-channels equations. In particular, the first-order DWBA result may be identically obtained by removing all couplings except for keeping the diagonal potentials in the elastic (α_i) and final (α) channel (U_1), and keeping the couplings *from* the elastic channel *to* the final channel (U_2).

Allowing for a distinct potential U_i in the entrance channel, this modified coupled-channels set is

$$[T_{\alpha_i} + U_i - E_{\alpha_i}]\psi_{\alpha_i} = 0$$
$$[T_\alpha + U_1 - E_\alpha]\psi_\alpha + U_2\,\psi_{\alpha_i} = 0. \qquad (3.3.65)$$

These equations may be equivalently solved in three ways:

 (i) By solving the differential equations according to the exact methods of subsection 6.3.2.
 (ii) By solving the inhomogeneous differential equation (3.3.65) by the iterative method of subsection 6.3.3. The elastic channel wave function will be just the uncoupled solution $\psi_{\alpha_i} = \chi_{\alpha_i}$, and the final channel will be populated after the first iteration.
 (iii) By evaluating the T matrix for the final channel by the DWBA integral of Eq. (3.3.60).

For the restricted set of equations (3.3.65), which do not have the back couplings from the second to the first channel, these three methods will give identical results.

3.4 Identical particles

The *spin-statistics theorem* is a fundamental result in quantum field theory. This states that spin-half objects obey Fermi-Dirac statistics and have antisymmetric wave functions, whereas integer-spin objects obey Bose-Einstein statistics and have symmetric wave functions. An (anti)symmetry of the collective wave functions means that, on interchanging the coordinates of any two particles, the wave function

---

[20] This necessary post-prior agreement of first-order T- (or S-) matrix elements in all partial waves yields a good check on the accuracy of calculations.

changes sign (antisymmetric for fermions), or remains the same (symmetric, for bosons).

Up to now we have not taken into account the Pauli principle for identical fermions, such as for protons and for neutrons. This principle is most universally followed if the wave function for a set of identical particles is *antisymmetric* under the interchange of the coordinates of any pair of these particles.

We first note that if two groups of nucleons are sufficiently far apart that their wave functions do not overlap, then the exchange of one or several nucleons between the groups cannot affect any observable. This means that antisymmetrization between the projectile and target tends to be less significant in scattering theory than in nuclear structure theory, because during scattering the phase spaces occupied by the projectile and by the target nucleons are more likely to be apart both in distance and in energy. The Pauli principle can thus be treated most often as a perturbative correction.

The only exception about antisymmetrization is exchange of *all* nucleons for identical nuclei, where there are now both direct and exchange amplitudes that add coherently, so such cases will be considered below, in subsection 3.4.2. We will discuss the structure issues again in subsection 5.3.2. First, however, we see how to treat protons and neutrons as almost identical particles.

3.4.1 Isospin

The proton and neutron are almost the same in mass ($\Delta m/m = 1.4 \times 10^{-3}$), and their different electric potentials and magnetic moments make only a small difference compared with the strong nuclear forces. This suggests treating them as identical in some sense, and including the electromagnetic effects (etc.) by perturbation theory.

The only sense that the neutron and proton need to be distinguished, then, is for the Pauli principle: to make antisymmetric wave functions. For this purpose (alone) we have the option of following the standard derivation of two 'states' for a *nucleon*, to be regarded as in one of two states depending on an *isobaric variable* τ_z. This distinguishes a neutron ($\tau_z = 1$) from a proton ($\tau_z = -1$).[21] Thus a full specification of a nucleon depends on position $\mathbf{r}$, spin z-component σ_z, *and* now also isospin z-component τ_z as $\psi(\mathbf{r}, \sigma_z, \tau_z)$.

If we define the neutron state vector as $|n\rangle = \begin{pmatrix} 1 \\ 0 \end{pmatrix}$ and $|p\rangle = \begin{pmatrix} 0 \\ 1 \end{pmatrix}$, then the $\hat{\tau}_z$ operator $\begin{pmatrix} 1 & 0 \\ 0 & -1 \end{pmatrix}$ has as eigenstates the above vectors:

$$\hat{\tau}_z|n\rangle = +1\,|n\rangle \quad \text{and} \quad \hat{\tau}_z|p\rangle = -1\,|p\rangle. \tag{3.4.1}$$

---

[21] This assignment is a matter of conventional choice.

The $\hat{\tau}_z$ may be supplemented by $\hat{\tau}_x$ and $\hat{\tau}_y$ which follow the familiar algebra for components of a spin-1/2 particle, namely

$$\hat{\tau}_x = \begin{pmatrix} 0 & 1 \\ 1 & 0 \end{pmatrix}; \quad \hat{\tau}_y = \begin{pmatrix} 0 & -i \\ i & 0 \end{pmatrix} \tag{3.4.2}$$

satisfying the commutators $[\tau_p, \tau_q] = 2i\tau_r$ for (pqr) cyclic permutations of (xyz), by analogy with the Pauli spin operators. This justifies the name of *isospin*. In terms of the original $|n\rangle$ and $|p\rangle$ states, the isospin operators may be written

$$\hat{\tau}_x = |p\rangle\langle n| + |n\rangle\langle p|$$
$$\hat{\tau}_y = i[|p\rangle\langle n| - |n\rangle\langle p|]$$
$$\hat{\tau}_z = |n\rangle\langle n| - |p\rangle\langle p|. \tag{3.4.3}$$

From the $\hat{\tau}_i$ matrix operators, we obtain the *isotopic spin operators*

$$\hat{\mathbf{t}} = \frac{1}{2}\hat{\boldsymbol{\tau}} = \frac{1}{2}(\hat{\tau}_x, \hat{\tau}_y, \hat{\tau}_z). \tag{3.4.4}$$

Since $\hat{t}^2 = 3/4$, the nucleon has total isospin $t = 1/2$, and z-components $m_t = t_z = +1/2$ for the neutron and $-1/2$ for the proton. The operators $\hat{t}_+ = t_x + it_y$ and $\hat{t}_- = t_x - it_y$ are the raising and lowering operators respectively, as from Eq. (3.4.3) we see that $\hat{t}_+ = \frac{1}{2}|n\rangle\langle p|$ and $\hat{t}_- = \frac{1}{2}|p\rangle\langle n|$.

Composite systems

For systems of two or more nucleons ($k = 1, 2, \ldots$), the isospins may be coupled to a total isospin

$$\hat{\mathbf{T}} = \sum_k \hat{\mathbf{t}}_k \tag{3.4.5}$$

with z-component $T_z = \frac{1}{2}(N - Z)$ for N neutrons and Z protons. For even numbers of nucleons, $T = 0, 1, \ldots$, and for odd numbers $T = \frac{1}{2}, \frac{3}{2}, \ldots$. Two neutrons have $T_z = 1$ and two protons have $T_z = -1$, and this is only possible if $T = 1$. A neutron and a proton together, by comparison, have $T_z = 0$, and hence either $T = 0$ or 1.

Antisymmetrization can be applied to neutrons and to protons separately, but we can make for nucleons what Bohr and Mottelson [12] call a *generalized antisymmetrization principle*, that the wave function for a set of fermions is antisymmetric with respect to the interchange of *all* coordinates (space, spin, *and* isospin) of any pair of nucleons.

Consider the case of two nucleons with coordinates $(\mathbf{r}_i, \sigma_{zi}, \tau_{zi})$, $i = 1, 2$. Antisymmetry requires $\Psi(1, 2) = -\Psi(2, 1)$. Let the spins s_1, s_2 be coupled to

S, the isospins t_1, t_2 to T, the relative angular momentum be L, and $\mathbf{L} + \mathbf{S} = \mathbf{J}$ the total spin. Then

$$\Psi_J(1, 2) = [Y_L(\mathbf{R}) \otimes [s_1 \otimes s_2]_S]_J \, [t_1 \otimes t_2]_T, \qquad (3.4.6)$$

which gives, on interchanging the particles 1 and 2,

$$\begin{aligned}
\Psi_J(2, 1) &= [Y_L(-\mathbf{R}) \otimes [s_2 \otimes s_1]_S]_J \, [t_2 \otimes t_1]_T \\
&= (-1)^L (-1)^{S - s_1 - s_2} (-1)^{T - t_1 - t_2} \Psi_J(1, 2) \\
&= (-1)^{L + S + T} \Psi_J(1, 2)
\end{aligned} \qquad (3.4.7)$$

as all $s_i = t_i = 1/2$. The factors like $(-1)^{S - s_1 - s_2}$ come from the symmetry properties of the Clebsch-Gordon coefficients:

$$\langle s_1 \mu_1, s_2 \mu_2 | S \mu \rangle = (-1)^{S - s_1 - s_2} \langle s_2 \mu_2, s_1 \mu_1 | S \mu \rangle. \qquad (3.4.8)$$

Antisymmetrization for two nucleons as fermions then requires that they be in a state where $L + S + T$ is *odd*.

For more complex nuclei composed of antisymmetric pairs of nucleons, T is still a good quantum number even though there are greater effects of Coulomb forces, which depend on the projections t_z. We find that isobaric sets of nuclei (those with constant A) have many sets of energy levels that are similar in absolute energy when $T > 0$ is a constant, and varying slightly only because of the additional Coulomb repulsion when a neutron is replaced by a proton. These sets of levels, illustrated for $T = 0$, 1 and 2 in Fig. 3.7, are called *isobaric analogue states*, and have very similar internal nuclear structure. There is little mixing between T values, only slight energy shifts that hardly change the structure.

3.4.2 *Direct and exchange amplitudes in elastic scattering*

The scattering analysis so far has dealt with reactions of the form $p + t \to p' + t'$ for projectile p, target t, ejectile p' and residue t' that are considered distinguishable nuclei. In the light of the previous section, however, we have to consider the cases of

(a) Scattering of identical fermions: $p = t$ of odd baryon number;
(b) Scattering of identical bosons: $p = t$ of even baryon number; and
(c) Exchange scattering: $p' = t$ and $t' = p$, and p is distinguishable from t,

where p and t may be clusters, not just individual nucleons. This is a way of taking into account one important overall property of clusters without considering their

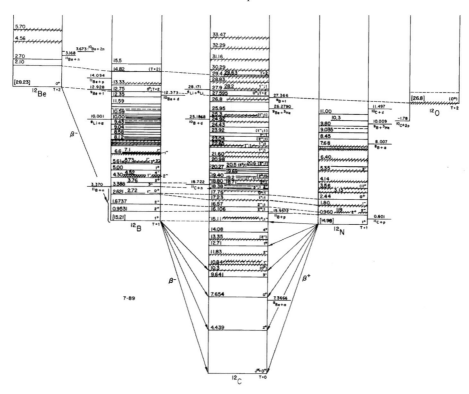

Fig. 3.7. Spectra for A = 12 isobars: the dashed lines indicate some of the isobaric analogue states in the A = 12 nuclear multiplet, from the compilation [13]. Each state is labeled by $J^\pi; T$ for total isospin T, where these are known. Reprinted from F. Ajzenberg-Selove, *Nucl. Phys. A* **506** (1990) 1. Copyright (1990), with permission from Elsevier.

internal structure in detail. If the nuclei were distinguishable, the wave function $\Psi^{M_{\text{tot}}}_{xJ_{\text{tot}}}(\mathbf{R}_x, \xi_p, \xi_t)$ for their combination in a system would simply be the product of their separate states ϕ_{I_p} and ϕ_{I_t} with angular momentum and orbital factors as in subsection 3.2.1. Now, however, we cannot always use such a product wave function if we are to satisfy the Pauli principle.

Let us define an 'exchange index' $\varepsilon = +1$ for boson-boson and $\varepsilon = -1$ for fermion-fermion collisions, so the (a) and (b) cases may be considered with one formalism. Define the operator $\hat{P}_{pt}$ which exchanges projectile and target coordinates in the wave function of the combined system. The spin-statistics theorem implies that $\varepsilon = (-1)^{2I_p}$, where I_p is the overall spin of the projectile. A complete wave function Ψ_ε should therefore satisfy $\hat{P}_{pt}\Psi_\varepsilon(\mathbf{R}) = \varepsilon\Psi_\varepsilon(-\mathbf{R})$, or

$$\hat{P}_{pt}\Psi^{M_{\text{tot}}}_{xJ_{\text{tot}}}(\mathbf{R}_x, \xi_p, \xi_t) = \varepsilon\Psi^{M_{\text{tot}}}_{xJ_{\text{tot}}}(-\mathbf{R}_x, \xi_t, \xi_p). \tag{3.4.9}$$

Identical spinless scattering

Consider first the simple case of spinless scattering,[22] but keep explicit reference to ε. The asymptotic form for non-identical scattering on spherical potentials is given by Eq. (2.4.12), namely

$$\psi^{\text{asym}}(\mathbf{R}) = A\left[e^{ikz} + f(\theta)\frac{e^{ikR}}{R}\right],$$

so a suitable wave function for identical particles is

$$\Psi_{\varepsilon}^{\text{asym}}(\mathbf{R}) = \psi^{\text{asym}}(\mathbf{R}) + \varepsilon\,\psi^{\text{asym}}(-\mathbf{R}). \tag{3.4.10}$$

We now calculate the initial flux j_i and final angular flux $\hat{j}_f$, in order to calculate the cross section $\sigma = \hat{j}_f/j_i$. The incident wave is

$$\Psi_{\varepsilon}^{\text{inc}}(\mathbf{R}) = A[e^{ikz} + \varepsilon\,e^{-ikz}] \tag{3.4.11}$$

and the scattered outgoing radial wave is

$$\Psi_{\varepsilon}^{\text{out}}(\mathbf{R}) = A[f(\theta) + \varepsilon f(\pi - \theta)]\frac{e^{ikR}}{R} \equiv Af_{\varepsilon}(\theta)\frac{e^{ikR}}{R}. \tag{3.4.12}$$

The total flux in $\Psi_{\varepsilon}^{\text{inc}}(\mathbf{R})$ is numerically zero. Goldberger and Watson give more detailed analysis using wave packets [14, §4.3], which shows that the incident flux is a combination of a projectile flux $\frac{\hbar k}{\mu}|A|^2$ in the $+z$ direction, and an (identical) target flux in the $-z$ direction. The projectile flux in the beam is therefore $j_i = \frac{\hbar k}{\mu}|A|^2$. The scattered flux is $\hat{j}_f = \frac{\hbar k}{\mu}|A|^2|f_{\varepsilon}(\theta)|^2$, so the needed cross section for identical particle scattering is

$$\sigma(\theta) = |f_{\varepsilon}(\theta)|^2 = |f(\theta) + \varepsilon f(\pi - \theta)|^2 \tag{3.4.13}$$

$$= |f(\theta)|^2 + |f(\pi - \theta)|^2 + 2\varepsilon\,\text{Re}\,f(\theta)^*f(\pi - \theta). \tag{3.4.14}$$

Figure 3.8 shows two kinds of semiclassical paths that could interfere in this way, corresponding to scattering angles θ and $\pi - \theta$.

Since $P_L(\cos(\pi - \theta)) = (-1)^L P_L(\cos\theta)$, the new amplitude has the partial wave expansion

$$f_{\varepsilon}(\theta) = \frac{1}{k}\sum_{L=0}^{\infty}(2L+1)P_L(\cos\theta)\mathbf{T}_L[1 + \varepsilon(-1)^L], \tag{3.4.15}$$

---

[22] This section is thus strictly applicable only to boson scattering, but the exercise is useful.

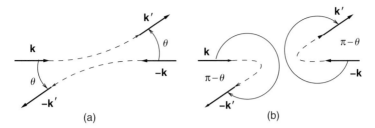

Fig. 3.8. Direct (a) and exchange (b) paths which interfere with each other, in the scattering of identical particles from c.m. momenta $\pm\mathbf{k}$ to $\pm\mathbf{k}'$ or $\mp\mathbf{k}'$, after scattering by angles θ or $\pi - \theta$.

where $\mathbf{T}_L$ is the scattering T matrix that would be obtained if the particles were distinguishable. This implies for the scattering of identical bosons, where $\varepsilon = 1$, that odd partial waves should be removed, and that the *amplitude* for the remaining even partial waves should be doubled.

Identical particles with spin

The above analysis for $I_p = I_t = 0$ is necessarily restricted to $\varepsilon = 1$ for identical bosons. For identical fermions it is essential to include spin. This is easiest in the S (channel spin) partial-wave basis, where the effect of the $\hat{P}_{pt}$ exchange operator is

$$\hat{P}_{pt}|L(I_p, I_t)S; J_{\mathrm{tot}}x\rangle = (-1)^L(-1)^{S-I_p-I_t}|L(I_t, I_p)S; J_{\mathrm{tot}}x\rangle \qquad (3.4.16)$$

with the radial wave function itself, $\psi_\varepsilon(R_x)$, not being affected. The $(-1)^L$ factor comes from reversing the direction of the radius vector in the spherical harmonic, and $(-1)^{S-I_p-I_t}$ comes from reordering the coupling of particle spins to make the channel-spin. The expression like Eq. (3.4.15) for the scattering amplitude is therefore different for different channel-spin values S:

$$f_S(\theta) = \frac{1}{k}\sum_{L=0}^{\infty}(2L+1)P_L(\cos\theta)\mathbf{T}_L[1 + \varepsilon(-1)^{L+S-I_p-I_t}]$$

$$= \frac{1}{k}\sum_{L=0}^{\infty}(2L+1)P_L(\cos\theta)\mathbf{T}_L[1 + (-1)^{L+S}] \qquad (3.4.17)$$

since $I_p = I_t$ and $\varepsilon = (-1)^{2I_p}$ for two identical particles. (This formula holds for bosons if we set $S = 0$ in that case.) The partial waves that give non-zero scattering are those with $L+S$ even. We say that the partial waves with $L+S$ odd are *blocked* by the Pauli principle. Nucleon-nucleon scattering, for example, is thus different in singlet ($S = 0$) and triplet ($S = 1$) states, and gives rise to the characteristic

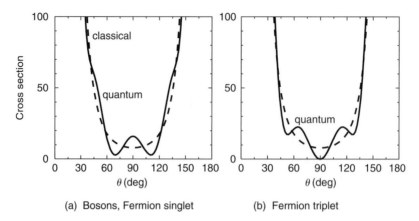

Fig. 3.9. Fermion singlet (a) and triplet (b) nucleon-nucleon scattering cross sections, assuming pure Coulomb scattering with $\eta = 5$. Case (a) also applies for boson scattering. The cross section is in units of $\eta^2/4k^2$.

interference patterns shown in Fig. 3.9. The singlet scattering on the left is the same as would be obtained for the scattering of two bosons.

We can repeat the above analysis using a total isospin T for the combined projectile and target system, and we will find that the allowed partial waves are those with $L + S + T$ odd. Two colliding neutrons or protons require $T = 1$, so this is the same conclusion as above.

If there are spin-dependent forces (such as the spin-orbit or tensor forces to be defined in the next chapter) then full multi-channel scattering theory must be used, not the simplified theory above where the T matrix only depends on L.

3.4.3 Integrated cross sections

The definition of angle-integrated cross sections has to be specified carefully for identical particles, since for every nucleus removed from the beam, the elastic scattering produces *two* identical nuclei to be detected. The differential cross section $\sigma(\theta) = |\tilde{f}_S(\theta)|^2$ measures the outgoing probability of detecting a specified product no matter how it was produced, but now the angular integral of this can be more than the flux leaving the initial beam! By convention, we resolve this discrepancy by *defining* the elastic σ_{el} as the cross section for removal from the *beam*. Then, using Eq. (3.4.17) in a derivation like that for Eq. (3.1.50),

$$\sigma_{\text{el}} = \frac{1}{2} \int_0^{2\pi} d\phi \int_0^{\pi} d\theta \sin\theta \, \sigma(\theta) \tag{3.4.18}$$

$$= 2\frac{\pi}{k^2} \sum_{L+S \text{ even}} (2L+1)|1 - \mathbf{S}_L|^2. \tag{3.4.19}$$

The reaction cross section is given by a similar expression

$$\sigma_R = 2\frac{\pi}{k^2} \sum_{L+S \text{ even}} (2L+1)[1 - |\mathbf{S}_L|^2]. \tag{3.4.20}$$

The same definition is used for the non-elastic products from the reaction of two identical nuclei. The cross sections $\sigma_{xpt:x_ip_it_i}$ from incoming channel $x_ip_it_i$ to exit channel xpt are defined as the flux from the beam of *projectiles* that goes to the exit channel. If the projectile and target are identical, then, in order that individual cross sections add up to this above total, the relation between the transition S-matrix elements and the angle-independent cross sections must be

$$\sigma_{xpt:x_ip_it_i} = \frac{\pi}{k_i^2} \frac{1 + \delta_{p_it_i}}{(2I_{p_i}+1)(2I_{t_i}+1)} \sum_{J_{\text{tot}}\pi\alpha\alpha_i} (2J_{\text{tot}}+1)|\tilde{\mathbf{S}}_{\alpha\alpha_i}^{J_{\text{tot}}\pi}|^2. \tag{3.4.21}$$

The generalized extra factor $1 + \delta_{p_it_i}$ doubles the cross section when the projectile and the target are identical *and* in the same excited state. This equation assumes that S-matrix elements have been set to zero for those partial waves blocked by the Pauli principle in either the entrance channel α_i or the exit channel α (or, suitable linear combinations set to zero, depending on the choice of partial-wave basis).

We still have the detailed balance $|\tilde{\mathbf{S}}_{\alpha\alpha_i}|^2 = |\tilde{\mathbf{S}}_{\alpha_i\alpha}|^2$, so the reverse cross section should now be related not by Eq. (3.2.61) but by

$$\sigma_{x_ip_it_i:xpt} = \frac{k_i^2}{k^2} \frac{(1 + \delta_{pt})}{(1 + \delta_{p_it_i})} \frac{(2I_{p_i}+1)(2I_{t_i}+1)}{(2I_p+1)(2I_t+1)} \sigma_{xpt:x_ip_it_i} \tag{3.4.22}$$

to the forward cross section $\sigma_{xpt:x_ip_it_i}$. The new factors will appear again in Chapter 12 when we use the cross sections to find reaction rates in gases.

3.4.4 Exchange transfer

This occurs in transfer reactions such as ^{6}He $+ \,^4$He $\rightarrow \,^4$He $+ \,^6$He, where the transfer process adds coherently to the elastic scattering amplitude, but with revised relative coordinate. In general, consider $p+t \rightarrow t'+p'$ reactions where the primes indicate the transfer channels. We construct a total wave function

$$\Psi_\varepsilon = \psi_{pt} + \varepsilon\hat{P}_{p't'}\psi_{t'p'} \tag{3.4.23}$$

where $\varepsilon = 1$ for $p \neq t$ and $\varepsilon = (-1)^{2I_p}$ for $p = t$, and where again $\hat{P}_{p't'}$ is the operator which exchanges the p' and t' nuclear coordinates. In the S basis, we saw that $\hat{P}_{pt}\psi_{\text{tp}}$ generates a diagonal factor $(-1)^{L+S-I_p-I_t}$. In the J basis, it generates a linear combination of $|\alpha\rangle$ basis states. The effect of elastic transfer is generally to give a backward-angle peak in the elastic scattering distribution.

3.5 Electromagnetic channels

3.5.1 *Maxwell equations and photon channels*

The Maxwell equations for the magnetic $\mathbf{H}$ and electric $\mathbf{E}$ field vectors are, in Gaussian units,

$$\nabla \times \mathbf{H} = \frac{4\pi}{c}\mathbf{j}_q + \frac{1}{c}\frac{d\mathbf{E}}{dt}$$

$$\nabla \times \mathbf{E} = -\frac{1}{c}\frac{d\mathbf{H}}{dt}$$

$$\nabla \cdot \mathbf{H} = 0$$

$$\nabla \cdot \mathbf{E} = 4\pi\rho_q, \tag{3.5.1}$$

where the source terms $\mathbf{j}_q$ and ρ_q are the charge current and charge density, respectively. These describe the classical electromagnetic field surrounding a nucleus. At low velocities, this is predominantly the electrostatic field arising from ρ_q, the charge density of the protons in the nucleus.

In addition to this electrostatic field, there can be radiative photons of higher energies produced by capture reactions, and similar photons that may lead to the breakup of nuclei. To a first approximation, these need only be considered one at a time, since (unlike a coherent laser field), radiative photons will generally react individually. In this section, therefore, we consider the coupling of *one radiative photon* with the charges in a nucleus that may lead to its excitation, and also the reverse: the production of a photon by a decaying nuclear state. The higher energy states in a nucleus may be in the continuum, in which case these reactions are photo-disintegration and photo-production reactions, respectively. The photon by itself will have a specific energy $E_\gamma = \hbar\omega$ and momentum $p_\gamma = \hbar k_\gamma = E_\gamma/c$.

Vector potential

We define, in the standard manner, the *vector potential* $\mathbf{A}(t)$ and *scalar potential* $\phi(t)$ such that

$$\mathbf{H} = \nabla \times \mathbf{A}(t), \tag{3.5.2a}$$

$$\mathbf{E} = -\nabla\phi(t) - \frac{1}{c}\frac{d\mathbf{A}}{dt}. \tag{3.5.2b}$$

Substituting these into the Maxwell equations (3.5.1) we find

$$\nabla^2 \mathbf{A}(t) + \frac{1}{c^2} \frac{\mathrm{d}^2 \mathbf{A}}{\mathrm{d}t^2} = -\frac{4\pi}{c} \mathbf{j}_q + \nabla(\nabla \cdot \mathbf{A}(t)) + \frac{1}{c} \nabla \frac{\mathrm{d}\phi}{\mathrm{d}t} \qquad (3.5.3)$$

$$\nabla^2 \phi(t) = -4\pi \rho_q(t) - \frac{1}{c} \frac{\mathrm{d}}{\mathrm{d}t} (\nabla \cdot \mathbf{A}(t)), \qquad (3.5.4)$$

where we have used the identity $\nabla \times (\nabla \times \mathbf{A}) = \nabla(\nabla \cdot \mathbf{A}) - \nabla^2 \mathbf{A}$.

The potentials $\mathbf{A}(t)$ and $\phi(t)$ are not fixed, but may be changed according to some arbitrary spatial *gauge* function $\chi(t)$ by

$$\mathbf{A}'(t) = \mathbf{A}(t) + \nabla \chi(t) \qquad (3.5.5)$$

$$\phi'(t) = \phi(t) + \partial \chi(t)/\partial t. \qquad (3.5.6)$$

This means that we are free to choose the gauge $\chi(t)$, for example so that

$$\nabla \cdot \mathbf{A}(t) = 0, \qquad (3.5.7)$$

which is called the *Coulomb* or *transverse gauge*. In this gauge, the scalar potential satisfies Poisson's equation

$$\nabla^2 \phi(t) = -4\pi \rho_q(t) \qquad (3.5.8)$$

and the vector potential satisfies an inhomogeneous wave equation

$$\nabla^2 \mathbf{A}(t) + \frac{1}{c^2} \frac{\mathrm{d}^2 \mathbf{A}}{\mathrm{d}t^2} = -\frac{4\pi}{c} \mathbf{j}_q + \frac{1}{c} \nabla \frac{\mathrm{d}\phi}{\mathrm{d}t}. \qquad (3.5.9)$$

We now make our physical *one-photon* approximation, separating the electrostatic field caused by the charges ρ_q from the radiative photon that couples to the current $\mathbf{j}_q$. In this case, ϕ becomes time-independent like ρ_q, and the electromagnetic wave equation for a radiative single photon becomes

$$\nabla^2 \mathbf{A}(t) + \frac{1}{c^2} \frac{\mathrm{d}^2 \mathbf{A}}{\mathrm{d}t^2} = -\frac{4\pi}{c} \mathbf{j}_q. \qquad (3.5.10)$$

For a fixed monochromatic photon energy $E_\gamma = \hbar\omega$, we can write the time-dependent vector potential $\mathbf{A}(t)$ in terms of a time-independent $\mathbf{A}$ and current $\mathbf{j}$ as

$$\mathbf{A}(t) = \mathbf{A} e^{-i\omega t} + \mathbf{A}^* e^{i\omega t} \qquad (3.5.11)$$

$$\mathbf{j}_q(t) = \mathbf{j} e^{-i\omega t} + \mathbf{j}^* e^{i\omega t}. \qquad (3.5.12)$$

The wave equation for the stationary-state $\mathbf{A}$-vector potential is thus

$$\nabla^2 \mathbf{A} + k_\gamma^2 \mathbf{A} = -\frac{4\pi}{c} \mathbf{j}. \qquad (3.5.13)$$

3.5.2 Coupling photons and particles

This classical description of the electromagnetic field will now be connected to a quantum description of the particle dynamics, by using for the source flux **j** the quantum mechanical matrix element, at each spatial position **r**, of the electric form of the general current operator of Eq. (3.1.101)

$$\mathbf{j}(\mathbf{r}) = \langle \Phi | \hat{\mathbf{j}}_q(\mathbf{r}) | \Psi \rangle, \tag{3.5.14}$$

where both $\Phi(\mathbf{r}_i)$ and $\Psi(\mathbf{r}_i)$ are wave functions depending on vector positions such as $\mathbf{r}_i$ for each charged particle i, and the bra-ket implies integration over all the $\mathbf{r}_i$. Both **r** and all the $\mathbf{r}_i$ are actual positions in the center-of-mass frame, not relative coordinates. The bra and ket wave functions are allowed to be different, giving off-diagonal matrix elements whereby electromagnetic effects of particle *transitions* may be described.

In the free-field case this matrix element is that of $\hat{\mathbf{j}}_{\text{free}}^q$,

$$\langle \Phi | \hat{\mathbf{j}}_{\text{free}}^q | \Psi \rangle = \frac{\hbar q}{2 i \mu} \left[\Phi^*(\nabla \Psi) - (\nabla \Phi)^* \Psi \right]. \tag{3.5.15}$$

Capture (photo-production)

Photo-nuclear couplings may work in two directions. A current of charged particles may produce photons, for example in the *photo-production* or *capture* reaction $^7\mathrm{Be}(\mathrm{p},\gamma)^8\mathrm{B}$, important for solar neutrino production. At other times, the photons may cause the movement of charged particles and the breakup of bound states as in $^8\mathrm{B}(\gamma,\mathrm{p})^7\mathrm{Be}$: this is called *photo-disintegration* and described in the following subsection.

The coupling from particle current to photon production, for particle initial scattering state Ψ and final bound state Φ_b, is therefore described by

$$\nabla^2 \mathbf{A} + k_\gamma^2 \mathbf{A} = -\frac{4\pi}{c} \langle \Phi_b | \hat{\mathbf{j}}_{\text{free}} | \Psi \rangle. \tag{3.5.16}$$

It is convenient to multiply this equation by $-\hbar c / k_\gamma$, so that the coefficient of **A** becomes $\hbar c / k_\gamma \times k_\gamma^2 = \hbar c k_\gamma = E_\gamma$, and has the same units of energy as in a Schrödinger equation:

$$\left[-\frac{\hbar c}{k_\gamma} \nabla^2 - E_\gamma \right] \mathbf{A}(\mathbf{r}) = \frac{4\pi \hbar}{k_\gamma} \langle \Phi_b | \hat{\mathbf{j}}_{\text{free}}(\mathbf{r}) | \Psi \rangle. \tag{3.5.17}$$

This equation has the form of a particle-to-photon coupled equation, linking the initial particle state Ψ to an outgoing photon field **A** with an 'interaction Hamiltonian' operator $H_{\gamma p} = \frac{4\pi \hbar}{k_\gamma} \langle \Phi_b | \hat{\mathbf{j}}_{\text{free}} |$.

Photo-disintegration

The standard *minimal gauge coupling* describing the influence of an electromagnetic field on particle motion is found by transforming the Schrödinger equation $[\hat{\mathbf{p}}^2/2m + V - E]\Psi = 0$ by the 'minimal' replacement

$$\hat{\mathbf{p}} \rightarrow \hat{\mathbf{p}} - \frac{q}{c}\mathbf{A}, \qquad (3.5.18)$$

where $\mathbf{A}$ is the vector potential for the electromagnetic field. Expanding the square we have

$$\left[\frac{1}{2\mu}\left(\hat{\mathbf{p}}^2 - \frac{q}{c}\{\mathbf{A}\cdot\hat{\mathbf{p}} + \hat{\mathbf{p}}\cdot\mathbf{A}\} + \frac{q^2}{c^2}|\mathbf{A}|^2\right) + V - E\right]\Psi = 0. \qquad (3.5.19)$$

Neglecting the $|\mathbf{A}|^2$ term according to our one-photon approximation, and, using $\hat{\mathbf{p}} = \hbar/i\,\vec{\nabla}$, this particle equation becomes

$$
\begin{aligned}
[\hat{T} + V - E]\Psi &= \frac{1}{2\mu}\frac{q}{c}\frac{\hbar}{i}\{\mathbf{A}\cdot\vec{\nabla} + \vec{\nabla}\cdot\mathbf{A}\}\Psi \\
&= \frac{1}{2\mu}\frac{q}{c}\frac{\hbar}{i}\{\mathbf{A}\cdot\vec{\nabla} - \overleftarrow{\nabla}\cdot\mathbf{A}\}\Psi \\
&= \frac{1}{c}\int d\mathbf{r}\,\mathbf{A}(\mathbf{r})\cdot\hat{\mathbf{j}}^q_{\text{free}}(\mathbf{r})\,\Psi, \qquad (3.5.20)
\end{aligned}
$$

where in the second step we use integration by parts, assuming that in any matrix element the product of both the initial and final wave functions goes to zero asymptotically, which is true for transitions to or from bound states. In the final step we used the electric current operator $\hat{\mathbf{j}}^q_{\text{free}}(\mathbf{r})$ defined in Eq. (3.1.99), with a charge q factor.

If the potential V in the Schrödinger equation is non-local, then it will be momentum-dependent, and strictly we should perform the minimal replacement of Eq. (3.5.18) in these operators as well. Without knowing the details of the non-locality we cannot give a general derivation, so for now we will assume that the effect of these non-localities is to supplement the free current operator in Eq. (3.5.20) by the 'two-body current' terms $\hat{\mathbf{j}}_2$ discussed in subsection 3.1.4. This gives the more general photo-disintegration result of

$$[\hat{T} + V - E]\Psi - \frac{1}{c}\int d\mathbf{r}\,\mathbf{A}(\mathbf{r})\cdot\hat{\mathbf{j}}(\mathbf{r})\,\Psi = 0. \qquad (3.5.21)$$

For a photo-disintegration reaction with particle initial state Φ_b and final scattering state Ψ, the main contributor to the second term will be the wave function of the bound state:

$$[\hat{T} + V - E]\Psi - \frac{1}{c}\int d\mathbf{r}\, \mathbf{A}(\mathbf{r}) \cdot \hat{\mathbf{j}}(\mathbf{r})\, \Phi_b = 0, \qquad (3.5.22)$$

where we have an effective 'interaction Hamiltonian' $H_{\text{int}} = -\frac{1}{c}\mathbf{A} \cdot \hat{\mathbf{j}}$ that couples between two particle states.

Equally, if Eq. (3.5.20) can be written as

$$[\hat{T} + V - E]\Psi - \frac{1}{c}\int d\mathbf{r}\, (\hat{\mathbf{j}}(\mathbf{r})\Phi_b) \cdot \mathbf{A}(\mathbf{r}) = 0, \qquad (3.5.23)$$

for fixed initial particle state Φ_b, then this again appears as an equation coupling the incoming photon field $\mathbf{A}$ to the particle state Ψ in the continuum. The coupling operator is $H_{p\gamma} = -\frac{1}{c}\int d\mathbf{r}\, \hat{\mathbf{j}}(\mathbf{r})\Phi_b \cdot$.

3.5.3 Photon cross sections

When the photon outgoing wave is $\mathbf{A}$ in direction $\mathbf{k}$, so $\mathbf{A}(\mathbf{r}) = \mathbf{a}e^{i\mathbf{k}\cdot\mathbf{r}}$, the cross section depends on the *number flux* of photons. In order to determine this from the magnitude of $\mathbf{A}$, we calculate the Poynting vector, which is the energy flux, and then divide by the energy of each photon.

Given $\mathbf{A}$, the physical vector potential of Eq. (3.5.11) is

$$\mathbf{A}(t) = 2\text{Re}(\mathbf{A}e^{-i\omega t})$$
$$= 2\text{Re}(\mathbf{a}e^{i(\mathbf{k}\cdot\mathbf{r}-\omega t)})$$
$$= 2\mathbf{a}\cos(\mathbf{k} \cdot \mathbf{r} - \omega t). \qquad (3.5.24)$$

The electric and magnetic fields are therefore

$$\mathbf{E}(t) = 2k_\gamma \mathbf{a}\sin(\mathbf{k} \cdot \mathbf{r} - \omega t) \qquad (3.5.25)$$
$$\text{and} \quad \mathbf{H}(t) = 2(\mathbf{k} \times \mathbf{a})\sin(\mathbf{k} \cdot \mathbf{r} - \omega t). \qquad (3.5.26)$$

The Poynting vector for the energy flux is constructed as

$$\mathbf{S}_P = \frac{c}{4\pi}\mathbf{E} \times \mathbf{H}. \qquad (3.5.27)$$

Far from the source, where the radiative field has $\mathbf{E}$ and $\mathbf{H}$ mutually perpendicular, the magnitudes $|\mathbf{H}|$ and $|\mathbf{E}|$ become equal, so

$$|\mathbf{S}_P| \simeq \frac{c}{4\pi}|\mathbf{E}|^2 = \frac{c}{4\pi}|\mathbf{H}|^2. \qquad (3.5.28)$$

As $\langle|\sin(t)|^2\rangle = 1/2$, the time-averaged $\langle|\mathbf{E}(t)|^2\rangle = \frac{1}{2}|2k_\gamma\mathbf{a}|^2 = 2k_\gamma^2|\mathbf{a}|^2$, and the time-averaged energy flux is

$$\langle|\mathbf{S}_P(t)|\rangle = \frac{k_\gamma^2 c}{2\pi}|\mathbf{a}|^2. \tag{3.5.29}$$

Dividing by the photon energy $E_\gamma = \hbar k_\gamma c$, we find the outgoing photon number flux

$$j_\gamma = \frac{k_\gamma}{2\pi\hbar}|\mathbf{a}|^2. \tag{3.5.30}$$

Because this flux formula has factors additional to the expression of Eq. (2.4.1) in quantum mechanics, it is now convenient to define a vector *photon wave function*

$$\mathbf{Z}(\mathbf{r}) = \sqrt{\frac{k_\gamma}{2\pi\hbar c}}\,\mathbf{A}(\mathbf{r}) \tag{3.5.31}$$

so that photon number density is simply $|\mathbf{Z}(\mathbf{r})|^2$, and the photon number flux now appears as should be expected:

$$j_\gamma = c|\mathbf{Z}(\mathbf{r})|^2. \tag{3.5.32}$$

By Eq. (2.4.1) this is appropriate for a quantum mechanical object moving at the speed of light c.

In terms of this new photon wave function, the photo-production 'coupled equation' (3.5.17) becomes

$$\left[-\frac{\hbar c}{k_\gamma}\nabla^2 - E_\gamma\right]\mathbf{Z}(\mathbf{r}) = \sqrt{\frac{k_\gamma}{2\pi\hbar c}}\frac{4\pi\hbar}{k_\gamma}\langle\Phi_b|\hat{\mathbf{j}}(\mathbf{r})|\Psi\rangle$$

$$= 2\sqrt{\frac{2\pi\hbar}{k_\gamma c}}\langle\Phi_b|\hat{\mathbf{j}}(\mathbf{r})|\Psi\rangle$$

$$= 2\sqrt{h/\omega}\langle\Phi_b|\hat{\mathbf{j}}(\mathbf{r})|\Psi\rangle, \tag{3.5.33}$$

and the photo-disintegration equation (3.5.23) becomes

$$[\hat{T} + V - E]\Psi = \sqrt{\frac{2\pi\hbar c}{k_\gamma}}\frac{1}{c}\int d\mathbf{r}\,\hat{\mathbf{j}}(\mathbf{r})\Phi_b \cdot \mathbf{Z}$$

$$= \sqrt{h/\omega}\int d\mathbf{r}\,\hat{\mathbf{j}}(\mathbf{r})\Phi_b \cdot \mathbf{Z}. \tag{3.5.34}$$

These two equations (3.5.33, 3.5.34) have *almost* the same kind of coupling interaction, namely the operators

$$V_{\gamma p} = 2\sqrt{h/\omega}\langle\Phi_b|\hat{\mathbf{j}}(\mathbf{r}), \tag{3.5.35a}$$

$$V_{p\gamma} = \sqrt{h/\omega}\int d\mathbf{r}\,\hat{\mathbf{j}}(\mathbf{r})\Phi_b\cdot, \tag{3.5.35b}$$

respectively, using the bound particle state Φ_b in each case. The factor of 2 difference in magnitude of these two couplings arises from the relativistic kinematics for photons, as explained in the Appendix on page 126.

The electromagnetic coupling operators of Eq. (3.5.35) are written in their *current* form, as they use the derivative current operator of Eq. (3.1.99). This is commonly judged as too complicated for everyday use, so in Section 4.7 an approximate form for the couplings will be derived. That form will be valid for long photon wavelengths (low-energy photons).

3.5.4 Partial waves and vector spherical harmonics

The photon vector potential $\mathbf{A}(\mathbf{r})$ and its normalized form $\mathbf{Z}(\mathbf{r})$ are described by a three-dimensional vector at every spatial position $\mathbf{r}$. These three coordinates may be equivalently mapped onto the three m-state amplitudes $\mu = -1, 0, +1$ for the photon as a spin-1 object.[23] Let ξ_μ be three (complex) unit vectors in 3D space, such that the complex coefficients Z_μ in $\mathbf{Z} = \sum_\mu Z_\mu\xi_\mu$ in this basis transform as the components of a spin-1 vector. One choice for the ξ_μ is

$$\xi_0 = \hat{\mathbf{z}} \quad \text{and} \quad \xi_{\pm 1} = \mp(\hat{\mathbf{x}} \pm i\hat{\mathbf{y}})/\sqrt{2}. \tag{3.5.36}$$

The vector field $\mathbf{Z}(\mathbf{r})$ can therefore be expanded as the coupling of a spatial angular momentum Λ with photon spin 1 to form a total spin J_γ. In a capture or disintegration reaction, this is coupled with the spin J_b of the bound state to give J_{tot}, the total angular momentum which must equal that from the particle scattering channel, $|\alpha\rangle = |(LIJ_p)J_p, I_t; J_{\text{tot}}\rangle$. In the normal manner of subsection 3.2.1 we use partial wave expansions to write the total wave function as a sum of photon (γ) and particle (p) channels as

$$\begin{aligned}
\Psi^{M_{\text{tot}}}_{J_{\text{tot}}} &= \Psi^{M_{\text{tot}}}_{\gamma J_{\text{tot}}} + \Psi^{M_{\text{tot}}}_{p J_{\text{tot}}} \\
&= \sum_{\Lambda J_\gamma b} |(\Lambda 1)J_\gamma, J_b; J_{\text{tot}}\rangle\zeta_\gamma(r)/r + \sum_{LJxpt} |(LIJ_p)J_p, I_t; J_{\text{tot}}\rangle\psi_\alpha(R)/R \\
&\equiv \sum_\gamma |\gamma\rangle\zeta_\gamma(r)/r + \sum_\alpha |\alpha\rangle\psi_\alpha(R)/R, \tag{3.5.37}
\end{aligned}$$

---

[23] The requirement of gauge-invariance and the reduction to two components will be presented in subsection 3.5.5.

where $|\gamma\rangle = \{(\Lambda 1)J_\gamma, J_b\}$ refers to the set of photon partial waves, as well as the coupling of the spin J_b of the bound state $\Phi_b(\mathbf{R})$. Both the $|\gamma\rangle$ and $|\alpha\rangle$ contain just the *angular* part of the basis states, with $\zeta_\gamma(r)$ and $\psi_\alpha(R)$ being the respective radial parts, and depend on J_{tot} implicitly. As usual, the particle channels are written using the relative separation R instead of the distance r from the center of mass, but these are linearly proportional.

The coupled state $|(\Lambda 1)J_\gamma\rangle$ in Eq. (3.5.37) is often called a *vector spherical harmonic*, explicitly the basis component

$$\mathbf{Y}_{\Lambda J}^M(\hat{\mathbf{r}}) = |(\Lambda 1)J\rangle = \sum_{m\mu} \langle \Lambda m, 1\mu | JM \rangle \, \boldsymbol{\xi}_\mu \, Y_\Lambda^m(\hat{\mathbf{r}}) \qquad (3.5.38)$$

that includes the photon vector states $\boldsymbol{\xi}_\mu$. It is defined for each $\Lambda = J_\gamma - 1$, J_γ, $J_\gamma + 1$, and has spatial parity $(-1)^\Lambda$. The vector spherical harmonics transform under coordinate rotations by the normal rotation matrices $D_{MM'}^{J_\gamma}(\mathcal{R})$ for a spin-J_γ object.

The electromagnetic field $\mathbf{Z}(\mathbf{r})$ can therefore be expanded in vector spherical harmonics, with a sum over a range of Λ and J_γ values. The J_γ value gives the total spin or *multipole* of the photon, and the spatial form of that multipole is determined by the Λ value in the range $|J_\gamma - 1| \le \Lambda \le J_\gamma + 1$.

Using Eq. (3.5.37), the coupled equations of Eqs. (3.5.33) and (3.5.34) may be rewritten in terms of radial functions for each partial wave $|\gamma\rangle$ and $|\alpha\rangle$:

$$[T_\Lambda - E_\gamma]\,\zeta_\gamma(r) = \sum_\alpha \langle \gamma | V_{\gamma p} | \alpha \rangle \, \psi_\alpha(R)$$

$$[T_\alpha + V_\alpha - E_\alpha]\,\psi_\alpha(R) = \sum_\gamma \langle \alpha | V_{p\gamma} | \gamma \rangle \, \zeta_\gamma(r), \qquad (3.5.39)$$

and solved to get S (or T) matrices from which we have cross sections following the pattern established for particle channels. When a photon T-matrix element is defined as the asymptotic amplitude $\mathbf{T}_{\gamma\alpha_i}^{J_{\text{tot}}\pi}$ of the outgoing solution of Eq. (3.5.39) for an incoming particle channel α_i and for system parity π, the photo production (capture) cross section

$$\sigma_{\text{cap}} = \frac{4\pi}{k_i^2} \frac{1}{(2I_{p_i}+1)(2I_{t_i}+1)} \frac{c}{v_i} \sum_{J_{\text{tot}}\pi\alpha_i} (2J_{\text{tot}}+1) \, |\mathbf{T}_{\gamma\alpha_i}^{J_{\text{tot}}\pi}|^2 \qquad (3.5.40)$$

since the outgoing photon speed is that of light, c.

The T-matrix elements can also be found by the equivalent integral expression

$$\mathbf{T}_{\gamma\alpha_i}^{J_{\text{tot}}\pi} = -\frac{1}{k_\gamma} \frac{k_\gamma}{\hbar c} \sum_\alpha \langle \zeta_\gamma^{(-)}(r) \langle \gamma | V_{\gamma p} | \alpha \rangle \psi_{\alpha\alpha_i}(R) \rangle, \qquad (3.5.41)$$

where the inner matrix element is an angular integral and the outer one a radial one. Since there is no potential in the photon wave equation, the $\zeta_\gamma^{(-)}(r)$ are simply the free-field solutions, giving

$$\mathbf{T}_{\gamma\alpha_i}^{J_{\text{tot}}\pi} = -\frac{1}{\hbar c} \sum_\alpha \langle F_\Lambda(0,k_\gamma r) \langle \gamma | V_{\gamma p} | \alpha \rangle \psi_{\alpha\alpha_i}(R) \rangle, \tag{3.5.42}$$

$$= -\frac{1}{\hbar c} \sum_\alpha \int d\mathbf{r} \, \frac{1}{rR} \, F_\Lambda(0,k_\gamma r) \, \phi_\gamma^* \, V_{\gamma p} \, \phi_\alpha \, \psi_{\alpha\alpha_i}(R), \tag{3.5.43}$$

where the radial and angular integrals have been recombined into one integral over all space, and ϕ_γ, ϕ_α are the wave functions of the respective basis functions.

We can further calculate the vector-dependent T-matrix amplitude $\mathbf{T}(\mathbf{k}_\gamma, \mathbf{k}_i)$ from the $\mathbf{T}_{\gamma\alpha_i}^{J_{\text{tot}}\pi}$ using Eq. (3.3.44). Since the photon wave functions in Eq. (3.5.42) are components of plane waves, we have the integral

$$\mathbf{T}^{\mu\mu_b:\mu_{p_i}\mu_{t_i}}(\mathbf{k}_\gamma, \mathbf{k}_i) = \langle \xi_\mu e^{i\mathbf{k}_\gamma \cdot \mathbf{r}} | V_{\gamma p} | \Psi_{x_i p_i t_i}^{\mu_{p_i}\mu_{t_i}}(\mathbf{R}; \mathbf{k}_i) \rangle \tag{3.5.44}$$

over $\mathbf{r}$, remembering that $V_{\gamma p}$ already contains an integral with respect to the particle positions $\mathbf{r}_i$, and contains the bound-state wave function $\Phi_b^{\mu_b}$. The scattering amplitude in the photon exit channel is then

$$f_{\mu\mu_b:\mu_{p_i}\mu_{t_i}}(\mathbf{k}_\gamma : \mathbf{k}_i) = -\frac{1}{4\pi} \frac{k_\gamma}{\hbar c} \mathbf{T}^{\mu\mu_b:\mu_{p_i}\mu_{t_i}}(\mathbf{k}_\gamma, \mathbf{k}_i). \tag{3.5.45}$$

Following the usual pattern of Eq. (3.2.17), the m-states are summed over the exit μ, μ_b and averaged over the entrance μ_{p_i}, μ_{t_i} to obtain the cross section for unpolarized beams.

A general photon field is therefore a linear combination of vector spherical harmonics with mixed parities and multipoles. The parity of the field is $(-1)^\Lambda$, and this parity is used to distinguish electric from magnetic multipole components of the electromagnetic field. The *electric* part of a field $\mathbf{A}$ (or $\mathbf{Z}$) is defined as that with parity $(-1)^{J_\gamma+1}$, and the *magnetic* part as that with parity $(-1)^{J_\gamma}$. That is, *the electric part has $\Lambda = J_\gamma \pm 1$, and the magnetic part has $\Lambda = J_\gamma$.* Both components are included on an equal footing in Eqs. (3.5.39) and (3.5.41), etc.

All the analysis of this subsection, however, has ignored the requirements of gauge invariance. The above theory assumes *three* m-states for the photon, $\mu = -1, 0, +1$, which is incorrect. We next show, therefore, what must be done to satisfy

- Nuclear transitions, we will see in general in Chapter 4, are classified as dipole, quadrupole, etc.
- Electric and magnetic transitions are classified not according to their spatial part Λ but their multipolarity: the total angular momentum J_γ that is transferred to or from the nucleus by the E1 and M1 photons have $J_\gamma = 1$, whereas the E2 and M2 have $J_\gamma = 2$, etc.
- When the additional parity change from the current operator is taken into account, electric transitions change nuclear parities according to $(-1)^{J_\gamma}$, and magnetic transitions by $(-1)^{J_\gamma - 1}$. Parities are not changed by M1, E2 and M3 transitions, for example, but *are* changed by E1 and M2 multipoles.
- M0 and E0 transitions are not allowed by angular momentum selection rules.

Box 3.5 **Electromagnetic and nuclear multipoles**

a specific Coulomb gauge condition. This procedure will remove some components from the plane wave $\xi_\mu e^{i\mathbf{k}_\gamma \cdot \mathbf{r}}$ in Eq. (3.5.44).

3.5.5 *Electric and magnetic parts in the Coulomb gauge*

If the coordinate system of Eq. (3.5.36) is not the laboratory system, but is specifically chosen so that the z axis coincides with $\hat{\mathbf{k}}_\gamma$, the *local* direction of photon propagation, the Coulomb gauge condition

$$\nabla \cdot \mathbf{A} = \nabla \cdot \mathbf{Z} = 0 \tag{3.5.46}$$

is simply $A_0 = Z_0 = 0$, as the polarization vectors $\boldsymbol{\xi}_{\pm 1}$ cover all the allowed transverse directions. In order to satisfy this condition, we first construct a suitable plane wave, and then reuse this gauged plane wave in the T-matrix integral (3.5.44) for the general photon cross section.

Consider first a general plane-wave vector field traveling in the $+z$ direction, $\mathbf{A}_\mu = \boldsymbol{\xi}_\mu e^{ikz}$. The polarization states allowed by the Coulomb gauge condition (3.5.46) are $\mu = \pm 1$, so we may set $\mathbf{A}_0 = 0$. Following the standard specification of Eq. (3.2.7) for coupling a spin-1 object to a plane wave in the direction of $\mathbf{k}$,

$$\xi_{\mu_i} e^{i\mathbf{k} \cdot \mathbf{r}} = \sum_{JM \, \Lambda m \mu} i^L Y_\Lambda^m(\mathbf{r}) \, \xi_\mu \, \langle \Lambda m, 1\mu | JM \rangle \, F_\Lambda(0, kr)/r$$

$$\times \frac{4\pi}{k} \sum_{\Lambda_i m_i} Y_{\Lambda_i}^{m_i}(\mathbf{k})^* \langle \Lambda_i m_i, 1\mu_i | JM \rangle. \tag{3.5.47}$$

We now specialize to $\mathbf{k}$ in the +z direction so $m = 0$, use the definition of the vector spherical harmonics above, note that in this free-field case there is no spin-flip so $\mu = \mu_i$, and rename Λ_i as Λ. This gives

$$\boldsymbol{\xi}_\mu e^{ikz} = \sqrt{4\pi} \sum_{J\Lambda} \sqrt{2\Lambda+1} \; i^L \mathbf{Y}_{\Lambda J}^\mu(\mathbf{r}) \; \langle \Lambda 0, 1\mu | J\mu \rangle F_\Lambda(0, kr)/(kr). \quad (3.5.48)$$

Now the electric parts (those that have $J = \Lambda \pm 1$) need to be grouped together. We therefore reassemble the above equation as

$$\boldsymbol{\xi}_\mu e^{ikz} = \sqrt{2\pi} \sum_J \sqrt{2J+1} \; i^J$$

$$\times \left[\sum_\Lambda \mathbf{Y}_{\Lambda J}^\mu(\mathbf{r}) \; i^{\Lambda-J} \sqrt{\frac{2\Lambda+1}{2J+1}} \langle \Lambda 0, 1\mu | J\mu \rangle \; F_\Lambda(0, kr)/(kr) \right]. \quad (3.5.49)$$

Let us split the $[\cdots]$ into two terms, and using the analytical forms of the $\langle \Lambda 0, 1\mu | J\mu \rangle$, define the $\Lambda = J$ magnetic part as

$$\mathbf{A}_{JM}(\mathbf{r}; \mathcal{M}) = (kr)^{-1} F_J(0, kr) \; \mathbf{Y}_{JJ}^M(\hat{\mathbf{r}}), \quad (3.5.50)$$

and electric part as the sum of the $\Lambda = J \pm 1$ terms

$$\mathbf{A}_{JM}(\mathbf{r}; \mathcal{E}) = \sqrt{\frac{J+1}{2J+1}} (kr)^{-1} F_{J-1}(0, kr) \; \mathbf{Y}_{J-1,J}^M(\hat{\mathbf{r}})$$

$$- \sqrt{\frac{J}{2J+1}} (kr)^{-1} F_{J+1}(0, kr) \; \mathbf{Y}_{J+1,J}^M(\hat{\mathbf{r}}). \quad (3.5.51)$$

Finally, we multiply the electric part by μ^2 so ensure that it is zero when $\mu = 0$, and combine the electric and magnetic parts to give the complete field $\boldsymbol{\xi}_\mu e^{ikz}$ satisfying the gauge condition:

$$\mathbf{A}_\mu = \boldsymbol{\xi}_\mu e^{ikz} = \mu\sqrt{2\pi} \sum_J \sqrt{2J+1} \; i^J [\mathbf{A}_{J\mu}(\mathbf{r}; \mathcal{M}) + i\mu \mathbf{A}_{J\mu}(\mathbf{r}; \mathcal{E})]. \quad (3.5.52)$$

Both these electric and magnetic basis components are normalized like plane waves,

$$\int \mathbf{A}_{JM}(\mathbf{r}; e) \cdot \mathbf{A}_{J'M'}(\mathbf{r}; e') \, d\mathbf{r} = \delta(k-k') \, \delta_{JJ'} \delta_{MM'} \delta_{ee'}, \quad (3.5.53)$$

where $e, e' = \mathcal{M}$ or $\mathcal{E}$, and they are related by

$$\mathbf{A}_{JM}(\mathbf{r}; \mathcal{M}) = \frac{1}{ik} \nabla \times \mathbf{A}_{JM}(\mathbf{r}; \mathcal{E}) \qquad (3.5.54)$$

$$\mathbf{A}_{JM}(\mathbf{r}; \mathcal{E}) = \frac{1}{ik} \nabla \times \mathbf{A}_{JM}(\mathbf{r}; \mathcal{M}), \qquad (3.5.55)$$

$$\text{so} \quad \nabla \cdot \mathbf{A}_{JM}(\mathbf{r}; e) = 0 \quad \text{for } e = \mathcal{M} \text{ or } \mathcal{E}. \qquad (3.5.56)$$

The Eq. (3.5.52) can be used directly for photon *entrance* channels, since then we do have a plane wave in the $+z$ direction. For *exit* channels, however, we need the expansion for a wave traveling in an arbitrary direction $\mathbf{k}$. We cannot use Eq. (3.5.47) directly, since the μ_i are components in the $+z$ direction, not in the $\mathbf{k}$ direction, and it is the $\mu = 0$ component *in the direction of travel* that the gauge condition requires to be zero. We therefore have to rotate Eq. (3.5.52) to obtain the μ-projections in the $\mathbf{k}$ direction as

$$\xi_\mu e^{i\mathbf{k}\cdot\mathbf{r}} = \mu\sqrt{2\pi} \sum_{JM} \sqrt{2J+1}\, i^J [\mathbf{A}_{JM}(\mathbf{r}; \mathcal{M}) + i\mu\mathbf{A}_{JM}(\mathbf{r}; \mathcal{E})] D^J_{M\mu}(\mathcal{R}_k), \quad (3.5.57)$$

where $\mathcal{R}_k$ is the rotation taking the z axis to the direction of $\mathbf{k}$. Because the laboratory z axis is not now the direction of motion, all M values contribute to this superposition, even in the Coulomb gauge.

These components can now be used to construct a scattering amplitude for photons in the Coulomb gauge if Eq. (3.5.57) is inserted into Eq. (3.5.44), yielding a sum of magnetic and electric parts

$$\mathbf{T}^\mu(\mathbf{k}_\gamma, \mathbf{k}_i; \mathcal{M}) = \mu\sqrt{2\pi} \sum_{JM} \hat{J}\, i^{-J} D^J_{M\mu}(\mathcal{R}_{k_\gamma})^* \langle \mathbf{A}_{JM}(\mathbf{r}; \mathcal{M}) | V_{\gamma p} | \Psi(\mathbf{R}; \mathbf{k}_i) \rangle$$

$$(3.5.58)$$

$$\mathbf{T}^\mu(\mathbf{k}_\gamma, \mathbf{k}_i; \mathcal{E}) = -i\mu^2\sqrt{2\pi} \sum_{JM} \hat{J}\, i^{-J} D^J_{M\mu}(\mathcal{R}_{k_\gamma})^* \langle \mathbf{A}_{JM}(\mathbf{r}; \mathcal{E}) | V_{\gamma p} | \Psi(\mathbf{R}; \mathbf{k}_i) \rangle,$$

$$(3.5.59)$$

suppressing the particle m-state labels and using $\hat{J} = \sqrt{2J+1}$. The $\langle \ldots \rangle$ matrix elements appear in

$$\mathbf{T}^{(e)}_{JM} = \langle \sqrt{4\pi}\sqrt{2J+1}\, i^J\, \mathbf{A}_{JM}(\mathbf{r}; (e)) | V_{\gamma p} | \Psi(\mathbf{R}; \mathbf{k}_i) \rangle, \qquad (3.5.60)$$

and so may be calculated by normal integral or coupled-channels methods as shown in the next chapter, and the results rotated to obtain the μ-projections for the photon in its direction of travel as

$$\mathbf{T}^\mu(\mathbf{k}_\gamma,\mathbf{k}_i;\mathcal{M}) = \frac{\mu}{\sqrt{2}}\sum_{JM} D^J_{M\mu}(\mathcal{R}_{k_i\to k_\gamma})^* \mathbf{T}^{(\mathcal{M})}_{JM}$$

$$\mathbf{T}^\mu(\mathbf{k}_\gamma,\mathbf{k}_i;\mathcal{E}) = -\frac{i\mu^2}{\sqrt{2}}\sum_{JM} D^J_{M\mu}(\mathcal{R}_{k_i\to k_\gamma})^* \mathbf{T}^{(\mathcal{E})}_{JM}. \tag{3.5.61}$$

This method is therefore equivalent to calculating the photon amplitudes $\mathbf{T}_{JM}$ for quantum numbers JM from forward scattering ($0°$), and then rotating the amplitudes to other angles according to the various JM values.

Finally, we may define a *longitudinal* part of the field, another combination of the $\Lambda = J \pm 1$ terms, as

$$\mathbf{A}_{JM}(\mathbf{r};\text{long}) \equiv \frac{1}{k}\nabla\left(\frac{1}{kr}F_J(0,kr)Y^M_J(\hat{\mathbf{r}})\right)$$

$$= \sqrt{\frac{J}{2J+1}}\frac{1}{kr}F_{J-1}(0,kr)\mathbf{Y}^M_{J-1,J}(\hat{\mathbf{r}})$$

$$+ \sqrt{\frac{J+1}{2J+1}}\frac{1}{kr}F_{J+1}(0,kr)\mathbf{Y}^M_{J+1,J}(\hat{\mathbf{r}}). \tag{3.5.62}$$

Though $\nabla \cdot \mathbf{A}_{J_\gamma M_\gamma}(\mathbf{r};\text{long}) \neq 0$, and hence does not satisfy the transverse gauge condition, this longitudinal component will be useful later since it differs from Eq. (3.5.51) only by the coefficients of its two terms. If the second terms in both Eqs. (3.5.51) and (3.5.62) could be neglected in some circumstances, for example when $kr \ll 1$, then

$$\mathbf{A}_{JM}(\mathbf{r};\mathcal{E}) \approx \sqrt{\frac{J+1}{2J+1}}(kr)^{-1}F_{J-1}(0,kr)\,\mathbf{Y}^M_{J-1,J}(\hat{\mathbf{r}})$$

$$= \sqrt{\frac{J+1}{J}}\sqrt{\frac{J}{2J+1}}(kr)^{-1}F_{J-1}(0,kr)\mathbf{Y}^M_{J-1,J}(\hat{\mathbf{r}})$$

$$\approx \sqrt{\frac{J+1}{J}}\mathbf{A}_{JM}(\mathbf{r};\text{long}). \tag{3.5.63}$$

This approximation will be used in the next chapter to simplify Eq. (3.5.59).

Appendix

S-matrix symmetry with relativistic kinematics

For low-energy nuclear reactions, Newtonian non-relativistic kinematics is sufficiently accurate. Later we will discuss breakup reactions at intermediate energies, but photons themselves, with zero rest mass, are necessarily relativistic, so here we apply the relativistic kinematics of subsection 2.3.3 to reactions that couple photon and particle channels.

In a coupled-channels scheme $[T_\alpha - E_\alpha]\psi_\alpha + \sum_\beta V_{\alpha\beta}\psi_\beta = 0$, the channel kinetic energy is $E_\alpha = E_{\text{tot}} - E_0$ for total relativistic energy $E_{\text{tot}}^2 = (\hbar k c)^2 + E_0^2$ and rest energy $E_0 = m_0 c^2$.

The kinetic energy operator is in general $T_\alpha = -t_\alpha \nabla^2$ with coefficient

$$t_\alpha = (E_{\text{tot}} - E_0)/k^2 = E_0(\gamma - 1)/k^2 \tag{3.5.64}$$

for $k = m_0 \gamma v/\hbar$ with $\gamma = (1 - v^2/c^2)^{-1/2}$. In the non-relativistic limit, $t_\alpha \to \hbar^2/2\mu_0$.

The T-matrix integrals of Eq. (3.3.23) can be also written more generally, for elastic channel α_i, as

$$\mathbf{T}_{\alpha\alpha_i} = -\frac{1}{t_\alpha k_\alpha} \int \tilde{F}_\alpha \sum_\beta V_{\alpha\beta} \psi_\beta \mathrm{d}R, \tag{3.5.65}$$

from which the symmetric S-matrix form of Eq. (3.2.12) is

$$\tilde{\mathbf{S}}_{\alpha\alpha_i} = 2\mathrm{i} \sqrt{\frac{v_\alpha}{v_{\alpha_i}}} \mathbf{T}_{\alpha\alpha_i} \tag{3.5.66}$$

$$= -\frac{2\mathrm{i}}{\hbar \sqrt{v_\alpha v_{\alpha_i}}} \frac{\hbar v_\alpha}{t_\alpha k_\alpha} \int \tilde{F}_\alpha \sum_\beta V_{\alpha\beta} \psi_\beta \mathrm{d}R. \tag{3.5.67}$$

Non-relativistically, the quantity $w_\alpha \equiv \hbar v_\alpha/t_\alpha k_\alpha = 2$, the same for all channels. The general relativistic expression, however, is $w_\alpha = 1 + (1 - v_\alpha^2/c^2)^{1/2}$, namely $w_\alpha = 2$ for $v_\alpha \ll c$, but $w_\alpha = 1$ for $v_\alpha = c$ with photons. For a symmetric S matrix $\tilde{\mathbf{S}}$ we must have $w_\alpha V_{\alpha\beta} = w_\beta V_{\beta\alpha}$. This implies that the photon and particle coupling interactions of Eq. (3.5.35) indeed satisfy

$$V_{\gamma p} = 2 V_{p\gamma}, \tag{3.5.68}$$

in contrast to non-relativistic particle-particle couplings which must all be symmetric.

Exercises

3.1 Consider that the neutron interacting potential with a ^{4}He target can be approximated by a square well potential of depth $-40\,\text{MeV}$ and radius $1.5\,\text{fm}$. Calculate the scattering phase shift for $L = 0$ and 1, for energies below $5\,\text{MeV}$. Are there any resonances?

3.2 The experiment of ^{14}O on protons at $8\,\text{MeV/u}$ was performed to determine states in the unbound system ^{15}F. Take a Woods-Saxon potential with parameters $V_{ws} = -49\,\text{MeV}$,

$r_{ws} = 1.25\,\text{fm}$ and $a_{ws} = 0.75\,\text{fm}$ and a spin orbit force with $V_{so} = 4.5\,\text{MeV}$, $r_{so} = 1.25\,\text{fm}$ and $a_{so} = 0.75$.

  (a) Calculate the S matrix and the phase shifts for the $s_{1/2}$ and $d_{5/2}$ channels.

  (b) Compare the values for the energies and widths of the virtual state and resonance obtained through the poles of the S matrix with those obtained through the derivative of the phase shifts.

  (c) Compare your results with those extracted from the data [15].

3.3 Single neutron transfer (d,p) reactions on ^{12}C have been measured many times. There is also a number of data for the inverse reaction (p,d) on ^{13}C. Perform a data search and find two matching reactions for which you can check the validity of detailed balance.

References

[1] M. Abramowitz and I. A. Stegun 1964, *Handbook of Mathematical Functions*, Washington: National Bureau of Standards.

[2] E. Merzbacher 1997, *Quantum Mechanics*, New York: Wiley.

[3] A. R. Barnett, *Comp. Phys. Comm.* **27** (1982) 147.

[4] I. J. Thompson and A. R. Barnett, *Comp. Phys. Comm.* **36** (1985) 363.

[5] J. R. Taylor 1972, *Scattering Theory*, New York: Wiley.

[6] L. Eisenbud 1948, Ph.D. dissertation, Princeton University (unpublished); E. P. Wigner, *Phys. Rev.* **98** (1955) 145.

[7] R. G. Sachs, *Phys. Rev.* **74** (1948) 433.

[8] D. O. Riska, *Physica Scripta* **31** (1985) 471.

[9] L. E. Marcucci, M. Viviani, R. Schiavilla, A. Kievsky and S. Rosati, *Phys. Rev. C* **72** (2005) 0140001.

[10] J. Gómez-Camacho and R. C. Johnson 2002, 'Polarisation in nuclear reactions', Chapter 3.1.5 in: *Scattering*, eds. Roy Pike and Pierre Sabatier, New York: Academic Press.

[11] G. R. Satchler 1983, *Direct Nuclear Reactions*, Oxford: Clarendon.

[12] A. Bohr and B. R. Mottelson 1975, *Nuclear Structure, Vol. 1*, Reading, MA: Benjamin.

[13] F. Ajzenberg-Selove, *Nucl. Phys. A* **506** (1990) 1.

[14] M. L. Goldberger and K. M. Watson 1992, *Collision Theory*, New York: Dover.

[15] W. A. Peters *et al.*, *Phys. Rev. C* **68** (2003) 034607.

4

Reaction mechanisms

An expert is a man who has made all the mistakes which can be made in
a very narrow field.

Niels Bohr

This chapter shows how the interactions in and between channels may be calculated
on the basis of some *potential model* for a few interacting bodies. That is, a
Hamiltonian is defined whose matrix elements give rise to channel couplings,
also known as *transition potentials*. The parameters in this Hamiltonian may be
found either from structure models (Chapter 5), or from fitting data (Chapter 15).
It is also possible to directly fit to experiment the *effects* of these couplings on the
asymptotic amplitudes of the wave functions, and this is the basis of the R-matrix
phenomenology discussed in Chapter 10.

4.1 Optical potentials

Before we can discuss more detailed reaction mechanisms, we need to see the typical
kinds of potentials used for elastic scattering, and also the binding potentials needed
to reproduce the usual single-particle structures of nucleons within a nucleus. We
will start by describing the most commonly used optical potentials for elastic
scattering, expanding on the introduction in Box 3.4.

4.1.1 Typical forms

The interaction potential between a nucleon and a spherical nucleus is usually
described by an attractive nuclear well of depth V_r with a radius R_r slightly larger
than the nuclear matter radius, and a diffuse nuclear surface. Most commonly we

use the 'Saxon-Woods' shape[1]

$$V(R) = -\frac{V_r}{1 + \exp\left(\frac{R - R_r}{a_r}\right)}. \tag{4.1.1}$$

As summarized in Box 3.4, the central depth V_r is typically between 40 and 50 MeV, and the diffuseness a_r about 0.6 fm. The radius R_r is proportional to the size of the nucleus, and is commonly around $R_r = r_r A^{1/3}$ for a nucleus of A nucleons, with $r_r \approx 1.2$ fm. Similar potentials can be used for the interaction between two nuclei with mass numbers A_1 and A_2, if the radii are scaled instead as $R_r = r_r(A_1^{1/3} + A_2^{1/3})$, since this is proportional to the sum of the individual radii.

Charged particles experience an additional Coulomb potential. This is different at small radii from the point-Coulomb potential $V_c(R)$ of Eq. (3.1.70) because of the non-zero mean radius for the protons in the nuclei. If their charge Ze is uniformly distributed over a radius of $R_{\text{Coul}} = r_{\text{Coul}} A^{1/3}$ for a nucleus of A nucleons, then for an incident nucleon of charge $Z_p e$, the Coulomb contribution to the potential is

$$V_{\text{Coul}}(R) = Z_p Z e^2 \times \begin{cases} \left(\dfrac{3}{2} - \dfrac{R^2}{2R_{\text{Coul}}^2}\right) \dfrac{1}{R_{\text{Coul}}} & \text{for } R \leq R_{\text{Coul}} \\ \dfrac{1}{R} & \text{for } R \geq R_{\text{Coul}}. \end{cases} \tag{4.1.2}$$

The nuclear and Coulomb potentials are usually combined with an imaginary and a spin-orbit part. The imaginary part, which is present at higher scattering energies as discussed in subsection 3.1.5, is also often given by a Woods-Saxon form

$$W(R) = -\frac{V_i}{1 + \exp\left(\frac{R - R_i}{a_i}\right)} \tag{4.1.3}$$

for a similar geometry $R_i \gtrsim R_r$ and $a_i \approx a_r$, and a depth V_i fitted to experiments giving $V_i \sim 10{-}20$ MeV depending on energy. Sometimes a surface-peaked imaginary contribution is also included, with a shape like the derivative of Eq. (4.1.3).

All the parameters (depth, radii and diffusenesses) should come from some model, or from fitting elastic scattering angular distributions. Usually the radii for the imaginary parts will be slightly larger than the real radii, reflecting the absorption that occurs from direct reactions just at and outside the nuclear surface.

The real parts of optical potential generally get weaker with increasing laboratory energy, with $\partial V_r/\partial E \approx -0.3$ for low energies up to about 20 MeV, but less rapidly at higher energies. At around 300 MeV the real part passes through zero, and becomes repulsive at higher energies, where the scattering tends to be dominated by the imaginary part.

---

[1] Also called a 'Woods-Saxon' (WS) or 'Fermi' shape.

The depth V_r for neutrons is typically less than for protons, and this difference increases for targets with large neutron excess $N - Z$, as neutrons attract protons more than neutrons or protons attract each other. This effect may be parametrized in terms of the projectile isospin operator t_z as

$$V(R) = V_0(R) + \frac{1}{2}t_z\frac{N-Z}{A}V_T(R) \qquad (4.1.4)$$

for an attractive 'isoscalar' component $V_0(R)$ and a positive 'isovector' component $V_T(R)$. In this way, the central depths V_p for protons and V_n for neutrons differ by

$$V_p - V_n \approx 50\left(\frac{N-Z}{A}\right)\text{MeV}. \qquad (4.1.5)$$

There will also be *spin-orbit* forces that couple the nucleon spin to its orbital motion, of the kind to be discussed on page 139.

4.1.2 Global optical potentials

If experimental data for a projectile scattering on a range of nuclei A at many incident energies are simultaneously fitted by an optical potential with coefficients that vary slowly with target mass number and energy, then a *global optical potential* is found that will be useful for interpolation (and sometimes extrapolation) to new reactions and energies.

The best-known global potentials for the scattering of protons and neutrons on nuclei are listed in Box 4.1, and those for the scattering of deuterons, tritons, ^{3}He and alpha particles are listed in Box 4.2.

Tests have shown [2] that in general the reproduction of elastic scattering data is worse with global potentials than using the specifically fitted potentials that appear in compilations. It may be argued, however, that optical potentials should only ever attempt to describe some simplified *average* features of scattering, and that hence global potentials are to be preferred. Global potentials, for example, would not be perturbed by any localized resonance phenomena, or the possible existence of scaling errors in the data from a specific experiment. The theory of averaging needed to derive an optical potential will be discussed in Section 11.5.

4.1.3 Folding potentials

Sometimes, instead of specific or global potential, a density distribution of the target nucleus is known from electron scattering or from structure calculations. In this case, folding with, for example, the JLM potential [23] is one way of obtaining a nucleon-nucleus potential. The theory for folding will be given in Section 5.2.

The compilation of Perey and Perey [1] is a useful listing of individual optical potentials for the many elastic scattering reactions fitted up to the year 1974.

The RIPL-2 Handbook [2] contains a list of the most important global potentials for both incident neutrons and protons:

- Koning and Delaroche [3] ($Z = 12$–83, $A = 27$–209, $E = 0.001$–200 MeV), and
- Madland [4] ($Z = 6$–82, $A = 12$–208, $E = 50$–400 MeV).

Other older global potentials that cover both incident neutrons and protons are those of

- Becchetti and Greenlees [5] ($Z = 20$–92, $A = 40$–238, $E = 10$–50 MeV),
- Walter and Guss [6] ($Z = 26$–82, $A = 54$–208, $E = 10$–80 MeV), and
- Varner *et al.* [7] ($Z = 20$–83, $A = 40$–209, $E = 16$–65 MeV).

Older global potentials developed exclusively for incident neutrons are those of

- Moldauer [8] ($Z = 20$–83, $A = 40$–209, $E = 0.001$–5 MeV),
- Wilmore and Hodgson [9] ($Z = 20$–92, $A = 40$–238, $E = 0.01$–25 MeV),
- Engelbrecht and Friedeldey [10] ($Z = 20$–83, $A = 40$–210, $E = 0.001$–155 MeV),
- Strohmaier *et al.* [11] ($Z = 23$–41, $A = 50$–95, $E = 0.001$–30 MeV).

For incident protons alone, there are the potentials of

- Menet *et al.* [12] ($Z = 6$–82, $A = 12$–208, $E = 30$–60 MeV),
- Perey [13] ($Z = 16$–49, $A = 30$–100, $E = 0.01$–22 MeV), and
- Patterson *et al.* [14] ($Z = 20$–82, $A = 48$–208, $E = 25$–45 MeV).

Box 4.1 **Compilations and global optical potentials for nucleon-nucleus scattering**

4.2 Single-nucleon binding potentials

Before we can discuss the transfer of nucleons from one interacting nucleus to another, we need to specify how they are bound (or unbound) in their initial and final states. Nucleons are bound inside nuclei, and as a first approximation we can consider them bound by the average attraction of all the other nucleons. This average potential will be very similar to the optical potentials we used above for scattering, but only the central and spin-orbit parts, without the imaginary component(s). (We discuss below a possible significance for the imaginary parts.) The eigenstates in the average potential are called *single-particle levels* since they ignore higher-order interactions and correlations. We will use **r** to describe the coordinates of the nucleons *within* a nucleus, along with the previous **R** to describe the relative coordinate of two scattering nuclei, and the respective angular momenta will be ℓ and L. Many kinds of forces are equally useful in both circumstances.

For incident deuterons, the RIPL-2 compilation [2] lists the global potentials from

- Bojowald *et al.* [15] ($Z = 6$–82, $A = 12$–208, $E = 20$–$100\,\mathrm{MeV}$),
- Daehnick *et al.* [16] ($Z = 20$–82, $A = 40$–208, $E = 11$–$90\,\mathrm{MeV}$),
- Lohr and Haeberli [17] ($Z = 20$–83, $A = 40$–209, $E = 8$–$13\,\mathrm{MeV}$), and
- Perey and Perey [18] ($Z = 20$–82, $A = 40$–208, $E = 11$–$27\,\mathrm{MeV}$).

For incident tritons and ^{3}He particles, the global potentials of Becchetti and Greenlees [19] ($Z = 20$–82, $A = 40$–208, $E = 1$–$40\,\mathrm{MeV}$) are available.

In the case of incident alpha particles, there are global potentials from

- McFadden and Satchler [20] ($Z = 8$–82, $A = 16$–208, $E = 1$–$25\,\mathrm{MeV}$),
- Avrigeanu *et al.* [21] ($Z = 8$–96, $A = 16$–250, $E = 1$–$73\,\mathrm{MeV}$),
- Huizenga and Igo [22] ($Z = 10$–92, $A = 20$–235, $E = 1$–$46\,\mathrm{MeV}$), and
- Strohmaier *et al.* [11] ($Z = 20$–45, $A = 40$–100, $E = 1$–$30\,\mathrm{MeV}$).

Box 4.2 **Global optical potentials for scattering of deuterons, tritons, ^{3}He and alpha particles**

4.2.1 Neutron and proton single-particle states in nuclei

Let us take a typical combination of central and spin-orbit potentials. The central potentials are given as in Eq. (4.1.1), and details of the spin-orbit forces will be given in subsection 4.3.2. Figure 4.1 then shows the eigenenergies of single-particle levels for the fixed potential parameters given in the caption, plotting the results as a function of nuclear mass A, by scaling both the radii as $A^{1/3}$. The multiple levels depend on the quantum numbers for motion in the $\mathbf{r}$ coordinate: angular momentum ℓ, intrinsic nucleon spin $s = 1/2$, and total spin $j = l \pm s$. The levels can be labeled by nj^{π} for parity $\pi = (-1)^{\ell}$, or equivalently by $n\ell_{j}$,[2] and where $n = 1, 2, \ldots$ is the number of nodes in the radial wave function.[3] The individual nucleons have wave functions $\phi^{m}_{\ell sj;b}(\mathbf{r})$ for the bound state b, namely angular and radial factors as

$$\phi^{m}_{\ell sj;b}(\mathbf{r}) = [Y_{\ell}(\hat{\mathbf{r}}) \otimes X_{s}]_{jm}\, u_{\ell sj;b}(r)/r, \qquad (4.2.4)$$

with X_{s} the state of the nucleon's intrinsic spin. They have boundary conditions $u_{\ell sj;b}(0) = 0$ at the origin, and go to zero as

$$u_{\ell sj;b}(r) =_{r>R_{n}} C_{\ell} W_{-\eta,\ell+\frac{1}{2}}(-2k_{I}r) \to_{r \gg \rho_{\mathrm{tp}}} C_{\ell}\mathrm{e}^{-k_{I}r} \to_{r \to \infty} 0 \qquad (4.2.5)$$

---

[2] Following atomic spectroscopy, we use names *s, p, d, f, g* etc., for $\ell = 0, 1, 2, 3, 4$ respectively. The lowest three levels in Fig. 4.1 are then equivalently named $1s_{1/2}$, $1p_{3/2}$ and $1p_{1/2}$.

[3] Here we include the node at the origin, so start counting with $n = 1$. Another convention is to include only nodes at $r > 0$, and therefore to start at $n = 0$.

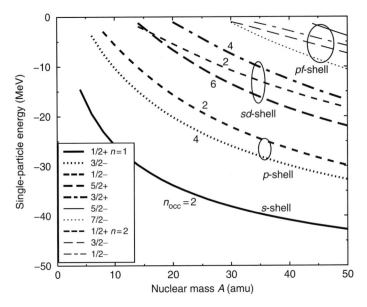

Fig. 4.1. Neutron eigenstates nj^π in a Saxon-Woods binding potential, for spin j, parity π and number of radial nodes n. The chosen potential has $V_r = 56.6\,\text{MeV}$, $r_r = 1.17\,\text{fm}$, $a_r = 0.75\,\text{fm}$, $V_{so} = 6.2\,\text{MeV}$, $r_{so} = 1.01\,\text{fm}$ and $a_{so} = 0.58\,\text{fm}$. Both radii scale with A as $R = rA^{1/3}$. The occupation numbers are $n_{\text{occ}} = 2j+1$.

outside the maximum radius R_n of the nuclear potentials, for some constant C_ℓ. The $W_{\kappa,\mu}(z)$ are the Whittaker functions defined in Box 4.3. The bound states are normalized as

$$\int |\phi^m_{\ell sj;b}(\mathbf{r})|^2 d\mathbf{r} = \int_0^\infty |u_{\ell sj;b}(r)|^2 dr = 1 \qquad (4.2.6)$$

for all magnetic substates m.

We see that the lowest level is for $\ell = 0$, labeled $1/2^+$ (or $s_{1/2}$), followed by a closely-spaced pair of $\ell = 1$ states $3/2^-$ and $1/2^-$. This pair is called the *p*-shell. Higher in energy is the *sd*-shell with three levels $5/2^+$, $1/2^+$ and $3/2^+$, and in heavier nuclei there is a *pf*-shell higher again. The second $1/2^+$ level, in the *sd*-shell, will have an additional radial node in its wave function, compared with the first such level in the *s*-shell. Single-particle levels for protons will be similar to these, but with a modified depth according to Eq. (4.1.4), and with their levels raised in energy by the average Coulomb repulsion.

The number N of neutrons and Z of protons to *occupy* these single-particle levels depends on the isotope. The maximum occupation numbers for each level are the values $n_{\text{occ}} = 2j + 1$ in Fig. 4.1. A ^{16}O nucleus, for example, has $N = Z = 8$, so each species of nucleon fills its *s*- and *p*-shells, if we adopt the simplest filling order from lowest to highest energies, and neglect changes caused by more complicated

---

Whittaker functions

Bound states are solutions of a Schrödinger equation for negative energies E that correspond to wave numbers $k = \sqrt{-2\mu E/\hbar^2}$ that are positive imaginary (see Fig. 3.4). That is, bound states have $k = \mathrm{i}k_I$ with imaginary part $k_I > 0$, and have Sommerfeld parameter of Eq. (3.1.71) that is also imaginary: $\eta_I = -Z_1 Z_2 e^2 \mu/(\hbar^2 k_I)$.

The Coulomb function needed for bound states is the outgoing function $H_L^+(\eta, \rho)$, as bound wave functions can be matched to this form everywhere outside the nuclear potentials. This H^+ function is not suitable for all bound states however, as it is not everywhere defined, such as whenever $1 + L + \mathrm{i}\eta = 0$ (this occurs for bound states in a purely Coulomb attractive well).

It is therefore common to use a *Whittaker function* that is H^+ rescaled as

$$W_{\mathrm{i}\eta,\ell+\frac{1}{2}}(2\mathrm{i}\rho) = \mathrm{e}^{-\pi\eta/2}\mathrm{e}^{-\mathrm{i}(\pi\ell/2+\sigma_L(\eta))}H_\ell^+(\eta,\rho). \tag{4.2.7}$$

Such a Whittaker function, from Eq. (3.1.69), has asymptotic form

$$W_{\mathrm{i}\eta,\ell+\frac{1}{2}}(2\mathrm{i}\rho) \to_{\rho \gg \rho_{\mathrm{tp}}} \mathrm{e}^{\mathrm{i}\rho-\pi\eta/2-\mathrm{i}\eta\ln(2\rho)}, \tag{4.2.8}$$

and, for bound states with $\rho = \mathrm{i}k_I r$, this becomes

$$W_{-\eta_I,\ell+\frac{1}{2}}(-2k_I r) \to_{\rho \gg \rho_{\mathrm{tp}}} \mathrm{e}^{-k_I r+\eta_I \ln(2k_I r)}. \tag{4.2.9}$$

For neutral particles this is the familiar exponential decay, and for charged particles the Whittaker function has a logarithmic variation arising from the Coulomb potential.

Box 4.3 **Coulomb functions for bound states**

mixing and correlations. We often use this single-particle filling prescription as a first approximation.

Nucleons occupying states in a nucleus can be seen in *pickup* reactions that remove one of the low-energy nucleons. It will leave a hole, which, if the hole is below other occupied levels, amounts to an excitation energy of the residual nucleus.

The energy of the highest occupied level is called the *Fermi energy* E_F, and levels above this are unoccupied. These unoccupied levels can be seen in *transfer* reactions that bring another nucleon into the nucleus.[4] If a newly occupied level after a transfer reaction is above another unoccupied level, then it also represents an excited state in the residual nucleus.

In nature, nucleon bound states will never be exactly at the single-particle levels shown in the figure, because of higher-order effects which go beyond the spherical potential model. Single-particle levels may be shifted, and also *fragmented* among

---

[4] Sometimes these are called 'stripping' reactions, but that word is often used for pickup reactions as well.

all the true eigenstates of the *A*-body system. The effects of this fragmentation will be described by *overlap functions*, the square norms of which are the *spectroscopic factors* (to be defined later), which will in general be different from unity.

To accommodate the energy shifts, it is normal to fine-tune the precise strength V_r of the central potential to reproduce the experimental binding energies. This guarantees the correct exponential tail of the wave function (necessary, we will see, for sub-Coulomb transfer reactions), and also the post-prior equivalence that should be obtained for first-order DWBA transfer cross sections.

When the binding potentials are too weak to bind a given single-particle level, then it will become unbound, and while it is still not too far above threshold it will be visible as a resonance or a virtual state. As discussed on page 66, states with $\ell = 0$ and no Coulomb barrier will become virtual states, and the others will become resonances. We generally assume that the potentials for the continuum are the same as for bound states, apart from the energy dependence mentioned in page 131.

In between resonances there are still scattering states, called the *non-resonant continuum*, since a Hamiltonian has eigenstates at *every positive* scattering energy $E = \hbar^2 k^2 / 2\mu$, as described in the previous chapter. We may label the corresponding wave functions by $\phi_{\ell sj}(\mathbf{r}; k)$, or equivalently using E rather than k. The wave functions in the one-channel case are factorizable as

$$\phi_{\ell sj}^m(\mathbf{r}; k) = [Y_\ell(\hat{\mathbf{r}}) \otimes X_s]_{jm} \, u_{\ell sj}(r; k)/r, \qquad (4.2.10)$$

where the $u_{\ell sj}(r; k)$ are externally normalized as usual for scattering wave functions: proportionally to $\frac{i}{2}[H^- - \mathbf{S}H^+]$. This asymptotic normalization gives the overlap integral of two radial functions as

$$\int_0^\infty u_{\ell sj}(r; k)^* u_{\ell sj}(r; k') \mathrm{d}r = \frac{\pi}{2} \delta(k - k'). \qquad (4.2.11)$$

It is therefore common to renormalize continuum states using

$$\hat{u}_{\ell sj}(r; k) = \sqrt{2/\pi} \, u_{\ell sj}(r; k) \qquad (4.2.12)$$

so that we now have an orthonormal set

$$\int_0^\infty \hat{u}_{\ell sj}(r; k)^* \hat{u}_{\ell sj}(r; k') \mathrm{d}r = \delta(k - k'). \qquad (4.2.13)$$

Other renormalizations are possible to give δ-functions in energy E. Using the $\hat{u}(r)$ wave functions, we can give the completeness relation when all the bound and scattering states are combined:

$$\sum_b |u_{\ell sj;b}(r)\rangle \langle u_{\ell sj;b}(r')| + \int_0^\infty \mathrm{d}k \, |\hat{u}_{\ell sj}(r; k)\rangle \langle \hat{u}_{\ell sj}(r'; k)| = \delta(r - r'), \qquad (4.2.14)$$

for each partial wave ℓsj.

Resonances, in this basis, are just concentrations of strength at particular energies, as measured by the magnitudes of $u_{\ell s j}(r; k)$ in the nuclear interior, so resonances should *not* be included as explicit terms in addition. A sum over resonances can only be given, Berggren [24] explains, if we deform the complex contour of the k integration in Eq. (4.2.14) so that it goes *below* some of the narrower resonances that are near the real k axis.

4.2.2 Optical potentials extended to bound states

In the previous section, we described the binding potentials giving all the single-particle levels for occupation in a simple uncorrelated structure model of the nucleus. Consider now a *full* nucleus, where all the levels up to the Fermi energy are occupied. The occupied states *above* the Fermi energy are candidates for further nucleons that can be transferred *in*, and the occupied states *below* the Fermi energy are candidates for where nucleons can be picked *out* in a removal reaction. That is, in this 'full nucleus' case, we interpret single-particle states below the Fermi level to refer to nuclear states in the $A - 1$ system as specific holes in the A-nucleon state, and interpret single-particle states with $E > E_F$ to refer to nuclear states in the $A + 1$ system as specific particles added to the A-nucleon state.

Holes or transfer states that are away from the Fermi level, and hence describe excited nuclei, will of course decay with some decay lifetime τ. This may occur by γ emission, or by ejection of other nucleons in the system, and the finite lifetime gives a *spreading width* to the excited state by $\Gamma \approx \hbar/\tau$, as for a resonance. It is called a 'spreading width' because the initial single-particle excitation is spreading into other modes of nuclear excitation.

This decay lifetime may also be described by an imaginary potential $W(R) < 0$, according to the factor $\mathrm{e}^{-2|W||t|/\hbar}$ in Eq. (3.1.109). In this way, therefore, we may extend the complex optical potential to bound states at negative energies, to describe the loss of flux from excited nucleonic states to more general nuclear excitations, such as statistical compound-nucleus modes. We choose the optical potential to give the correct decay times for excited states away from the Fermi level. Clearly this extended optical potential will have an imaginary part which goes to zero at and near the Fermi energy, and will have an absorptive component that will grow as $|E - E_F|$ increases.

4.3 Coupling potentials

Now that we have defined central potentials for scattering and bound states between nucleons and nuclei, we are able to give details of the *off-diagonal* potentials that couple together different partial waves within a multi-channel set. These coupling potentials describe processes that usually transfer angular momentum and energy from one interacting body to another, so we begin with an analysis of their multipole structure.

4.3.1 Multipole analysis of transition potentials

We are here concerned with an interaction part H_{intr} of the Hamiltonian that is responsible for the coupling between distinct states of a nucleus, and its orbital motion, as the nucleus interacts with another. The Hamiltonian itself is a scalar, which may be constructed as a sum of transition operators of multipoles λ for both the nucleus and its spatial orbit. For each λ value, let the orbital angular transition operator be the spherical harmonic $Y_\lambda^m(\hat{\mathbf{R}})$ and the nuclear transition operator be $T_\lambda(\xi)$ for the necessary internal coordinates ξ of the nucleus. The λ-multipole part of H_{intr} is the scalar product of these:

$$H_{\text{intr}}^\lambda(\xi, \mathbf{R}) = \sqrt{4\pi}\, \mathcal{F}_\lambda(R) \sum_{m=-\lambda}^{+\lambda} T_\lambda^m(\xi)^* Y_\lambda^m(\hat{\mathbf{R}})$$

$$\equiv \sqrt{4\pi}\, \mathcal{F}_\lambda(R)\, T_\lambda(\xi) \cdot Y_\lambda(\hat{\mathbf{R}}), \qquad (4.3.1)$$

where $\sqrt{4\pi}\mathcal{F}_\lambda(R)$ expresses the as-yet-unknown radial dependence. The $\sqrt{4\pi}$ is a constant chosen so that a monopole equation gives $H_{\text{intr}}^0(\xi, \mathbf{R}) = \mathcal{F}_0(R)$ when we adopt a normalization of $T_0 = 1$.

The quantity λ is called the *multipolarity of the reaction*, or *angular momentum transfer*. The $\lambda = 0$ case describes *monopole* (scalar) transitions, which are either diagonal (like the optical potential), or off-diagonal couplings between states of the same spin. The $\lambda = 1$ describes *dipole* (vector) transitions and 2 the *quadrupole* (tensor) processes.

Most often, a given reaction mechanism changes the spin state of only one of the interacting pair of nuclei, the spin of the other nucleus not being dynamically coupled. In this simpler case, the J and S coupling schemes of subsection 3.2.1 both reduce to $|(LI)J_{\text{tot}}\rangle$. In this basis, the R-dependent matrix element of any coupling Hamiltonian (4.3.1) may be found by integrating over all variables except R, obtaining

$$\langle (L_f I_f) J_{\text{tot}} | H_{\text{intr}}^\lambda(\xi, \mathbf{R}) | (L_i I_i) J_{\text{tot}} \rangle = \sqrt{4\pi}\, \mathcal{F}_\lambda(R)$$

$$\times (-1)^{\lambda + J_{\text{tot}} + L_i + I_f} \begin{Bmatrix} L_i & I_i & J_{\text{tot}} \\ I_f & L_f & \lambda \end{Bmatrix} \langle L_f \parallel Y_\lambda \parallel L_i \rangle \langle I_f \parallel T_\lambda \parallel I_i \rangle \qquad (4.3.2)$$

by Eq. (1A-72) of Bohr and Mottelson [25]. The reduced matrix element of the spherical harmonic, with $\hat{x} \equiv \sqrt{2x+1}$, is

$$\langle L_f \parallel Y_\lambda \parallel L_i \rangle = (4\pi)^{-1/2}\hat{\lambda}\hat{L}_i \langle L_i 0, \lambda 0 | L_f 0 \rangle \qquad (4.3.3)$$

when using the Wigner-Eckhart theorem with the normalization of [25]:[5]

$$\langle j_f m_f | \hat{O}_{jm} | j_i m_i \rangle = \frac{\langle j_i m_i, jm | j_f m_f \rangle}{\hat{j}_f} \langle j_f || \hat{O}_j || j_i \rangle. \qquad (4.3.4)$$

We now define a *transition potential* to include the nuclear reduced matrix element for a specific pair of states:

$$V_{fi}^\lambda(R) = \mathcal{F}_\lambda(R) \langle I_f \parallel \mathcal{T}_\lambda \parallel I_i \rangle \qquad (4.3.5)$$

for some $\mathcal{F}_\lambda(R)$ called a *form factor*, so we henceforth use the general form

$$\langle (L_f I_f) J_{\text{tot}} | H_{\text{intr}}^\lambda(\xi, \mathbf{R}) | (L_i I_i) J_{\text{tot}} \rangle$$

$$= V_{fi}^\lambda(R)(-1)^{\lambda + J_{\text{tot}} + L_i + I_f} \hat{\lambda} \hat{L}_i \langle L_i 0, \lambda 0 | L_f 0 \rangle \begin{Bmatrix} L_i & I_i & J_{\text{tot}} \\ I_f & L_f & \lambda \end{Bmatrix} \qquad (4.3.6)$$

$$\equiv V_{fi}^\lambda(R)(-1)^{\lambda + J_{\text{tot}} - I_i - L_f} \hat{\lambda} \hat{L}_i \langle L_i 0, \lambda 0 | L_f 0 \rangle W(L_i I_i L_f I_f; J_{\text{tot}} \lambda).$$

in terms of the 6J or Racah coefficients W of angular-momentum theory. The total coupling between two states is the sum over all multipoles of the terms in Eq. (4.3.6).

When there are spins in the set of coupled angular momenta that are *spectators*, in the sense of not participating in the dynamical transitions, rearrangements of the coupling order must be used in order to isolate those spins actually involved. Expressions are then obtained that are similar to Eqs. (4.3.6), but are more complicated, and contain sums over additional intermediate quantum numbers.

In later sections, we calculate the transition potentials $V_{fi}^\lambda(R)$ using the $\mathcal{F}_\lambda(R)$ form factors that may be common to several states i, f. These functions of the internuclear separation R are determined by the physical reaction mechanism, and will be different for different target nuclei, and different for single-particle and collective processes. For transfer reactions we will find that the form factors will be non-local.

4.3.2 Spin-dependent potentials

The monopole diagonal potential $\lambda = 0$ has the simplest form $\mathcal{F}_0(R)$, and could be one of the folded or optical potentials that we have used to describe the interaction of two nuclei before any non-elastic reactions occur. Typical optical potentials have already been discussed, and folded potentials will be described in Section 5.2.

When nuclei are interacting that have non-zero spins, non-zero multipole components may have an effect even if the nuclei stay in their ground state.

---

[5] Note that reduced matrix elements are defined differently from this in [26].

Typically, $\lambda = 1$ *vector* potentials and $\lambda = 2$ *tensor* potentials may be present, up to $\lambda \leq 2s$, which is twice the modulus of the sum $\mathbf{s} = \sum_i \mathbf{s}_i$ of the spins $\mathbf{s}_i$ of the interacting bodies.[6] Individual nucleons have spins 1/2, so $\lambda = 1$ multipoles may contribute to the scattering of nucleons on any other body. Pairs of nucleons may have total spin $s = 0$ or 1. The deuteron ground state has $s = 1$, for example, and in general $\lambda = 2$ tensor forces may operate during the scattering of any two nucleons in an $s = 1$ state. Neither of these vector or tensor forces changes the energy eigenstate of a particle or nucleus.

Vector forces

The most common vector potential is the *spin-orbit force*. This $\lambda = 1$ force can couple a single $s = 1/2$ nucleon with its orbital motion with $\mathcal{T}_1(\xi) = \mathbf{s}$, the spin operator itself. This gives

$$V_{so} = \mathcal{F}_1^{so}(R)\, 2\mathbf{L}\cdot\mathbf{s} \tag{4.3.7}$$

of which the matrix elements may be calculated in the J basis $\mathbf{J} = \mathbf{L} + \mathbf{s}$ as

$$\langle (Ls)J|2\mathbf{L}\cdot\mathbf{s}|(Ls)J \rangle = J(J{+}1) - L(L{+}1) - s(s{+}1)$$

$$= \begin{cases} +L & \text{for } J = L{+}1/2 \\ -L{-}1 & \text{for } J = L{-}1/2, \end{cases} \tag{4.3.8}$$

so only the diagonal $\langle (Ls)J|V_{so}|(Ls)J \rangle = \mathcal{F}_1^{so}(R)[J(J{+}1) - L(L{+}1) - s(s{+}1)]$ elements are non-zero. In the S basis of Eq. (3.2.2), the spin-orbit force gives off-diagonal couplings.[7]

In comparison with electrons in atoms, for which the spin-orbit potential is (Thomas [27])

$$\mathcal{F}_e^{so}(R) = -\frac{1}{2m_e^2 c^2} \frac{1}{R} \frac{dV(R)}{dR} \tag{4.3.9}$$

in terms of the central (electrostatic) potential $V(R)$, the nuclear spin-orbit force is much stronger. If this formula is used directly for the nuclear spin-orbit forces, it is found to be too weak, requiring an amplification factor commonly taken as 25.

In practice, negative spin-orbit form factors $\mathcal{F}_1^{so}(R)$ are still taken as having this 'Thomas form' as derivatives of the form of a nuclear central potential, but with a scaling factor that is somewhat arbitrarily fixed in terms of the pion mass m_π:

$$\mathcal{F}_1^{so}(R) = \left(\frac{\hbar}{m_\pi c}\right)^2 \frac{1}{R} \frac{d}{dR} \frac{V_{so}}{1 + \exp(\frac{R-R_{so}}{a_{so}})}, \tag{4.3.10}$$

---

[6] We prefer to use s for the spin of a single nucleon or a very light cluster, and I for the spin of a composite nucleus.
[7] This is one small but useful advantage of the J-basis formulation.

so $\hbar^2/(m_\pi c)^2 = 2.00\,\text{fm}^2$, and $\mathcal{F}_1^{\text{so}}(R)$ has the same units as V_{so}. This requires a potential factor of the order of $V_{so} = 5$–$8\,\text{MeV}$ for nucleons.[8]

Another simple spin-dependent force possible with two nuclei both with spins (I_p and I_t) is the *spin-spin force* $V_{ss} = \mathcal{F}_1^{\text{ss}}(R)\mathbf{I}_p \cdot \mathbf{I}_t$. In the S (channel spin) basis of Eq. (3.2.2) the spin-spin force is diagonal, with matrix elements

$$\langle L(I_p, I_t)S; J_{\text{tot}} | V_{ss} | L(I_p, I_t)S; J_{\text{tot}} \rangle$$
$$= \mathcal{F}_1^{\text{ss}}(R)\tfrac{1}{2}[S(S+1) - I_p(I_p+1) - I_t(I_t+1)]. \qquad (4.3.11)$$

In the J basis (Eq. 3.2.1) there are off-diagonal couplings from the spin-spin force.

Tensor forces

An $s = 1$ cluster can be influenced in its orbital motion by a $\lambda = 2$ (rank two) tensor force, as can the motion between two $s = 1/2$ nucleons when they are together in an $S = 1$ (triplet) state. These tensor forces do couple together different $L \neq L'$, in contrast to vector forces (rank-one tensors).

For an object of spin s, Satchler [28] showed that there are three generic kinds of rank-two operators that could be used for tensor forces. These couple the spin with either the radius, momentum or orbital angular-momentum vectors, yielding forms conventionally written as

$$T_r = (\mathbf{s} \cdot \hat{\mathbf{R}})^2 - \tfrac{2}{3} \qquad (4.3.12)$$

$$T_p = (\mathbf{s} \cdot \mathbf{p})^2 - \tfrac{2}{3}p^2 \qquad (4.3.13)$$

$$T_L = (\mathbf{s} \cdot \mathbf{L})^2 - \tfrac{1}{2}\mathbf{S} \cdot \mathbf{L} - \tfrac{2}{3}\mathbf{L}^2. \qquad (4.3.14)$$

where $\hat{\mathbf{R}}$ is the unit vector in the direction of $\mathbf{R}$. All of these quadrupole forms need to be combined with some radial form factor $\mathcal{F}_2(R)$, which may be complex.

4.4 Inelastic couplings

An inelastic reaction is one where a nucleus changes its energy eigenstate, from state i to state f, when interacting with another nucleus. The inelastically excited nucleus changes its energy from ϵ_i to ϵ_f, determined according to the eigen-equations of Eq. (3.2.39). The energy Q-value for the reaction is therefore $Q = \epsilon_i - \epsilon_f$, as this is the energy released, and is only positive if the nucleus is de-excited with $\epsilon_f < \epsilon_i$.

An inelastic reaction will usually change the spin of the nucleus, by some transfer of angular momentum that is the multipolarity λ of the reaction. This angular

---

[8] Note that sometimes spin-orbit forces are defined with $\mathbf{L} \cdot \mathbf{s}$ in place of the $2\mathbf{L} \cdot \mathbf{s}$ in Eq. (4.3.7), and/or sometimes without the $\hbar^2/(m_\pi c)^2$, and these redefine the numerical strength factor V_{so}.

momentum is usually transferred from the orbital angular momentum of relative motion, but sometimes directly from the other interacting nucleus. If the initial and final nuclear spins are I_i and I_f respectively, then there holds a triangular relationship $\Delta(\lambda, I_i, I_f)$, or

$$|I_i - I_f| \leq \lambda \leq I_i + I_f. \tag{4.4.1}$$

Thus monopole transitions keep $I_i = I_f$, and dipole transitions ($\lambda=1$) change I_i by 1 at most.

Normal parity transitions change the nuclear parity only for odd λ multipoles, and all transitions caused by central forces are *normal* in this sense. Other 'non-normal' transitions may occur when spin-dependent forces change nuclear spin orientations during a reaction, but these are generally weaker processes for relative velocities $v \ll c$ (the speed of light).

Inelastic transitions in a nucleus are caused by the interaction with a second nucleus, which may act by electromagnetic and/or nuclear forces. The nuclear forces may be central, or may couple to the spins of individual nucleons. The electromagnetic forces are almost entirely electric at low and medium energies, and are therefore principally the Coulomb forces which couple to the charges of the protons.

If Coulomb forces contribute to the inelastic transition, then for R outside the radii R_n of the nuclear potential, the inelastic form factors have characteristic forms as inverse powers of $R^{\lambda+1}$. We will in Eq. (4.4.25) be defining a *Coulomb reduced matrix element* $\langle I_f \parallel E\lambda \parallel I_i \rangle$, in terms of which

$$V_{fi}^\lambda(R) \overset{R \gg R_n}{\longrightarrow} \frac{\langle I_f || E\lambda || I_i \rangle}{R^{\lambda+1}} \frac{\sqrt{4\pi} e^2 Z_t}{2\lambda+1}, \tag{4.4.2}$$

where we factorize out $Z_t e$, the charge of the second nucleus.

The quantity

$$B(E\lambda, I_i \rightarrow I_f) = \frac{1}{2I_i+1} |\langle I_f \parallel E\lambda \parallel I_i \rangle|^2 \tag{4.4.3}$$

is called the *reduced transition probability*, and the rate for peripheral reactions depends only on $B(E\lambda, I_i \rightarrow I_f)$, and no other structural property. These may be low-energy photo-nuclear reactions, gamma decay processes, or very forward angle (or low-energy) Coulomb excitation reactions. The value of $B(E\lambda)$ is thus sometimes a good contact point between theory and experiment, at least for Coulomb-dominated processes.

The reduced matrix element, as defined, will be symmetric,

$$\langle I_f \parallel E\lambda \parallel I_i \rangle = \langle I_i \parallel E\lambda \parallel I_f \rangle, \tag{4.4.4}$$

so

$$(2I_i+1) \, B(E\lambda, I_i \to I_f) = (2I_f+1) \, B(E\lambda, I_f \to I_i), \tag{4.4.5}$$

and the 'up' and 'down' $B(E\lambda)$ values will be different (except when $I_i = I_f$).

To find $V_{fi}^\lambda(R)$ at smaller radii, we need a potential model for the cause of the reaction. This model could be a rotational or vibrational collective model, a single-particle excitation model, or a more microscopic calculation of the many-body nuclear states and how they interact with the second nucleus. The collective and single-particle models are now described in turn, while the use of microscopic models is discussed in Chapter 5.

4.4.1 Collective inelastic processes

The most common collective models for nuclei consider the rotation of a deformed nuclear shape, or the deformational vibration of a nucleus that is initially spherical.

Rotational inelastic excitation

The simplest rotational model defines a deformed surface of a nucleus, and constructs the nuclear interaction potential as a function of the distance to the surface along a radial line. This gives natural parity transitions among members of a given *rotational band* of a nucleus. These bands are a set of states of spin I that start from some *bandhead $I = K$*, and have excitation energies typically like

$$\epsilon_I = \frac{\hbar^2}{2\mathcal{M}} \left[I(I+1) - K(K+1) \right], \tag{4.4.6}$$

where $\mathcal{M}$ is the moment of inertia of the nucleus. The moment $\mathcal{M}$ will be constant for the rotational band of a rigid rotor, with value

$$\mathcal{M}_{\text{rigid}} = \frac{2}{5} m_u A R_0^2 \tag{4.4.7}$$

for average surface radius R_0 and m_u the unit atomic mass (amu). The value of K is the projection of the intrinsic nuclear angular momentum on the rotational axis, and hence the smallest possible value of rotating spin I. Rotational bands with $K = 0$ will consist of the set of even levels $I = 0, 2, 4, \ldots$, whereas bands for $K > 0$ will have all levels $I = K, K+1, K+2$, etc.

A deformed nucleus is considered to have an *intrinsic state* in a *body-fixed frame of reference* (r', θ', ϕ') that can be defined by placing the z' axis along some axis of deformation symmetry. The varying radius $\tilde{R}$ of the surface of a deformed nucleus can always be expanded using spherical harmonics in terms of polar angles θ' and ϕ' in the body-fixed frame as

$$\tilde{R}(\theta', \phi') = R_0 + \sum_{q=2}^{q_{max}} \sum_{\mu=-q}^{+q} d_{q\mu} Y_q^\mu(\theta', \phi'), \qquad (4.4.8)$$

where q_{max} is sufficiently large for the desired accuracy. We are only interested in $q \geq 2$ as $q = 0$ can be included in R_0, and $q = 1$ moves the center of mass. Requiring $\tilde{R}(\theta', \phi')$ real means that $d_{q\mu} = (-1)^\mu d_{q-\mu}$, so not all the $d_{q\mu}$ are independent. The R_0 is the average surface radius of the potential.

The most common $d_{q\mu}$ is $q = 2$ for quadrupole deformations. Axially deformed nuclei only have a_{20} non-zero, whereas a triaxially deformed nucleus has $a_{22} = a_{2-2}$ also non-zero. The $d_{q\mu}$ have units of length, and d_{q0} is usually called the *deformation length* δ_q. The fractional deformation is

$$\beta_q = \delta_q / R_0. \qquad (4.4.9)$$

A prolate deformed nucleus has $\beta_2 > 0$, and oblate deformation occurs with $\beta_2 < 0$.

In the simplest rotational model for a deformed surface, we describe the interaction V with the second nucleus as depending on the distance to the surface as in Fig. 4.2:

$$V(R, \theta', \phi') = U(R - \tilde{R}(\theta', \phi') + R_0) \qquad (4.4.10)$$

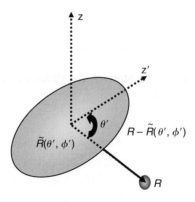

Fig. 4.2. A nucleus is rotating about the **z** laboratory axis, and is itself a deformed object with axis of symmetry along the **z**′ axis that continuously varies with respect to **z**. The interaction potential of another body at R with a deformed object of radius $\tilde{R}(\theta', \phi')$ may depend on the radial distance $R - \tilde{R}(\theta', \phi')$ to its surface.

for some suitable nuclear potential function $U(R)$ such as an optical potential. The first step is to expand the potential of (4.4.10) into a sum of tensor components like Eq. (4.3.1). A simple approximate procedure is to expand $U(R - \tilde{R}(\theta', \phi') + R_0)$ to first order in the $d_{q\mu}$, which should be satisfactory for small deformations. This gives

$$V(R, \theta', \phi') \approx U(R) - U'(R) \sum_{q\mu} d_{q\mu} Y_q^\mu(\theta', \phi'), \qquad (4.4.11)$$

where the first term $U(R)$ is the diagonal optical potential.

For an axially-deformed rotator, the first-order result gives multipoles $\lambda = q$:

$$V(R, \theta', \phi') = U(R) - U'(R) \sum_{\lambda=2}^{q_{\max}} \delta_\lambda Y_\lambda^0(\theta', \phi'). \qquad (4.4.12)$$

Now (θ', ϕ') are the angles between the laboratory vector $\hat{\mathbf{R}}$ and the axis $\hat{\mathbf{z}}'$ of the body-fixed coordinate system, so (as $\mu = 0$ here)

$$Y_\lambda^0(\theta', \phi') = \frac{\hat{\lambda}}{\sqrt{4\pi}} P_\lambda(\cos\theta') = \frac{\hat{\lambda}}{\sqrt{4\pi}} \frac{4\pi}{2\lambda+1} \sum_m Y_\lambda^m(\hat{\xi}) Y_\lambda^m(\hat{\mathbf{R}})^*, \qquad (4.4.13)$$

so the dependence of the potential V is now

$$V(R, \hat{\xi}, \hat{\mathbf{R}}) = U(R) - U'(R) \sum_\lambda \delta_\lambda \frac{\sqrt{4\pi}}{\hat{\lambda}} \sum_m Y_\lambda^m(\hat{\xi}) Y_\lambda^m(\hat{\mathbf{R}})^*. \qquad (4.4.14)$$

This interaction can change the state of the deformed nucleus by the non-zero matrix elements $\langle \phi_{I_f} | V(R, \hat{\xi}, \hat{\mathbf{R}}) | \phi_{I_i} \rangle$. Here, I_i is the initial spin and I_f the final spin of the rotating nucleus. We may construct the rotationally excited states IM as the intrinsic state ϕ_K operated on by a rotation matrix $D_{MK}^I(\omega)$ where ω is the set of Euler angles that transforms (θ', ϕ') in the body-fixed frame to (θ, ϕ) in the laboratory frame. From the normalization property of the $D_{MK}^I(\omega)$, a normalized state of such a rotor is

$$\phi_{IM} = \frac{\hat{I}}{\sqrt{8\pi^2}} D_{MK}^I(\omega)^* \phi_K. \qquad (4.4.15)$$

We are now able to calculate the matrix elements of the potential (4.4.14) between rotational states (4.4.15), using

$$\langle I_f M_f | Y_\lambda^m(\hat{\xi})^* | I_i M_i \rangle$$

$$= \frac{\hat{I}_i \hat{I}_f}{8\pi^2} \int d\omega \langle \phi_K | D_{M_f K}^{I_f}(\omega) Y_\lambda^{m*} D_{M_i K}^{I_i}(\omega)^* | \phi_K \rangle$$

$$= \frac{\hat{I}_i \hat{I}_f}{8\pi^2} \frac{\hat{\lambda}}{\sqrt{4\pi}} \int d\omega D_{M_f K}^{I_f}(\omega) D_{-m0}^\lambda(\omega) D_{-M_i -K}^{I_i}(\omega) \ (-1)^{K-m-M_i}$$

$$= \frac{\hat{\lambda}}{\sqrt{4\pi}} \hat{I}_i \langle I_i K, \lambda 0 | I_f K \rangle \hat{I}_f^{-1} \langle I_i M_i, \lambda m | I_f M_f \rangle, \qquad (4.4.16)$$

since $Y_\lambda^m = \hat{\lambda}/\sqrt{4\pi} D_{m0}^\lambda(\omega)^*$. Using the adopted definition of a reduced matrix element, we find that the integral over the core angle gives the factor

$$\langle I_f \parallel Y_\lambda \parallel I_i \rangle = \frac{\hat{\lambda}}{\sqrt{4\pi}} \hat{I}_i \langle I_i K, \lambda 0 | I_f K \rangle. \qquad (4.4.17)$$

These reduced matrix elements give a transition potential in the rotational model of

$$V_{fi}^\lambda(R) = -\frac{\delta_\lambda}{\sqrt{4\pi}} U'(R) \hat{I}_i \langle I_i K, \lambda 0 | I_f K \rangle, \qquad (4.4.18)$$

which enters into Eq. (4.3.6) to complete the partial-wave matrix element.

The Coulomb interaction between a deformed charge density of Z_p protons and a second nucleus of Z_t protons does not depend on just the distance to the surface as the nuclear force does, but on a more global integral. Let the combined proton probability density distribution $\rho_q(\mathbf{r})$ be normalized as $\int \rho_q(\mathbf{r}) d\mathbf{r} = Z_p$. The electrostatic potential from these protons is then

$$V_C(\mathbf{R}) = e \int d\mathbf{r} \frac{\rho_q(\mathbf{r})}{|\mathbf{R} - \mathbf{r}|}, \qquad (4.4.19)$$

where these coordinates are in the laboratory frame. Using

$$\frac{1}{|\mathbf{R} - \mathbf{r}|} = \sum_{\lambda\mu} \frac{4\pi}{2\lambda+1} Y_\lambda^\mu(\mathbf{r})^* Y_\lambda^\mu(\mathbf{R}) f(R, r) \qquad (4.4.20)$$

with the 'near field' and 'far field' forms

$$f(R, r) \equiv \begin{cases} R^\lambda/r^{\lambda+1} & \text{for } R \leq r \\ r^\lambda/R^{\lambda+1} & \text{for } R \geq r \end{cases} \qquad (4.4.21)$$

respectively, then

$$
H_{\text{intr}} = eZ_t V_C(\mathbf{R})
$$
$$
= \sum_{\lambda\mu} \frac{4\pi Z_t e^2}{2\lambda+1} Y_\lambda^\mu(\mathbf{R}) \int Y_\lambda^\mu(\mathbf{r})^* f(R,r)\rho_q(\mathbf{r})\mathrm{d}\mathbf{r}. \tag{4.4.22}
$$

Comparing this with Eq. (4.3.1), we find that the rotational form factor is

$$
\mathcal{F}_\lambda(R)\, \mathcal{T}_\lambda^\mu(\xi) = \frac{\sqrt{4\pi}Z_t e^2}{2\lambda+1} \int Y_\lambda^\mu(\mathbf{r}) f(R,r)\rho_q(\mathbf{r})\mathrm{d}\mathbf{r}, \tag{4.4.23}
$$

so the matrix elements for transitions from nuclear spin I_i to I_f are

$$
V_{fi}^\lambda(R) = \langle I_f \parallel \mathcal{F}_\lambda \mathcal{T}_\lambda \parallel I_i \rangle = \frac{\sqrt{4\pi}Z_t e^2}{2\lambda+1} \langle I_f \parallel Y_\lambda^\mu(\mathbf{r}) f(R,r)\rho_q(\mathbf{r}) \parallel I_i \rangle
$$
$$
\xrightarrow{R\gg r} \frac{\sqrt{4\pi}e^2 Z_t}{2\lambda+1} \frac{\langle I_f \parallel Y_\lambda^\mu(\mathbf{r}) r^\lambda \rho_q(\mathbf{r}) \parallel I_i \rangle}{R^{\lambda+1}}. \tag{4.4.24}
$$

We define the numerator of the second term as the *Coulomb reduced matrix element*

$$
\langle I_f \parallel E\lambda \parallel I_i \rangle = \langle I_f \parallel Y_\lambda^\mu(\mathbf{r}) r^\lambda \rho_q(\mathbf{r}) \parallel I_i \rangle. \tag{4.4.25}
$$

For a deformed sphere of constant internal density with deformation length δ_λ and mean radius R_c, the rotational picture gives an approximate value for this matrix element of

$$
\langle I_f \parallel E\lambda \parallel I_i \rangle = \frac{3Z_p \delta_\lambda R_c^{\lambda-1}}{4\pi} \hat{I}_i \langle I_i K, \lambda 0 | I_f K \rangle. \tag{4.4.26}
$$

The accuracy may be improved by taking the Taylor series beyond Eq. (4.4.11), to second order in the $d_{q\mu}$. This gives rise to tensor products $Y_q \otimes Y_{q'}$, which couple up to λ satisfying $|q - q'| \leq \lambda \leq q + q'$. This means that second-order terms contribute also to the monopole parts, changing the diagonal distorting potential. There are also now $\lambda = 2q$ multipoles, allowing direct excitation of, for example, $0^+ \to 4^+$ transitions for $q = 2$ quadrupole deformations.

For best accuracy, a result accurate to all orders in the $d_{q\mu}$ may be found by carrying out a numerical angular integration in the body-fixed frame for the overlap $\langle Y_\lambda(\hat{\mathbf{R}}')|U(R-\tilde{R}(\theta',\phi')+R_0)\rangle$. For axial deformations, this is an integral over θ' only:

$$
\mathcal{F}_\lambda(R) = \frac{\hat{\lambda}}{2} \int_{-1}^{+1} U(R-\tilde{R}(\theta')-R_0)P_\lambda(\cos\theta')\mathrm{d}(\cos\theta'), \tag{4.4.27}
$$

and here

$$R - \tilde{R}(\theta') + R_0 = R - \sum_{q=2}^{q_{max}} \delta_q \sqrt{\frac{2q+1}{4\pi}} P_q(\cos\theta') \qquad (4.4.28)$$

for deformation lengths δ_q.

Sometimes, particular information about specific transitions is available from structure calculations, or from other experiments such as lifetime measurements. Including this knowledge will require changing by hand the values of $V_{fi}^\lambda(R)$ for specific $I_i \to I_f$ transitions. This is a way in which previous measurements of $B(E\lambda, I_i \to I_f)$ may be taken into account.

Vibrational models of nuclei give rise to inelastic transition potentials with forms similar to those from the rotational model. The excitation of quadrupole one-phonon states in a nucleus will give a 2_1^+ state, and then two-phonon states consist of two phonons of energy which can be coupled together to give a triplet of 0^+, 2_2^+ and 4^+ states at approximately twice the first 2_1^+ energy. Octupole phonons also may occur, generating a 3^- excited state with one phonon of energy.

The theory of vibrational nuclei may be developed along the same lines as above for rotational couplings, but now the $\hat{a}_{q\mu}$ are *operators* which change the state of the nucleus by exciting phonon states, with one or more phonons. Vibrational models are described in detail in, for example, Eisenberg and Greiner [29].

For a given $B(E\lambda)$, the vibrational model yields the same off-diagonal couplings as does the rotational model, but the *diagonal* re-orientation couplings are zero. This reflects the fact that a purely vibrational 2^+ state, for example, has zero static quadrupole moment. Real nuclei are usually between the pure limits, and have some small static quadrupole moments even for vibrational excited states.

4.4.2 Single-particle inelastic processes

Sometimes the state of a nucleon (or cluster of nucleons) in a nucleus changes because of differential tidal forces when interacting with a second nucleus. Consider the change of state of a valence nucleon or cluster v from initial state $\phi_{\ell_i s j_i}(\mathbf{r})$ to $\phi_{\ell_f s j_f}(\mathbf{r})$, so that, when bound to a core c of spin I_c, the inelastic transition is from $(I_c j_i)I_i$ to $(I_c j_f)I_f$.

The potential model to generate this transition describes a three-body system with core c, valence v and target t. The model is defined by the three masses m_c, m_v and m_t respectively, and the three pair-wise potentials $V_{cv}(\mathbf{r})$, $V_{vt}(\mathbf{r}_{vt})$ and $V_{ct}(\mathbf{r}_{ct})$, where the coordinates $\mathbf{r}_{vt}$ and $\mathbf{r}_{ct}$, as shown in the diagram, are related to $\mathbf{R}$ and $\mathbf{r}$ by

$$\mathbf{r}_{vt} = \mathbf{R} + \frac{m_c}{m_p}\mathbf{r}$$

$$\mathbf{r}_{ct} = \mathbf{R} - \frac{m_v}{m_p}\mathbf{r}, \tag{4.4.29}$$

where $m_p = m_c + m_v$, the total mass of the nucleus being excited. The transition arises as a tidal effect because neither V_{vt} nor V_{ct} act just on the center of mass at R. If we consider the principal central parts of these two potentials, then the core spin I_c, nucleon spin s and target spin I_t are all spectators. The inelastic transition potential is therefore

$$V_{fi}(R) = \langle (L_f J_f)\Lambda | V_{vt} + V_{ct} | (L_i J_i)\Lambda \rangle, \tag{4.4.30}$$

where L_i, L_f are the initial and final orbital angular momenta of relative motion, and $\Lambda = \mathbf{L} + \mathbf{J}$ is their sum which is a conserved angular momentum. The sum $\mathbf{J}$ is to be combined with $\mathbf{s}$, $\mathbf{I}_c$ and $\mathbf{J}_t$ to give the total spin J_{tot} for the whole system.

With V_{vt} and V_{ct} depending only on the length square roots of

$$r_{vt}^2 = R^2 + \frac{m_c^2}{m_p^2}r^2 + \frac{2m_c}{m_p}Rrz$$

$$r_{ct}^2 = R^2 + \frac{m_v^2}{m_p^2}r^2 - \frac{2m_v}{m_p}Rrz, \tag{4.4.31}$$

where $z = \hat{\mathbf{R}} \cdot \hat{\mathbf{r}}$, the cosine of the angle between $\mathbf{R}$ and $\mathbf{r}$, all the effects of V_{vt} and V_{ct} are in the two-variable multipole function

$$\mathcal{F}_\lambda(R, r) = \frac{1}{2}\int_{-1}^{+1}[V_{vt}(r_{vt}) + V_{ct}(r_{ct})]P_\lambda(z)\mathrm{d}z. \tag{4.4.32}$$

In terms of this $\mathcal{F}_\lambda(R, r)$ the potentials in the matrix element (4.4.30) are

$$V_{vt} + V_{ct} = \sum_\lambda (2\lambda+1)\mathcal{F}_\lambda(R, r)P_\lambda(z)$$

$$= \sum_\lambda (2\lambda+1)\mathcal{F}_\lambda(R, r)\sum_m Y_\lambda^m(\hat{\mathbf{r}})^* Y_\lambda^m(\hat{\mathbf{R}}). \tag{4.4.33}$$

If the single-particle states have angular and radial parts given by Eq. (4.2.10), then the transition potential for multipole component λ is

$$V_{fi}^\lambda(R) = \int_0^\infty u_{\ell_f s j_f}(r)\mathcal{F}_\lambda(R, r)u_{\ell_i s j_i}(r)\mathrm{d}r \; \hat{\lambda}\hat{\ell}_i \; \langle \ell_i 0, \lambda 0 | \ell_f 0 \rangle. \tag{4.4.34}$$

4.5 Particle rearrangements

4.5.1 Transfer reactions

In a transfer reaction, a valence nucleon (or cluster) is transferred from the projectile to the target, or vice versa. When this is transferred from the projectile it is often called *stripping*, and when it is added to the projectile from the target this is *pickup*.[9]

For transfer processes, as for single-particle inelastic excitations, we use $\mathbf{R}$ to refer to the difference of the projectile and target positions, and $\mathbf{r}$ to the relative coordinates of the valence cluster to its core. For mass-transfer reactions, there will be distinct pairs of vectors in the initial state $\mathbf{R}, \mathbf{r}$, and in the final state $\mathbf{R}', \mathbf{r}'$ because, by recoil effects, the vectors $\mathbf{R} \neq \mathbf{R}'$ as shown in Fig. 4.3.

Let us consider the specific case of stripping, that is removing a nucleon from the projectile. For now we deal with the simple case that ignores the core structure and treats its spin as a spectator, and assumes that the composite nucleus is uniquely composed by a single bound state. The initial bound state of the projectile $\phi_p(\mathbf{r})$ and the final bound state of the residual nucleus $\phi_t(\mathbf{r}')$ then satisfy the eigenvalue equations

$$[H_p - \varepsilon_p]\phi_p(\mathbf{r}) = 0 \quad \text{where} \quad H_p = T_\mathbf{r} + V_p(\mathbf{r})$$
$$[H_t - \varepsilon_t]\phi_t(\mathbf{r}') = 0 \quad \text{where} \quad H_t = T_{\mathbf{r}'} + V_t(\mathbf{r}'). \quad (4.5.1)$$

The binding potentials $V_p(\mathbf{r})$ and $V_t(\mathbf{r}')$ are usually fitted so the eigenenergies agree with experimental separation energies, using the methods described in Section 6.4.

The Q-value for the reaction, the amount of energy released during the transfer, is $Q = \varepsilon_p - \varepsilon_t$. In terms of the *binding energies $B \equiv -\varepsilon$* which are positive for bound states, the Q-value is $Q = B_t - B_p$, so $B_t = B_p + Q$.

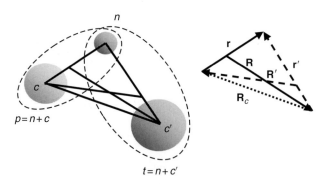

Fig. 4.3. Coordinates used in a one-particle transfer reaction.

---

[9] Note that sometimes the word 'stripping' is used to refer to *any* removal process from the projectile, including say breakup and more complicated reactions, but in this section it refers only to the transfer to a specific bound state around the target.

The dynamical details of the transfer coupling arise from the matrix elements of the Hamiltonian for the three bodies involved: the initial and final cores, and the valence particle. This Hamiltonian is

$$H = T_{\mathbf{r}} + T_{\mathbf{R}} + V_p(\mathbf{r}) + V_t(\mathbf{r}') + U_{c'c}(\mathbf{R}_c), \qquad (4.5.2)$$

where $U_{c'c}(\mathbf{R}_c)$ is the *core-core* optical potential. The pair of kinetic energy terms can equivalently be written $T_{\mathbf{r}} + T_{\mathbf{R}} = T_{\mathbf{r}'} + T_{\mathbf{R}'}$, so there will henceforth be two ways of expanding the Hamiltonian as we have already seen in subsection 3.2.2. These are called *prior* and *post*:

$$
\begin{aligned}
H = H_{\text{prior}} &= T_{\mathbf{R}} + U_i(R) + H_p(\mathbf{r}) + \mathcal{V}_i(\mathbf{R}, \mathbf{r}) \\
&= H_{\text{post}} = T_{\mathbf{R}'} + U_f(R') + H_t(\mathbf{r}') + \mathcal{V}_f(\mathbf{R}', \mathbf{r}'), \qquad (4.5.3)
\end{aligned}
$$

where the $U_{i,f}$ are the entrance and exit diagonal potentials respectively, so the *interaction terms* $\mathcal{V}_i$ and $\mathcal{V}_f$ to be used must be

$$
\begin{aligned}
\mathcal{V}_i(\mathbf{R}, \mathbf{r}) &= V_t(\mathbf{r}') + U_{c'c}(\mathbf{R}_c) - U_i(R) \\
\text{or} \quad \mathcal{V}_f(\mathbf{R}', \mathbf{r}') &= V_p(\mathbf{r}) + U_{c'c}(\mathbf{R}_c) - U_f(R'). \qquad (4.5.4)
\end{aligned}
$$

It is these interaction terms which cause the transition from one bound state to another, since the other terms in the post and prior Hamiltonians are diagonal with respect to transfer channels. The first part of the $\mathcal{V}_{i,f}$ is the binding potential, and the remaining two terms are called the *remnant terms*, which, as they are often similar in magnitude and contain complex potentials, are sometimes neglected for convenience. The post and prior forms give exactly the same results in first-order DWBA, as was shown on page 104. With sufficient numerical accuracy this equality should hold also in practice, and this equality requires that the remnant terms be always included.

The matrix elements of these interaction terms are now found for the case of two transfer channels. Using the model space[10]

$$\Psi = \left[\phi_p(\mathbf{r}) \otimes Y_L(\hat{\mathbf{R}}) \right]_{\Lambda} \psi_i(R)/R + \left[\phi_t(\mathbf{r}') \otimes Y_{L'}(\hat{\mathbf{R}}') \right]_{\Lambda} \psi_f(R')/R', \qquad (4.5.5)$$

we need to find transition matrix elements such as

$$V_{fi}^o(R', R) = R' \langle [\phi_t(\mathbf{r}') \otimes Y_{L'}(\hat{\mathbf{R}}')] \mid \mathcal{V}_o \mid [\phi_p(\mathbf{r}) \otimes Y_L(\hat{\mathbf{R}})]_{\Lambda} \rangle R^{-1}, \qquad (4.5.6)$$

with $o = i, f$ for prior and post interactions respectively in Eq. (4.5.4).

---

[10] We omit the two core states ϕ_c and $\phi_{c'}$ since here they are spectators with no dynamical role.

Finite-range transfers

This matrix element is a non-local integral operator, as it operates on the function $\psi_i(R)$ to produce a function of R'. We therefore derive the non-local kernel $V_{fi}^o(R', R)$ so that the matrix element operation on a wave function, which initially involves a five-dimensional integral over $\mathbf{r}'$ and $\hat{\mathbf{R}}'$, may be calculated by means of a one-dimensional integral over R:

$$\Omega_f(R') = \int V_{fi}^o(R', R)\psi_i(R)\mathrm{d}R. \tag{4.5.7}$$

Such a source term may be used, for example, in the Green function integral methods of Chapter 3.

Note that when the initial and final single-particle states are real, then the kernel function is symmetric

$$V_{fi}^o(R', R) = V_{if}^o(R, R'), \tag{4.5.8}$$

whether o is post or prior.

Since the $\mathbf{r}$ and $\mathbf{r}'$ are linear combinations of the channel vectors $\mathbf{R}$ and $\mathbf{R}'$, we have

$$\mathbf{r} = p\mathbf{R}' + q\mathbf{R} \quad \text{and} \quad \mathbf{r}' = p'\mathbf{R}' + q'\mathbf{R}, \tag{4.5.9}$$

where $p = -\omega$, $q = v_t\omega$, $p' = -v_p\omega$, and $q' = \omega$, with $v_p = m_c/m_p$, $v_t = m_{c'}/m_{t'}$, and $\omega = (1-v_pv_t)^{-1}$. The 'core-core' vector is $\mathbf{R}_c = \mathbf{r}'-\mathbf{r} = (p'-p)\mathbf{R}'+(q'-q)\mathbf{R}$.

To calculate the matrix element of Eq. (4.5.6), we first convert the bound states $\phi_t(\mathbf{r}')$ and $\phi_p(\mathbf{r})$ into functions of $\mathbf{R}$ and $\mathbf{R}'$ using Eq. (4.5.9). To do this, we must transform the spherical harmonics $Y_\ell(\hat{\mathbf{r}})$ and $Y_{\ell'}(\hat{\mathbf{r}}')$ into linear combinations of the spherical harmonics $Y_n(\hat{\mathbf{R}})$ and $Y_{n'}(\hat{\mathbf{R}}')$. This is done by means of the Moshinsky solid-harmonic expansion [30]:

$$Y_\ell^m(\hat{\mathbf{r}}) = \sqrt{4\pi} \sum_{n=0}^\ell \sum_{\lambda=-n}^n c(\ell, n)\frac{(pR')^{\ell-n}(qR)^n}{r^\ell}$$
$$\times\ Y_{\ell-n}^{m-\lambda}(\hat{\mathbf{R}}')Y_n^\lambda(\hat{\mathbf{R}})\langle\ell-n\,m-\lambda, n\lambda|\ell m\rangle, \tag{4.5.10}$$

$$\text{where}\quad c(\ell, n) = \left(\frac{(2\ell + 1)!}{(2n + 1)!(2(\ell-n) + 1)!}\right)^{1/2}. \tag{4.5.11}$$

Considering only the case where the remnant part of the interaction $\mathcal{V}_o$ of Eq. (4.5.6) contains just scalar potentials, we perform the Legendre expansion

$$\mathcal{V}_o\frac{u_{\ell'}(r')}{r'^{\ell'+1}}\frac{u_\ell(r)}{r^{\ell+1}} = \sum_{T=0}^{T_{\max}}(2T+1)\mathbf{q}_{\ell',\ell}^{T,o}(R', R)P_T(z), \tag{4.5.12}$$

where the $u_\ell(r)$ are the radial parts of the bound wave functions according to Eq. (4.2.10). The limit $T_{\max}$ is chosen large enough to generate all the couplings for partial waves to be used. Here, the Legendre polynomials $P_T(z)$ are functions of z, the cosine of the angle between $\mathbf{R}$ and $\mathbf{R}'$. According to Eq. (4.5.4) with scalar potentials, the $\mathcal{V}_o$ depends only on the lengths of the vectors R_c and $\{r', R\}$ or $\{r, R'\}$, all of which may be calculated in terms of $\{R, R'\}$ and z according to formulae such as $r = (p^2 R'^2 + q^2 R^2 + 2pqRR'z)^{1/2}$ in the numerical quadrature for the integral

$$\mathbf{q}_{\ell',\ell}^{T,o}(R',R) = \frac{1}{2}\int_{-1}^{+1}\mathcal{V}_o\frac{u_{\ell'}(r')}{r'^{\ell'+1}}\frac{u_\ell(r)}{r^{\ell+1}}P_T(z)\mathrm{d}z. \tag{4.5.13}$$

Using the Legendre expansion, the radial kernel function is

$$\mathcal{V}_{\ell'L':\ell L}^{\Lambda o}(R',R) = (-1)^{L+L'}\hat{L}\hat{L}'\sum_{TKK'}\begin{pmatrix} K & L & T \\ 0 & 0 & 0 \end{pmatrix}\begin{pmatrix} K' & L' & T \\ 0 & 0 & 0 \end{pmatrix}$$

$$\times \sum_\lambda W(\ell L\ell'L';\Lambda\lambda)\,W(KLK'L';T\lambda)\,\mathcal{F}_{\lambda,K'KT}^{\ell'\ell\Lambda o}(R',R), \tag{4.5.14}$$

where we use the non-local form factor

$$\mathcal{F}_{\lambda,K'KT}^{\ell'\ell\Lambda o}(R',R) = \frac{|q|^3}{2}\sum_{nn'}RR'(pR')^{\ell-n}(qR)^n(p'R')^{\ell'-n'}(q'R)^{n'}$$

$$\times (2T+1)(-1)^{\Lambda+T}\,\hat{\ell}\hat{\ell}'\,\widehat{(\ell-n)}\,\widehat{(\ell'-n')}\hat{n}\hat{n}'(2K+1)(2K'+1)$$

$$\times c(\ell,n)c(\ell',n')\begin{pmatrix} \ell-n & n' & K \\ 0 & 0 & 0 \end{pmatrix}\begin{pmatrix} \ell'-n' & n & K' \\ 0 & 0 & 0 \end{pmatrix}$$

$$\times (2\lambda+1)\begin{Bmatrix} \ell' & \lambda & \ell \\ n' & K & \ell-n \\ \ell'-n' & K' & n \end{Bmatrix}\mathbf{q}_{\ell',\ell}^{T,o}(R',R). \tag{4.5.15}$$

Combining these factors, we are able to calculate the non-local kernels $\mathcal{V}_{\ell'L':\ell L}^{\Lambda o}(R',R)$ to calculate transfer reactions using an exact treatment of the finite range of the potentials and all the recoil terms from the finite masses of the cores c and c'.

Zero-range transfers

When the wave functions $\phi_\ell(\mathbf{r})$ are all s-states ($\ell = 0$), when the remnant terms can be neglected, and when the interaction potential is approximately of zero range, $\mathcal{V}_o\phi(\mathbf{r}) \sim D_0\delta(\mathbf{r})$, then the form factor $\mathcal{V}_{\ell'L':\ell L}^{\Lambda}(R',R)$ of equation (4.5.14) can be simplified to

$$
V^L_{\ell'L':0L}(R',R) = D_0 \frac{(-1)^{L'-\ell'}}{\hat{L}} \frac{\hat{\ell}'\hat{L}\hat{L}'}{\sqrt{4\pi}} \begin{pmatrix} \ell' & L & L' \\ 0 & 0 & 0 \end{pmatrix}
$$

$$
\times \frac{1}{R'} u_{\ell'}(R') \frac{q^2}{p} \delta(pR' + qR). \tag{4.5.16}
$$

This can be made local by defining a new step size $h' = -ph/q \equiv v_t h$ for R' in the stripping channel f, and this considerably simplifies the problem of solving the coupled equations.

Zero-range approximations are primarily useful for transfers to or from s-shell nuclei such as deuterons, tritons and the α particle.

Local energy approximation

If the interaction potential is of small range, though not zero, and the projectile still contains only s-states, then a first-order correction may be made to the above form factor, and the accuracy can be considerably improved. This correction will depend on the rate of oscillation of the source wave function $\psi_i(R)$ within a 'finite-range effective radius' parametrized as ρ_{eff}. The rate of oscillation is estimated from the local energy in the entrance and exit channels, which is proportional to

$$
K_i(R)^2 = \frac{2\mu_p}{\hbar^2} (E_i - U_i), \tag{4.5.17}
$$

and the result [31] is to multiply the target bound state $u_{\ell'}(R')$ in Eq. (4.5.16) by a factor

$$
\left[1 + \rho_{\text{eff}}^2 \frac{2\mu_p}{\hbar^2} \left(U_{cc}(R') + V_{\ell'}(R') - U_f(R') + \varepsilon_p \right) \right], \tag{4.5.18}
$$

where μ_p is the reduced mass of the valence particle in the projectile, and $V_{\ell'}(R')$ is the binding potential for $u_{\ell'}(R')$.

At sub-Coulomb incident energies [32], the details of the nuclear potentials in Eq. (4.5.18) become invisible, and as the longer-ranged Coulomb potentials cancel in Eq. (4.5.18), the form factor can be simplified to

$$
u_{\ell'}(R') D_0 \left[1 + \rho_{\text{eff}}^2 \frac{2\mu_p}{\hbar^2} \varepsilon_p \right] = D u_{\ell'}(R'), \tag{4.5.19}
$$

where

$$D = (1 + k_p^2 \rho_{\text{eff}}^2)D_0 \qquad (4.5.20)$$

is the effective zero-range coupling constant for sub-Coulomb transfers, using the bound-state wave number $k_p^2 = 2\mu_p \varepsilon_p/\hbar^2$ for the projectile.

The parameters D_0 and D can be derived from the details of the projectile bound state $\phi_0(\mathbf{r})$. The zero-range constant D_0 is the integral

$$D_0 = \sqrt{4\pi} \int_0^\infty r V_0(r) u_0(r) \mathrm{d}r. \qquad (4.5.21)$$

The parameter D, on the other hand, reflects the asymptotic strength of the wave function $u_0(r)$ as $r \to \infty$, as it is the magnitude of this tail which is important in sub-Coulomb reactions:

$$u_0(r) =_{r \to \infty} \frac{2\mu_p}{\hbar^2} \frac{1}{\sqrt{4\pi}} D e^{-k_p r}. \qquad (4.5.22)$$

It may be also found, using Schrödinger's equation, from the integral

$$D = \sqrt{4\pi} \int_0^\infty \frac{\sinh(k_p r)}{k_p} V_0(r) u_0(r) \mathrm{d}r. \qquad (4.5.23)$$

From this equation we can see that as the range of the potential becomes smaller, D approaches D_0. The 'finite-range effective radius' ρ_{eff} of equation (4.5.20) is thus some measure of the mean radius of the potential $V_0(r)$.

Asymptotic Normalization Coefficient

The Asymptotic Normalization Coefficient (ANC) of any projectile or target bound-state wave function, of any partial wave ℓ, is defined as the asymptotic coefficient C_ℓ of the Whittaker function. That function is a decaying exponential for uncharged particles (see Box 4.3). The ANC is thus determined by

$$u_\ell(r) =_{r > R_n} C_\ell W_{-\eta, \ell+\frac{1}{2}}(-2k_p r) \approx C_\ell e^{-k_p r}, \qquad (4.5.24)$$

(and is therefore related to the above D value of s-wave states by $C_\ell = \frac{2\mu_p}{\hbar^2} \frac{1}{\sqrt{4\pi}} D$).

The ANC describes the strength of the exponential tail of a bound-state wave function, and hence capture and transfer reactions at low astrophysical energies are directly dependent on its value. Determining this value will be one of the objectives of nuclear spectroscopy, as discussed in Chapter 14.

4.5.2 Knockout reactions

A variety of different reactions are commonly called 'knockout reactions':

- any removal of a nucleon (cluster) from a nucleus, whether by transfer, breakup, or more complicated non-elastic reaction leaving one or more residual nuclei in excited states,
- *quasi-elastic knockout* such as the breakup reactions ^{12}C(p,pp)^{11}B with high-energy projectiles so the incident and bound protons scatter as if elastically to $\sim 90°$ relative angles in a three-body final state, and
- knockout of a cluster which is replaced in the target by the projectile, such as ^{14}C(p,α)^{11}B, to a two-body final state with the nuclei in specific energy levels.

This last kind of *knockout* (transfer) reaction is considered here, as it is the more important for low incident proton energies, and is usually exothermic ($Q > 0$).

A (p,α) reaction can be regarded as a superposition of amplitudes for two different mechanisms. The first is a triton transfer:

$$\left({}^{14}\text{C} = {}^{11}\text{B} + {}^{3}\text{H} \right) + \text{p} \rightarrow {}^{11}\text{B} + \left(\alpha = \text{p} + {}^{3}\text{H} \right), \qquad (4.5.25)$$

and the second is again a transfer, this time of a ^{10}Be cluster from ^{12}C to the proton:

$$\left({}^{14}\text{C} = {}^{10}\text{Be} + \alpha \right) + \text{p} \rightarrow \alpha + \left({}^{11}\text{B} = \text{p} + {}^{10}\text{Be} \right). \qquad (4.5.26)$$

This second mechanism, called *heavy-particle transfer*, produces the final two nuclei with the new projectile identical to the initial target, and vice versa, so its amplitude needs to be permuted by an operator $P_{\alpha,^{11}\text{B}}$ before combining with the first amplitude, as explained in subsection 3.4.2.

The heavy-particle stripping interaction terms, from Eq. (4.5.4), are

$$\text{prior } \mathcal{V}_i = V_{\text{p}-\text{Be}} + U_{\text{p}\alpha} - U_{\text{pC}}$$
$$\text{post } \mathcal{V}_f = V_{\alpha-\text{Be}} + U_{\text{p}\alpha} - U_{\alpha\text{B}}, \qquad (4.5.27)$$

of which the most important terms are not the binding potentials $V_{\text{p}-\text{Be}}$ or $V_{\alpha\text{Be}}$ as usual for normal transfers, but the second term $U_{\text{p}\alpha}$ describing the interaction between the incident proton and the knocked out α particle. This is the potential we would expect to be dominant if we picture the proton interacting with and knocking out the α particle from its bound state.

4.5.3 Breakup reactions

When a nucleus in a reaction breaks up into two or more pieces which are both detected separately, we have breakup reactions as defined in Chapter 2. Breakup reactions necessarily involve the *continuum final states* in some nucleus. We now consider two-body breakup.

One way of populating continuum states is to use the single-particle excitation mechanism of subsection 4.4.2 to move a nucleon from a bound state $u_{\ell s j;b}(r)$ to a continuum state $u_{\ell' s j'}(r; k')$ for some scattering momentum k'. This scattering state now extends to infinity in r, and hence allows the detection of one or both pieces moving with relative coordinate **r** and relative momentum **k**′. Ideally we will want to calculate the cross sections to all possible final states $\ell' s j'; k'$, so, to avoid omission or over-counting they should constitute a complete set. For that reason the renormalized wave functions $\hat{u}_{\ell' s j'}(r; k')$ of Eq. (4.2.12) are generally used, up to some maximum k' value judged suitable for each particular reaction.

Strictly speaking, there should also be couplings, not only from the ground state to the continuum, but also *between* all continuum states. The $\hat{u}_{\ell' s j'}(r; k')$ wave functions are suitable for the first purpose, but, because they extend to infinity, not for the second task of calculating *continuum-to-continuum* couplings. The Coupled Discretized Continuum Channels (CDCC) method has been devised to solve this problem and allow all physically important couplings to be included, and will be discussed in Chapter 8.

Transfer mechanisms can also populate the continuum, using the methods of subsection 4.5.1 respectively. Convergence is more difficult in this case, however, because of continuing small contributions from very large R values. Ideally, as mentioned also in subsection 8.3.4, the radial integrals should be deformed in the complex plane by the method of Vincent and Fortune [33] in order to obtain numerically stable results.

4.5.4 Capture reactions

Two nuclei may approach each other and fuse together, and the rate of fusion depends on the potential in the entrance channel as well as the capture mechanism. The important factor in the entrance channel is clearly the penetration of any Coulomb or centrifugal barrier present at middle distances. If the scattering wave comes in from large relative distances, then it will be attenuated by the time it has tunneled through any Coulomb or centrifugal barrier, and we will calculate a *penetrability factor* to describe this reduction.

There are different mechanisms for trapping the particles permanently, once they have come inside the Coulomb barrier. One way is for a γ-ray to be emitted, and the particles lose that energy and fall down into a bound state of relative motion. This is a *direct-capture* process, to be discussed further in Section 4.7. Another direct mechanism is for one of the nuclei to be pushed up to an excited state, and absorbing energy, which makes escape less likely. This mechanism can produce doorway resonances.

A third way is for the particles to be captured by some of the long-living resonances of the compound nucleus that is formed of both the nuclei together.

This will happen especially with heavier nuclei, where there is a high *level density* of these resonances: many per MeV. Compound-nucleus resonances are usually narrow (that is, long-lived), so they capture flux from the direct-reaction channels and release it later into other channels in ways to be discussed further in Chapter 11. These other channels may be neutron or γ emission, but they are called *statistical* or *evaporation* products, rather than direct reactions. The direct processes are most likely with light nuclei, where the level density of resonances is rather low. We will see in subsection 11.5.3 that the flux going to compound resonances in heavy nuclei can be simulated by an absorptive imaginary 'fusion part' to the optical potential, a part that is inside the Coulomb barrier.

To calculate the *penetrability factors* $P_L(E)$ for traversing a Coulomb barrier, we need to know the shape of the barrier: how it is composed of a Coulomb repulsion at medium and large distances, along with a nuclear attraction at short distances. This amounts to solving the Schrödinger equation for the relative motion. This can be done exactly by the methods of Chapter 3, or by parametrizing the wave functions at an intermediate boundary and using the R-matrix methods of Chapter 10. Hill and Wheeler [34] showed how to solve the Schrödinger equation exactly with a potential that has an inverted parabolic barrier. Alternatively, we could use the WKB methods of Chapter 7 to get approximate penetration factors.

Once we have the penetrability factors $P_L(E)$ for each partial wave L, then a cross section will depend on them according to

$$\sigma_X(E) = \frac{\pi}{k^2} \sum_L (2L+1) P_L(E) \mathcal{B}_L(X), \qquad (4.5.28)$$

where $\mathcal{B}_L(X)$ is a branching ratio to observing a specific final product X once the barrier has been passed. If X is fusion, there are many compound nuclear states and so $\mathcal{B}_L(X) \approx 1$, as capture is then with a large probability irreversible.

4.6 Isospin transitions

4.6.1 Charge-exchange reactions

These are reactions in which the participating nuclei keep their masses constant, but change their charge, such as the reaction $^{12}C(p,n)^{12}N$. Various mechanisms may contribute to such a transition:

- heavy-particle transfer of ^{11}C from $^{12}C = {}^{11}C + n$ to $^{12}N = {}^{11}C + p$, as discussed in subsection 4.5.2,
- two-step transfers via a $^{11}C + d$ intermediate state, the first transfer adding a neutron to the projectile and the second step removing a proton, and
- direct conversion of a proton to a neutron, for example by a meson π^+ being emitted from the proton, and absorbed by the ^{12}C where it changes a neutron there into a proton.

The first two mechanisms can be calculated by the transfer interactions described previously, but we need to combine these with the last direct *charge-exchange* mechanism, to be described in this subsection. It can be modeled by an isospin operator which raises the projectile isospin from $t_z = -1/2$ (p) to $+1/2$ (n), alongside lowering the target isospin from $T_z = 0$ for ^{12}C to -1 for ^{12}N. Such an effect would be caused by a component of the Hamiltonian that couples together the two isospins **t** and **T**, the isospin operators for the projectile and target respectively.

Lane [35] showed that the optical potential for the scattering of protons and neutrons on nuclei appears to have a **t** · **T** contribution. In heavy nuclei, for example, the additional attraction between protons and neutrons means that the optical potential for the scattering of protons on neutron-rich nuclei is more attractive than for neutron scattering. As discussed in subsection 4.1.1, this is commonly parametrized as

$$V(R) = V_0(R) + \frac{1}{2} t_z \frac{N-Z}{A} V_T(R), \tag{4.6.1}$$

where t_z is the projectile operator, and the target has proton and neutron numbers Z, N. Such a force implies a corresponding neutron–proton charge-exchange interaction, if we generalize the expression to an invariant form

$$V(R) = V_0(R) + \frac{\mathbf{t} \cdot \mathbf{T}}{A} V_T(R). \tag{4.6.2}$$

The $V_T(R)$ potential can therefore cause transitions to isobaric analogue states of the target (see subsection 3.4.1), as well as producing scalar shifts in the neutron and proton optical potentials.

Fermi transitions

Such Fermi transitions are the simplest form of isospin coupling, whereby the interaction term is

$$H_F = V_F(R)\mathbf{t} \cdot \mathbf{T}, \tag{4.6.3}$$

for **t** the isospin operator for the projectile, and **T** for the target. The operator

$$\mathbf{t} \cdot \mathbf{T} = t_x T_x + t_y T_y + t_z T_z$$
$$= \frac{1}{2}[t_+ T_- + t_- T_+] + t_z T_z, \tag{4.6.4}$$

where $t_\pm = t_x \pm i t_y$ is the raising (+) and lowering (−) operator for the projectile, and $T_\pm$ similarly for the target. The three terms in Eq. (4.6.4) have different effects: the first $t_+ T_-$ converts a projectile proton to a neutron and a target neutron to a proton, the second does the opposite, and the third makes no conversions of protons or neutrons. In general, the effect of the $T_{\pm}$ operators is

$$T_{\pm}|T, T_z\rangle = \sqrt{T(T+1) - T_z(T_z\pm1)}|T, T_z \pm 1\rangle. \qquad (4.6.5)$$

For arbitrary initial and final state of the target nucleus ϕ_i and ϕ_f, the magnitude of the charge-exchange transition depends on the matrix element $\langle\phi_f|\mathbf{T}|\phi_i\rangle$. The isospin raising and lowering parts of this matrix element will be used in Chapter 5 in connection with β-decay processes.

Gamow-Teller transitions

The Gamow-Teller operators are the next-simplest form of charge exchange. These involve a spin as well as an isospin transition:

$$H_{GT} = V_{GT}(R) \, (\mathbf{s} \cdot \mathbf{S}) \, (\mathbf{t} \cdot \mathbf{T}), \qquad (4.6.6)$$

where $\mathbf{s}$ and $\mathbf{S}$ are spin operators for the projectile and target respectively. These operators do not change the spatial configurations within a nucleus (they keep the same partial wave ℓ and preserve the radial wave functions). The H_{GT} operator contains a spin-spin coupling, so cannot, for example, couple 0^+ states together, in contrast to the Fermi operator H_F.

The single-particle reduced matrix elements of H_{GT} for one nucleus (projectile or target) are, for n$\rightarrow$p transitions for example:

$$\langle u_{\ell's j'}(r)p \parallel s t_- \parallel u_{\ell s j}(r)n\rangle$$

$$= \frac{1}{2}\langle u_{\ell's j'}(r) \parallel \mathbf{s} \parallel u_{\ell s j}(r)\rangle$$

$$= \frac{1}{2}\int_0^\infty u_{\ell's'j'}(r)^* u_{\ell s j}(r)dr$$

$$\times \hat{j}\hat{j'}(-1)^{\ell+s+j'+1} \begin{Bmatrix} \ell & s & j \\ 1 & j' & s \end{Bmatrix} \langle s \parallel \mathbf{s} \parallel s\rangle\delta_{\ell\ell'}, \qquad (4.6.7)$$

with $\langle s \parallel \mathbf{s} \parallel s\rangle = \sqrt{s(s+1)(2s+1)}$. The radial integral is a measure of the spatial similarity between the initial and final states.

The Gamow-Teller transition is not limited to one or a few isobaric analogue states like the Fermi transition, but will populate many spin and charge-exchange states over a range of energies, with varying transition rates. The GT measurements therefore depend on more details of nuclear structure, and measurements of GT transitions can be used to probe that structure.

The strength of $V_F(R)$ and $V_{GT}(R)$ can be found by fitting to experiment, or derived theoretically by folding nuclear wave functions with effective NN forces

V_{NN}. The $V_F(R)$ and $V_{GT}(R)$ strengths therefore depend directly on the isovector and spin-flip-isovector components of V_{NN} respectively.

4.6.2 Generalized multipole transitions

Nucleon-nucleon interactions V_{NN} are known to have in general many spatial-, spin- and isospin-transition components. When all of these are included for the transition potential that couples a given pair of initial and final nuclear states, we have a generalized set of multipole transitions. These can be classified according to spatial angular-momentum transfer (previously called λ, here also called L), spin transfer (here S), and isospin transfer T. As well, there may be dependence on the vector sum $\mathbf{L} + \mathbf{S} = \mathbf{J}$. The general form may therefore be written schematically as

$$H_{LSJT} = V_{LSJT}(R) \left[Y_L(\hat{\mathbf{R}}) \otimes \mathcal{T}_S(\mathbf{s}, \mathbf{S}) \right]_J \mathcal{T}_T(\mathbf{t}, \mathbf{T}), \qquad (4.6.8)$$

where $L \geq 0$, and $\mathcal{T}_S$, $\mathcal{T}_T$ are tensors of rank S and T composed of their vector arguments.

There are long descriptive names for each of these combinations: $S=1$: 'spin flip'; $T=0$: 'isoscalar' and $T=1$: 'isovector'; and $L=0$: 'monopole'; $L=1$: 'dipole'; $L=2$: 'quadrupole'. Thus the $LST = 210$ transition will be called the 'isoscalar spin-flip quadrupole' reaction.

4.7 Photo-nuclear couplings

4.7.1 Single-photon reactions

Subsection 3.5.1 derived the equations that couple a single photon of energy $E_\gamma = \hbar\omega$ with a particle of charge q moving to or from a bound state Φ_b. From Eqs. (3.5.33, 3.5.34) these are:

$$\left[-\frac{\hbar c}{k}\nabla^2 - E_\gamma \right] \mathbf{Z}(\mathbf{r}) = 2\sqrt{h/\omega}\langle\Phi_b|\hat{\mathbf{j}}_q(\mathbf{r})|\Psi(\mathbf{R})\rangle,$$

$$[\hat{T} + V - E_i]\Psi(\mathbf{R}) = \sqrt{h/\omega}(\hat{\mathbf{j}}_q\Phi_b) \cdot \mathbf{Z}(\mathbf{r}). \qquad (4.7.1)$$

Here, the coordinate $\mathbf{r}$ is the distance from the center of mass of the whole system, and the current operator $\hat{\mathbf{j}}(\mathbf{r})$ gives at each position $\mathbf{r}$ the current after integrating over all the charged-particle positions $\mathbf{r}_i$, which are again their distances from that center of mass. For the bound state Φ_b at energy E_b, and continuum state Ψ at E_i, the photon energy is $E_\gamma = E_i - E_b$. The $\hat{\mathbf{j}}_q$ is the electric current operator defined in subsection 3.1.4, and $Z(\mathbf{r}) = \sqrt{k/hc}\ \mathbf{A}$ is the normalized vector potential of Eq. (3.5.31) for the electromagnetic field.

The 3-dimensional wave functions of Eq. (4.7.1) have been expanded in partial waves in Eq. (3.5.37), from which the partial-wave T matrix is given by Eq. (3.5.42), and a vector-form T matrix is also given by Eq. (3.5.44). In order to satisfy the Coulomb gauge condition, subsection 3.5.5 explains how this vector-form T matrix is replaced by that of Eq. (3.5.60):

$$\mathbf{T}_{JM}^{(\mathcal{E},M)} = \langle \sqrt{4\pi}\sqrt{2J+1}\ \mathrm{i}^J\ \mathbf{A}_{JM}(\mathbf{r};(\mathcal{E},M))|V_{\gamma p}|\Psi(\mathbf{R};\mathbf{k}_i)\rangle \qquad (4.7.2)$$

which we must now evaluate. These matrix elements of the derivative current operator can be evaluated exactly (see for example [36] or [37]), but the complications of derivatives mean that simplified alternatives are often used. One minor detail is that the center-of-mass coordinates $\mathbf{r}$ for the position of the photon and $\mathbf{r}_i$ for the charged particles are different from $\mathbf{R}$, the relative separation of the two nuclei, so the couplings are spatially non-local. The most common simplifying approximations used are (a) the long-wavelength approximation, and (b) using Siegert's theorem to transform current into charge-density operators.

The *long-wavelength approximation* means that the lowest Λ partial wave in the allowed range $|J-1| \leq \Lambda \leq J+1$ is taken to be dominant. This is reasonable if the photon wavelengths $2\pi\hbar c/E_\gamma$ are much larger than the nuclear bound states: even large nuclear halo states of $\sim 20\,\mathrm{fm}$ in extent allow photon energies up to $E_\gamma \sim 60\,\mathrm{MeV}$ before the long-wavelength approximation is unusable.

Siegert's theorem [38, 39] uses the continuity equation satisfied by the particle density to allow integration by parts to transform the current operators into second derivatives, where they may be replaced by using the Schrödinger equation satisfied by the particle wave function $\Psi(\mathbf{R})$.

4.7.2 Electric transitions using the Siegert theorem

The Siegert theorem uses the long-wavelength approximation of $kr \ll 1$ to replace the electric plane-wave component by a derivative of the longitudinal plane wave. Since $F_\Lambda(0, kr) \propto (kr)^{\Lambda+1}$, this limit allows us to neglect the $\Lambda = J+1$ terms of both $\mathbf{A}_{JM}(\mathbf{r};\mathcal{E})$ and $\mathbf{A}_{JM}(\mathbf{r};\mathrm{long})$. The two remaining $\Lambda = J-1$ terms are thus proportional to each other, as explained on page 126, allowing the replacement

$$\mathbf{A}_{JM}(\mathbf{r};\mathcal{E}) \simeq \sqrt{\frac{J+1}{J}}\,\mathbf{A}_{JM}(\mathbf{r};\mathrm{long}) = \sqrt{\frac{J+1}{J}}\frac{1}{k}\nabla\frac{1}{kr}F_J(0,kr)Y_J^M(\hat{\mathbf{r}}) \qquad (4.7.3)$$

to a good approximation. If we use the abbreviations

$$s_{kJ} = \sqrt{4\pi}\sqrt{2J+1}\,\mathrm{i}^{-J}\frac{1}{k^2}\sqrt{\frac{J+1}{J}} \tag{4.7.4}$$

$$t_{kJ} = s_{kJ}\,2\sqrt{\frac{h}{\omega}} \tag{4.7.5}$$

$$f_{JM}(\mathbf{r}) = r^{-1}F_J(0,kr)Y_J^M(\hat{\mathbf{r}}), \tag{4.7.6}$$

the electric T matrix from Eq. (4.7.2) can be calculated as

$$\mathbf{T}_{JM}^{(\mathcal{E})} = s_{kJ}\,\langle\nabla f_{JM}(\mathbf{r})|V_{\gamma p}|\,\Psi(\mathbf{R};\mathbf{k}_i)\rangle \tag{4.7.7}$$

$$= t_{kJ}\,\langle\nabla f_{JM}(\mathbf{r})\,\Phi_b\,|\,\hat{\mathbf{j}}_q(\mathbf{r})\,|\,\Psi\rangle. \tag{4.7.8}$$

Remember that the $\langle|\ldots|\rangle$ notation implies integration over all the variables on the left: integration over both photon position $\mathbf{r}$ and particle positions $\mathbf{r}_i$ is implied.

We now use integration by parts to move the ∇ from acting on the f_J to act on the current operator. The limits of the integrand give zero since the bound state $\Phi_b(R) \to 0$ as $R \to \infty$, so

$$\mathbf{T}_{JM}^{(\mathcal{E})} = -t_{kJ}\,\langle\Phi_b\,f_{JM}(\mathbf{r})\,|\nabla\cdot\hat{\mathbf{j}}_q(\mathbf{r})|\,\Psi\rangle. \tag{4.7.9}$$

The divergence of the current $\nabla\cdot\hat{\mathbf{j}}_q(\mathbf{r})$ is then replaced using the continuity equation (3.1.105), yielding

$$\mathbf{T}_{JM}^{(\mathcal{E})} = -t_{kJ}\langle f_{JM}(\mathbf{r})\Phi_b|\frac{\mathrm{i}q}{\hbar}[\delta(\mathbf{r}-\mathbf{r}_i)\,H - H^\dagger\,\delta(\mathbf{r}-\mathbf{r}_i)]\,|\Psi\rangle$$

$$= -t_{kJ}\frac{\mathrm{i}q}{\hbar}\langle f_{JM}(\mathbf{r})|[\Phi_b^*H\Psi - (H\Phi_b)^*\Psi]\rangle$$

$$= -t_{kJ}\frac{\mathrm{i}q}{\hbar}(E_i-E_b^*)\,\langle\Phi_b f_{JM}(\mathbf{r})|\Psi\rangle. \tag{4.7.10}$$

In the last step we used the Schrödinger equations satisfied by Φ_b and Ψ, $[H - E_b]\Phi_b = [H - E_i]\Psi = 0$, and have rewritten the integral as a matrix element in the $\mathbf{r}$ coordinate. Since $E_i - E_b = E_\gamma = \hbar\omega = \hbar kc$ for the photon, and E_b is real for a bound state, the constant factors become

$$-t_{kJ}\frac{\mathrm{i}q}{\hbar}E_\gamma = -\sqrt{4\pi}\sqrt{2J+1}\,\mathrm{i}^{-J}\frac{1}{k^2}\sqrt{\frac{J+1}{J}}2\sqrt{\frac{h}{\omega}}\frac{\mathrm{i}q}{\hbar}E_\gamma$$

$$= -\mathrm{i}q\sqrt{4\pi}\sqrt{2J+1}\,\mathrm{i}^{-J}\sqrt{\frac{8\pi\,\hbar c(J+1)}{k^3 J}}. \tag{4.7.11}$$

The electric photon vector-form **T** matrix is thus

$$\mathbf{T}_{JM}^{(\mathcal{E})} = -i\sqrt{4\pi}\sqrt{2J+1}\sqrt{\frac{8\pi\hbar c(J+1)}{k^3 J}}\langle\Phi_b\, q\, i^J f_{JM}|\Psi\rangle. \tag{4.7.12}$$

In the long-wavelength approximation, we use Eq. (3.1.17) to give

$$F_J(0,kr) \simeq \frac{1}{(2J+1)!!}(kr)^{J+1} \text{ thus } f_{JM}(\mathbf{r}) \simeq \frac{k^{J+1}}{(2J+1)!!}r^J Y_J^M. \tag{4.7.13}$$

The Seigert theorem therefore gives us a matrix element that is proportional to the standard multipole operator of order J for the charge density of Eq. (4.4.25):

$$\langle\Phi_b|q f_{JM}|\Psi\rangle \simeq \frac{k^J}{(2J+1)!!}\langle\Phi_b|q r^J Y_J^M|\Psi\rangle. \tag{4.7.14}$$

The conjugation of f_{JM} and the sign of M in this equation depend on whether the photon is in the initial or the final state.

Reconstituting a local form factor for coupled equations

Having made the reduction to charge-density form, it is sometimes convenient for the uniform treatment of all reaction channels to reconstitute a coupled photon equation which gives the same partial-wave $\mathbf{T}_{\gamma\alpha_i}$-matrix element as Eq. (4.7.12). For photon partial wave γ as in Eq. (3.5.39),

$$\sqrt{4\pi}\sqrt{2J+1} = 4\pi Y_J^0(+z) = kA(\gamma;+z), \tag{4.7.15}$$

and when $\mathbf{T}_{JM}^{(\mathcal{E})}$ is the amplitude for the emission of photons in the $+z$ direction, using Eq. (3.3.44) gives

$$\mathbf{T}_{\gamma\alpha_i} = -\frac{1}{k}\frac{k}{\hbar c}\frac{q}{i}\sqrt{\frac{8\pi\hbar c(J+1)}{Jk}}\sum_\alpha\langle i^J F_J\phi_b|\psi_{\alpha\alpha_i}\rangle. \tag{4.7.16}$$

The same **T**-matrix numerical value can be obtained from the asymptotic solution of a reformulated photon channel equation

$$\left[-\frac{\hbar c}{k}\left(\frac{d^2}{dr^2}-\frac{J(J+1)}{r^2}\right)-E_\gamma\right]\tilde{\zeta}_\gamma(r)$$
$$= \frac{q}{i}\sqrt{8\pi\hbar c(J+1)/(Jk)}\,\phi_b\psi_\alpha(R). \tag{4.7.17}$$

The new photon wave functions $\tilde{\zeta}_\gamma(r)$ are *not* equal to the previous functions $\zeta_\gamma(r)$, to start with being in partial wave $\Lambda' = J$ rather than $\Lambda = J-1$, but they have been constructed to have the same asymptotic **T**-matrix amplitude.

Since the wave functions $\Psi(R)$ are written as functions of the two-body separation R and not on the distance r from their center of mass, Eq. (4.7.17) is a non-local equation. It may be approximated by a local equation if we can use the power series of Eq. (4.7.14) to include the scaling factor in $r = vR$:

$$\left[-\frac{\hbar c}{k} \left(\frac{d^2}{dR^2} - \frac{J(J+1)}{R^2} \right) - E_\gamma \right] \tilde{\zeta}_\gamma(R)$$
$$= v^{J+2} \frac{q}{i} \sqrt{8\pi \hbar c (J+1)/(Jk)} \phi_b \psi_\alpha(R). \tag{4.7.18}$$

This equation, valid in the long-wavelength approximation or when $v \simeq 1$, implies that the transition potential from particle to gamma channels is

$$V_\lambda^{\gamma p}(R) = v^{J+2} \frac{q}{i} \sqrt{8\pi \hbar c (J+1)/(Jk)} \; \phi_b(R) \tag{4.7.19}$$

for multipolarity λ equal to the bound-state orbital angular momentum.

From the **T**-matrix element calculated either by Eq. (4.7.12) or by Eq. (4.7.18), the capture cross section for photon emission is (as stated in Chapter 3)

$$\sigma_{cap}^J = \frac{4\pi}{k_i^2} \frac{1}{(2I_p+1)(2I_t+1)} \frac{c}{v_i} \sum_{J_{tot}} (2J_{tot}+1) \, |\mathbf{T}_{\gamma\alpha_i}^{J_{tot}\pi}|^2. \tag{4.7.20}$$

4.7.3 Combining multiple-particle and γ channels

The coupled-channels formalism that we have used as a general framework for reaction theory is designed to have only *two* bodies in relative motion in each partition. This means that if a two-body composite system or a continuum state decays by γ emission, the theory describes the relative motion of the emitted photon, and the remaining nucleus has to remain as one body, as effectively bound. Any further decays of that residual system will have to be described statistically with the Hauser-Feshbach methods given in Chapter 11.

What the coupled-channels framework can describe well is the production of a composite system, $A + B \rightarrow (AB)^*$, and then the couplings between all the decays of this system to two-body channels. Such channels include particle channels $C+D$, $C'+D'$, etc., as well as γ-decay channels $\gamma + E$, $\gamma' + E'$, etc., where all the C, D and E nuclei are effectively bound. The coupled-channels set for total spin and parity J_{tot}^π will have partial waves for each of the particle channels $C + D$, as well as a set of γ channels, one or more for each residual nucleus E after γ emission. Medium and heavy nuclei will typically have very many states E to which the composite system could decay by photon emission, and each state will contribute channels to the coupled-channels set.

If the composite (AB)* is a resonance, then the coupled-channels framework as presented here will correctly generate the branching ratios to all the particle and γ-decay channels. In the language to be developed in Chapter 10, it will generate all the *partial widths* for both kinds of exit channels. The partial widths to all the possible residual states E will add together, and combine with the particle partial widths for the C + D channels, to form the total width of the resonance that is observed. In Section 11.2 we will discuss some approximations which may help to simplify the problem when there are very many γ-emission channels in heavier nuclei.

4.7.4 Connecting photon cross sections and $B(EJ)$

Let us define the electric multipole operator

$$\mathcal{M}_{JM}^q = qr^J Y_J^M(\hat{\mathbf{r}}) \tag{4.7.21}$$

as that which appears in the earlier matrix element (4.4.25) for Coulomb inelastic scattering. Then the function defined above in Eq. (4.7.6) becomes in the long-wavelength approximation

$$qf_J(\mathbf{r}) \simeq \frac{k^{J+1}}{(2J+1)!!}\mathcal{M}_{JM}^q. \tag{4.7.22}$$

From Eq. (4.7.20), the capture cross section is

$$\sigma_{\text{cap}}^J = \frac{16\pi^3}{k_i^2} \frac{1}{(2I_p+1)(2I_t+1)} \frac{J+1}{J} \frac{k^{2J+1}}{[(2J+1)!!]^2} \frac{1}{\hbar v_i}$$

$$\times \sum_{m_b M M_i} \left| \langle \Phi_b^{m_b} | \mathcal{M}_{JM}^q | \sqrt{\frac{2}{\pi}} \Psi_i^{M_i} \rangle \right|^2. \tag{4.7.23}$$

The incoming scattering waves $\sqrt{2/\pi}\,\Psi_i$ are normalized as delta functions in k, and $\frac{d}{dk} = \hbar v_i \frac{d}{dE}$, so

$$\frac{1}{\hbar v_i} \sum_{m_b M M_i} \left| \langle \Phi_b^{m_b} | \mathcal{M}_{JM}^q | \sqrt{\frac{2}{\pi}} \Psi_i^{M_i} \rangle \right|^2 = (2J_i+1) \frac{d}{dE} B(EJ, i \to \gamma). \tag{4.7.24}$$

This gives the $\gamma \leftarrow p$ cross section

$$\sigma_{\text{cap}}^J = \frac{16\pi^3}{k_i^2} \frac{(2J_i+1)}{(2I_p+1)(2I_t+1)} \frac{J+1}{J} \frac{k^{2J+1}}{[(2J+1)!!]^2} \frac{dB(EJ, i \to \gamma)}{dE}. \tag{4.7.25}$$

Using the detailed balance for photo-nuclear reactions as given by Eq. (3.2.4), the reverse ($p \leftarrow \gamma$) photo-disintegration cross section is

$$\sigma_{\text{photo}}^J = \frac{k_i^2 (2I_p + 1)(2I_t + 1)}{k^2 2(2J_b + 1)} \sigma_{\text{cap}}^J$$

$$= \frac{(2\pi)^3 (J+1)}{J[(2J+1)!!]^2} \frac{2J_i + 1}{2J_b + 1} k^{2J-1} \frac{dB(EJ, p \to \gamma)}{dE}$$

$$= \frac{(2\pi)^3 (J+1)}{J[(2J+1)!!]^2} k^{2J-1} \frac{dB(EJ, \gamma \to p)}{dE}. \tag{4.7.26}$$

For transitions between *discrete* states, see Section 5.4.

Transitions for two-body nuclei

Consider a nucleus consisting of two bodies as in subsection 4.4.2, with a charge Z_c at $\mathbf{r} = -\frac{m_v}{m_p}\mathbf{R}$ and another Z_v at $\mathbf{r} = \frac{m_c}{m_p}\mathbf{R}$, where $\mathbf{R}$ is the vector separation between the clusters. If its initial scattering state is $\Psi_i(\mathbf{R})$ and final bound state is $\Phi_b(\mathbf{R})$, then the electric matrix element EJ is

$$\langle \Phi_b | \mathcal{M}_{JM}^E | \Psi_i \rangle = \int d\mathbf{R} \, \Phi_b^* \Psi_i \left[Z_c e \left(\frac{m_v}{m_p} R \right)^J Y_J^M(-\mathbf{R}) + Z_v e \left(\frac{m_c}{m_p} R \right)^J Y_J^M(\mathbf{R}) \right]$$

$$= e_J \int \Phi_b(\mathbf{R})^* R^J \Psi_i(\mathbf{R}) \, d\mathbf{R}, \tag{4.7.27}$$

where the *effective multipole charge* is

$$e_J = Z_c e \left(-\frac{m_v}{m_p} \right)^J + Z_v e \left(\frac{m_c}{m_p} \right)^J. \tag{4.7.28}$$

The spatial integral is

$$\int \Phi_b^* R^J \Psi_i \, d\mathbf{R} = \frac{1}{\sqrt{4\pi}} \frac{\hat{J}\hat{L}_i}{\hat{L}_f} \langle L_i M_i, JM | L_f M_f \rangle \langle L_i 0, J 0 | L_f 0 \rangle \langle \Phi_b | R^J | \psi_i \rangle, \tag{4.7.29}$$

and the reduced transition probabilities are

$$B(EJ, i \to b) = e_J^2 \frac{2J+1}{4\pi} \left| \langle L_i 0, J 0 | L_f 0 \rangle \langle \Phi_b(R) | R^J | \psi_i(R) \rangle \right|^2. \tag{4.7.30}$$

More general Racah algebra expressions may be obtained when the core and/or valence particles have spins, and these are spectators to the electric transition. When the core or valence particles themselves undergo electric transitions, further terms must be added that involve internal matrix elements for one or both of the bodies.

Limiting cases for neutrons

Consider $L_f = 0$ s-wave bound states of a neutron with a delta-function binding potential $V(R) = \delta(R)$ at the origin. The wave function is $\Phi_b(\mathbf{R}) = \sqrt{2\gamma} R^{-1} \exp(-\gamma R) Y_0^0(\hat{\mathbf{R}})$ for wave number $\gamma = \sqrt{2\mu|E_b|/\hbar^2}$, and the scattering states are p-wave plane waves. The dipole reduced transition probability for photo-disintegration is

$$\frac{dB(E1, 0^+ \to 1^-)}{dE} = \frac{2\hbar^2}{\pi^2\mu}\left(\frac{Z_c e}{A_c+1}\right)^2 \frac{\sqrt{E_b}E^{3/2}}{(E+E_b)^4}. \tag{4.7.31}$$

If instead $\Phi_b(\mathbf{R}) = C_0 R^{-1} \exp(-\gamma R) Y_0^0(\hat{\mathbf{R}})$ for asymptotic normalization coefficient (ANC) C_0 defined by Eq. (4.5.24), then this result will be scaled by $C_0^2/2\gamma$.

Limiting cases for protons

For low-energy capture of charged particles, the dominating feature is the Coulomb barrier in the entrance channel. At very low energies, if there are no resonances then the entrance phase shifts will be zero, and the incoming wave function will be essentially $\psi_{\alpha\alpha_i}(R) = \delta_{\alpha\alpha_i} F_L(\eta_i, k_i R)$. According to Box 3.2, in the $\eta \gg L_i$ and $k_i R \ll 1$ limits this becomes

$$F_{L_i}(\eta_i, k_i R) \to \frac{(\eta_i k_i)^{L_i}}{L_i!(2L_i+1)!}\sqrt{\frac{2\pi\eta_i}{e^{2\pi\eta_i}}} k_i R^{L_i+1}, \tag{4.7.32}$$

noting that $\eta_i k_i$ is independent of energy. Since now this elastic wave function is the only non-zero channel in Eq. (4.7.16), we get from Eq. (4.7.20) the cross section for a specific J_{tot}, entrance channel L_i and multipole J of

$$\begin{aligned}
\sigma_{\text{cap}}^J &= \frac{4\pi}{k_i^2} g_{J_{\text{tot}}} \frac{c}{v_i} \left|\frac{q}{\hbar c}\sqrt{\frac{8\pi\hbar c(J+1)}{Jk}}\langle F_J(0, kr)|\phi_b|F_{L_i}(\eta_i, k_i R)\rangle\right|^2 \\
&= \frac{e^{-2\pi\eta_i}}{E_i}\frac{32\pi^3\hbar c}{k} g_{J_{\text{tot}}} \frac{(\eta_i k_i)^{2L_i+1}}{[L_i!(2L_i+1)!]^2}\frac{(J+1)}{J}\frac{q^2}{\hbar c}\left|\langle F_J(0, kr)|R^{L_i+1}|\phi_b\rangle\right|^2,
\end{aligned} \tag{4.7.33}$$

omitting the angular-momentum coupling factors. At low proton energies E_i, the photon energy $E_\gamma = E_i - E_b$ will tend to a constant, so the factor governing the low-energy behavior here is just the $e^{-2\pi\eta_i}/E_i$, which is exactly that built into the definition of the astrophysical S-factor in Eq. (1.1.4). The same factor thus describes the energy dependence for all partial waves without resonances.

The integrand of the final matrix element will be dominated by the long-range behavior of the bound-state wave function $\phi_b(R)$, which can be characterized by its asymptotic normalization coefficient C_b defined according to Eq. (4.5.24). At low incident energies, therefore, the astrophysical S-factor will tend to be constant with energy, and the constant will be proportional to C_b^2. For more details, see references [40, 41, 42, 43].

4.7.5 Magnetic transitions

The coupled equations Eq. (4.7.1) describe the combined electric and magnetic contributions to the photon cross section. The separated electric part was calculated from Eq. (4.7.7), and now we look at MJ transitions which give overall parity changes of $(-1)^{J-1}$ as explained in Box 3.5. The *magnetic* part of the MJ transition integral depends only on the $\Lambda = J$ components, and is

$$\mathbf{T}_{JM}^{(\mathcal{M})} = \langle \sqrt{4\pi}\sqrt{2J+1}\, \mathrm{i}^J\, \mathbf{A}_{JM}(\mathbf{r};\mathcal{M})|V_{\gamma p}|\Psi(\mathbf{R};\mathbf{k}_i)\rangle, \qquad (4.7.34)$$

where, from Eq. (3.5.50), the magnetic plane-wave part is

$$\mathbf{A}_{JM}(\mathbf{r};\mathcal{M}) = (kr)^{-1}F_J(0,kr)\,\mathbf{Y}_{JJ}^M(\hat{\mathbf{r}})\,, \qquad (4.7.35)$$

and, from Eq. (3.5.35), the coupling operator is

$$
\begin{aligned}
V_{\gamma p} &= 2\sqrt{h/\omega}\langle\Phi_b|\hat{\mathbf{j}}_q(\mathbf{r})\rangle \\
&= 2\sqrt{\frac{2\pi\hbar}{kc}}\frac{\hbar q}{2\mathrm{i}\mu}\langle\Phi_b|\{\delta(\mathbf{r}-\mathbf{r}_i)\,\nabla - \nabla\,\delta(\mathbf{r}-\mathbf{r}_i)\}\,. \qquad (4.7.36)
\end{aligned}
$$

Now the vector spherical harmonic in Eq. (4.7.35) may be obtained also by acting on a spherical harmonic with the angular-momentum vector operator,

$$\mathbf{L}\,Y_J^M(\hat{\mathbf{r}}) = [J(J+1)]^{\frac{1}{2}}\mathbf{Y}_{JJ}^M(\hat{\mathbf{r}}), \qquad (4.7.37)$$

so

$$
\begin{aligned}
\mathbf{A}_{JM}(\mathbf{r};\mathcal{M}) &= [J(J+1)]^{-\frac{1}{2}}(kr)^{-1}F_J(0,kr)\,\mathbf{L}\,Y_J^M(\hat{\mathbf{r}}) \\
&= [J(J+1)]^{-\frac{1}{2}}\mathbf{L}\,(kr)^{-1}F_J(0,kr)\,Y_J^M(\hat{\mathbf{r}}) \qquad (4.7.38) \\
&\equiv [J(J+1)]^{-\frac{1}{2}}k^{-1}\mathbf{L}f_{JM}(\mathbf{r}), \qquad (4.7.39)
\end{aligned}
$$

using the fact that $\mathbf{L}$ commutes with any purely radial function, and then using the definition of $f_{JM}(\mathbf{r})$ in Eq. (4.7.4).

The T matrix is therefore

$$\mathbf{T}_{JM}^{(M)} = \sqrt{\frac{4\pi(2J+1)}{k^2 J(J+1)}} \, \langle \mathrm{i}^J \, \mathbf{L} f_{JM}(\mathbf{r}) | V_{\gamma p} | \Psi(\mathbf{R}; \mathbf{k}_i) \rangle \qquad (4.7.40)$$

$$= 2\sqrt{\frac{2\pi\hbar}{kc}} \frac{\hbar q}{2\mathrm{i}\mu} \sqrt{\frac{4\pi(2J+1)}{k^2 J(J+1)}} \mathrm{i}^{-J} I, \qquad (4.7.41)$$

where I is the integral

$$I = \int \mathrm{d}\mathbf{r} \, [\Phi_b^* \, \mathbf{L} f_{JM}(\mathbf{r})^* \cdot \nabla\Psi + \Psi \, \mathbf{L} f_{JM}(\mathbf{r})^* \cdot \nabla\Phi_b^*]. \qquad (4.7.42)$$

Now $\mathbf{L} \cdot \nabla = \mathbf{r} \times \mathbf{p} \cdot \nabla = 0$, so $[\mathbf{L} f(\mathbf{r})] \cdot \nabla = -[\nabla f(\mathbf{r})] \cdot \mathbf{L}$ for any spatial function $f(\mathbf{r})$. Thus

$$I = -\int \mathrm{d}\mathbf{r} \, [\Phi_b^* \, \nabla f_{JM}(\mathbf{r})^* \cdot \mathbf{L}\Psi + \Psi \, \nabla f_{JM}(\mathbf{r})^* \cdot \mathbf{L}\Phi_b^*], \qquad (4.7.43)$$

where from Eq. (3.5.62) we find that

$$\nabla f_{JM}(\mathbf{r}) = k^2 \mathbf{A}_{JM}(\mathbf{r}; \mathrm{long}) \qquad (4.7.44)$$

$$\approx \sqrt{\frac{J}{2J+1}} \frac{k}{r} F_{J-1}(0, kr) \mathbf{Y}_{J-1,J}^M(\hat{\mathbf{r}}) \qquad (4.7.45)$$

in the long-wavelength approximation which allows us to neglect the second term in Eq. (3.5.62). These steps remove all the derivative operators from the matrix element of I, giving

$$I = -\sqrt{\frac{J}{2J+1}} \frac{k}{r} \int \mathrm{d}\mathbf{r} F_{\Lambda'}(0, kr) [\Phi_b^* \mathbf{Y}_{\Lambda'J}^{M*} \cdot \mathbf{L}\Psi + \Psi \mathbf{Y}_{\Lambda'J}^{M*} \cdot \mathbf{L}\Phi_b^*], \qquad (4.7.46)$$

where we define $\Lambda' = J-1$. The photon spatial wave functions are now in partial wave Λ' rather than the original $\Lambda = J$. The parity change between Ψ and Φ_b is now simply $(-1)^{\Lambda'}$, which agrees with Eq. (4.7.46) as there are no longer derivative operators which reverse the parity.

Note that this transformation of photon partial waves is the opposite of that accomplished by the Siegert theorem, which transforms an original $\Lambda = J-1$ to $\Lambda' = J$ in order to remove the derivative operators in the electric matrix element. In both cases, the electromagnetic parity change is simply now $(-1)^{\Lambda'}$.

The angular parts of the matrix elements have reduced matrix elements

$$\langle L_b || \mathbf{Y}_{\Lambda'J}^{M*} \cdot \mathbf{L} || L_i \rangle = \frac{(2L_i+1)\hat{\Lambda}'\hat{J}\sqrt{L_i(L_i+1)}}{\sqrt{4\pi}} \langle L_b 0, \Lambda' 0 | L_i 0 \rangle W(L_i 1 L_b \Lambda'; L_i J).$$

$$(4.7.47)$$

We may now follow the same procedures as above for electric transitions, and derive transition potentials for use in coupled particle-photon equations, and also define a *magnetic multipole operator* $\mathcal{M}_{JM}^{\mathrm{mag}}$ to construct a reduced transition probability $dB(MJ, \gamma \to p)/dE$.

Exercises

4.1 Calculate the elastic scattering of protons on ^{112}Cd at 30 MeV using the global potential of Koning and Delaroche. Compare with the results obtained with Becchetti and Greenlees.

4.2 Work out the potentials for 10 MeV proton scattering on ^{112}Cd and compare with those obtained for 30 MeV.

4.3 The local energy approximation is very useful to simplify the non-local transfer kernel. Derive the expression for the form factor given in Eq. (4.5.19).

4.4 The zero-range approximation was widely used in the early days of nuclear reactions when analyzing (d,p) data. In this approximation, the transition matrix is proportional to D_0, the zero-range constant. Determine D_0 and D assuming a deuteron gaussian potential $V(r) = -72.15 \exp(-(r/1.484)^2)$ in MeV, for radii in fm. Find the effective radius ρ_{eff}.

4.5 $E1$ transitions connecting the ground state with the continuum are very important in capture reactions. If one takes the asymptotic limit, there is an analytic derivation for the $dB(E1)/dE$, the result of which is shown in Eq. (4.7.31). Assume the bound state is given by $\Phi_b(\mathbf{R}) = \sqrt{2\gamma}R^{-1}\exp(-\gamma R)Y_0^0(\hat{\mathbf{R}})$ and the final state is a plane wave. Derive Eq. (4.7.31).

References

[1] C. M. Perey and F. G. Perey, *At. Data Nucl. Data Tables* **13** (1974) 293.

[2] T. Belgya, O. Bersillon, R. Capote, T. Fukahori, G. Zhigang, S. Goriely, M. Herman, A. V. Ignatyuk, S. Kailas, A. Koning, P. Oblozhinsky, V. Plujko and P. Young. *Handbook for Calculations of Nuclear Reaction Data: Reference Input Parameter Library*, URL:www-nds.iaea.org/RIPL-2, IAEA:Vienna, 2005.

[3] A. J. Koning and J. P. Delaroche, *Nucl. Phys. A* **713** (2003) 231.

[4] D. G. Madland, 'Progress in the development of global medium-energy nucleon-nucleus optical-model potential', Proc. OECD/NEA Specialists Meeting on Nucleon-Nucleus Optical Model to 200 MeV, Bruyères-le-Châtel, France (1997) 129.

[5] F. D. Becchetti and G. W. Greenlees, *Phys. Rev.* **182** (1969) 1190.

[6] R. L. Walter and P. P. Guss, *Rad. Effects* **95** (1986) 73.

[7] R. L. Varner, W. J. Thompson, T. L. Mcabee, E. J. Ludwig and T. B. Clegg, *Phys. Rep.* **201** (1991) 57.

[8] P. A. Moldauer, *Nucl. Phys.* **47** (1963) 65.

[9] D. Wilmore and P. E. Hodgson, *Nucl. Phys.* **55** (1964) 673.

[10] C. A. Engelbrecht and H. Friedeldey, *Ann. Phys.* **42** (1967) 262.

[11] B. Strohmaier, M. Uhl and W. Reiter, 'Neutron cross section calculations for ^{52}Cr, ^{55}Mn, ^{56}Fe, and ^{58,60}Ni', INDC(NDS)-128, IAEA Vienna (1982).

[12] J. J. H. Menet, E. E. Gross, J. J. Malanify and A. Zucker, *Phys. Rev. C* **4** (1971) 1114.

[13] F. G. Perey, *Phys. Rev.* **131** (1963) 745.

[14] D. M. Patterson, R. R. Doering and A. Galonsky, *Nucl. Phys. A* **263** (1976) 261.

[15] J. Bojowald, H. Machner, H. Nann, W. Oelert, M. Rogge and P. Turek, *Phys. Rev. C* **38** (1988) 1153.

[16] W. W. Daehnik, J. D. Childs, and Z. Vrcelj, *Phys. Rev. C* **21** (1980) 2253.

[17] J. M. Lohr and W. Haeberli, *Nucl. Phys. A* **232** (1974) 381.

[18] C. M. Perey and F. G. Perey, *At. Data Nucl. Data Tables* **17** (1976) 1.

[19] F. D. Becchetti and G. W. Greenlees, *Annual Report*, J. H. Williams Laboratory, University of Minnesota (1969).

[20] L. McFadden and G. R. Satchler, *Nucl. Phys.* **84** (1966) 177.

[21] V. Avrigeanu, P. E. Hodgson and M. Avrigeanu, *Phys. Rev. C* **49** (1994) 2136.

[22] J. R. Huizenga and G. Igo, *Nucl. Phys.* **29** (1962) 462.

[23] J. -P. Jeukenne, A. Lejeune and C. Mahaux, *Phys. Rev. C* **16** (1977) 80.

[24] T. Berggren and P. Lind, *Phys. Rev. C* **47** (1993) 768.

[25] A. Bohr and B. R. Mottelson 1975, *Nuclear Structure, Vol. 1*, Reading, MA: Benjamin.

[26] D. Brink and G. R. Satchler 1993, *Angular Momentum*, Oxford: Clarendon Press.

[27] L. H. Thomas, *Nature* **117** (1926) 514.

[28] G. R. Satchler, *Nucl. Phys.* **11** (1960) 116.

[29] J. M. Eisenberg and W. Greiner 1987, *Nuclear Theory: Nuclear models, Vol. 1*, Amsterdam: North-Holland.

[30] M. Moshinsky, *Nucl. Phys.* **13** (1959) 104.

[31] P. J. A. Buttle and L. J. B. Goldfarb, *Proc. Phys. Soc.* (London) **83** (1964) 701; see also §6.14.1 of G. R. Satchler 1983, *Direct Nuclear Reactions*, Oxford: Clarendon Press.

[32] L. J. B. Goldfarb and E. Parry, *Nucl. Phys.* **A116** (1968) 309.

[33] C. M. Vincent and H. T. Fortune, *Phys. Rev. C* **2** (1970) 782.

[34] D. L. Hill and J. A. Wheeler, *Phys. Rev.* **89** (1953) 1102.

[35] A. M. Lane, *Nucl. Phys.* **35** (1962) 676.

[36] J. M. Lafferty and S. R. Cotanch, *Nucl. Phys.* **A373** (1982) 363.

[37] W. E. Parker, *et al.*, *Phys. Rev. C* **52** (1995) 252.

[38] A. J. F. Siegert, *Phys. Rev.* **52** (1937) 787.

[39] J. M. Eisenberg and W. Greiner 1988, *Nuclear Theory: Excitation Mechanisms of the Nucleus, Vol. 2*, Amsterdam: North-Holland.

[40] S. Typel and G. Baur, *Nucl. Phys. A* **759** (2005) 247.

[41] C. Forssen, N. B. Shulgina and M. V. Zhukov, *Phys. Lett.* **B549** (2002) 79.

[42] A. M. Mukhamedzhanov and F. M. Nunes, *Nucl. Phys. A* **708** (2002) 437.

[43] D. Baye, *Phys. Rev. C* **70** (2004) 015801.

5

Connecting structure with reactions

Winning the (Nobel) prize wasn't half as exciting as doing the work.

Maria Mayer

In the previous chapter we presented non-elastic mechanisms based on rotational or vibrational models, and on transfer, capture and knockout reactions based on the separation of a nucleus into a nucleon and a core cluster. In this chapter we see how the necessary properties of these structure models may be related to nuclear structure theories that should be more exact because they are microscopic and take *all* the many nucleons of the nuclei into account. It is beyond the scope of this book on reactions to discuss detailed methods and numerical examples using microscopic models, so we adopt the aim of establishing a 'common language' between structure and direct-reaction theories. In particular, we show how masses, sizes, folding potentials, overlaps and other matrix elements may be defined in terms of structure models and then used for reaction calculations.

5.1 Summary of structure models

A nucleus contains a number A of nucleons, all pairs ij of which interact with each other by a nucleon-nucleon potential $V^{(2)}(\mathbf{r}_i - \mathbf{r}_j)$ which is strongly repulsive at short distances ($r_{ij} \equiv |\mathbf{r}_i - \mathbf{r}_j| \lesssim 1$ fm), attractive at medium distances (1 fm $\lesssim r_{ij} \lesssim 4$ fm), while protons repel even at large distances. There may also be three-body forces between bodies ijk according to some form $V^{(3)}(\mathbf{r}_i - \mathbf{r}_j, \mathbf{r}_i - \mathbf{r}_k)$. In general, $V^{(2)}$ and $V^{(3)}$ depend also on the spin states of the interacting nucleons, and should therefore be expanded on a sufficiently complete set of vector and tensor operators.

The exact solution for the nuclear wave function will be specified in terms of the positions $\boldsymbol{\rho}_i = \mathbf{r}_i - \mathbf{S}$ of each nucleon with respect to the center of mass $\mathbf{S} = \sum_{i=1}^{A} \mathbf{r}_i / A$ of the whole nucleus. Only $A-1$ of these $\boldsymbol{\rho}_i$ coordinates will be linearly dependent as $\sum_{i=1}^{A} \boldsymbol{\rho}_i = 0$, so we write the internal wave function as

$\Phi(\boldsymbol{\rho}_1,\ldots,\boldsymbol{\rho}_{A-1})$ that is independent of $\mathbf{S}$. In this way we can specify the internal state of the nucleus in a manner which is translationally invariant.

Let $\Phi_{I\mu}$ be the eigensolution of the many-body Schrödinger equation[1] when the total nucleus has spin I and z-projection μ. It is the solution of a Schrödinger equation with E_I as the energy eigenvalue,[2] namely

$$H_A \Phi_{I\mu}(\boldsymbol{\rho}_1,\ldots,\boldsymbol{\rho}_{A-1}) = E_I \Phi_{I\mu} \tag{5.1.1}$$

where the Hamiltonian H_A is the sum of kinetic and potential energy parts for the A nucleons. The kinetic energy part is ideally written in terms of translationally invariant coordinates such as the $\{\boldsymbol{\rho}_i\}$, but a simple alternative is to write it as the total kinetic energy of all the A nucleons *minus* the kinetic energy of the motion of the center of mass. In this way, the Hamiltonian H_A can be given as

$$H_A = -\sum_{i=1}^{A} \frac{\hbar^2}{2m_i} \nabla_{\mathbf{r}_i}^2 + \frac{\hbar^2}{2M} \nabla_{\mathbf{S}}^2 + \sum_{i>j}^{A} V^{(2)}(\mathbf{r}_i-\mathbf{r}_j) + \sum_{i>j>k}^{A} V^{(3)}(\mathbf{r}_i-\mathbf{r}_j, \mathbf{r}_i-\mathbf{r}_k),$$

$$\tag{5.1.2}$$

where $M = \sum_i m_i$ is the total mass of the nucleus. The boundary conditions to be satisfied are that, for all i,

$$\lim_{\boldsymbol{\rho}_i \to \infty} \Phi_{I\mu}(\ldots,\boldsymbol{\rho}_i,\ldots) = 0 \tag{5.1.3}$$

for bound states, when E_I is below all the breakup thresholds. The wave function can then be normalized, usually as

$$\int \mathrm{d}\boldsymbol{\rho}_1 \ldots \int \mathrm{d}\boldsymbol{\rho}_{A-1} |\Phi_{I\mu}(\boldsymbol{\rho}_1,\ldots,\boldsymbol{\rho}_{A-1})|^2 = 1 \tag{5.1.4}$$

for any μ. When E_I is above some breakup threshold, then the exact eigensolution should be rather a linear combination of ingoing and outgoing scattering boundary conditions, as described in Chapter 3.

The solving of $(H_A - E_I)\Phi_{I\mu} = 0$ is a very difficult numerical problem, and has so far been achieved completely only for $A = 2-4$ by few-body methods, for $A \lesssim 12$ by time-intensive Monte Carlo techniques [1], and with scattering boundary conditions only up to $A \approx 5$ [2]. To date, the Green's function Monte Carlo

---

[1] In this chapter, we will use Φ for the wave function of a whole nucleus, and ϕ for the individual single-particle wave functions.
[2] The energy eigenvalue does not depend on μ if the Hamiltonian is purely internal and hence rotationally invariant.

method [1] is the only one that uses realistic NN forces directly in the many-body Schrödinger equation (5.1.1) with $A \geq 5$.

More commonly, various approximations are made to simplify the problem. For applications, we thus desire approximations which still reproduce as accurately as possible the features of $\Phi_{I\mu}$ influencing low-energy reactions. The approximation methods all reduce the allowed complexity of the wave function, and derive *effective interactions* to reformulate the Hamiltonian.

We can see this formally as dividing the Hilbert space for the wave functions into two parts by means of projection operators P for the simplified range, and $Q = 1 - P$ for the remaining part. Feshbach [3] shows the effective Hamiltonian for the P part

$$H_{\text{eff}}(P\Phi_{I\mu}) = E_I(P\Phi_{I\mu}) \qquad (5.1.5)$$

with

$$H_{\text{eff}} = PHP + PHQ\frac{1}{E - QHQ}QHP. \qquad (5.1.6)$$

The Green's operator $(E - QHQ)^{-1}$ should have specified scattering boundary conditions for any open channels. The Lee-Suzuki method [4] is another method of deriving a simplified Hamiltonian for the reduced model space.

The No-Core Shell Model (NCSM) [5] and $V_{\text{low-k}}$ [6] methods apply these reductions in a systematic way. The more traditional practice has been to define a reduced model space for $P\Phi_{I\mu}$, and then use *experimental* results to fit the set of parameters in the effective Hamiltonian H_{PP}.

We actually use this method – of finding an effective interaction appropriate for a reduced model space – to find the interactions needed for low-energy nuclear reactions. A wide range of model spaces will be used, such as shell models, cluster models, mean-field models, and collective models for the whole nucleus, in order of decreasing scope of the model space. In this sequence, therefore, progressively *fewer* nucleon-nucleon correlations are described, and the models become correspondingly simpler. Which model we should use will depend on which features determine the nuclear reactions of interest.

We now briefly describe the model spaces and features of this range of nuclear models.

5.1.1 Shell models

Shell-model wave functions are expanded using harmonic oscillator wave functions as the single-particle basis states, for some oscillator energy $\hbar\omega$. A single *Slater determinant* for A identical nucleons is the antisymmetrized product of A single-particle wave functions:

$$\Theta_{\mathcal{S}}(\mathbf{r}_1, \ldots, \mathbf{r}_A) = \mathcal{A} \prod_{i=1}^{A} \phi_{n_i \ell_i j_i}(\mathbf{r}_i, s_i) \tag{5.1.7}$$

for a given set $\mathcal{S} = \{n_i \ell_i j_i, \ i = 1 \cdots A\}$ of single-particle quantum numbers. The $\mathcal{A}$ is the antisymmetrization operator. Some shell models couple all the states to good total angular momentum I, whereas others use an m-state basis and only discriminate between different I after diagonalization.

The single-nucleon states $\phi(\mathbf{r}_i, \sigma_i)$ for spatial coordinate $\mathbf{r}_i$ and Pauli spin projections σ_i, labeled by quantum numbers $n\ell j$, are

$$\phi_{n\ell j}^m(\mathbf{r}, \sigma) = R_{n\ell}(r)/r \ [Y_\ell^\mu(\hat{\mathbf{r}}) \otimes \chi_s^\sigma]_{jm} \tag{5.1.8}$$

where $s = 1/2$ is the fixed nucleonic spin, $n = 0, 1, \ldots$ is the number of radial nodes (excluding the origin), and m is the angular-momentum projection on the z axis. The harmonic-oscillator radial forms are

$$R_{nl}(r) = \left[\frac{2\alpha \Gamma(n+1)}{\Gamma(n+l+\frac{3}{2})} \right]^{\frac{1}{2}} \mathrm{e}^{-\alpha^2 r^2/2} (\alpha r)^{l+1} L_n^{l+\frac{1}{2}}(\alpha^2 r^2) \tag{5.1.9}$$

for $\alpha = \sqrt{\omega m/\hbar}$, where

$$L_n^{l+\frac{1}{2}}(\alpha^2 r^2) = \sum_{k=0}^{n} \frac{\Gamma(n+l+\frac{3}{2})(-\alpha^2 r^2)^k}{\Gamma(n-k+1)\Gamma(l+k+\frac{3}{2})k!} \tag{5.1.10}$$

are associated Laguerre polynomials. The energies of these basis states are shown in Fig. 5.1 in the leftmost column. The second column shows the splittings expected in a finite well, and the third shows the effect of spin-orbit forces. If there were no further interactions, then these levels will fill up in sequences according to $n_{\mathrm{occ}}(j)$, and the gaps between various levels give rise to the closed-shell (magic) numbers when a shell is completely full.

In a specific shell-model calculation (including interactions), the many-body wave function $\Phi(\mathbf{r}_1, \ldots, \mathbf{r}_A)$ is necessarily expanded for a specific set of allowed $n_i \ell_i j_i$ states for some value of $\hbar\omega$ chosen according to the expected radial size of the final nucleus. The calculations may be classified by the number $\mathcal{N} = 2n_i + \ell_i$ of oscillator quanta for each nucleon. According to $\mathcal{N}$, we show in Table 5.1 the historical sequence of shell-models that have been solved. The size of the matrices needing to be diagonalized in the shell model rises very rapidly with $\mathcal{N}$ and the number of nucleons, limiting the range of applicable nuclei according to the computers available. Shell-model calculations are also classified as $0\hbar\omega$ or $1\hbar\omega$ according to the *excitations* $\Delta\mathcal{N}$ allowed above the initial $\mathcal{N}$.

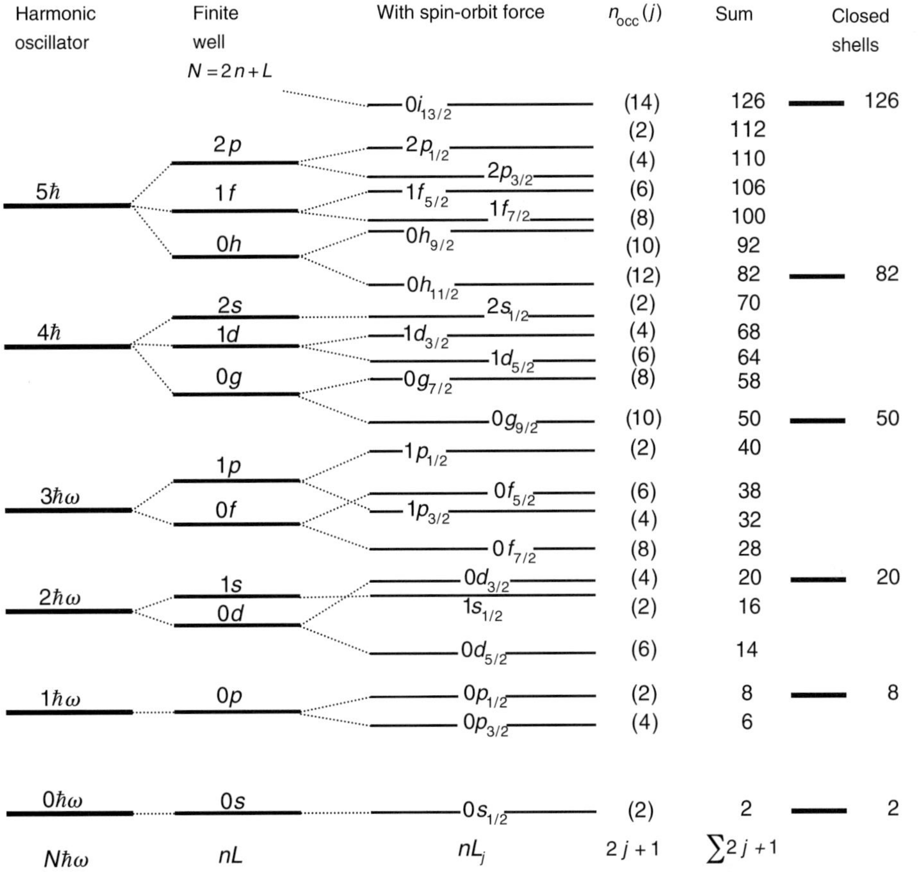

Harmonic oscillator	Finite well $N = 2n+L$	With spin-orbit force	$n_{occ}(j)$	Sum	Closed shells
		$0i_{13/2}$	(14)	126	126
	$2p$	$2p_{1/2}$	(2)	112	
$5\hbar$		$2p_{3/2}$	(4)	110	
	$1f$	$1f_{5/2}$	(6)	106	
		$1f_{7/2}$	(8)	100	
	$0h$	$0h_{9/2}$	(10)	92	
		$0h_{11/2}$	(12)	82	82
$4\hbar$	$2s$	$2s_{1/2}$	(2)	70	
	$1d$	$1d_{3/2}$	(4)	68	
		$1d_{5/2}$	(6)	64	
	$0g$	$0g_{7/2}$	(8)	58	
		$0g_{9/2}$	(10)	50	50
	$1p$	$1p_{1/2}$	(2)	40	
$3\hbar\omega$		$0f_{5/2}$	(6)	38	
	$0f$	$1p_{3/2}$	(4)	32	
		$0f_{7/2}$	(8)	28	
	$1s$	$0d_{3/2}$	(4)	20	20
$2\hbar\omega$	$0d$	$1s_{1/2}$	(2)	16	
		$0d_{5/2}$	(6)	14	
$1\hbar\omega$	$0p$	$0p_{1/2}$	(2)	8	8
		$0p_{3/2}$	(4)	6	
$0\hbar\omega$	$0s$	$0s_{1/2}$	(2)	2	2
$N\hbar\omega$	nL	nL_j	$2j+1$	$\sum 2j+1$	

Fig. 5.1. Single-particle levels are the basis states for the shell model.

The calculations in Table 5.1 also differ in how the angular momenta $\{\mathbf{l}_i, \mathbf{s}_i\}$ are coupled together. Traditional methods [7, 8] used sequences of *fractional parentage expansions* to define states $\Phi_{I\mu}$ of good angular momentum I, but more recent calculations [9, 10, 5] use the *m*-state scheme, wherein the expectation values of I are only calculated for each eigenstate after diagonalization.

The many-body wave function for the whole nucleus is constructed as a superposition of many such Slater determinants for different quantum number sets $\mathcal{S}$. For a given set of $\phi_{n_i \ell_i j_i}$ basis states, limited as above by a specific $\mathcal{N}$ value, the shell-model wave function is

$$\Phi_A^{SM}(\mathbf{r}_1, \ldots, \mathbf{r}_A) = \sum_{\mathcal{S}=1}^{\mathcal{S}_{\max}} a_{\mathcal{S}} \, \mathcal{A} \prod_{i=1}^{A} \phi_{n_i \ell_i j_i}(\mathbf{r}_i, \sigma_i) \qquad (5.1.11)$$

Table 5.1. *The range of shell-model calculations performed for*
nucleons in shells with range of quanta from $\mathcal{N}$ to $\mathcal{N} + \Delta\mathcal{N}$.

$\mathcal{N}$	shell	$\Delta\mathcal{N}$	models for
0	0s only	0	simplified ^{3}He and ^{4}He
1	0p	0	^{6}Li to ^{15}N: Cohen and Kurath [7]
2	sd	0	^{17}O to ^{39}Ca: Wildenthal [8]
1, 2	mixed p & sd	1	e.g. ^{11}Be: Brown [9]
3	pf	0	^{41}C and above: Caurier [10]
0, 1	sp	8	No-Core Shell Model [5] up to ^{16}O

using amplitudes a_S for each set S of single-particle states, where S_{max} is the total number of such sets in the model. The a_S amplitudes are finally found as the eigenvector components of the effective Hamiltonian H_{eff} in this basis, as

$$[H_{\text{eff}} - E_{\text{SM}}]\Phi_A^{\text{SM}}(\mathbf{r}_1, \ldots, \mathbf{r}_A) = 0. \qquad (5.1.12)$$

Because of the restricted range of $\mathcal{N}\hbar\omega$, the effective potentials that must be used in H_{eff} cannot include all the short-range repulsive part of the NN interaction. Either effective $\tilde{V}^{(2)}$ and $\tilde{V}^{(3)}$ are calculated explicitly, as in the NCSM [5], or the values of $\langle ab|V_{\text{NN}}^{(2)}|cd\rangle$ matrix elements of the form

$$\langle \phi_{n_a\ell_a j_a}(\mathbf{r})\phi_{n_b\ell_b j_b}(\mathbf{r}') \mid V_{\text{NN}}^{(2)}(\mathbf{r}-\mathbf{r}') \mid \phi_{n_c\ell_c j_c}(\mathbf{r})\phi_{n_d\ell_d j_d}(\mathbf{r}')\rangle \qquad (5.1.13)$$

are fitted so that the A-body eigenenergies approach those seen in nature.

States of a specific $\mathcal{N}\hbar\omega$ calculation all have the same parity: if there are n nucleons in a shell $\mathcal{N}$, the overall parity of the nucleons in that shell will be $(-1)^{n\mathcal{N}}$. In nature these states are mixed with others of opposite parity, which may come about in the model either by including *hole* states from the deeply bound filled shells, or by including *intruder* states from less-bound higher shells. The mixed p,sd-shell models of [9] attempt to include intruder states in a systematic way by mixing $\mathcal{N} = 1$ and 2 basis states.

The wave function of Eq. (5.1.11) for a state of A nucleons is a function of all coordinates $\{\mathbf{r}_i, i = 1 \cdots A\}$, and therefore includes some center-of-mass motion of the whole nucleus. In the shell model, if a complete set of harmonic-oscillator basis states is included up to some shell closure at $\mathcal{N} + \Delta\mathcal{N}$, it can be arranged that the center-of-mass motion is in its ground state $\Psi_{L=0}^{\text{cm}}(\mathbf{S})$ for the whole nucleus, and therefore factorized from the wave function. The true A-body wave function appearing in Eq. (5.1.4) is therefore

$$\Phi_{I\mu}(\boldsymbol{\rho}_1, \ldots, \boldsymbol{\rho}_{A-1}) = P_{I\mu}\Phi_A^{SM}(\mathbf{r}_1, \ldots, \mathbf{r}_A)/\Psi_{L=0}^{cm}(\mathbf{S}) \qquad (5.1.14)$$

where $P_{I\mu}$ is a projection operator that selects total spin state $I\mu$. In this way, translationally invariant properties may be extracted.

One significant feature of harmonic-oscillator basis states $R_{n\ell}(r)$ is that they all decay at large distances as $\exp(-\alpha^2 r^2/2)$, and so fail to accurately describe weakly bound states, which have $\exp(-\gamma r)$ behavior.[3]

The original harmonic oscillator basis states necessarily fail with unbound states, as these should oscillate to infinity. To meet this problem, there has been recent research on continuum shell models (e.g. [12, 13, 14, 15]) to describe these unbound states by either adding in, or matching to, explicit scattering wave functions with the correct asymptotic forms for positive (continuum) energies.

5.1.2 Cluster models

These models divide the nucleus into two or more groups, each group containing either one nucleon or a bound cluster of nucleons. The effective interactions *between* the clusters is used to find the dynamical behavior. The reduced dimensions of the model space allow both bound and scattering states to be calculated.

Examples of cluster models are

- *particle-core* models: $^{17}O = {}^{16}O + n$; $^8B = {}^7Be + p$
- *deuteron-core* models: $^6Li = \alpha + d$
- *triton-core* models: $^7Li = \alpha + t$
- *alpha-cluster* models: $^8Be = 2\alpha$; $^{12}C^* = 3\alpha$; $^{20}Ne = {}^{16}O + \alpha$
- *three-cluster* models: $^6He = \alpha + n + n$; $^8B = \alpha + {}^3He + p$

Cluster models simplify the many-body problem by assuming that each cluster is either frozen in its ground state, or is allowed to be in one of a small set of low-lying excited states.

The effective interactions used in cluster models are of two kinds. They may be (A) real optical potentials, with maybe spin-orbit or tensor components, or (B) otherwise derived from an NN effective force using fixed antisymmetrized wave functions for each cluster. Method (A) is less microscopic, but allows each

---

[3] A traditional 'fix' has been to use spectroscopic amplitudes from diagonalizations to multiply wave functions calculated in finite Woods-Saxon wells. These wave functions in finite wells cannot be used in the main shell-model calculation for the $\phi_{n\ell j}^m(\mathbf{r}, \sigma)$, as then the factorization of center-of-mass motion as in Eq. (5.1.14) would not work.

A more systematic approach is to use the 'transformed harmonic oscillator' (THO) basis [11], which redefines the meaning of the radius coordinate as to model some exponential tail more accurately.

cluster-cluster potential to be fine-tuned to known bound states and/or resonances. Method (B) is called the *resonating group method* (RGM) that uses traditionally a Gaussian central and spin-orbit NN interaction, with one adjustable parameter in its spin-dependence to fit some essential threshold energy [16].

Two-cluster models are easy to solve, using elementary bound or scattering techniques, as in Chapter 3. Three-cluster models, by contrast, lead to partial differential equations in the two *Jacobi coordinates*: two translationally invariant relative coordinate vectors. The partial differential equations may be converted into coupled one-dimensional *hyperspherical* equations using the methods of Chapter 9. Such models, for example for ^{6}He $= \alpha + $ n $ + $ n, describe the 'valence' n–n correlations with good accuracy, but not those of the neutrons inside the core. That is, the cluster-cluster correlations are now modeled precisely, at the expense of having fixed structures within the clusters.

5.1.3 Mean-field models

Medium and heavy nuclei are beyond the scope of shell-model computations, and have no simple cluster decomposition. For these, *mean-field models* are most commonly used [17]. These are now often called *density-functional methods*, because of the central role of the density $\rho(\mathbf{r})$.

In a mean-field model, the nucleons are taken as affected only by the *mean* or *average potential attraction* of all the other nucleons. If this mean field $V_m(\mathbf{r})$ is defined in a body-fixed coordinate system, the single-particle eigenstates $\phi_i(\mathbf{r})$ are found by solving

$$[\hat{T}_{\mathbf{r}} + V_m(\mathbf{r})]\phi_i(\mathbf{r}) = \varepsilon_i\phi_i(\mathbf{r}), \tag{5.1.15}$$

and then the levels $\varepsilon_1 < \varepsilon_2 < \cdots < \varepsilon_F$ are filled in order of increasing energy by $n_{\mathrm{occ}}(i)$ of the A nucleons while respecting the Pauli exclusion principle. This means that the number of nucleons in an orbital, the occupation number $n_{\mathrm{occ}}(i)$, is less than $2j_i+1$ for neutrons and protons separately. The energy ε_F of the highest occupied state is called the *Fermi energy*. The total nucleon density of the nucleus is thus

$$\rho(\mathbf{r}) = \sum_{i=1}^{F} n_{\mathrm{occ}}(i)|\phi_i(\mathbf{r})|^2 \tag{5.1.16}$$

where $\sum_{i=1}^{F} n_{\mathrm{occ}}(i) = A$. This density is a function of the vector $\mathbf{r}$ in the body-fixed coordinate frame, allowing a nucleus with intrinsic deformation to be described. The mean field is then calculated from the density $\rho(\mathbf{r})$ by folding with some effective NN interaction $V_{NN}(\mathbf{r})$ by

$$V_m(\mathbf{r}) = \int d\mathbf{r}' V_{NN}(\mathbf{r}-\mathbf{r}')\rho(\mathbf{r}'). \qquad (5.1.17)$$

If this is different from the starting mean field in Eq. (5.1.15), then the calculation of $\rho(\mathbf{r})$ and $V_m(\mathbf{r})$ from each other must be repeated until they are consistent with each other: the eigenstates in the potential produce that same potential when the density is folded with the effective interaction. The model is then called *self-consistent*.

In *Hartree* calculations, the overall nuclear wave function is a product of all the occupied orbitals, whereas in *Hartree-Fock* calculations the exchange terms are included, as calculated from a Slater determinant. The total wave function of the nucleus is in that case an antisymmetrized product of the single-particle states ϕ_i,

$$\Phi^{HF}(\mathbf{r}_1,\ldots,\mathbf{r}_A) = \mathcal{A} \prod_{p=1}^{A} \phi_{i_p}(\mathbf{r}_p,\mathbf{s}_p) \qquad (5.1.18)$$

where $1 \leq i_p \leq F$ is the orbital of Eq. (5.1.15) occupied by particle p. We note that in mean-field theories we do not have a clean factorization of the center-of-mass motion according to Eq. (5.1.14), but these models are more commonly used for heavier nuclei where there is less center-of-mass motion. Angular-momentum projection operators $P_{I\mu}$ can still be used to select composite states of good angular momentum.

Mean-field models may be improved with using a density-dependent effective interaction $V_{NN}(\mathbf{r};\rho)$, proportional for example to some power of the density ρ. The existence of these non-linearities makes it difficult to write down the specific Hamiltonian for the system, which is one reason why all these models are now more accurately classified as *density-functional theories*.[4]

Mean-field models are simpler for even–even nuclei, as then nucleons are in pairs of time-reversed orbits. In general, the $V_{NN}(\mathbf{r};\rho)$ may be finite range forces such as Gogny [18] or Volkov [19], or even simpler zero-range Skyrme-class forces [21]. In all cases, the parameters defining the force are *fitted* to give binding energies and charge radii of a small set of benchmark nuclei, and then used to extrapolate to a much wider range of nuclei approaching the driplines. One nucleon and pair separation energies may be predicted, as well as all the proton and neutron density distributions. Mean-field models may be used to predict bound-state wave functions for transfer or capture scattering calculations according to the orbital occupation numbers.

---

[4] We may further include *pairing effects* either by BCS superconductivity models, or by a Bogolyubov pairing field. With weakly bound levels ε_i, treatment of pairing requires a careful inclusion of positive-energy continuum states in order to get the needed complete set of paired states that are properly confined.

To calculate the *excited states* of nuclei in the mean-field approach, we can either use a time-dependent analysis of how the nucleus responds to perturbations, or use a linearized perturbation theory. There is the *time-dependent Hartree-Fock* (TDHF) [20] method, or the linear perturbation analysis called the *random-phase approximation* (RPA). The RPA generates excited states of the nucleus as a superposition of 'particle-hole doorway states' that can be reached from the ground state by one application of particle-hole operators. The doorway states can themselves be coupled to two-particle two-hole states, and this will further fragment each doorway state into many more levels by a spreading process. These 2p2h states should ideally also be included in structure calculations, in a 'second RPA' framework. The set of all compound-nucleus states is more numerous again, and in principle contains all excitations, including 3p3h and 4p4h operations, etc.

5.1.4 Collective nuclear-matter descriptions

The very simplest nuclear description is according to the density distribution $\rho(\mathbf{r})$. This may be derived from one of the structure models above, or we could attempt to immediately write down a plausible form for that function. This form describes the collective size and shape of a nucleus: spherical or deformed. The density, however derived, can now be used with a nucleon-nucleon potential to find the principal interaction potential of the nucleus with another.

We have a choice describing the densities $\rho(\mathbf{r})$ in the laboratory frame of reference (r, θ, ϕ), or $\rho(\mathbf{r}')$ in a body-fixed frame (r, θ', ϕ') (where that frame can be defined). The radii in the two frames are equal. Laboratory densities will depend on the angular momentum state $I\mu$ of the nucleus, whereas body-fixed densities are *instrinsic*, and depend only on the quantum number K, which is the projection of the instrinsic spin on the body-fixed $\mathbf{z}'$ axis. Mean-field, density-functional and collective rotational models give the nuclear wave function initially in the body-fixed frame, whereas shell-model wave functions are usually projected onto good angular-momentum states in the laboratory frame.

In terms of a wave function that depends on A coordinates, such as that of Eqs. (5.1.11) or (5.1.18), in either coordinate frame the single-particle density is the sum of the probability distribution of each nucleon:

$$\rho(\mathbf{r}) = \sum_{i=1}^{A} \int \cdots \int |\Phi(\mathbf{r}_1, \ldots, \mathbf{r}_A)|^2 \delta(\mathbf{r}-\mathbf{r}_i) \, d\mathbf{r}_1 \cdots d\mathbf{r}_A$$

$$\equiv \langle \Phi(\mathbf{r}_1, \ldots, \mathbf{r}_A)| \sum_{i=1}^{A} \delta(\mathbf{r}-\mathbf{r}_i) |\Phi(\mathbf{r}_1, \ldots, \mathbf{r}_A)\rangle. \qquad (5.1.19)$$

If the wave function carries quantum numbers $I\mu$ or K, so will the density. When these wave functions are antisymmetric, we need only look at the dependence on one of the coordinates, $\mathbf{r}_1$ say, and multiply by A:

$$\rho(\mathbf{r}) = A \langle \Phi(\mathbf{r}_1, \ldots, \mathbf{r}_A)| \delta(\mathbf{r}-\mathbf{r}_1) |\Phi(\mathbf{r}_1, \ldots, \mathbf{r}_A)\rangle. \qquad (5.1.20)$$

The density of Eq. (5.1.19) may include the effects of center-of-mass motion, and have therefore an inflated mean radius because the division of Eq. (5.1.14) has not been performed. For more accuracy, we should thus use the wave functions $\Phi(\boldsymbol{\rho}_1, \ldots, \boldsymbol{\rho}_{A-1})$ of Eq. (5.1.1), because these are translationally invariant, and we have $\boldsymbol{\rho}_A + \sum_{i=1}^{A-1} \boldsymbol{\rho}_i = 0$. The single-particle density that is translationally invariant is therefore

$$\rho(\mathbf{r}) = \langle \Phi(\boldsymbol{\rho}_1, \ldots) \left| \left[\sum_{i=1}^{A-1} \delta(\mathbf{r}-\boldsymbol{\rho}_i) + \delta(\mathbf{r}+\sum_{i}^{A-1} \boldsymbol{\rho}_i) \right] \right| \Phi(\boldsymbol{\rho}_1, \ldots)\rangle. \qquad (5.1.21)$$

However derived, the density distribution $\rho(\mathbf{r})$ for A nucleons is normalized by

$$\int \mathrm{d}\mathbf{r}\rho(\mathbf{r}) = A. \qquad (5.1.22)$$

The $\rho(\mathbf{r})$ will be spherically symmetric in the laboratory frame when $I = 0$, or when the nucleus is an unpolarized sum over all projection quantum numbers μ. In the spherical case, the density only depends on $r = |\mathbf{r}|$, and is normalized as

$$4\pi \int_0^\infty r^2 \mathrm{d}r\rho(r) = A. \qquad (5.1.23)$$

Its root mean square radius r_{rms}, measured from the center of mass of the whole nucleus, is defined by

$$r_{\mathrm{rms}}^2 = \langle r^2 \rangle = \frac{1}{A} \int \mathrm{d}\mathbf{r} \, r^2 \rho(\mathbf{r}) = \frac{4\pi}{A} \int_0^\infty r^4 \mathrm{d}r\rho(r). \qquad (5.1.24)$$

Now each nucleon has itself an intrinsic density $\rho_{\mathrm{int}}(\mathbf{r})$ corresponding to a radius of about 0.8 fm. This implies that the density of nucleonic matter $\rho_{\mathrm{mat}}(\mathbf{r})$ is the convolution of this finite size with the point-density distribution:

$$\rho_{\mathrm{mat}}(\mathbf{r}) = \int \mathrm{d}\mathbf{r}' \rho_{\mathrm{int}}(\mathbf{r} - \mathbf{r}')\rho(\mathbf{r}'). \qquad (5.1.25)$$

The rms radius of ρ_{mat} is larger according to $r_{\mathrm{mat}}^2 = r_{\mathrm{rms}}^2 + r_{\mathrm{int}}^2$. Usually in nuclear physics, to avoid ambiguity, densities are communicated using the point nuclear density $\rho(\mathbf{r})$.

Separate densities $\rho_p(\mathbf{r})$ and $\rho_n(\mathbf{r})$ may be defined for protons and neutrons, respectively, to allow for distinct charge and matter radii. At radii where $\rho_n(r) \gg \rho_p(r)$, we would have what is called a *neutron skin*.

We now describe some density distributions of simple shapes:

Sphere of uniform density

A sphere of uniform density ρ_0 and radius R satisfies $\frac{4}{3}\pi R^3 \rho_0 = A$ so

$$\rho_0 = \frac{3A}{4\pi R^3} \tag{5.1.26}$$

and has rms radius $r_{\mathrm{rms}} = \sqrt{\frac{3}{5}}R$.

Spherical sphere of Fermi (Woods-Saxon) distribution

A spherically symmetric sphere with density

$$\rho(r) = \rho_0 \left[1 + e^{\frac{r-R}{a}}\right]^{-1} \tag{5.1.27}$$

must be normalized to A nucleons by

$$\rho_0 = \frac{3A}{4\pi R^3}\left[1 + \left(\frac{\pi a}{R}\right)^2 + 6\left(\frac{a}{R}\right)^3 e^{-R/a} + \cdots\right] \tag{5.1.28}$$

and has rms radius

$$r_{\mathrm{rms}} \simeq \left[\frac{3}{5}\left(R^2 + \frac{7}{3}\pi^2 a^2\right)\right]^{1/2}. \tag{5.1.29}$$

Spherical sphere of Gaussian distribution

A spherically symmetric sphere with density

$$\rho(r) = \rho_0\, e^{-(r/R)^2} \tag{5.1.30}$$

must be normalized to A nucleons by

$$\rho_0 = \frac{A}{\pi^{3/2}R^3} \tag{5.1.31}$$

and has rms radius

$$r_{\mathrm{rms}} = \sqrt{\frac{3}{2}}R. \tag{5.1.32}$$

Deformed spheroidal density

In the body-fixed coordinate system attached to the nuclear shape, a deformed spheroidal density $\rho(\mathbf{r}')$ will depend on the polar angles θ', ϕ' just as the potential did on page 144. A simple form of this dependence may vary as before with some 'surface radius' $\tilde{R}(\theta', \phi')$ according to

$$\rho(r, \theta', \phi') = \rho_0(r - \tilde{R}(\theta', \phi') + R_0), \qquad (5.1.33)$$

for some angle-independent density profile function $\rho_0(R)$ that could be one of the shapes given above. For a sphere of uniform density, $\rho_0(R)$ would be a step function that is unity for $R < R_0$ and zero for $R > R_0$, and for the Fermi or Gaussian distributions to it would be the functions given in equations (5.1.27) or (5.1.30). Now, we allow these Fermi or Gaussian distributions to become deformed. The $\tilde{R}(\theta', \phi')$ is defined so it is the half-density radius in the case of the sharp cutoff or the Fermi distribution.

We can always expand the angular dependence of $\tilde{R}(\theta', \phi')$ in spherical harmonics

$$\tilde{R}(\theta', \phi') = R_0 + \sum_{\lambda=0}^{\lambda_{max}} \sum_{\mu=-\lambda}^{+\lambda} \delta_{\lambda\mu} Y_\lambda^\mu(\theta', \phi'), \qquad (5.1.34)$$

where λ_{max} is sufficiently large for the desired accuracy. Here, the $\delta_{\lambda\mu}$ are called *matter deformation lengths*, and $\beta_{\lambda\mu} = \delta_{\lambda\mu}/R_0$ are the fractional matter deformations. This equation is similar to Eq. (4.4.8), but there we were talking of *potential* deformations, while here of *matter* deformations.

Multipole moments

The multipole moments of a deformed distribution are defined using operators similar to the electric multipole operators of Eq. (4.7.21):

$$\mathbf{Q}_{\lambda\mu} = \sum_k r_k^\lambda Y_\lambda^\mu(\hat{\mathbf{r}}_k) \qquad (5.1.35)$$

where the index k runs over all nucleons for the matter distribution. The electric multipole moment would be obtained if the sum was over all protons, and if a factor e is included for the proton charge.[5]

In the laboratory frame, the *quadrupole moment* of a nucleus in a quantum mechanical state $\Phi_{I\mu}$ is defined as the matrix element of $\mathbf{Q}_{20}$ evaluated in the

---

[5] Very often electric moments are given numerically without this factor, as for example $3\,e\mathrm{fm}^2$. This does *not* mean the moment has units of $e.\mathrm{fm}^2$, but *rather* that $3e$, if $e = 1.2\,\mathrm{MeV}^{1/2}\,\mathrm{fm}^{1/2}$ is substituted, has units of fm^2.

$\mu = I$ magnetic substate:

$$Q = \sqrt{\tfrac{16\pi}{5}} \langle \Phi_{II} | \mathbf{Q}_{20} | \Phi_{II} \rangle, \tag{5.1.36}$$

where the $\sqrt{16\pi/5}$ factor is conventional. Since, by the Wigner-Eckhart theorem, $\langle \Phi_{II} | \mathbf{Q}_{20} | \Phi_{II} \rangle = \hat{I}^{-1} \langle II, 20 | II \rangle \langle \Phi_I \parallel \mathbf{Q}_2 \parallel \Phi_I \rangle$, we have a formula for the quadrupole moment,

$$Q = \sqrt{\tfrac{16\pi}{5}} \hat{I}^{-1} \langle II, 20 | II \rangle \langle \Phi_I \parallel \mathbf{Q}_2 \parallel \Phi_I \rangle, \tag{5.1.37}$$

in terms of reduced matrix elements as defined in Eq. (4.4.25): the only difference is the unit charge factor e in the charge density $\rho(\mathbf{r}_p)$ appearing in Eq. (4.4.25). This moment for a specific quantum state of spin I is called a *spectroscopic quadrupole moment*.

In the body-fixed frame there is no need for the Wigner-Eckhart theorem, as the moments may be calculated directly by quadrature of the density $\rho(r, \theta', \phi')$ in Eq. (5.1.33). For an axially symmetric quadrupole-deformed density $q = 2$ and $\mu = 0$, so the *intrinsic quadrupole moment* is

$$Q = \sqrt{\tfrac{16\pi}{5}} \int d\mathbf{r}' \, r'^2 \rho(\mathbf{r}') Y_2^0(\hat{\mathbf{r}}'). \tag{5.1.38}$$

A first-order expansion of Eq. (5.1.33) for small δ_2 gives

$$Q = \frac{3R_0 A}{\sqrt{5\pi}} \delta_2. \tag{5.1.39}$$

The A can be replaced by Z to give the moment for a charge instead of matter density of point nucleons.

If a nucleus consists of a spherical core plus an unpaired single nucleon in a state $\phi_{\ell sj}(\mathbf{r})$ of angular momentum $\boldsymbol{\ell} + \mathbf{s} = \mathbf{j}$, then its single-particle quadrupole moment will be from the single nucleon. By Eq. (5.1.36), this is

$$Q_{\text{sp}} = -\frac{2j - 1}{2(j+1)} \langle r^2 \rangle, \tag{5.1.40}$$

where the mean square radius in the state is $\langle r^2 \rangle = \int r^2 |\phi_{\ell sj}(\mathbf{r})|^2 d\mathbf{r}$. Note that Q_{sp} is zero for $j = \tfrac{1}{2}$, as it is a rank-2 tensor operator, and that it is negative (oblate) for all other j values.

5.2 Folded potentials

Having defined density distributions that are spherical or deformed in some reference frame, we now wish to calculate the corresponding interaction potentials with a scattering nucleon or another scattering nucleus. We may do this if we know the interaction between each nucleon in the density distribution and the other scattering body, and resulting potential is given by a convolution integral. This integral is also called a *folding integral*, so the potentials obtained are called *folded potentials*. If the other scattering body is structureless, we have a *single-folded* potential, and if we have also to integrate over the other body's constituent nucleons, then we have a *double-folded* potential.

Consider first a target nucleus with density $\rho_t(\mathbf{r}_t)$ of A_t nucleons interacting with a structureless projectile p at position $\mathbf{r}$ by means of a potential $V_{pN}(\mathbf{r} - \mathbf{r}_t)$ between p and each nucleon N of the target. A simple estimate of the overall potential between the projectile and the target is the convolution integral

$$U_{pt}(\mathbf{r}) = \int d\mathbf{r}_t \, V_{pN}(\mathbf{r} - \mathbf{r}_t) \, \rho_t(\mathbf{r}_t). \tag{5.2.1}$$

This is the single-folded potential, since there is just one integration coordinate $\mathbf{r}_t$ as in Fig. 5.2.

If the projectile itself consists of many nucleons, with distribution $\rho_p(\mathbf{r}_p)$, each of which interacts with a target nucleon by a nucleon-nucleon potential $V_{NN}(\mathbf{r}_p - \mathbf{r}_t)$, then we need to calculate what is called the *double-folded potential*. The coordinates are illustrated in Fig. 5.3, in terms of which the integral is

$$U_{pt}(\mathbf{r}) = \int d\mathbf{r}_t \int d\mathbf{r}_p \, \rho_t(\mathbf{r}_t)\rho_p(\mathbf{r}_p) \, V_{NN}(\mathbf{r} + \mathbf{r}_p - \mathbf{r}_t). \tag{5.2.2}$$

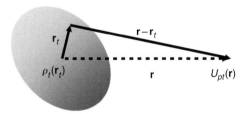

Fig. 5.2. Coordinates for single-folding integrals.

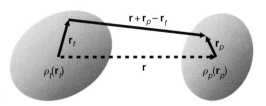

Fig. 5.3. Coordinates for double-folding integrals.

5.2.1 Fourier methods

These density-independent folded potentials are most conveniently calculated using Fourier transforms, so that the above convolution integrals become products of Fourier densities. Following Satchler [22, App. C], we normalize our Fourier transforms by defining

$$\tilde{f}(\mathbf{k}) = \int d\mathbf{r} \, e^{i\mathbf{k}\cdot\mathbf{r}} f(\mathbf{r}) \tag{5.2.3}$$

so that the inverse transform is

$$f(\mathbf{r}) = \frac{1}{(2\pi)^3} \int d\mathbf{k} \, e^{-i\mathbf{k}\cdot\mathbf{r}} \tilde{f}(\mathbf{k}), \tag{5.2.4}$$

keeping $\int d\mathbf{r} \, e^{i\mathbf{k}\cdot\mathbf{r}} = (2\pi)^3 \delta(\mathbf{k})$.[6]

Consider the projectile and target densities with respective Fourier expansions

$$\rho_p(\mathbf{r}_p) = (2\pi)^{-3} \int d\mathbf{k}_p \, e^{-i\mathbf{k}_p\cdot\mathbf{r}_p} \, \tilde{\rho}_p(\mathbf{k}_p)$$

$$\rho_t(\mathbf{r}_t) = (2\pi)^{-3} \int d\mathbf{k}_t \, e^{-i\mathbf{k}_t\cdot\mathbf{r}_t} \, \tilde{\rho}_t(\mathbf{k}_t), \tag{5.2.5}$$

and the NN force with transform given by $V_{\text{NN}}(\mathbf{s}) = (2\pi)^{-3} \int d\mathbf{k} \, e^{-i\mathbf{k}\cdot\mathbf{s}} \tilde{V}_{\text{NN}}(\mathbf{k})$. The Fourier transform of the folded potential is therefore

$$\tilde{U}_{pt}(\mathbf{k}) = \tilde{\rho}_p(\mathbf{k}) \, \tilde{\rho}_t(\mathbf{k}) \, \tilde{V}_{\text{NN}}(\mathbf{k}), \tag{5.2.6}$$

from which we extract the needed spatial dependence of Eq. (5.2.2) as

$$U_{pt}(\mathbf{r}) = \frac{1}{(2\pi)^3} \int d\mathbf{k} \, e^{-i\mathbf{k}\cdot\mathbf{r}} \, \tilde{U}_{pt}(\mathbf{k}). \tag{5.2.7}$$

Effective interactions $V_{\text{NN}}(r)$ that are derived as a spatial potential function are often written as sums of analytic Yukawa, Gaussian and zero-range forms. These have analytic Fourier transforms for $\tilde{V}_{\text{NN}}(\mathbf{k})$:

$V_{\text{NN}}(r)$	$\tilde{V}_{\text{NN}}(k)$
$V_0 \dfrac{e^{-\mu r}}{\mu r}$	$\dfrac{4\pi V_0}{\mu} \dfrac{1}{k^2 + \mu^2}$
$V_0 e^{-(r/b)^2}$	$V_0 \pi^{3/2} b^3 e^{-b^2 k^2/4}$
$V_0 \delta(\mathbf{r})$	V_0

---

[6] Note that other definitions may have $(2\pi)^{3/2}$ factors placed differently in these equations.

5.2.2 Deformed densities

If we follow for the density the same expansions as used on p. 138 for potentials, we define density multipoles $\rho_{\lambda\mu}(r)$ by

$$\rho(\mathbf{r}) = \sum_{\lambda\mu} \rho_{\lambda\mu}(r)\sqrt{4\pi}\, Y_\lambda^\mu(\hat{\mathbf{r}}), \tag{5.2.8}$$

and similarly for the Fourier transform

$$\tilde{\rho}(\mathbf{k}) = \sum_{\lambda\mu} \tilde{\rho}_{\lambda\mu}(k)\sqrt{4\pi}\, Y_\lambda^\mu(\hat{\mathbf{k}}). \tag{5.2.9}$$

These spatial and spectral multipoles are related using Eqs. (5.2.3) and (3.1.24), giving

$$\tilde{\rho}_{\lambda\mu}(k) = 4\pi\, \mathrm{i}^\lambda \int_0^\infty \frac{r}{k} F_\lambda(0, kr)\rho_{\lambda\mu}(r)\mathrm{d}r, \tag{5.2.10}$$

so, for example,

$$\tilde{\rho}_{00}(0) = \int \mathrm{d}\mathbf{r}\, \rho(\mathbf{r}) = 4\pi \int_0^\infty r^2 \rho_{00}(r)\mathrm{d}r = A. \tag{5.2.11}$$

One useful limit is for $\lambda = 0$, $F_0(0, kr) = \sin(kr)$, and for small k, $F_0(0, kr) \simeq kr - k^3 r^3/6$ so

$$\tilde{\rho}_{00}(k) \simeq 4\pi \int_0^\infty r^2 \left(1 - \frac{k^2 r^2}{6}\right)\rho_{00}(r)\mathrm{d}r$$

$$= A\left(1 - \frac{k^2}{6}\langle r^2 \rangle\right). \tag{5.2.12}$$

Here, the decrease of density in momentum space is proportional to the mean square radius.

If just one of ρ_p and ρ_t is deformed with multipole λ, then $U_{pt}(\mathbf{r})$ will also carry an order-λ angular dependence. Supposing that the target is the deformed nucleus, we have

$$\tilde{U}_{pt}^{\Lambda\mu}(k) = \tilde{\rho}_p(k)\tilde{\rho}_t^{\Lambda\mu}(k)\tilde{V}_{\mathrm{NN}}(k), \tag{5.2.13}$$

from which the radial multipoles of the deformed potential are

$$U_{pt}^{\Lambda\mu}(r) = \frac{1}{2\pi^2}\mathrm{i}^{-\lambda} \int_0^\infty \frac{k}{r} F_\lambda(0, kr)\tilde{U}_{pt}^{\Lambda\mu}(k)\mathrm{d}k. \tag{5.2.14}$$

These expressions for the multipoles of the potential between two nuclei can be used more generally than the rotational models of subsection 4.4.1. Those models assumed that the interaction potentials depended on the distance from the 'surface' of a nucleus that depended on angle, whereas now we have the means to calculate the potential for any angle-dependent density distribution, by folding that distribution with the interaction between the constituent nucleons and the other scattering body.

5.2.3 Typical forms of effective interactions

The folding methods of this section use effective interactions $V_{NN}(r)$. It is beyond the scope of this book to discuss all the effective interactions that have been proposed and tested in the literature, so we mention two of the most widely used.

One commonly used $V_{NN}(r)$ effective interaction for folding is the M3Y interaction, determined by Bertsch *et al.* [23] as a simple parametrization derived from realistic forces. This real interaction consists of a sum of two Yukawa terms, as well as a zero-range delta function term to simulate exchange effects. The isospin-average form of the M3Y potential is

$$V_{M3Y}(\mathbf{r}) = 7999\frac{e^{-4r}}{4r} - 2135\frac{e^{-2.5r}}{2.5r} - 252\,\delta(\mathbf{r}), \qquad (5.2.15)$$

in units of MeV. This gives only the real part of a nucleus-nucleus potential, so the imaginary part of the nucleus-nucleus potential that is needed when there are open channels would have to be found or fitted independently.

For nucleon-nucleus scattering, another effective interaction widely used is the JLM potential of Jeukenne, Lejeune and Mahaux [24], as mentioned in subsection 4.1.3. This contains both real and imaginary components with parameters that are energy and density dependent.

5.3 Overlap functions

In order to calculate transfers, captures, or single-particle excitations involving a nucleus B, we need to know what will be the quantum states of a nucleon that could be removed from B, as well as the corresponding states of the core A that remains. This information is given by knowledge of the one-body *overlap functions* $\phi = \langle \Phi_A | \Phi_B \rangle$ of the many-body wave functions of Φ_A and Φ_B.

5.3.1 Non-antisymmetrized theory

When discussing transfer reactions on p. 150, we considered a simple case by ignoring the core structure and its spin, and by assuming that the composite nucleus was uniquely composed by a single bound state. In general, however, nuclei are more complicated, the bound nucleon may be identical to others in the core, and the removal of a nucleon may leave the core in distinct states with different probability amplitudes.

We take into account the identity of nucleons in the next section. For now we ignore the Pauli principle, and consider the transfer of a 'valence' particle v to a core A to form a composite nucleus $B = A + v$. We define the core in spin state I_A to have wave function $\Phi^A_{I_A}(\xi_A)$, and the composite nucleus in spin state I_B to have $\Phi^B_{I_B}(\xi_B)$, where ξ_A refers to the internal A coordinates and $\xi_B = \{\xi_A, \mathbf{r}\}$ is the composite set of core + particle coordinates.

We may then need to calculate and use the *overlap functions*

$$\phi_{I_A:I_B}(\mathbf{r}) = \langle \Phi^A_{I_A}(\xi_A) | \Phi^B_{I_B}(\xi_A, \mathbf{r}) \rangle, \qquad (5.3.1)$$

of which there will be many, one for each $\{I_A, I_B\}$ pair. The parity of $\phi_{I_A:I_B}(\mathbf{r})$ is $\pi_B \pi_A \pi_v$, where π_v is the internal parity of the valence particle v. The integration in this equation is over the core coordinates ξ_A. The magnitude of $\phi_{I:J}(\mathbf{r})$ measures the amplitude for removing a nucleon at $\mathbf{r}$ from composite state I_B and leaving the core in state I_A. Each composite state I_B may in turn be reconstructed as a superposition over all the mutually orthogonal core states:

$$\Phi^B_{I_B}(\xi_A, \mathbf{r}) = \sum_{I_A} \phi_{I_A:I_B}(\mathbf{r}) \Phi^A_{I_A}(\xi_A). \qquad (5.3.2)$$

The individual overlaps $\phi_{I_A:I_B}(\mathbf{r})$ are *not* normalized to unity, as only their summed normalization, for each I_B value separately, satisfies

$$\sum_{I_A} \int d\mathbf{r} |\phi_{I_A:I_B}(\mathbf{r})|^2 = 1. \qquad (5.3.3)$$

This follows from Eq. (5.3.2) and the unit normalizations of all the $\Phi^A_{I_A}$ and $\Phi^B_{I_B}$ of Eq. (5.1.4). Remember that we are so far ignoring the possibility of identical particles.

For practical calculations of transfer reactions, etc., we need to expand overlap functions in partial waves. Let the transferred particle v have spin s, and we keep the core and composite angular momenta as I_A and I_B respectively. After choosing a particular coupling order such as $|(\ell s)j, I_A; I_B\rangle$, the overlap Eq. (5.3.2) is equivalently written as a composition for nucleus B in terms of A and a valence particle:[7]

$$\Phi^B_{I_B}(\xi_A, \mathbf{r}) = \sum_{I_A, \ell s j} u^{jI_A I_B}_{\ell s j}(r)/r \left[\left[Y_\ell(\hat{\mathbf{r}}) \otimes \chi_s \right]_j \otimes \Phi^A_{I_A}(\xi_A) \right]_{I_B}, \qquad (5.3.4)$$

---

[7] Since the coordinates on both sides are internal or relative coordinates, there is no center-of-mass motion on either side of this equation.

for some single-particle radial overlap wave functions $u_{\ell sj}^{jI_A I_B}(r)$.[8] By the composition of parities, we require $(-1)^{\ell}\pi_A\pi_v = \pi_B$. In this theory, which is not antisymmetrized, the radial wave functions are normalized to unity for each I_B, according to

$$\sum_{\ell sj I_A} \int_0^\infty dr |u_{\ell sj}^{jI_A I_B}(r)|^2 = 1. \qquad (5.3.5)$$

For nucleons, only the value $s = \frac{1}{2}$ enters into these summations.

Spectroscopic factors

This non-unit normalization of the *individual* $u_{\ell sj}^{jI_A I_B}(r)$ allows us to write them non-trivially as a product of normalized wave function and an amplitude, called the *spectroscopic amplitude*, or the *coefficient of fractional parentage*:

$$u_{\ell sj}^{jI_A I_B}(r) = A_{\ell sj}^{jI_A I_B}\, v_{\ell sj}^{jI_A I_B}(r) \qquad (5.3.6)$$

where $v_{\ell sj}^{jI_A I_B}(r)$ is a single-particle wave function normalized to unity as $||v_{\ell sj}^{jI_A I_B}(r)|| = 1$. We may explicitly construct these as

$$v_{\ell sj}^{jI_A I_B}(r) = \frac{u_{\ell sj}^{jI_A I_B}(r)}{\left[\int_0^\infty |u_{\ell sj}^{jI_A I_B}(r')|^2 dr'\right]^{1/2}} \qquad (5.3.7)$$

for each set of quantum numbers for which the denominator is non-zero.

This gives a fractional parentage expansion for the composite nucleus of

$$\Phi_{I_B}^B(\xi_A, \mathbf{r}) = \sum_{I_A, \ell sj} A_{\ell sj}^{jI_A I_B}\frac{1}{r} v_{\ell sj}^{jI_A I_B}(r) \left[\left[Y_\ell(\hat{\mathbf{r}}) \otimes \chi_s\right]_j \otimes \Phi_{I_A}^A(\xi_A)\right]_{I_B}. \qquad (5.3.8)$$

The square modulus of $A_{\ell sj}^{jI_A I_B}$ is

$$S_{\ell sj}^{jI_A I_B} = |A_{\ell sj}^{jI_A I_B}|^2, \qquad (5.3.9)$$

and is called the *spectroscopic factor*. The spectroscopic factor may be interpreted as a probability, namely the probability of finding core state I_A within a composite state I_B when removing a nucleon in partial-wave state ℓsj, because

$$\sum_{\ell sj I} |A_{\ell sj}^{jI_A I_B}|^2 = \sum_{\ell sj I} S_{\ell sj}^{jI_A I_B} = 1. \qquad (5.3.10)$$

---

[8] We keep the quantum number j in two places here, so both ℓsj and $jI_A I_B$ describe sets of quantum numbers satisfying triangular coupling relations.

Note, however, that the Eqs. (5.3.8) and (5.3.10) must be replaced when antisymmetrization is taken into account, as described in subsection 5.3.2.

Isospin C coefficients

It may be that the nuclear states $\Phi_{I_A}^A$ and $\Phi_{I_B}^B$ have been defined with isospin quantum numbers T and z-component $T_z = \frac{1}{2}(N-Z)$ as on page 107. In this case, we will have constructed amplitudes $\tilde{A}_{\ell sj}^{jI_A I_B}$, which in general are different from the $A_{\ell sj}^{jI_A I_B}$. If the transferred v is a nucleon ($t = \frac{1}{2}$) and is definitely a neutron ($t_z = \frac{1}{2}$) or a proton ($t_z = -\frac{1}{2}$), then these spectroscopic amplitudes are related by the isospin Clebsch-Gordan coefficient $C = \langle T_A T_{zA}, t t_z | T_B T_{zB} \rangle$, as

$$A_{\ell sj}^{jI_A I_B} = C \tilde{A}_{\ell sj}^{jI_A I_B}. \tag{5.3.11}$$

The relation between the corresponding spectroscopic factors is[9]

$$S_{\ell sj}^{jI_A I_B} = C^2 \tilde{S}_{\ell sj}^{jI_A I_B}. \tag{5.3.12}$$

Phases of spectroscopic amplitudes

The sign of the spectroscopic amplitudes $A_{\ell sj}^{jI_A I_B}$ depends on

(i) the overall phases of both $\Phi_{I_A}^A$ and $\Phi_{I_B}^B$, which must be kept fixed for all the overlaps calculated from them,

(ii) the coupling scheme and order of coupling, which is $\boldsymbol{\ell} + \mathbf{s} = \mathbf{j}$, $\mathbf{j} + \mathbf{I}_A = \mathbf{I}_B$ here. Reversing a coupling order, for example using $\mathbf{s} + \boldsymbol{\ell} = \mathbf{j}$, changes the phase by $(-1)^{j-\ell-s}$. Using another coupling scheme, such as the channel-spin scheme of $\mathbf{s} + \mathbf{I}_A = \mathbf{S}$, $\boldsymbol{\ell} + \mathbf{S} = \mathbf{I}_B$ generates another set of coefficients $B_{sI_A S}^{\ell SI_B}$ related by the Racah algebra transformations seen in Eq. (3.2.3), and

(iii) the sign of the normalized radial wave functions $v_{\ell sj}^{jI_A I_B}(r)$: whether for example these are defined to be positive at infinity, or positive near the origin. FRESCO single-particle wave functions are defined to be positive near the origin (see Appendix B).

.

5.3.2 Antisymmetrized theory

The analysis of the previous subsection 5.3.1 assumes that the transferred particle v is distinguishable from all other nucleons within the core A. We know, however, that neutrons and protons are not distinguishable among themselves, and that, as discussed in Section 3.4, microscopic wave functions must be antisymmetric under

---

[9] Note that Satchler [22] and others have used a reverse notation, whereby $\tilde{S} = C^2 S$.

exchanging any two neutrons, or any two protons. This means that on, say, removing a neutron from a nucleus B, there may be $n > 1$ neutrons that could be in the same orbital and removed in the same way. The cross section will therefore be increased by the factors n. It would be desirable to include this effect in reaction calculations if we could just change the spectroscopic amplitudes and factors by multiplicative constants of $\sqrt{n}$ and n respectively. To determine whether this is possible requires a more detailed analysis of reaction theory, by using antisymmetrized wave functions for all the nucleons present.

It has been shown by Macfarlane [25] and Austern [26] that *inert* groups of nucleons have no influence on the spectroscopic factors. An inert group is one whose configuration is common to both $\Phi^A_{I_A}$ and $\Phi^B_{I_B}$, and could be, for example, deeply bound closed shells. Only the remaining 'active' nucleons need to be considered for antisymmetrization.

If the isospin formalism is used for constructing $\Phi^A_{I_A}$ and $\Phi^B_{I_B}$, then we do not need to distinguish neutrons from protons. Otherwise, antisymmetrization and counting n identical particles is to be done separately for neutrons and for protons. Another formalism that is often recommended for identical particles is that of second quantization. As Austern [26, p. 61] points out, however, the results should be the same, and, if we can consider particle labels correctly, we can avoid the sizeable formal apparatus of second quantization. We shall see next that often it is better, for dynamical reasons, to separately consider direct and exchange terms, especially since exchange terms are more often non-local.

The exchange amplitudes that should in principle be included in a complete reaction model are exactly analogous to the additional reaction mechanisms that possibly occur if the nucleons were not identical. Many exchange amplitudes are analogous to knockout-exchange reactions, so they are small when knockout is small, for the same dynamical reasons. They should both be small, for example, in reactions at high energies, or when there are large momentum transfers.

What is most important, however, is to at least calculate the correct magnitudes of the direct terms, by, for example, counting the number of equivalent amplitudes. The detailed derivation of the correct magnitude of the direct terms in a given reaction is discussed in Austern [26, pp. 58–61], in Towner [27, §7.2] and in Satchler [22, §2.11.3]. The result of this theory is that the correct results are obtained by changing the construction of spectroscopic amplitudes as follows.

In a reaction $a + A \rightarrow b + B$ by the transfer of nucleon or cluster v, let

$$n^B_A = \binom{N_B}{v} \equiv \frac{N_B!}{N_A!v!} \tag{5.3.13}$$

be the number of nucleons in B that are in the same orbital, and identical to that valence nucleon(s) v transferred. Here, N_B is the number of such *valence* nucleons in nucleus B, and $N_A = N_B - v$ is the number of such identical nucleons remaining in core nucleus A.

Instead of Eq. (5.3.8), we now define the spectroscopic amplitudes by the overlap equation

$$\Phi^B_{I_B}(\xi_A, \mathbf{r}) = \frac{1}{\sqrt{n^B_A}} \sum_{I_A, \ell s j} A^{j I_A I_B}_{\ell s j} \frac{1}{r} v^{j I_A I_B}_{\ell s j}(r) \left[\left[Y_\ell(\hat{\mathbf{r}}) \otimes \chi_s \right]_j \otimes \Phi^A_{I_A}(\xi_A) \right]_{I_B}. \quad (5.3.14)$$

Since Φ^B, Φ^A and $v^{j I_A I_B}_{\ell s j}(r)$ are all normalized to unity, this gives

$$\sum_{\ell s j I_A} |A^{j I_A I_B}_{\ell s j}|^2 = \sum_{\ell s j I_A} S^{j I_A I_B}_{\ell s j} = n^B_A \quad (5.3.15)$$

as the sum rule to be satisfied in place of Eq. (5.3.10). The spectroscopic factors S are now to be interpreted as a probability multiplied by the number of available valence nucleons, so that one-step cross sections are still proportional to these spectroscopic factors.

5.3.3 Cluster overlaps

The general theory presented above still holds if the transferred particle v is a cluster of nucleons such as a deuteron, two neutrons, or an alpha particle. The intrinsic state χ_s for the particle v of spin s is now the entire intrinsic wave function of the cluster, and equations (5.3.14, 5.3.15) still hold.

Reactions involving the transfer of a cluster can either be treated as the movement of a single 'particle', or more microscopically as a combination of simultaneous and sequential transfers.

The bound states of clusters treated as one particle are commonly found in a Saxon-Woods potential with the same geometry parameters as are found for low-energy scattering of the same nuclei. The number of radial nodes is chosen by means of a shell-model counting of oscillator quanta, to accommodate the Pauli exclusion principle, as follows.

Suppose that the valence nucleon shell has $\mathcal{N}_i = 2n_i + \ell_i$ quanta, that is $\mathcal{N} = 1$ for the p-shell, 2 for the sd-shell, etc. If the transferred valence cluster is taken as composed of v such nucleons, then the total energy of the cluster nucleons will be $v\mathcal{N}_i \hbar \omega$. If the *internal* cluster configuration has $\mathcal{N}_{\text{int}} = 2n_{\text{int}} + \ell_{\text{int}}$ quanta (0 for s-shell clusters), then the cluster-core motion in partial wave L has N nodes where

$$2N + L + \mathcal{N}_{\text{int}} = v\mathcal{N}_i. \quad (5.3.16)$$

Usually L is fixed by the spin of the composite state, so N may be determined. Note that $N = 0, 1, \ldots$ is the number of nodes *excluding* the origin.[10]

5.4 General matrix elements

5.4.1 Coulomb and nuclear transitions

For initial Ψ_i and final Ψ_f states of a nucleus, the transition may in general be governed by reduced matrix elements $\langle \Psi_f \| \hat{V} \| \Psi_i \rangle$, with the operator $\hat{V}$ representing the relevant Coulomb, nuclear or charge-exchange processes that either are spontaneous, or are induced by another nucleus during a collision.

The Coulomb and charge-exchange matrix elements (when squared) may often be written as proportional to a reduced transition probability $B(i \to f)$ that depends on the particular nuclear states, and other factors that depend on the interacting partner. The electric quantities $B(E\lambda, i \to f)$ of Eq. (4.4.3) and the magnetic quantities $B(M\lambda, i \to f)$ are of this kind. Their definition allows them to be related to electromagnetic decay lifetimes, so measurements of these lifetimes can tell us the magnitude of the matrix elements needed in various reactions, and vice versa. (The *phases* of the matrix elements are not determined, only their absolute magnitudes.)

For transitions between *discrete* states we may derive results, analogous to those of subsection 4.7.4, that give now not cross sections but *transition rates* $T(E\lambda, i \to f)$:

$$T(E\lambda, i \to f) = \frac{8\pi(\lambda+1)}{\lambda[(2\lambda+1)!!]^2} \frac{1}{\hbar} k^{2\lambda+1} B(E\lambda, i \to f), \qquad (5.4.1)$$

where k is the photon wave number. The half-life is $t_{\frac{1}{2}} = (\ln 2)/T$. The same equation relates $T(M\lambda)$ to $B(M\lambda)$ for magnetic transitions.

5.4.2 Allowed β-decays

For allowed β-decay transitions, the only contributing matrix elements are those of the Fermi and Gamow-Teller operators, as defined in Section 4.6. We define weak-interaction reduced matrix elements by analogy to the electric and magnetic elements above. The Fermi reduced transition probability is defined, following Bohr

---

[10] The NN parameter for FRESCO bound states includes the origin, so $NN = N + 1 \geq 1$.

and Mottelson [28, App. D], using a sum over all nucleons k:

$$B(F, i \to f) = \frac{\delta_{I_i I_f}}{2I_i + 1} \left| \sum_k \langle \Psi_f \parallel Y_0^0 g_V t_-(k) \delta(\mathbf{r} - \mathbf{r}_k) \parallel \Psi_f \rangle \right|^2$$

$$= \frac{\delta_{I_i I_f}}{2I_i + 1} \frac{g_V^2}{4\pi} \left| \sum_k \langle \Psi_f \parallel t_-(k) \parallel \Psi_f \rangle \right|^2 \quad (5.4.2)$$

where $t_-(k)$ is defined on p. 107 as the isospin lowering operator for the k'th nucleon. From experiments, the vector coupling constant g_V is found to be

$$g_V = (1.40 \pm 0.02) 10^{-49} \, \text{erg.cm}^3 \quad (5.4.3)$$

$$= (1.36 \pm 0.02) 10^{-3} \frac{e^2 \hbar^2}{m_p^2 c^2}. \quad (5.4.4)$$

The Gamow-Teller reduced transition probability is correspondingly

$$B(GT, i \to f) = \frac{1}{2I_i + 1} \frac{g_A^2}{4\pi} \left| \sum_k \langle \Psi_f \parallel 2s_k t_-(k) \parallel \Psi_f \rangle \right|^2 \quad (5.4.5)$$

where the axial-vector coupling constant satisfies $g_A / g_V = -1.73 \pm 0.01$.

The actual β-decay rate depends on the electric field of the nucleus and its influence on the behaviors of the electrons after they have been emitted at various energies, with the corresponding opposite variations in the neutrino energies. For these reasons, a dimensionless quantity f is usually calculated to represent the electron and neutrino phase spaces, and is found by integrals over the electron spectra. The β-decay results are then expressed as the product $f t_{\frac{1}{2}}$, for half-life $t_{\frac{1}{2}}$. The $f t_{\frac{1}{2}}$ product (usually written just ft) then reflects nuclear structure by

$$f t_{\frac{1}{2}} [B(F) + B(GT)] = \frac{\pi^2 \hbar^7 \ln 2}{2 m_e^5 c^4} \equiv \frac{g_V^2}{4\pi}. \quad (5.4.6)$$

The transition decay rate $T = (\ln 2)/t_{\frac{1}{2}}$ is thus proportional to $B(F) + B(GT)$.

Exercises

5.1 Consider a simple two-body description of the halo nucleus ^{15}C as a single neutron and an inert ^{14}C core. The effective nuclear interaction between the neutron and ^{14}C is often taken to be a Woods-Saxon form of Eq. (4.1.1), with radius $R_r = 1.2 A^{1/3}$ fm (A being the mass of the core), and diffuseness $a_r = 0.65$ fm. The depth of the interaction V_r is fitted to the binding energy $S_n = 1.2$ MeV.

(a) In what orbital do you expect to find the neutron in ^{15}C(g.s.)? Considering the standard shell structure, what excited states would you expect to find? How do your expectations compare to experiment?

(b) Determine the depth of the interaction that produces the correct binding energy of ^{15}C, assuming first no spin-orbit and next a spin-orbit strength of 7 MeV. Compare the radial wave functions.

(c) Study the energy dependence of the n-^{14}C phase shifts $\delta_l(E)$ for $l = 0, 1, 2$ and energies from 0.1 to 4.0 MeV. Take the same scattering interaction as for the ground state (with zero spin-orbit force) for all partial waves. Are there any resonances in the system? If yes, what are their characteristic quantum numbers, and the corresponding energies and widths?

(d) What is the effect of the spin-orbit interaction on the phase shifts calculated in (c)?

5.2 Consider that the low-lying levels in ^{40}Ca can be described within a vibrational model. Based on the experimental B(E2) strength connecting the first 2^+ state with the ground state, determine the deformation length of the nucleus.

5.3 Estimate the potential between a neutron and ^{16}O with the single-folding model. Use the M3Y NN effective interaction and assume a Gaussian distribution for the nucleons in ^{16}O with the radius of 2.5 fm. Compare with the result for the JLM interaction.

References

[1] S. C. Pieper and R. B. Wiringa, *Ann. Rev. Nucl. and Part. Sci.* **51** (2001) 53.
[2] K. M. Nollett, S. C. Pieper, R. B. Wiringa, J. Carlson and G. M. Hale, *Phys. Rev. Letts.* **99** (2007) 022502.
[3] H. Feshbach, *Ann. Phys. (NY)* **5** (1958) 357; **19** (1962) 287.
[4] R. Suzuki and S. Y. Lee, *Prog. Theor. Phys.* **64** (1980) 2019.
[5] P. Navrátil and B. R. Barrett, *Phys. Rev. C* **54** (1996) 2986; **57** (1998) 3119.
[6] S. K. Bogner, T. T. S. Kuo and A. Schwenk, *Phys. Rep.* **386** (2003)1.
[7] S. Cohen and D. Kurath, *Nucl. Phys.* **73** (1965) 1.
[8] B. H. Wildenthal, *Prog. Part. Nucl. Phys.* **11** (1984) 5.
[9] E. K. Warburton and B. A. Brown, *Phys. Rev. C* **46** (1992) 923.
[10] E. Caurier, A. P. Zuker and A. Poves, *Phys. Rev. C* **50** (1994) 225.
[11] M. V. Stoitsov and I. Zh. Petkov, *Ann. Phys. (NY)* **184** (1988) 121; M. V. Stoitsov, J. Dobaczewski, P. Ring and S. Pittel, *Phys. Rev. C* **61** (2000) 034311.
[12] K. Bennaceur, F. Nowacki, J. Okolowicz and M. Ploszajczak, *J. Phys. G.* **24** (1998) 1631.
[13] J. Okolowicz, M. Ploszajczak and I. Rotter, *Phys. Rep.* **374** (2003) 271.
[14] A. Volya and V. Zelevinsky, *Phys. Rev. C* **67** (2003) 054322.
[15] P. Descouvemont, *Phys. Rev. C* **38** (1988) 2397; *Nucl. Phys. A* **615** (1997) 261.
[16] D. R. Thompson, M. Lemere and Y. C. Tang, *Nucl. Phys. A,* **286** (1977) 53.
[17] M. Bender, P.-H. Heenen and P.-G. Reinhard, *Rev. Mod. Phys.* **75** (2003) 121.
[18] J. Decharge and D. Gogny, *Phys. Rev. C* **21** (1980) 1568.
[19] A. B. Volkov, *Nucl. Phys.* **74** (1965) 33.
[20] P. Bonche, S. E. Koonin and J. N. Negele, *Phys. Rev. C* **13** (1976) 1226.
[21] T. H. R. Skyrme, *Proc. Roy. Soc. London. Series A* **260** (1961) 127; D. Vautherin and D. M. Brink, *Phys. Rev. C* **5** (1972) 626.
[22] G. R. Satchler 1983, *Direct Nuclear Reactions*, Oxford: Clarendon Press.

[23] G. Bertsch, J. Borysowicz, H. McManus and W. G. Love, *Nucl. Phys.* **A284** (1977) 399.

[24] J.-P. Jeukenne, A. Lejeune and C. Mahaux, *Phys. Rev. C* **16** (1977) 80.

[25] M. H. Macfarlane and J. B. French, *Rev. Mod. Phys.* **32** (1960) 567.

[26] N. Austern 1970, *Direct Nuclear Reaction Theories*, New York: Wiley-Interscience.

[27] I. Towner 1977, *A Shell Model Description of Light Nuclei*, Oxford: Clarendon Press.

[28] A. Bohr and B. R. Mottelson 1975, *Nuclear Structure, Vol. 1*, Reading, MA: Benjamin.

6

Solving the equations

It is nice to know that the computer understands the problem.
But I would like to understand it too.

Eugene Wigner

In order to find the S-matrix elements and hence the cross sections, we have first to solve the coupled equations that represent the Schrödinger equation in the partial wave basis that we use. For elastic scattering we have a single second-order differential equation, and in Section 6.1 we expand on the details of the solution method outlined on page 55. We subsequently discuss iterative and exact solution methods for both bound state and scattering problems, and in Section 6.5 we see how R-matrix methods are generally suitable for solving coupled-channels problems.

6.1 Elastic scattering

For each partial wave L, we need to numerically solve the one-channel radial Schrödinger equation of Eq. (3.1.10):

$$\left[-\frac{\hbar^2}{2\mu} \left(\frac{d^2}{dR^2} - \frac{L(L+1)}{R^2} \right) + V(R) - E \right] \chi_L(R) = 0, \qquad (6.1.1)$$

with boundary conditions of

$$\chi_L(0) = 0 \qquad (6.1.2)$$

$$\chi_L(a) = \tfrac{i}{2} [H_L^-(\eta, ka) - \mathbf{S}_L H_L^+(\eta, ka)] \qquad (6.1.3)$$

$$\chi_L'(a) = \tfrac{i}{2} [H_L^{-\prime}(\eta, ka) - \mathbf{S}_L H_L^{+\prime}(\eta, ka)] \qquad (6.1.4)$$

for matching radius a. The derivatives H' are with respect to R.

The differential equation may be rewritten as

$$\chi_L''(R) = f(R)\chi_L(R) \tag{6.1.5}$$

$$\text{for } f(R) = \frac{2\mu}{\hbar^2}[V(R) - E] + \frac{L(L+1)}{R^2}.$$

Trial solutions

This one-channel problem may be solved as follows. We use a trial solution $y(R)$ with initial conditions $y(0) = 0$ and $y'(0) = p$ for some non-zero constant p. The differential equation (6.1.5) may then be integrated numerically, to determine $y(R)$ for all R. The desired solution $\chi_L(R)$ has some other initial derivative, and is thus proportional to this trial solution:

$$c\, y(R) = \chi_L(R) \tag{6.1.6}$$

for some unknown scaling coefficient c. Its value may be found by solving at $R = a$ the two simultaneous equations

$$c\, y(a) = \text{i}/2\, [H_L^-(\eta, ka) - \mathbf{S}_L H_L^+(\eta, ka)] \tag{6.1.7}$$

$$c\, y'(a) = \text{i}/2\, [H_L^{-\prime}(\eta, ka) - \mathbf{S}_L H_L^{+\prime}(\eta, ka)] \tag{6.1.8}$$

for the two unknowns c and $\mathbf{S}_L$. If only $\mathbf{S}_L$ is required then it is perhaps simpler to form $\mathbf{R} = y(a)/(ay'(a))$ and then find $\mathbf{S}_L$ by Eq. (3.1.30).

Numerical integration

To produce the trial solution, the numerical integration of Eq. (6.1.5) generally proceeds by defining a radial step size h, so the wave functions are defined at a grid of points as $y_n = y(nh)$ for $n = 0, 1, 2 \ldots, m_a$ up to $m_a h = a$, the matching point. The initial conditions are $y_0 = 0$ and $y_0' = p$ some non-zero initial derivative. Equation (6.1.5) is a second-order differential equation and may easily be converted to a pair of coupled first-order equations for $y(R)$ and $x(R) \equiv y'(R)$. This would allow, for example, Runge-Kutta methods to be used, producing eventually y_m, and also $x_m = y_m'$, for use in Eq. (6.1.8).

Because Eq. (6.1.5) has no first-order derivatives, and such derivatives rarely occur in nuclear scattering calculations, a more efficient alternative method is that of Numerov. It uses as starting conditions the wave function values at two adjacent radial points: $y_0 = 0$ and $y_1 = ph$, and propagates the solution, for $n = 1, 2 \ldots$, by

$$y_{n+1} = [(2 + 10\tilde{f}_n)y_n - y_{n-1}(1 - \tilde{f}_{n-1})]/(1 - \tilde{f}_{n+1}), \tag{6.1.9}$$

where $\tilde{f}_n = h^2 f_n/12$ with $f_n = f(nh)$. The division here sometimes gives problems: it requires a special treatment of $n = 1$ for $L = 1$ to avoid a division by zero, and does not generalize efficiently to coupled-channels problems. A more convenient method is therefore the 'modified Numerov method' [1], which integrates an auxiliary wave function z_n by

$$z_{n+1} = 2z_n - z_{n-1} + h^2 f_n z_n, \tag{6.1.10}$$

from which the true wave function is found by $y_n = (1 - \tilde{f}_n)z_n$. The z_n integration may be started by $z_0 = 0$ and $z_1 = ph$. The S matrix is now found by satisfying the boundary conditions of Eq. (6.1.7) for y_n at two distinct radii, say $n = m_a$ and, say, $m_a - 5$, since the first derivative is not available for direct matching. (For this, we must have chosen $(m_a - 5)h > R_n$, the maximum radius of the potential.)

The accuracy of Runge-Kutta or Numerov methods requires that the step size h becomes smaller as the local wave numbers $k(R) = \sqrt{(E - V(R))2\mu/\hbar^2}$ increase, so that there are a sufficient number of radial points within each oscillation of the wave solution. Attractive potential wells increase the local $k(R)$ in the interior as $V(R) \ll 0$, but when imaginary optical potentials are used the absorption means that the interior enhancement is less important. A practical guideline is that the product $kh \lesssim 0.2$ for at least the asymptotic k that appears in Eq. (6.1.7). Further 'enhanced Numerov methods' have been proposed [2], which are more accurate for high-energy scattering with slowly varying potentials as they allow larger kh products.

Another influence on numerical accuracy is the large centrifugal barrier $L(L + 1)/R^2$ for small R. Again, a practical guideline is that in the Numerov method, the starting $R = R_{\min}$ point should be of the order of $R_{\min} \gtrsim 2.0Lh$. This is to move the starting point increasingly away from the origin as L increases, so the centrifugal barrier ($\sim L^2/R^2$) does not become too large.

6.2 Classifications

The feasible methods of solving the partial-wave equations are very different for different kinds of couplings. In particular, they depend on whether the couplings are local or non-local in the radii R_x for all the partitions x. Non-locality is another name for the case that $R_x \neq R_{x'}$. Let us classify these cases, and see where they might typically occur, because non-local couplings make solving the coupled equations more difficult.

6.2.1 Local and non-local couplings

The coupling operators $\hat{V}^o_{\alpha\alpha'} = R_x \langle \alpha | V_o | \alpha' \rangle R_{x'}^{-1}$ as defined in subsection 3.2.2 couples *to* channel α *from* channel α', and may be local or non-local according to the physical process being described with the interaction potential V_o.

Typically, **inelastic couplings** of Section 4.4 with $x = x'$ and $pt \neq p't'$ are local, which means that $R_x \equiv R'_x$. The effect of operating with $\hat{V}$ on a channel wave function at radius R_x is therefore to give a resulting source function at the same radius:

$$\hat{V}^o_{\alpha\alpha'}\psi_{\alpha'}(R_x) = V^{inel}_{\alpha\alpha'}(R_x)\psi_{\alpha'}(R_x). \tag{6.2.1}$$

The 'inelastic form factor' $V^{inel}_{\alpha\alpha'}(R_x)$ is thus a function of one variable R_x.

For **transfer couplings** of subsection 4.5.1, however, the prior and post mass partitions and radius vectors are different, and the dynamics of recoil give rise to *non-local* couplings. Such couplings are described by specifying a two-variable *kernel function* $V^{transf}_{\alpha\alpha'}(R_x, R_{x'})$, as for example in Eq. (4.5.14), so that

$$\hat{V}^o_{\alpha\alpha'}\psi_{\alpha'}\big|_{R_x} = R_x\langle\alpha|V_o|\alpha'\rangle R^{-1}_{x'}\psi_{\alpha'}(R_{x'}) \tag{6.2.2}$$

may be written explicitly as

$$\hat{V}^o_{\alpha\alpha'}\psi_{\alpha'}\big|_{R_x} = \int dR_{x'}\, V^{o,transf}_{\alpha\alpha'}(R_x, R_{x'})\psi_{\alpha'}(R_{x'}). \tag{6.2.3}$$

The $N_{\alpha'\alpha}$ operators of Eq. (3.2.50) are also non-local for transfer couplings. Because local couplings are much easier to specify and include numerically, sometimes we may consider using the 'zero range approximation' to a non-local coupling of Eq. (4.5.16), so that

$$V^{o,transf}_{\alpha\alpha'}(R_x, R_{x'}) \simeq \delta(R_x - R_{x'})\, V^{ZR}_{\alpha\alpha'}(R_x). \tag{6.2.4}$$

Elastic couplings with $x = x'$ and $pt = p't'$ are usually diagonal (occasionally with a small non-local term added). If local, then

$$\hat{V}_{\alpha\alpha}\psi_\alpha(R_x) = U_\alpha(R_x)\psi_\alpha(R_x) \tag{6.2.5}$$

for some *diagonal monopole potential* $U_\alpha(R_x)$.

6.2.2 Simplified solutions

We saw in subsection 3.2.2, Eq. (3.2.52), that a set of N coupled equations have the general form

$$[H_\alpha - E_\alpha]\psi_{\alpha\alpha_i}(R_\alpha) + \sum_{\beta \neq \alpha}^N \hat{V}_{\alpha\beta}\psi_{\beta\alpha_i}(R_\beta) = 0, \tag{6.2.6}$$

where α_i identifies the incoming elastic channel. Let us consider various ways in which this set of equations $\alpha = 1, \ldots, N$ may be applied in practice, and also sometimes simplified.

Kinds of channels

If transfer channels in different partitions are coupled, then we call the set **coupled reaction channels**, or CRC. If breakup channels in the continuum are included with some discretization method, then we call the set **coupled discretized continuum channels**, or CDCC. If no transfer or breakup couplings are included, just bound inelastic states, then we often refer to plain **coupled channels**, or CC, sets.

Simplifications

Elastic scattering: If $N = 1$ then there is only one uncoupled channel, α_i, with equation $[H_{\alpha_i} - E_{\alpha_i}]\psi_{\alpha_i\alpha_i}(R_{\alpha_i}) = 0$. Within a larger set, this is the channel which is sure to be strong, because of the incoming beam.

One-step DWBA: If an $N = 2$ set consists of an elastic channel α_i and another channel β with only coupling from α_i to β via $\hat{V}_{\beta\alpha_i}$:

$$\left[H_{\alpha_i} - E_{\alpha_i}\right]\psi_{\alpha_i\alpha_i}(R_{\alpha_i}) = 0$$
$$\left[H_\beta - E_\beta\right]\psi_{\beta\alpha_i}(R_\beta) + \hat{V}_{\beta\alpha_i}\psi_{\alpha_i}(R_{\alpha_i}) = 0, \qquad (6.2.7)$$

then subsection 3.3.5 shows how this is equivalent to the first-order *distorted-wave Born approximation*, or DWBA.

N-step DWBA: For a general set of $N+1$ channels, we may consider the sequence of N steps of couplings $\alpha = 1 \rightarrow 2 \rightarrow \cdots N+1$, we set $\alpha_i = 1$. These will be considered in more detail on page 208.

Coupled-channels Born approximation: If the coupled-channels collection is divided into two sets, it is possible to solve the full coupled-channels problem *within* each set, but only include to first order the couplings *between* the two sets. This may be useful, for example, when there are strong collective excitations for both the initial and final nuclei, but weaker transfer couplings that connect them. This mixture of coupled channels and first-order Born approximation is called *coupled-channels Born approximation*, or CCBA.

6.3 Multi-channel equations

6.3.1 Alternate methods

This chapter discusses the methods used to solve the coupled reaction channels equations (3.2.51), when there are both local couplings $V_{\alpha\alpha'}^\lambda(R_x)$ *and* non-local kernels $V_{\alpha\alpha'}(R_x, R_{x'})$ in general. We will see in the next subsection that a group of

M equations can be solved 'exactly' (subject only to radial discretization errors) by finding a set of M linearly independent groups of solutions, and taking a linear combination of these which satisfies the required boundary conditions. This 'close-coupling' method is only practicable, however, if there are not too many equations (the numerical effort can rise as M^3), and if there are *only local couplings*. With local couplings the independent solutions can be found in a single outward 'sweep' of some M^2 radial trial functions. Non-local couplings mean, however, that the source terms at a given radius depend on the wave functions at other radii both larger and smaller, so that this 'exact' method becomes impractical.

In many cases of interest in nuclear physics, nevertheless, because the non-local couplings are not too strong, they can be treated as successive perturbations. They can then be applied iteratively until further applications have progressively smaller effects, and the solutions have converged (to some preset criterion of accuracy). Some failures of convergence can be remedied by the use of Padé approximants, as discussed on page 214, otherwise R-matrix methods may be used as described in Section 6.5.

6.3.2 Close-coupling methods for local couplings

The coupled equations of Eq. (3.2.52) for a given J_{tot}^π may be rewritten to appear like Eq. (6.1.5), but now with a *vector* of channel wave functions $\psi_{\alpha\alpha_i}$, and a *matrix* of coupling potentials $\mathsf{V}(R)$. Again we integrate a trial solution, now a matrix Y, as the solution of

$$\mathsf{Y}''(R) = \mathsf{F}(R)\mathsf{Y}(R) \tag{6.3.1}$$

$$\text{for } \mathsf{F}(R) = \frac{2\mu_\alpha}{\hbar^2}[\mathsf{V}(R) - E_\alpha] + \frac{L_\alpha(L_\alpha+1)}{R^2},$$

where the energies E_α and centrifugal barriers are diagonal matrices.

These coupled differential equations can be solved, following the method of [3], by forming the linearly independent solution sets $\{\mathsf{Y}_{\alpha\beta}(R)\}$, where the β'th solution consists of a set of all channels ($\alpha = 1 \ldots M$) that is made linearly independent of the other sets by having a distinctive starting value at $R = R_{\min}$:

$$\mathsf{Y}_{\alpha\beta}(R_{\min} - h) = 0, \quad \mathsf{Y}_{\alpha\beta}(R_{\min}) = \delta_{\alpha\beta}\frac{(k_\alpha R_{\min})^{L_\alpha+1}}{(2L_\alpha+1)!!} \tag{6.3.2}$$

for the initial conditions in the radial integration of equations (6.3.1). The factor of $\delta_{\alpha\beta}$ may be varied. These solutions together form a matrix, where the row index of Y refers to the channel within the coupled channels, and the column index refers to a linearly independent solution. The boundary conditions of Eq. (3.2.10) are to be satisfied by the $\psi_{\alpha\alpha_i}$ solution, for incoming plane wave in channel α_i.

The solutions $\psi_{\alpha\alpha_i}$ are linear combinations of the $Y_{\alpha\beta}(R)$

$$\psi_{\alpha\alpha_i} = \sum_{\beta=1}^{M} c_{\beta\alpha_i} Y_{\alpha\beta}(R), \qquad (6.3.3)$$

again by satisfying the boundary conditions of Eq. (3.2.10) at $R = a$ and say $R = a - 5h$. The S-matrix elements are a product of the matching procedure.

If only the S matrix is required, and not the wave functions $\psi(R)$, then calculation of the R matrix may be useful. In place of the one-channel definition Eq. (3.1.28) we now use the matrix expression

$$\mathbf{R} = Y(a)[aY(a)']^{-1}, \qquad (6.3.4)$$

which is independent of the starting derivatives in Eq. (6.3.2). The S matrix may then be found by a multi-channel generalization of Eq. (3.1.30):

$$\mathbf{S} = [\mathbf{H}^+ - a\mathbf{R}\mathbf{H}^{+\prime}]^{-1}[\mathbf{H}^- - a\mathbf{R}\mathbf{H}^{-\prime}], \qquad (6.3.5)$$

where the matrices $\mathbf{H}^{\pm}$ are composed with the Coulomb functions $H_\alpha^{\pm}$ along the diagonals.

For cross sections, the S matrix and wave functions $\psi_{\alpha\alpha_i}$ are needed only for those columns α_i corresponding to one of the incoming channels for fixed total spin J_T and parity π. If the sum of the incoming projectile and target spins is greater than one, then there will be several such α_i.

Tolsma and Veltkamp [4] point out one difficulty with this method, which is that if the couplings $V(R)$ have strong off-diagonal components, then the linear independence of the columns of $Y_{\alpha\beta}(R)$ will be reduced as R increases through a classically forbidden region. This is because the components with negative local kinetic energy will generally consist of an exponentially growing part and an exponentially decreasing part. The former are responsible for a tendency to destroy the initially generated linear independence of the solution vectors. The longer the integration continues through a classically forbidden region, the stronger this tendency toward instability will be; for instance, it will occur in scattering problems of nuclear physics with energies near or below the Coulomb barrier. For large matching radii a, maintaining linear dependence at low energies becomes rather difficult.

Tolsma *et al.* [4] propose a stabilization procedure to monitor and if necessary re-orthogonalize the solution vectors. If this were not done, there would be large cancellations in the sum of equation (6.3.3), resulting if severe in complete loss of accuracy of the S-matrix elements and the solution wave functions. A similar problem occurs at small radii, inside all the centrifugal barriers, and this may be remedied by increasing the starting radius $R_{\min}$ at which the radial integrations begin, as already discussed for one-channel solutions. In the multi-channel case it is *more* necessary, for reasons of stability at small radii, to have a minimum radius proportional to some angular momentum $\bar{L}$ typical of the coupled-channels set:

$$R_{\min} \geq c_L \bar{L} h \tag{6.3.6}$$

for some constant $c_L \sim 2.0$. If this remedial action is insufficient, then the R-matrix methods of Section 6.5 may be needed.

6.3.3 Iterative solutions

Iterative methods may be used as one way of solving the coupled equations, and are often necessary when non-local (e.g. transfer) couplings are present. Moreover, there is also a physical relation between successive iterations and the N-step DWBA.

Let us begin again with a model Hamiltonian H for the coupled system, as we did in subsection 3.2.2. This Hamiltonian is projected as before onto the partial-wave basis states $\phi_\alpha = |\alpha\rangle$. If E_α is the asymptotic kinetic energy in the α channel, then a channel-projected Hamiltonian H_α satisfies

$$H_\alpha - E_\alpha = \langle \phi_\alpha | H - E | \phi_\alpha \rangle \tag{6.3.7}$$

and will be composed of a kinetic energy term and a diagonal optical potential. The 'interaction potential' $\mathcal{V}_\alpha$ is then defined to be everything in H not included in H_α, so

$$H_\alpha - E_\alpha + \mathcal{V}_\alpha = H - E. \tag{6.3.8}$$

The kinetic energy in each channel is $E_\alpha = E - \epsilon_\alpha$, depending on the sum ϵ_α of the internal excitation energies, which is the same as the energy E_{xpt} appearing in Eq. (3.2.45). This construction gives $\mathcal{V}_\alpha$ with vanishing diagonal matrix elements $\langle \phi_\alpha | \mathcal{V}_\alpha | \phi_\alpha \rangle = 0$. These are the same definitions as before in subsection 4.5.1, and differ from those in subsection 3.2.2 by the movement here of the diagonal optical potential from V_x into $\mathcal{V}_\alpha$.

Coupled equation set for N basis states

If we take the model Schrödinger equation $[H - E]\psi = 0$, and project separately onto the different basis states ϕ_α, we derive the set of equations

$$[E_\alpha - H_\alpha] \psi_\alpha(R_\alpha) = \sum_{\beta \neq \alpha} \langle \phi_\alpha | H - E | \phi_\beta \rangle \psi_\beta(R_\beta). \tag{6.3.9}$$

which couple together the unknown wave functions $\psi_\alpha(R_\alpha)$. The matrix element $\langle \phi_\alpha | H - E | \phi_\beta \rangle$ has two different forms, as we saw in subsection 4.5.1, depending on whether we expand

$$H - E = H_\alpha - E_\alpha + \mathcal{V}_\alpha \quad \text{(the 'post' form)}$$
$$= H_\beta - E_\beta + \mathcal{V}_\beta \quad \text{(the 'prior' form)}.$$

Thus

$$\langle \phi_\alpha | H - E | \phi_\beta \rangle = \mathcal{V}_{\alpha\beta}^{\text{post}} + [H_\alpha - E_\alpha] N_{\alpha\beta} \quad \text{(post)} \tag{6.3.10}$$

$$\text{or} = \mathcal{V}_{\alpha\beta}^{\text{prior}} + N_{\alpha\beta} [H_\beta - E_\beta] \quad \text{(prior)}$$

where

$$\mathcal{V}_{\alpha\beta}^{\text{post}} \equiv \langle \phi_\alpha | \mathcal{V}_\alpha | \phi_\beta \rangle, \quad \mathcal{V}_{\alpha\beta}^{\text{prior}} \equiv \langle \phi_\alpha | \mathcal{V}_\beta | \phi_\beta \rangle, \quad N_{\alpha\beta} \equiv \langle \phi_\alpha | \phi_\beta \rangle. \tag{6.3.11}$$

The overlap function $N_{\alpha\beta} = \langle \phi_\alpha | \phi_\beta \rangle$ in equation (6.3.10) is the same as defined in Eq. (3.2.50), and arises from the non-orthogonality between the basis states ϕ_α and ϕ_β if these are in different mass partitions. We will see below that this *non-orthogonality term* has no effect in first-order DWBA, and can be made to disappear in second-order DWBA, if the first and second steps use the prior and post interactions respectively.

N-step DWBA

If the coupling interactions $\mathcal{V}_\alpha$ in equation (6.3.10) are weak, or if the back-coupling effects of these interactions are already included in the optical potentials of the prior channel, then it becomes reasonable to use a multi-step distorted-wave Born approximation (DWBA) as introduced in subsection 6.2.2. Let us consider such a multi-step approximation which labels the channels so that it always feeds flux 'forwards' in the sequence $\alpha = 1 \rightarrow 2 \rightarrow \cdots \rightarrow N+1$, neglecting the back couplings. In the elastic channel α_i (here $\alpha_i = 1$), the wave function is governed by the optical potential defined there, and the wave function in the α channel is driven by couplings from all the previous channels:

$$[E_\alpha - H_\alpha] \psi_\alpha(R_\alpha) = \sum_{\beta=1}^{i-1} \langle \phi_\alpha | H - E | \phi_\beta \rangle \psi_\beta(R_\beta). \tag{6.3.12}$$

Initial channel:

$$[E_1 - H_1] \psi_1(R_1) = 0. \tag{6.3.13}$$

Second channel:

$$[E_2 - H_2] \psi_2(R_2) = \langle \phi_2 | H - E | \phi_1 \rangle \psi_1(R_1). \tag{6.3.14}$$

If the *prior* interaction is used, the right-hand side becomes

$$= \langle \phi_2 | \mathcal{V}_1 | \phi_1 \rangle \psi_1 + \langle \phi_2 | \phi_1 \rangle [H_1 - E_1] \psi_1$$

$$= \langle \phi_2 | \mathcal{V}_1 | \phi_1 \rangle \psi_1, \text{ as } \psi_1 \text{ satisfies Eq. (6.3.13)} \quad (6.3.15)$$

$$= \mathcal{V}_{21}^{\text{prior}} \psi_1. \quad (6.3.16)$$

Final channel: $(c = N + 1)$

$$[E_c - H_c] \, \psi_c(R_c) = \sum_{\beta=1}^{c-1} \langle \phi_c | H - E | \phi_\beta \rangle \, \psi_\beta(R_\beta). \quad (6.3.17)$$

If the *post* interaction had been used for all the couplings to this last channel, then

$$[E_c - H_c] \, \psi_c(R_c) = \sum_{\beta=1}^{c-1} \langle \phi_c | \mathcal{V}_c | \phi_\beta \rangle \psi_\beta + [H_c - E_c] \sum_{\beta=1}^{\beta=c-1} \langle \phi_c | \phi_\beta \rangle \psi_\beta, \quad (6.3.18)$$

so

$$[E_c - H_c] \, \chi_c(R_c) = \sum_{\beta=1}^{c-1} \mathcal{V}_{cj}^{\text{post}} \psi_\beta, \quad (6.3.19)$$

where we have defined

$$\chi_c(R_c) = \psi_c + \sum_{\beta=1}^{c-1} \langle \phi_c | \phi_\beta \rangle \psi_\beta$$

$$= \langle \phi_c | \psi \rangle.$$

Note that, as all the ϕ_β are square-integrable and hence decay faster than R^{-1} at large radii, the ψ_c and χ_c are the same asymptotically. They differ only at finite radii, and hence yield the same asymptotic scattering amplitudes. The equation for χ_c has no non-orthogonality terms once the *post* interaction is used in the final channel.

These results imply that in N-step DWBA, some non-orthogonality terms can be made to disappear if 'prior' interactions are used for the first step, and/or if 'post' interactions are used for the final step. This means that the non-orthogonality term never appears in the first-order DWBA, irrespective of the choice of prior or post forms. In second-order DWBA, the *prior-post* combination must be chosen to avoid the non-orthogonality terms [5]. It should be also clear that non-orthogonality terms will have to be evaluated if the DWBA is continued beyond second order.

Full solution by iteration

There are a number of different ways of solving the CRC equations with the non-orthogonality terms present: for discussions of different approaches see [6], [7] and the survey of [8, ch. 3].

There are schemes available which can iterate all channels with an arbitrary choice of post or prior interactions for all the couplings. Define

$$\theta_{\alpha\beta} = 0 \text{ or } 1 : \text{presence of post in a } \beta \to \alpha \text{ coupling}, \tag{6.3.20}$$

$$\text{so } 1 - \theta_{\alpha\beta} = 1 \text{ or } 0 : \text{presence of prior.}$$

The following iterative scheme [9] ($n = 1, 2, \ldots$) on convergence then solves the CRC equations (6.3.9):

For $n = 0$, start with

$$\psi_\alpha^{(0)} = \delta_{\alpha,\alpha_i} \psi_{\text{elastic}}$$

$$\delta\Omega_\alpha^{(0)} = \delta\psi_\alpha^{(0)} = 0. \tag{6.3.21}$$

For $n = 1 \to N$ (for N-step DWBA) solve

$$[H_\alpha - E_\alpha]\chi_\alpha^{(n)} + \Omega_\alpha^{(n-1)} = 0 \tag{6.3.22}$$

with

$$\Omega_\alpha^{(n-1)} = \sum_\beta [\theta_{\alpha\beta} V_{\alpha\beta}^{\text{post}} + (1 - \theta_{\alpha\beta}) V_{\alpha\beta}^{\text{prior}}] \psi_\beta^{(n-1)} - \delta\Omega_\alpha^{(n-1)}, \tag{6.3.23}$$

then calculate for subsequent iterations

$$\delta\psi_\alpha^{(n)} = \sum_\beta \theta_{\alpha\beta} \langle\phi_\alpha|\phi_\beta\rangle \psi_\beta^{(n-1)} \tag{6.3.24}$$

$$\delta\Omega_\alpha^{(n)} = \sum_\beta (1 - \theta_{\alpha\beta}) \langle\phi_\alpha|\phi_\beta\rangle [\Omega_\beta^{(n-1)} + [H_\beta - E_\beta]\delta\psi_\beta^{(n)}] \tag{6.3.25}$$

$$\psi_\alpha^{(n)} = \chi_\alpha^{(n)} - \delta\psi_\alpha^{(n)}. \tag{6.3.26}$$

This scheme avoids numerical differentiations except in an higher-order correction to $\delta\Omega_\alpha$ that arises in some circumstances.

When the non-orthogonality terms are included properly, it becomes merely a matter of numerical convenience whether post or prior couplings are used, for one, two, and multi-step calculations. The equivalence of the two coupling forms can be confirmed in practice [10], and used as one check on the accuracy of the numerical methods employed.

6.3.4 Numerical iterations

If the non-local interactions $\mathcal{V}_{\alpha\alpha'}(R, R')$ in the CRC equations (3.2.52) are present, then, without R-matrix methods, it will always be necessary to solve the coupled channels by iteration. To solve Eq. (6.3.22), the coupled radial equations of Eq. (6.3.1) become inhomogeneous, having an additional source term Ω

$$\mathsf{Y}''(R) = \mathsf{F}(R)\mathsf{Y}(R) + \Omega(R) \qquad (6.3.27)$$

with $\mathsf{F}(R)$ defined as before in Eq. (6.3.1).

Now the trial solution matrix $\mathsf{Y}(R)$ needs to include also a particular *inhomogeneous* solution. Let this particular solution be denoted by $\mathsf{Y}_{\alpha\beta}(R)$ for $\beta = 0$, and for this solution we may for convenience choose the zero starting conditions

$$\mathsf{Y}_{\alpha 0}(R_{\min}) = \mathsf{Y}'_{\alpha 0}(R_{\min}) = 0, \qquad (6.3.28)$$

in contrast to the *homogeneous* $\mathsf{Y}'_{\alpha\beta}$ for $\beta \geq 1$ which have non-zero starting derivatives.

Then again the needed solutions $\psi_{\alpha\alpha_i}$ are linear combinations of the $\mathsf{Y}_{\alpha\beta}(R)$

$$\psi_{\alpha\alpha_i} = \sum_{\beta=0}^{M} c_{\beta\alpha_i}\mathsf{Y}_{\alpha\beta}(R) \qquad (6.3.29)$$

where $c_{\beta\alpha_i} = 1$ always if $\beta = 0$. The whole iteration procedure needs to be repeated for each required incoming boundary condition α_i as the $\Omega_\alpha^{(n-1)}$ will be different, but the same $\mathsf{Y}_{\alpha\beta}(R)$ for $\beta \geq 1$ are used, and should be stored after their first generation.

6.3.5 Convergence of iterative methods

The iterative method of solving the CRC equations (3.2.52, 6.3.9) will converge if the couplings are sufficiently small. The procedure will however diverge if the couplings are too large, or if the system is too near a resonance. On divergence, the successive wave functions $\psi_{\alpha\alpha_i}^{(n)}$ will become larger and larger as n increases, and not converge to any fixed limit. Unitarity will of course be violated as the S-matrix elements will become much larger than unity.

Improving the convergence rate

There are several ways of improving the rate of convergence, especially when there are non-local potentials:

(A) Solving some partial number of the local couplings exactly by the methods of subsection 6.3.2, and iterating only on the non-local couplings and the remaining local couplings.
(B) Use Padé approximants to accelerate the convergence of the sequence of S-matrix elements.
(C) Find a separable expansion for the non-local kernels, so that they can be included exactly in the coupled-channels solution.
(D) Expand the wave functions with a range of square-integrable basis states, and take the coefficients of the wave functions in this basis as the unknowns in a system of linear equations.
(E) Use R-matrix eigenvalue expansions in an interior radial region.

We discuss each of these in turn:

(A) Partial iteration: If the non-local interactions $V_{\alpha\alpha'}(R, R')$ in the CRC equations (3.2.52) are present, then it will always be necessary to solve the coupled channels by iteration. With the local couplings, however, we have a choice whether to iterate, or to include them in the exact solutions of the close-coupling method. A simple option is to allow a specifiable number N_{cc} of channels to be coupled exactly, with the remainder only being fed after one or more iterations. This would be useful, for example, if the channels for the low-lying states of a highly deformed target were included in this block of N_{cc} channels, and if the remaining channels (e.g. for transfers) were not fed by more than 2 or 3 steps beyond this initial block. Restricting the number of these iterations to one is equivalent to solving a *coupled-channels Born approximation* (CCBA) model.

(B) Padé acceleration: It is very useful to be able to iterate the coupled equations in a conventional manner, as then 1-, 2- and 3- step DWBA results (etc.) can be recovered by stopping the iterations short of full convergence. Using Padé acceleration, as described in the next subsection and used in [11, 12, 9], has the advantages that it need only be employed if ordinary iterations are seen to diverge, and that it transforms the previously divergent results with little new computational effort.

(C) Separable expansions: The separable expansion method works when a non-local potential has a separable approximation $\hat{V} = |v\rangle\langle v|$ for some vector $|v\rangle$ not necessarily normalized. This means that the kernel function factorizes as

$$V(R, R') = v(R)v(R').$$
(6.3.30)

A separable approximation allows a radial equation such as (6.1.1) to be solved using a superposition of homogeneous and inhomogeneous solutions. If we want to solve

$$[H - E]|\chi\rangle + |v\rangle\langle v|\chi\rangle = 0 \qquad (6.3.31)$$

then, as well as trial solution $y(R)$ of Eq. (6.1.6), we find also the inhomogeneous solution $[H - E]y^{(0)}(R) + v(R) = 0$. The desired solution is now the linear combination $\chi(R) = cy(R) + c_0 y^{(0)}(R)$ that satisfies Eq. (6.3.31), which leads to the requirement $(\langle v|y^{(0)}\rangle - 1)c + \langle v|y\rangle c_v = 0$. This equation, in combination with the existing Eq. (6.1.7), enables all the c and $\mathbf{S}_L$ to be found. The method can be generalized to a linear combination of separable terms

$$V(R, R') = \sum_{\lambda=1}^{\Lambda} v_\lambda(R) v_\lambda(R'). \qquad (6.3.32)$$

This is useful in nucleonic few-body reactions, but becomes unsatisfactory for handling the non-locality in transfer reactions involving heavier nuclei. This is because if the masses of the initial and final nuclei become large relative to the mass of the transferred particle, the form factor for the transfers becomes more nearly local. As we approach the no-recoil limit where the form factors would be exactly local, the separable expansion of Eq. (6.3.32) of a nearly local kernel requires more and more terms. In the limit of a local form factor, the separable expansion will require the same number of terms Λ as there are radial grid points.

(D) Basis expansions: The method of expanding the wave functions in Coulomb, Gaussian [13], Lagrange-mesh [14], Airy [15] functions has been used. This method is practical provided the characteristic spatial widths of the basis states were chosen in accordance with the wave numbers k_α in the respective channels. This requirement is less severe with light-ion reactions, where the wavelengths are typically $\gtrsim 5$ fm. For some reactions, however, the oscillation rates are much larger, and a more economical method would be to expand in terms of sinusoidal or Airy functions that depend explicitly on the local wave number over some radial region.

(E) R-matrix eigenvalue expansions: This method, described in more detail below in Section 6.5, is essentially a variant of method (D), where the basis functions are now eigensolutions of some part of the physical Hamiltonian, namely the diagonal potentials. This method works even at higher scattering energies, since for the basis we can select a subset of the single-channel eigensolutions with eigenenergies in the neighborhood of the actual scattering energy. The off-diagonal couplings are then diagonalized in this selected basis.

Padé approximants for sequence extrapolation

A given sequence $S_0, S_1, \ldots$ of S-matrix elements that result from iterating the CRC equations can be regarded as the successive partial sums of the polynomial

$$f(z) = S_0 + (S_1 - S_0)z + (S_2 - S_1)z^2 + \cdots \tag{6.3.33}$$

evaluated at $z = 1$. This polynomial will clearly converge for z sufficiently small, but will necessarily diverge if the analytic continuation of the $f(z)$ function has any pole or singularities inside the circle $|z| > 1$ in the complex z-plane. The problem that Padé approximants solve is that of finding a computable approximation to the analytic continuation of the $f(z)$ function. This is accomplished by finding a rational approximation

$$P_{[n,m]}(z) = \frac{p_0 + p_1 z + p_2 z^2 + \cdots + p_n z^n}{1 + q_1 z + q_2 z^2 + \cdots + q_m z^m} \tag{6.3.34}$$

which agrees with the $f(z)$ function in the region where the latter does converge, as tested by matching the coefficients in the polynomial expansion of $P_{[n,m]}(z)$ up to and including the coefficient of z^{n+m}.

There are many different ways [16] of evaluating the coefficients $\{p_m, q_n\}$, but for the present problem we can use Wynn's ε algorithm [17], which is a method of evaluating the upper right half of the Padé table at $z = 1$ directly in terms of the original sequence $S_0, S_1, \ldots$.

Wynn's epsilon algorithm

Initializing $\varepsilon_0^{(j)} = S_j$ from the given divergent sequence S_j, and $\varepsilon_{-1}^{(j)} = 0$, we may generate an array using the rule

$$\varepsilon_{k+1}^{(j)} = \varepsilon_{k-1}^{(j+1)} + (\varepsilon_k^{(j+1)} - \varepsilon_k^{(j)})^{-1}. \tag{6.3.35}$$

All the $\varepsilon_k^{(j)}$ then map onto the transposed upper right half of the Padé table, including the diagonal, according to

$$\varepsilon_{2k}^{(j)} = P_{[k,k+j]}(1). \tag{6.3.36}$$

Experience has shown that for typical sequences the most accurate Padé approximants are those near the diagonal of the Padé table, and these are just the right-most $\varepsilon_{2k}^{(0)}$ in the ε table.

When accelerating a *vector* of S-matrix elements $\mathbf{S}_j$, with a component for each coupled channel, then it is important to accelerate the vector as a whole. Wynn [18] pointed out that this can be done using the Samuelson inverse $\mathbf{x}^{-1} = (\mathbf{x} \cdot \mathbf{x}^*)^{-1}\mathbf{x}^*$ where $\mathbf{x}^*$ is the complex conjugate of $\mathbf{x}$. Otherwise there will be problems when iterating (say) a two-channel system with alternating backwards and forwards coupling, because of zero divisors in the ε algorithm.

6.4 Multi-channel bound states

As well as multi-channel scattering solutions, sometimes coupled-channels bound states are also needed. The bound state of a particle in a deformed potential will have several channels with different angular momenta coupled together, and such states may be needed for calculation of transfers to or from deformed nuclei.

There are various techniques for calculating the wave functions of these bound states: for a review see reference [19]. The Sturmian expansion [20] or R-matrix eigenvalue (Section 6.5) methods can be used, or the coupled equations can be solved iteratively. The expansion methods have the advantage that *all* solutions in the deformed potential are found, where sometimes the iterative method has difficulty in converging to a particular solution if there are other permitted solutions near in energy.

The iterative method has the advantage that the radial wave functions (once found) are subject only to the discretization error for the Schrödinger's equation, and are not dependent on a more time-consuming diagonalization of large matrices. Iterative solutions satisfy the correct boundary conditions independently of the size of a basis-state set, and are anyway needed to find the Sturmian or channel eigensolutions for constructing the basis states of those methods.

6.4.1 Coupled-channels eigenvalue problem

If a particle is bound at negative energy E around a (say) rotational core, its wave function may be found as the eigensolution of a given potential:

$$[T_\ell(r) + V(r) + \epsilon_I - E]u_{\ell sjI}(r) + \sum_{\ell' j' I'} V_{\ell sjI:\ell' sj'I'}(r)u_{\ell' sj'I'}(r) = 0 \qquad (6.4.1)$$

with boundary conditions $u_{\ell sjI}(0) = 0$ and $u_{\ell sjI}(r) \propto W_{-\eta,\ell+\frac{1}{2}}(-2k_I r)$ as $r \geq a$, where $W_{-\eta,\ell+\frac{1}{2}}(\rho)$ is the Whittaker function and $k_I^2 = 2\mu|E - \epsilon_I|/\hbar^2$ is the asymptotic wave number. The total spin J index has been suppressed. The couplings $V(r)$ may be calculated using the theory of subsection 4.4.1.

If the core cannot be excited, then these coupled equations reduce to one uncoupled equation, but solving this equation can still be regarded as a special case of the coupled bound-state problem. Eigensolutions can be found by solving either for the bound-state energy E, or by varying the depth of the binding potential for a specified eigenenergy E. In general, furthermore, we could choose to vary any multipole of any part of the binding potentials (except the Coulomb component).

All of these eigenvalue problems may be regarded as special cases of finding eigensolutions of a set of M coupled-channels equations, represented as the question of finding an eigenvalue ω to satisfy the equations

$$\left[\frac{d^2}{dr^2} - \frac{\ell_i(\ell_i+1)}{r^2}\right]\phi_i(r) + \sum_{j=1}^{M}\left[U_{ij}(r) + \omega V_{ij}(r)\right]\phi_j(r) = 0 \qquad (6.4.2)$$

with boundary conditions

$$\phi_i(a) = p_i W_{-\eta_i,\ell_i+\frac{1}{2}}(-2k_i R) \qquad (6.4.3)$$

$$\phi_i(a+\delta R) = p_i W_{-\eta_i,\ell_i+\frac{1}{2}}(-2k_i(a+\delta R)) \qquad (6.4.4)$$

$$\phi_i(0) = 0, \qquad (6.4.5)$$

and $k_i^2 \equiv \kappa_i^2 + \theta\omega$ and $\eta_i \equiv \nu_i/(2k_i)$, for given partial waves ℓ_i, fixed potentials $U_{ij}(r)$, variable potentials $V_{ij}(r)$, matching radius a, and Coulomb proportionality constants $\nu_i = 2\mu Z_1 Z_2 e^2/\hbar^2$. The p_i are as yet unknown coefficients.

The asymptotic energy constants κ_i^2 do not explicitly appear in Eq. (6.4.2) because we absorb them for convenience into the diagonal components $U_{ii}(r)$. Let θ be the asymptotic component of all the diagonal $V_{ii}(r)$.

The solution begins by constructing the trial integration functions for a trial value of ω, on either side of an intermediate matching point $r = r_m$:

$f_{i:j}^{\text{in}}(r)$ for r from h to r_m, starting with $f_{i:j}^{\text{in}}(h) = \delta_{ij} h^{\ell_i+1}/(2\ell_i+1)!!$, and

$f_{i:j}^{\text{out}}(r)$ for r from a in to r_m, starting with $f_{i:j}^{\text{out}}(a) = \delta_{ij} W_{-\eta_i,\ell_i+\frac{1}{2}}(-2k_i a)$.

The intermediate matching point $r = r_m$ should be chosen where the wave functions are still oscillatory, to avoid having to integrate outwards into the classically forbidden region. This is because, as discussed in subsection 6.3.2, we should in such regions always integrate in the direction where the desired solution is increasing.

The solution is therefore

$$\phi_i(r) = \begin{cases} \sum_j b_j f_{i:j}^{\text{in}}(r) & \text{for } r < r_m \\ \sum_j c_j f_{i:j}^{\text{out}}(r) & \text{for } r \geq r_m, \end{cases} \qquad (6.4.6)$$

and the matching conditions are the equality of the two expressions and their derivatives at $r = r_m$. The normalization is still arbitrary, so we may fix $c_1 = 1$. In general the equations (6.4.2) have no solution as ω is not exactly an eigenvalue. The method therefore uses the discrepancy in the matching conditions to estimate how ω should be changed to $\omega + \delta\omega$ to reduce that discrepancy, and iterates this process to reduce $\delta\omega$.

Thus at each iteration we first solve as simultaneous equations the $2M - 1$ of the matching conditions

$$\sum_j b_j f_{i;j}^{\text{in}}(r_m) = \sum_j c_j f_{i;j}^{\text{out}}(r_m) \text{ for all } i \tag{6.4.7}$$

$$\sum_j b_j f_{i;j}^{\text{in}}(r_m)' = \sum_j c_j f_{i;j}^{\text{out}}(r_m)' \text{ for all } i \neq 1 \tag{6.4.8}$$

along with $c_1 = 1$ for the $2M$ unknowns b_i, c_i. If the function $\phi_i(r)$ is then constructed using equation (6.4.6), there will be a discrepancy as

$$\phi_{\text{in}}' \equiv \phi_1'(r)|_{r<r_m} \neq \phi_{\text{out}}' \equiv \phi_1'(r)|_{r>r_m}, \tag{6.4.9}$$

and this difference will generate $\delta\omega$ via

$$\delta\omega \sum_{ij=1}^M \int_0^R \phi_i(r) V_{ij}(r) \phi_j(r) \mathrm{d}r = \phi_1(r_m)[\phi_{\text{out}}' - \phi_{\text{in}}']. \tag{6.4.10}$$

It is necessary while iterating in this manner to monitor the *number of nodes* in one or more selected components of the wave function. When the iterations have converged to some accuracy criterion on the size of $\delta\omega$, the set of wave functions can be normalized in the usual manner:

$$\sum_{i=1}^M \int_0^\infty |\phi_i(r)|^2 \mathrm{d}r = 1. \tag{6.4.11}$$

6.5 R-matrix methods

The R-matrix method, proposed by Wigner and Eisenbud [21] and promulgated in detail by Lane and Thomas [22], uses an orthonormal basis expansion in the interior of some R-matrix radius a, using eigenfunctions of the diagonal parts of the Hamiltonian as basis states. With the diagonalized interior wave functions it constructs the R matrix of Eq. (6.3.4) to match to asymptotic Coulomb functions as before. The novel feature of this method is that it uses a fixed logarithmic derivative at the R-matrix radius for all the basis states to render the kinetic energy Hermitian on a finite region, and hence make the basis states orthonormal.

The radial stepping methods of solving the coupled equations only allow local couplings to be treated properly, and non-local couplings from transfers have to be included iteratively. The present R-matrix method is an equivalent way of solving the coupled equations, and has the advantages of being more stable numerically, and also allowing non-local components of the Hamiltonian in an interior region to be included to all orders. It has recently been revived in nuclear physics applications [14, 23] for these reasons. Both transfer and non-orthogonality non-localities may be included non-perturbatively, and resonances and bound states may both be described without difficulty.

6.5.1 One-channel R-matrix expansions

We begin by describing how to solve the one-channel Schrödinger equation of Eq. (6.1.1) with the new expansion method. Even though it is more easily solved using the stepping methods of Section 6.1, the one-channel case here is instructive. It also turns out that correcting the errors in the one-channel expansion, by the method of Buttle [24], is sufficient to correct most of the multi-channel errors.

Consider again the one-channel equation (6.1.1). This has a solution for all continuum energies E, and because its scattering wave functions over a finite range $[0, a]$ are not orthogonal to each other they cannot be used as orthonormal basis states for an eigenvalue expansion. Another statement of that fact is that the kinetic energy operator is not Hermitian over a finite radial range, unless all the wave functions satisfy some special boundary conditions to ensure the Hermiticity. The R-matrix basis states are therefore chosen to ensure Hermiticity of the kinetic energy and hence the orthogonality of the eigensolutions over a finite interval. This is done by requiring them all to have fixed logarithmic derivatives

$$\beta = \frac{\mathrm{d}}{\mathrm{d}R} \ln w(R) \equiv \frac{w'(R)}{w(R)} \tag{6.5.1}$$

at $R = a$. They also satisfy $w(0) = 0$, and are solutions of

$$\left[-\frac{\hbar^2}{2\mu} \left(\frac{\mathrm{d}^2}{\mathrm{d}R^2} - \frac{L(L+1)}{R^2} \right) + V(R) - \varepsilon_n \right] w_n(R) = 0 \tag{6.5.2}$$

for some eigenvalues ε_n, $n = 1, 2, \ldots$, labeling the distinct eigensolutions according to the number of radial nodes in the wave function. We assume that the $V(R)$ is real.

At present β and a are both free input parameters, and the same results should be obtained on convergence for any value of β, and for any matching radius outside the potentials.

Orthonormal basis

Suppose $w_n(R)$ and $w_m(R)$ are two solutions of Eq. (6.5.2) with eigenenergies $\varepsilon_n \neq \varepsilon_m$. If we multiply the w_n equation by w_m, and subtract from the exchanged equation, we obtain

$$-\frac{\hbar^2}{2\mu}[w_m w_n'' - w_n w_m''] + (\varepsilon_m - \varepsilon_n)w_n w_m = 0. \qquad (6.5.3)$$

Integrating this equation by parts with limits $R = 0$ to a gives

$$-\frac{\hbar^2}{2\mu}[w_m(a)w_n'(a) - w_n(a)w_m'(a)] + (\varepsilon_m - \varepsilon_n)\int_0^a w_n w_m dR = 0, \qquad (6.5.4)$$

in which the first term is zero since both $w_n'/w_n = w_m'/w_m = \beta$ at a. This implies that, since $\varepsilon_n \neq \varepsilon_m$, distinct eigenstates are orthogonal. We now assume that the w_n basis states are all normalized to unity, so they may all be taken as an orthonormal set satisfying

$$\int_0^a w_n(R)w_m(R)dR = \delta_{mn}. \qquad (6.5.5)$$

Expansion of the scattering wave function

We now wish to construct the scattering solution $\chi(R)$ of Eq. (6.5.2) at general energy E as a linear combination of N states w_n according to

$$\chi(R) = \sum_{n=1}^N A_n w_n(R) \qquad (6.5.6)$$

for coefficients $A_n = \int_0^a w_n(R)\chi(R)dR$ to be found in agreement with the usual scattering boundary conditions of Eq. (6.1.2). We expect to approach convergence as N increases.

Note that we can *not* simply match the wave functions and derivatives of each side of Eq. (6.5.6), as might be expected from elementary quantum theory. If we do this, then, since all the $w_n(R)$ have the same logarithmic derivative β, so will their sum on the right side of the equation. A scattering wave function $\chi(R)$ certainly does not have a fixed logarithmic derivative! This discrepancy comes about from the manner of convergence of the sum (6.5.6) as a function of N. The convergence is only with respect to the values of the wave functions, and *not* their derivatives. Exactly how this convergence proceeds will be illustrated in Fig. 6.1 below. The lack of convergence of derivatives means that we must instead use an integral expression to find the expansion coefficients.

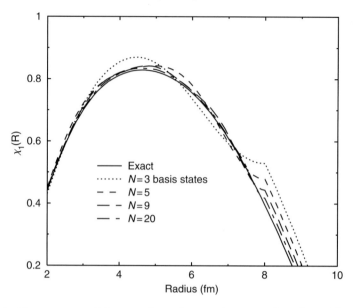

Fig. 6.1. Convergence of the one-channel scattering wave function with varying numbers of basis states, for $a = 8\,\text{fm}$ and $\beta = 0$. We plot the real part of $p_{1/2}$ neutron scattering wave function on ^{4}He at 5 MeV.

If we repeat the procedure of the previous paragraph with the pair of solutions $w_n(R)$ and $\chi(R)$, then in place of Eq. (6.5.4) we have

$$-\frac{\hbar^2}{2\mu}\left[\chi(a)w'_n(a) - w_n(a)\chi'(a)\right] + (E - \varepsilon_n)A_n = 0, \qquad (6.5.7)$$

which gives

$$A_n = \frac{\hbar^2}{2\mu}\frac{1}{E - \varepsilon_n}[\chi(a)w'_n(a) - w_n(a)\chi'(a)]$$

$$= \frac{\hbar^2}{2\mu}\frac{w_n(a)}{\varepsilon_n - E}[\chi'(a) - \beta\chi(a)]. \qquad (6.5.8)$$

If we use these expressions in Eq. (6.5.6) evaluated at $R = a$, we obtain useful information about the logarithmic derivative of the unknown solution $\chi(R)$. We get

$$\chi(a) = \sum_{n=1}^{N}\frac{\hbar^2}{2\mu}\frac{w_n(a)}{\varepsilon_n - E}[\chi'(a) - \beta\chi(a)]w_n(a), \qquad (6.5.9)$$

so

$$\frac{\chi(a)}{\chi'(a) - \beta\chi(a)} = \sum_{n=1}^{N}\frac{\hbar^2}{2\mu}\frac{w_n(a)^2}{\varepsilon_n - E}. \qquad (6.5.10)$$

If β were zero here, the left side would be just $\chi(a)/\chi'(a)$, which is exactly the R matrix defined in Eqs. (3.1.28, 6.3.4). Because we now want freedom to choose β for other reasons, we now define a 'generalized' R matrix for a wave function as

$$\mathbf{R} = \frac{1}{a} \frac{\chi(R)}{\chi'(R) - \beta\chi(R)}, \quad (6.5.11)$$

in terms of which the S matrix, instead of Eqs. (3.1.30, 6.3.5), is

$$\mathbf{S} = \frac{H^- - a\mathbf{R}(H'^- - \beta H^-)}{H^+ - a\mathbf{R}(H'^+ - \beta H^+)}. \quad (6.5.12)$$

It is therefore into this equation that we substitute the R-matrix expression

$$\mathbf{R} = \sum_{n=1}^{N} \frac{\hbar^2}{2\mu a} \frac{w_n(a)^2}{\varepsilon_n - E}. \quad (6.5.13)$$

We now define for each pole n the *reduced width amplitudes*

$$\gamma_n = \sqrt{\frac{\hbar^2}{2\mu a}} \, w_n(a), \quad (6.5.14)$$

in terms of which the R matrix has the familiar simple form

$$\mathbf{R} = \sum_{n=1}^{N} \frac{\gamma_n^2}{\varepsilon_n - E}. \quad (6.5.15)$$

The γ_n^2 are called the *reduced widths*.

The scattering wave function in the same approximation is

$$\chi(R) = \sum_{n=1}^{N} \frac{\hbar^2}{2\mu a} \frac{w_n(a)}{\varepsilon_n - E} [\chi'(a) - \beta\chi(a)] w_n(R). \quad (6.5.16)$$

The convergence with N of this wave function is illustrated in Fig. 6.1. We see that even though the function of (6.5.16) has zero derivative at the matching radius, it still converges closer and closer to the correct wave function with non-zero derivative.

6.5.2 The multi-channel R matrix

This method for multi-channel problems uses as basis states the eigenfunctions of the real part of the diagonal potential in each channel. If the boundary condition of Eq. (6.5.1) is used, these form an orthonormal basis set. In that way, diagonal real potentials within the coupled-channels set are treated more accurately, while off-diagonal and all imaginary potentials are treated via their matrix elements in this basis.

This means that the basis functions should now be written as $w_\alpha^n(R_\alpha)$ for the n'th basis state in channel α, with radial coordinate R_α to follow the notation for the coupled equations (6.3.9). It is most convenient to take them as all the eigensolutions of the diagonal real potential U_α in each separate channel:

$$\left[T_{\alpha L}(R_\alpha) + U_\alpha(R_\alpha) - \varepsilon_{n\alpha} \right] w_\alpha^n(R_\alpha) = 0 \qquad (6.5.17)$$

for eigenenergies $\varepsilon_{n\alpha}$, with the basis functions again all having fixed logarithmic derivatives $\beta = \mathrm{d}\ln w_\alpha^n(R_\alpha)/\mathrm{d}R_\alpha$ at a.

The wave functions of the coupled problem (6.3.9) can now be solved completely over the interior range $[0, a]$, by using the orthonormal basis set of the $\{w_\alpha^n(R_\alpha)\}$ with coefficients to be determined. The coefficients are found in two stages: first by finding all the eigensolutions $g_\alpha^p(R_\alpha)$ of the equations (6.3.9) using the above orthonormal basis, and then expanding the scattering wave functions in terms of these $g_\alpha^p(R_\alpha)$.

In the traditional R-matrix method, the diagonalization of the M-channel Hamiltonian in equation (6.3.9) proceeds by finding the radial wave eigenfunctions $g_\alpha(R_\alpha)$ in that basis. This is to solve

$$[T_{\alpha L} + U_\alpha + \epsilon_\alpha] g_\alpha(R_\alpha) + \sum_{\alpha' \neq \alpha} \hat{V}_{\alpha\alpha'} \, g_{\alpha'}(R_{\alpha'}) = e \, g_\alpha(R_\alpha), \qquad (6.5.18)$$

where $\hat{V}_{\alpha\alpha'}$ refers to all the *off*-diagonal couplings, local or non-local, since the diagonal real potentials U_α already appear, and the energies ϵ_α are the core excitation energies in each channel. This yields $P = NM$ eigenenergies e_p with corresponding multi-channel eigenstates

$$g_\alpha^p(R_\alpha) = \sum_{n=1}^{N} c_\alpha^{pn} w_\alpha^n(R_\alpha). \qquad (6.5.19)$$

Eigenstates here with $e_p < 0$ are close to the bound states, while solutions with $e_p > 0$ contribute to the scattering solutions. Certain of the $e_p > 0$ solutions may correspond to low-lying resonances if those are present, but the majority of the positive eigenenergies have no simple physical interpretation. These $g_\alpha^p(R_\alpha)$ form another orthonormal basis in the interior region if the Hamiltonian is Hermitian.

The coefficients c_α^{pn} and energies e_p in Eq. (6.5.19) satisfy matrix equations

$$(\varepsilon_{n\alpha} + \epsilon_\alpha) c_\alpha^{pn} + \sum_{n'\alpha'} \langle w_\alpha^n \mid \hat{V}_{\alpha\alpha'} \mid w_{\alpha'}^{n'} \rangle c_{\alpha'}^{pn'} = e_p c_\alpha^{pn} \qquad (6.5.20)$$

for each eigenstate p. These are eigenvalue equations of the matrix form

$$\mathbf{Hc} = e\mathbf{c}. \qquad (6.5.21)$$

Because our coupled equations may have different reduced masses in different transfer channels, we now define a third form of the R matrix by

$$t_\alpha^{1/2} \psi_\alpha(R_\alpha) = a \sum_{\alpha'} \mathbf{R}_{\alpha\alpha'}(E)\Big[\psi_{\alpha'}'(R_\alpha) - \beta\psi_{\alpha'}(R_\alpha)\Big]t_{\alpha'}^{1/2}. \qquad (6.5.22)$$

Here the factors $t_\alpha \equiv \hbar^2/2\mu_\alpha$ are placed to render the R matrix symmetric even for transfer channels. The progressively more generalized R-matrix definitions are summarized in Box 6.1.

For scattering states at arbitrary energy E, the coupled solutions are then expanded in terms of the multi-channel eigenstates as $\psi_{\alpha\alpha_i} = \sum_p A_{\alpha\alpha_i}^p g_\alpha^p$. The R matrix for this wave function, using Eq. (6.5.22), can be calculated [22, 25] from the $g_\alpha^p(a)$ eigenfunctions by the standard methods, which are similar to that given in detail for the one-channel case leading to Eq. (6.5.10). We find

$$\mathbf{R}_{\alpha\alpha'}(E) = \sqrt{\frac{t_\alpha}{a}} \sum_{p=1}^P \frac{g_\alpha^p(a)g_{\alpha'}^p(a)}{e_p - E} \sqrt{\frac{t_{\alpha'}}{a}}. \qquad (6.5.23)$$

The *reduced width amplitudes* for each channel α and pole p,

$$\gamma_{p\alpha} = \sqrt{\frac{t_\alpha}{a}}\, g_\alpha^p(a) = \sqrt{\frac{\hbar^2}{2\mu_\alpha a}}\, g_\alpha^p(a), \qquad (6.5.24)$$

in terms of which the R matrix has the familiar form

$$\mathbf{R}_{\alpha\alpha'}(E) = \sum_{p=1}^P \frac{\gamma_{p\alpha}\gamma_{p\alpha'}}{e_p - E}. \qquad (6.5.25)$$

Linear equations for the R matrix

There is an alternative method [26, 14] for finding the $\mathbf{R}_{\alpha\alpha'}$, which does not diagonalize the matrix on the left side of Eq. (6.5.21), but solves a set of linear equations. We need the solution of $(\mathbf{H} - E)\mathbf{x} = \mathbf{w}(a)$ for the right-hand side

For partial wave L, Eq. (3.1.28) at radius a defines the matrix element

$$\mathbf{R}_L = \frac{1}{a}\frac{\chi_L(a)}{\chi'_L(a)}.$$
(6.5.26)

For a matrix of coupled-channels solutions, Eq. (6.3.4) defines the matrix

$$\mathbf{R} = \frac{1}{a}\mathsf{Y}(a)[\mathsf{Y}'(a)]^{-1}.$$
(6.5.27)

For one channel, with logarithmic boundary condition β, Eq. (6.5.11) defines the matrix element

$$\mathbf{R} = \frac{1}{a}\frac{\chi(a)}{\chi'(a) - \beta\chi(a)}.$$
(6.5.28)

For coupled transfer channels, with $\beta \neq 0$ and variable reduced masses μ_α, Eq. (6.5.22) defines a matrix $\mathbf{R}_{\alpha\alpha'}$ by

$$t_\alpha^{1/2}\psi_\alpha(a) = a\sum_{\alpha'}\mathbf{R}_{\alpha\alpha'}\left[\psi'_{\alpha'}(a) - \beta\psi_{\alpha'}(a)\right]t_{\alpha'}^{1/2},$$
(6.5.29)

using $t_\alpha \equiv \hbar^2/2\mu_\alpha$. In matrix form with t as a matrix of diagonal elements t_α,

$$\mathsf{t}^{\frac{1}{2}}\,\mathsf{Y}(a) = a\,\mathbf{R}\,[\mathsf{Y}'(a) - \beta\mathsf{Y}(a)]\,\mathsf{t}^{\frac{1}{2}},$$
(6.5.30)

so

$$\mathbf{R} = \frac{1}{a}\,\mathsf{t}^{\frac{1}{2}}\,\mathsf{Y}(a)\,\mathsf{t}^{-\frac{1}{2}}\,[\mathsf{Y}'(a) - \beta\mathsf{Y}(a)]^{-1}.$$
(6.5.31)

Box 6.1 **Progressively more flexible definitions of the R matrix**

consisting of the values of the basis functions at the R-matrix boundary. Then we can solve directly

$$\mathbf{R} = a^{-1}\,\mathsf{t}^{\frac{1}{2}}\,\mathbf{w}^T(a)\,(\mathbf{H} - E)^{-1}\,\mathbf{w}(a)\,\mathsf{t}^{\frac{1}{2}},$$
(6.5.32)

using the notation of Eq. (6.5.31). This method has the advantage of naturally continuing the R-matrix method to complex potentials, avoiding the diagonalization of non-Hermitian matrices.

Scattering S matrix and wave functions

Using Eqs. (3.2.10) and (6.5.22), and writing the Coulomb functions $\mathbf{H}^{\pm}$ as diagonal matrices as we did before at Eq. (6.3.5), the scattering S matrix is given in terms of $\mathbf{R}$ by

$$\mathbf{S} = [\mathbf{t}^{\frac{1}{2}}\mathbf{H}^+ - a\mathbf{R}\,\mathbf{t}^{\frac{1}{2}}(\mathbf{H}^{+\prime} - \beta\mathbf{H}^+)]^{-1}[\mathbf{t}^{\frac{1}{2}}\mathbf{H}^- - a\mathbf{R}\,\mathbf{t}^{\frac{1}{2}}(\mathbf{H}^{-\prime} - \beta\mathbf{H}^-)]. \quad (6.5.33)$$

The scattering states at the arbitrary energy E and incoming wave in channel α_i are linear combinations of the multi-channel eigenstates of Eq. (6.5.19):

$$\psi_{\alpha\alpha_i}(R_\alpha) = \sum_p A^p_{\alpha_i}\,g^p_\alpha(R_\alpha), \quad (6.5.34)$$

with expansion coefficients for each eigenstate p and incoming channel α_i of

$$A^p_{\alpha_i} = \frac{\hbar^2}{2\mu_{\alpha_i}}\frac{1}{e_p - E}\sum_{\alpha'} g^p_{\alpha'}(a)\Big[\delta_{\alpha'\alpha_i}(H^{-\prime}_L(k_{\alpha'}a) - \beta H^-_L(k_{\alpha'}a))$$

$$- \mathbf{S}_{\alpha'\alpha_i}(H^{+\prime}_L(k_{\alpha'}a) - \beta H^+_L(k_{\alpha'}a))\Big]. \quad (6.5.35)$$

Buttle corrections

Even the one-channel R-matrix values are only correct in the limit of an infinite number of basis states N. In the one-channel case, however, we can easily calculate the exact solution by the methods of Section 6.1, even if the diagonal potentials are complex, so it is straightforward to calculate the errors arising from having a finite basis in this case. The proposal of Buttle [24] is to calculate the one-channel corrections to the R matrix, and apply them additively to the *diagonal* terms of the many-channel R matrix. This is clearly exact in the limit of zero couplings, and is surprisingly efficient even when the couplings are stronger. The Buttle correction can be used not only for real diagonal potentials, but also for complex optical potentials, in order to correct for finite-basis effects in the weak coupling limit.

The method, therefore, to improve the accuracy of calculations with finite N (and hence finite P), is to add a 'Buttle correction' to the right-hand side of Eqs. (6.5.23, 6.5.32). This modifies the diagonal terms $\mathbf{R}_{\alpha\alpha}(E)$ to reproduce for each uncoupled problem the exact scattering solution $\chi_\alpha(R_\alpha)$ after this has been integrated separately. From the definition of the energy eigenstates $w^n_\alpha(R_\alpha)$, the R-matrix sum from (6.5.23) for each uncoupled channel is

$$\mathbf{R}^u_\alpha(E) = \frac{\hbar^2}{2\mu_\alpha a}\sum_{n=1}^N\frac{w^n_\alpha(a)^2}{\varepsilon_{n\alpha} - E} \quad (6.5.36)$$

and the exact one-channel R matrix is $\mathbf{R}_\alpha^0(E) = a^{-1}\chi_\alpha(a)/(\chi_\alpha'(a) - \beta\chi_\alpha(a))$. The Buttle-corrected full R matrix to be used in Eq. (6.5.33) is then

$$\mathbf{R}_{\alpha\alpha'}^c(E) = \mathbf{R}_{\alpha\alpha'}(E) + \delta_{\alpha\alpha'}\left[\mathbf{R}_\alpha^0(\tilde{E}) - \mathbf{R}_\alpha^u(\tilde{E})\right]. \qquad (6.5.37)$$

The energy $\tilde{E}$ can be equal to E, or chosen just near to it if necessary to avoid the poles in Eq. (6.5.36), since the Buttle correction (in the square brackets) varies smoothly with energy.

CRC matrix elements

The solution of the CRC equations (6.3.9) with all the non-orthogonality terms in Eq. (6.3.10) requires in Eq. (6.5.20) the matrix element integrals of the form

$$\langle w_\alpha^n|V_{\alpha\alpha'}|w_{\alpha'}^{n'}\rangle = \langle w_\alpha^n \mid \langle \Phi_\alpha|H_m - E|\Phi_{\alpha'}\rangle \mid w_{\alpha'}^{n'}\rangle \qquad (6.5.38)$$

for $m = \alpha$ (post) or $m = \alpha'$ (prior). In the post form, H_m contains $T_\alpha + U_\alpha$, and since w_α^n is just the eigenfunction of this operator with eigenvalue $\varepsilon_{n\alpha}$, we can operate to the left to obtain

$$\langle w_\alpha^n|V_{\alpha\alpha'}|w_{\alpha'}^{n'}\rangle_{\text{post}} = \langle w_\alpha^n\Phi_\alpha|V_\alpha|\Phi_{\alpha'}w_{\alpha'}^{n'}\rangle + (\varepsilon_{n\alpha} - E_\alpha)\langle w_\alpha^n\Phi_\alpha|\Phi_{\alpha'}w_{\alpha'}^{n'}\rangle, \quad (6.5.39)$$

with the similar prior form

$$\langle w_\alpha^n|V_{\alpha\alpha'}|w_{\alpha'}^{n'}\rangle_{\text{prior}} = \langle w_\alpha^n\Phi_\alpha|V_{\alpha'}|\Phi_{\alpha'}w_{\alpha'}^{n'}\rangle + \langle w_\alpha^n\Phi_\alpha|\Phi_{\alpha'}w_{\alpha'}^{n'}\rangle(\varepsilon_{n'\alpha'} - E_{\alpha'}).$$
$$(6.5.40)$$

The wave function overlaps in the second term $\langle\Phi_\alpha|\Phi_{\alpha'}\rangle$ go to zero asymptotically, and may be assumed small when R_α, $R_{\alpha'} > a$. The standard R-matrix theory therefore still applies in the surface region $R_\alpha \sim a$.

6.6 Coupled asymptotic wave functions

In the above descriptions, radial or expansion methods have been used to solve the Schrödinger equation in the interior up to some matching radius $R = a$. The R matrix is then found at this point, and matching there with asymptotic Coulomb wave functions $H^\pm$ allows the S matrix to be found to predict the cross sections. Sometimes, however, there are couplings outside the radius a that have large radial extents, and are not easily treated by the interior methods. Such couplings could be, for example, the Coulomb multipoles of subsection 4.4.1, or they could be from photo-nuclear couplings to weakly bound states, such as the proton states bound by 0.137 MeV in ^{8}B, or by 22 keV in an ^{15}O excited state.

We have two ways of including such long-range couplings. Either we can propagate the R matrix from a to some large 'asymptotic radius' R_a where the uncoupled Coulomb wave functions $H^{\pm}$ can be used, or else we propagate the Coulomb wave functions $H^{\pm}$ inwards to the smaller a radius, during which process they become coupled. The first R-matrix propagation method is extensively used in electron and atomic scattering [27], using the algorithms of Light and Walker [28] or of Burke *et al.* [29], whereas the second inward wave-function propagation can be performed either using the asymptotic expansions of Burke and Shey [30, 31], Gailitis [32], or Christley *et al.* [15].

In FRESCO, the method of Christley *et al.* [15] has been implemented, which uses expansions on Airy functions, as these are exact single-channel solutions for piecewise linear potentials. These expansions are used to integrate inward the asymptotic uncoupled $H^{\pm}$ from a large radius R_{asym} to the matching radius a, where they become now *coupled* Coulomb wave functions, and are represented by *full* matrices $\mathbf{H}^{\pm}$. These $\mathbf{H}^{\pm}$ and their radial derivatives can be used directly for matching in Eqs. (3.2.10) and (6.5.33).

Exercises

6.1 What method similar to that for Eq. (6.3.31) provides a scattering solution χ orthogonal to a given ϕ, that is $\langle\phi|\chi\rangle = 0$? This might be needed if ϕ is a Pauli forbidden state, and we want to implement Pauli blocking. Under what conditions does this orthogonality requirement have no effect at any energy?

6.2 Consider elastic scattering from a potential $V(R) = V_1(R) + \lambda V_2(R)$ in a specific partial wave, and iterate on $V_2(R)$ for $\lambda = 1$.

    (a) Examine the rate of numerical convergence, with and without Padé acceleration, and determine the form of the error series of powers of λ.

    (b) Examine the PWBA case when $V_1 = 0$, for V_2 being purely real, or being purely imaginary.

    (c) Examine a case when $V_1 \neq 0$ and near a resonance, to see if the Padé acceleration is successful in this case. Determine from the λ series where would be the pole in the complex λ-plane.

6.3 What is the meaning of the energies where the Buttle correction is zero? What should we do to make it zero at some specific energy, at for example a resonance?

References

[1] M. A. Melkanoff, T. Sawada and J. Raynal, *Math. Comput. Phys.* **6** (1966) 1.
[2] A. E. Thorlacius and E. D. Cooper, *J. of Comp. Physics* **72** (1987) 70.
[3] B. Buck, A. P. Stamp and P. E. Hodgson, *Phil. Mag.* **8** (1963) 1805.
[4] L. D. Tolsma and G. W. Veltkamp, *Comput. Phys. Commun.* **40** (1986) 233.
[5] T. Udagawa, H. H. Wolter and W. R. Coker, *Phys. Rev. Letts.* **31** (1973) 1507.
[6] B. Imanishi, M. Ichimura and M. Kawai, *Phys. Lett.* **52B** (1974) 267.
[7] S. R. Cotanch and C. M. Vincent, *Phys. Rev. C* **14** (1976) 1739.

[8] G. R. Satchler 1983, *Direct Nuclear Reactions*, Oxford: Clarendon Press.

[9] I. J. Thompson, *Comput. Phys. Rep.* **7** (1988) 167.

[10] J. S. Lilley, B. R. Fulton, D. W. Banes, T. M. Cormier, I. J. Thompson, S. Landowne and H. H. Wolter, *Phys. Letts.* **128B** (1983) 153; J. S. Lilley, M. A. Nagarajan, D. W. Banes, B. R. Fulton and I. J. Thompson, *Nucl. Phys.* **A463** (1987) 710.

[11] J. Raynal 1972, p. 281 in *Computing As a Language of Physics*, Vienna: IAEA.

[12] M. Rhoades-Brown, M. H. Macfarlane and S. C. Pieper, *Phys. Rev. C* **21** (1980) 2436.

[13] M. Kawai, M. Kamimura, Y. Mito and K. Takesako, *Prog. Theor. Phys.* **59** (1978) 674, 676. M. Kamimura, *Prog. Theor. Phys. Suppl.* **62** (1977) 236.

[14] D. Baye, M. Hesse, J.-M. Sparenberg and M. Vincke, *J. Phys. B* **31** (1998) 3439; *Nucl. Phys. A* **640** (1998) 37.

[15] J. A. Christley and I. J. Thompson, *Comp. Phys. Comm.* **79** (1994) 143.

[16] P. R. Graves-Morris 1973, *Padé Approximants*, Bristol: Institute of Physics.

[17] P. Wynn, *Numer. Math.* **8** (1966) 264; see also A. Genz, p. 112 in [16].

[18] P. Wynn, *Math. Comput.* **16** (1961) 23.

[19] W. Ogle, S. Wahlborn, R. Piepenbring and S. Frederiksson, *Rev. Mod. Phys.* **43** (1971) 424.

[20] J. M. Bang and J. S. Vaagen, *Z. Physik A* **297** (1980) 223.

[21] E. P. Wigner and L. Eisenbud, *Phys. Rev.* **72** (1947) 29.

[22] A. M. Lane and R. G. Thomas, *Rev. Mod. Phys.* **30** (1958) 257.

[23] I. J. Thompson, B. V. Danilin, V. D. Efros, J. S. Vaagen, J. M. Bang and M. V. Zhukov, *Phys. Rev. C* **61** (2000) 24318.

[24] P. J. A. Buttle, *Phys. Rev.* **160** (1967) 719.

[25] P. G. Burke and M. J. Seaton, *J. Phys.* **B17** (1984) L683.

[26] D. H. Glass, P. G. Burke, H. W. van der Hart and C. J. Noble, *J. Phys. B* **30** (1997) 3801.

[27] V. M. Burke and C. J. Noble, *Comp. Phys. Commun.* **85** (1995) 471.

[28] J. C. Light and R. B. Walker, *J. Chem. Phys.* **65** (1976) 4272; E. B. Stechel, R. B. Walker and J. C. Light, *J. Chem. Phys.* **69** (1978) 3518.

[29] K. L. Baluja, P. G. Burke and L. A. Morgan, *Comput. Phys. Commun.* **27** (1982) 299.

[30] P. G. Burke, D. D. McVicar and K. Smith, *Proc. Phys. Soc.* **83** (1964) 397.

[31] F. Rösel, *J. Phys. G.* **3** (1977) 613.

[32] M. Gailitis, *J. Phys. B* **9** (1976) 843.

7

Approximate solutions

> If you meet an operator walking down Guildford High Street, you can't
> tell if it is Hermitean just by looking at it. You've got to know the environ-
> ment it is living in.
>
> *Ron Johnson*

Approximations in physics are often very useful. In nuclear reactions in particular, depending on the specific regime, some approximations may offer a very large simplification of the problem and still provide great accuracy. Of course, to some extent, all methods in nuclear reactions are approximate, but let us consider that the solution methods discussed in Chapter 6 are the starting point to which approximations can be considered.

One idea appears when there are variables with distinct timescales, as then an adiabatic approximation can be made. Another idea is based on classical arguments. By taking a certain trajectory for the projectile, we can separate out the dynamics of the reaction and treat just what is happening to the projectile within quantum mechanics (semiclassical methods). For instance, at very high energies, the projectile's trajectory is hardly bent and the straight-line approximation is valid (called the eikonal approximation). For cases where the reaction is Coulomb dominated, taking a Rutherford trajectory for the center of mass of the projectile may be adequate, and if so provides useful simplifications of the problem. When the potential is very smooth and for slow reactions (low energies), we may make use of the WKB approximation. In this chapter we summarize the most important approximations used in reactions and discuss their limits of validity.

7.1 Few-body adiabatic scattering

The main idea behind any adiabatic model is that we can separate a fast variable and a slow variable, such that during the reaction the slow variable does not change from its initial value. Here we consider the adiabatic high-energy approximation, also known as the sudden approximation, where the internal motion of the projectile is considered slow compared to the motion of the c.m. of the projectile relative to the target. In the energy domain, the energy scales associated with the slow variable should be much smaller than those associated with the fast variable, and so we can assume the projectile excitations are practically degenerate with the ground state, by comparison with the beam energy.

7.1.1 Three-body adiabatic model

Let us consider a two-body projectile composed of core and valence clusters ($p = v+c$) incident on a target t. The three-body Schrödinger equation for the problem is:

$$[\hat{T}_R + H_{\text{int}}(\mathbf{r}) + U_{ct} + U_{vt} - E]\Psi(\mathbf{r}, \mathbf{R}) = 0, \tag{7.1.1}$$

where the internal Hamiltonian of the projectile is $H_{\text{int}} = T_r + V_{vc}(r)$ such that $H_{\text{int}}\phi_i = \epsilon_i\phi_i$ defines for $i = 0$ the ground-state wave function ϕ_0 and its energy ϵ_0. We neglect the spin and intrinsic Hamiltonian of the target, and use the coordinates $(\mathbf{r}, \mathbf{R})$ describing the relative motion between $c+v$ and $p+t$ respectively, as shown in Fig. 7.1. The full three-body scattering wave function can be expanded as

$$\Psi = \phi_0(\mathbf{r})\chi_0(\mathbf{R}) + \sum_{i>0} \phi_i(\mathbf{r})\chi_i(\mathbf{R}), \tag{7.1.2}$$

where the first term is the elastic component, and the others describe the projectile inelastic and breakup states. Using this in Eq. (7.1.1), we obtain

$$[\hat{T}_R + \epsilon_0 + U_{ct} + U_{vt} - E]\phi_0(\mathbf{r})\chi_0(\mathbf{R})$$
$$+ \sum_{i>0}[\hat{T}_R + \epsilon_i + U_{ct} + U_{vt} - E]\phi_i(\mathbf{r})\chi_i(\mathbf{R}) = 0. \tag{7.1.3}$$

We now assume that the relative motion of the projectile coordinate $\mathbf{r}$ is slow compared to the motion of the projectile relative to the target $p + t$ (coordinate $\mathbf{R}$),

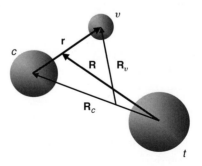

Fig. 7.1. Coordinates for adiabatic models.

which is to say that $|\epsilon_i - \epsilon_0| \ll E$. Under this assumption, $H_{int}\Psi \approx \epsilon_0\Psi$, so the excitation energies of the projectile can be replaced by a constant ϵ_0 by comparison to E. This implies that, to a good approximation,

$$(\hat{T}_R + U_{ct} + U_{vt} - (E - \epsilon_0))\Psi(\mathbf{r}, \mathbf{R}) \approx 0. \tag{7.1.4}$$

The *adiabatic model* consists in exactly solving the equation

$$(\hat{T}_R + U_{ct} + U_{vt} - (E - \epsilon_0))\Psi^{ad}(\mathbf{r}, \mathbf{R}) = 0, \tag{7.1.5}$$

with the same incident boundary conditions, so that $\Psi(\mathbf{r}, \mathbf{R}) \approx \Psi^{ad}(\mathbf{r}, \mathbf{R})$ at least for r smaller than the size of the projectile. The adiabatic wave function is also a superposition of elastic and breakup channels,

$$\Psi^{ad}(\mathbf{r}, \mathbf{R}) = \phi_0(\mathbf{r})\chi_0^{ad}(\mathbf{R}) + \sum_{i>0} \phi_i(\mathbf{r})\chi_i^{ad}(\mathbf{R}), \tag{7.1.6}$$

for some new 'adiabatic' radial functions $\chi_i^{ad}(\mathbf{R})$. The scattering boundary conditions for incident momentum $\mathbf{K}_0$ are that for $R > R_n$ the adiabatic wave function contains the incoming elastic channel $\phi_0(\mathbf{r})\chi_c(\mathbf{K}_0, \mathbf{R})$ (of Eq. (3.1.72)), plus all outgoing waves.

The important advantage of the adiabatic model is that in Eq. (7.1.5) the value of $\mathbf{r}$ appears only parametrically (without $\mathbf{r}$ derivatives), so the solutions for each value of $\mathbf{r}$ can be obtained independently of those for other $\mathbf{r}$ values. This is the general form of the adiabatic equation, so we do not use Eq. (7.1.6) to find the solution, but solve a separate coupled-channels problem for each value of r. We therefore find adiabatic partial-wave functions $\chi_{L'\ell';L\ell}^J(R; r)$ satisfying

$$[T_{L'}(R) - (E - \epsilon_0)]\chi_{L'\ell';L\ell_0}^J(R; r)$$
$$= -\sum_{L''\ell''} \langle(L'\ell')J|U_{ct} + U_{vt}|(L''\ell'')J\rangle\chi_{L''\ell'';L\ell_0}^J(R; r), \tag{7.1.7}$$

for each separate r, with the usual boundary conditions at $R \geq R_n$ of

$$\chi_{L'\ell';L\ell_0}^J(R; r) \to F_{L'}(K_0R)\delta_{\ell_0\ell'}\delta_{L'L} + \mathbf{T}_{L'\ell';L\ell_0}^+(r)H_{L'}^+(K_0R), \tag{7.1.8}$$

so the T-matrix element depends on the parameter r.

The complete wave function in terms of these wave functions is [1]

$$\Psi_{K_0,m}^{ad}(\mathbf{r}, \mathbf{R}) = \frac{4\pi}{K_0} \sum_{JM_JLM} \sum_{L'M'\ell'm'} \langle LM, \ell_0m_0|JM_J\rangle\langle L'M', \ell'm'|JM_J\rangle$$
$$\times i^{L'}\frac{u_{\ell_0}(r)}{r}Y_{\ell_0}^{m_0*}(\hat{\mathbf{r}})\frac{\chi_{L'\ell';L\ell_0}^J(R; r)}{R}Y_{L'}^{M'}(\hat{\mathbf{R}})Y_{\ell'}^{m'}(\hat{\mathbf{r}})Y_L^{M*}(\hat{\mathbf{K}}_0), \tag{7.1.9}$$

where $u_{\ell_0}(r)$ is the radial ground-state wave function of the projectile (neglecting all intrinsic spins as well as the target spin). The elastic scattering amplitude can be obtained from $\mathbf{T}(r)$ by averaging it over the ground-state density $|u_{\ell_0}(r)|^2$:

$$f_{m:m_0}^{\text{el}}(\mathbf{K}) = f_c(\theta) + \frac{4\pi}{K_0} \sum_{LL'J} \langle L'M', \ell_0 m | Jm_0 \rangle \langle L0, \ell_0 m_0 | Jm_0 \rangle e^{i(\sigma_{L'}(\eta) + \sigma_L(\eta))}$$

$$\times\, Y_L^0(\hat{\mathbf{K}}_0) Y_{L'}^{M'}(\hat{\mathbf{K}}) \int_0^\infty dr\, |u_{\ell_0}(r)|^2\, \mathbf{T}_{L'\ell_0;L\ell_0}^J(r), \qquad (7.1.10)$$

with $|\mathbf{K}| = K_0$ and $\mathbf{K}_0$ in the $+\hat{\mathbf{z}}$ direction, where we have added a point Coulomb contribution as in subsection 3.1.2. The adiabatic model includes the effects of breakup intermediate states on elastic scattering.

7.1.2 The Johnson and Soper potential for transfer reactions

Given the small binding energy of the deuteron, in (d,p) reactions it is likely that deuteron breakup is an intermediate step in transfer processes. An adiabatic prescription has been developed for (d,p) reactions by Johnson and Soper [2], whereby deuteron breakup is included in a simple way. The exact post-form T matrix for a transfer outgoing channel, found from the two-potential Eq. (3.3.51) applied with the post transfer interaction of Eq. (4.5.4), is

$$\mathbf{T}_{fi} = -\frac{2\mu_f}{\hbar^2 K_f} \langle \psi_f^{(-)} \phi_n | \mathcal{V}_f | \Psi(r, R) \rangle, \qquad (7.1.11)$$

where $\Psi(r, R)$ is the exact wave function for the deuteron-target system, ϕ_n is a neutron bound state, and ψ_f is a proton exit wave function. For the deuteron, the dominating term in the interaction $\mathcal{V}_f$ is $V_{\text{np}}(r)$, the neutron–proton binding potential, which has a short range of $\rho_{\text{eff}} \sim 1$ or 2 fm.

In DWBA, the wave function Ψ is approximated by an elastic component $\phi_0(r)\chi_0(R)$, with $\phi_0(r)$ being the deuteron ground state, which means that the breakup components of the wave function are neglected except for their influence on $\chi_0(R)$. This influence appears through the optical potentials, since these are determined from fitting the elastic scattering cross section. In transfer reactions, however, the χ_0 is used around the surface region and sometimes in the interior of the nucleus, whereas the elastic scattering for the fit is a purely asymptotic property. It is thus not clear whether U_{opt} from fitting elastic scattering can characterize $\chi_0(R)$ in the region where it is needed for transfer reactions.

The important realization by Johnson and Soper [2] was that, for deuteron transfer reactions, the wave function $\Psi(r, R)$ is only required in the specific region where $r \lesssim \rho_{\text{eff}}$: where r is smaller than the range of the potential. For an $\ell = 0$ deuteron bound in the limit of a zero-range potential V_{np}, we thus only need $r = 0$, a part which is very easily obtained in the adiabatic model. For $r = 0$, we simply solve Eq. (7.1.5) for that value of the parameter r:

$$(\hat{T}_R + U_{ct} + U_{nt} - (E - \epsilon_0))\Psi^{\mathrm{ad}}(0, \mathbf{R}) = 0. \qquad (7.1.12)$$

For this r value, the proton- and neutron-target potentials are evaluated at the same position R, so the above equation is equivalent to

$$(\hat{T}_R + U_{\mathrm{ad}}(R) - (E - \epsilon_0))\tilde{\chi}(\mathbf{R}) = 0, \qquad (7.1.13)$$

where we have an effective 'Johnson-Soper' potential

$$U_{\mathrm{ad}}(R) = U_{ct}(R) + U_{vt}(R). \qquad (7.1.14)$$

A new 'adiabatic scattering wave function' $\tilde{\chi}(\mathbf{R})$ defined by

$$\phi_0(0)\tilde{\chi}(\mathbf{R}) = \Psi^{\mathrm{ad}}(0, \mathbf{R}) \qquad (7.1.15)$$

therefore includes contributions from all the breakup components, as in the expansion of Eq. (7.1.6).

Note that the new distorting optical potential U_{ad} is just the sum of the neutron-target and proton-target optical potentials evaluated at the same point. This is called the *adiabatic* or *Johnson-Soper* potential. It no longer fits the elastic scattering, but is intended to produce a distorted wave that includes deuteron breakup effects in the region of configuration space where transfer takes place. In many applications, the adiabatic optical potential improves the description of the transfer angular distribution, whether using the zero-range or the finite-range transfer interactions.

A finite-range version of the adiabatic model was developed by Johnson and Tandy [3], who obtain an averaged adiabatic potential by folding with weight factor $V_{np}(r)|u_{\ell_0}(r)|^2$. For practical purposes, we can also use simplified prescriptions (e.g. [4]).

In order for the adiabatic approximation to be accurate, the beam energy should be well above the typical range of ϵ_i for deuteron excitation into the continuum, which for nuclear breakup is up to $\epsilon_i = 10 \, \mathrm{MeV}$. Thus, we expect this prescription to be adequate for beam energies above $E_d \gg 20 \, \mathrm{MeV}$.

7.1.3 The Johnson special three-body model

The three-body wave function $\Psi^{\mathrm{ad}}(\mathbf{r}, \mathbf{R})$ of Eq. (7.1.5) satisfies the boundary condition of an incoming plane wave $\phi_0(\mathbf{r})e^{i\mathbf{K}_0 \cdot \mathbf{R}}$ plus outgoing spherical waves. Johnson [5] showed that a further simplification is possible when the target-valence interaction is negligible ($U_{vt} = 0$). This is typically the case for neutron halo systems: exactly so for Coulomb breakup, and approximately for nuclear breakup.

The equation to be solved becomes a single-potential equation,

$$\left[-\frac{\hbar^2}{2\mu_{pt}} \nabla_R^2 + U_{ct}(\mathbf{R}_c) - (E - \epsilon_0) \right] \Psi_{\mathbf{K}_0}^{sp}(\mathbf{r}, \mathbf{R}) = 0, \qquad (7.1.16)$$

with $\mathbf{R} = \mathbf{R}_c + m_v/m_p \mathbf{r}$. We can exactly solve this one-potential equation, if we transform the radial variable from $\mathbf{R}$ to that appearing in the potential, namely $\mathbf{R}_c$, and correct the position shift with an exponential factor as in

$$\Psi_{\mathbf{K}_0}^{sp}(\mathbf{r}, \mathbf{R}) = \phi_0(\mathbf{r}) e^{im_v/m_p \mathbf{K}_0 \cdot \mathbf{r}} \chi_{\mathbf{K}_0}(\mathbf{R}_c), \qquad (7.1.17)$$

to obtain an equation for $\chi_{\mathbf{K}_0}(\mathbf{R}_c)$:

$$\left[-\frac{\hbar^2}{2\mu_{pt}} \nabla_{R_c}^2 + U_{ct}(\mathbf{R}_c) - (E - \epsilon_0) \right] \chi_{\mathbf{K}_0}(\mathbf{R}_c) = 0. \qquad (7.1.18)$$

The adiabatic wave function of Eq. (7.1.17) satisfies by construction the boundary conditions, and the distorted waves $\chi_{\mathbf{K}_0}(\mathbf{R}_c)$ describe the scattering of a point-like projectile with an interaction U_{ct} with the target. Note the important fact that Eq. (7.1.18) has a kinetic energy operator expressed in terms of the core-target coordinate $\mathbf{R}_c$, but instead of containing the reduced mass for just the $c + t$ system, it contains the reduced mass for the full projectile with the target, $\mu_{pt} = m_p m_t / (m_p + m_t)$.[1]

The exact elastic-scattering transition amplitude is given by

$$\mathbf{T}_{el}^{exact}(\mathbf{K}') = \langle \phi_0(\mathbf{r}) e^{i\mathbf{K}' \cdot \mathbf{R}} | U_{ct} | \Psi^{exact}(\mathbf{r}, \mathbf{R}) \rangle, \qquad (7.1.19)$$

but in the special model, $\Psi^{exact}(\mathbf{r}, \mathbf{R})$ is replaced by Eq. (7.1.17), so

$$\mathbf{T}_{el}^{sp}(\mathbf{K}') = \langle \phi_0(\mathbf{r}) e^{i\mathbf{K}' \cdot \mathbf{R}_c} e^{im_v/m_p \mathbf{K}' \cdot \mathbf{r}} | U_{ct} | \chi_{\mathbf{K}_0}(\mathbf{R}_c) \phi_0(\mathbf{r}) e^{im_v/m_p \mathbf{K}_0 \cdot \mathbf{r}} \rangle. \qquad (7.1.20)$$

The integrals over $\mathbf{R}_c$ and $\mathbf{r}$ are independent, so we can define a form factor dependent on $\mathbf{Q} = \mathbf{K}' - \mathbf{K}_0$ as the Fourier transform of the ground-state density (*not* wave function):

$$F(\mathbf{Q}) = \int d^3 r |\phi_0(\mathbf{r})|^2 e^{im_v/m_p \mathbf{Q} \cdot \mathbf{r}}. \qquad (7.1.21)$$

This multiplies a U_{ct}-only scattering transition amplitude

$$\mathbf{T}_{el}^{point} = \langle e^{i\mathbf{K}' \cdot \mathbf{R}_c} | U_{ct} | \chi_{\mathbf{K}_0}(\mathbf{R}_c) \rangle, \qquad (7.1.22)$$

---

[1] This is distinct from the impulse multiple scattering formulations [6] where μ_{ct} would be used instead. It has been shown [7] that both adiabatic and impulse multiple scattering predict the same results in the high-energy limit.

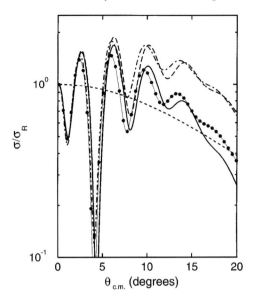

Fig. 7.2. Results from [5] for the Johnson special model elastic scattering of ^{11}Be on ^{12}C at 49.3 MeV/u: the dashed line shows the point projectile with ^{10}Be–^{12}C optical potential, the dot-dashed shows the results using the folded potential, the solid line is the result of the special model and is compared to the exact adiabatic prediction of Eq. (7.1.10), the filled circles. Reprinted with permission from R. C. Johnson *et al.*, *Phys. Rev. Lett.* **79** (1997) 2771. Copyright (1997) by the American Physical Society.

to obtain the final transition amplitude

$$\mathbf{T}_{el}^{ad} = F(\mathbf{Q}) \, \mathbf{T}_{el}^{point}. \tag{7.1.23}$$

This convenient expression resembles that of electron elastic scattering in the plane-wave Born approximation, but here all the breakup continuum is being taken into account. The Johnson special model has been applied to the elastic scattering of ^{11}Be on ^{12}C at 49.3 MeV/u [5], giving the angular distribution of the cross section shown in Fig. 7.2. Limitations of the model have been studied in [7].

7.1.4 The adiabatic wave function for breakup

The adiabatic wave functions $\Psi_{\mathbf{K}_0}^{ad}(\mathbf{r}, \mathbf{R})$ and $\Psi_{\mathbf{K}_0}^{sp}(\mathbf{r}, \mathbf{R})$ contain breakup to all orders, but not to all distances. To obtain breakup cross sections, it is important to remember that Ψ^{sp} is not accurate when $r \to \infty$, since the bound state $\phi_0(r)$ in Eq. (7.1.17) forces Ψ^{ad} to zero. The breakup amplitudes for final-state momenta $(\mathbf{K}_0, \mathbf{k}_v, \mathbf{k}_c)$ should rather be constructed by using the wave function in a post-form

T-matrix integral, from the special model for example [8]:

$$\mathbf{T}_{bu}^{sp}(\mathbf{K}_0, \mathbf{k}_v, \mathbf{k}_c) = \langle \chi_{\mathbf{k}_c}^{(-)}(\mathbf{R}_c) e^{i\mathbf{k}_v \cdot \mathbf{R}_v} | V_{vc} | \Psi_{\mathbf{K}_0}^{sp}(\mathbf{r}, \mathbf{R}) \rangle. \tag{7.1.24}$$

This integral only uses the wave function at small $r < \rho_{\text{eff}}$ distances, where ρ_{eff} is the radial range of the V_{vc} potential. Substituting Eq. (7.1.17) into Eq. (7.1.24), and factorizing the integrals in $\mathbf{r}$ and $\mathbf{R}_c$, we obtain

$$\mathbf{T}_{bu}^{ad}(\mathbf{K}_0, \mathbf{k}_v, \mathbf{k}_c) = \langle e^{i\mathbf{q}_v \cdot \mathbf{r}} | V_{vc}(\mathbf{r}) | \phi_0(\mathbf{r}) \rangle \langle \chi_{\mathbf{k}_c}^{(-)}(\mathbf{R}_c) e^{i\beta_c \mathbf{k}_v \cdot \mathbf{R}_c} | \chi_{\mathbf{K}_0}(\mathbf{R}_c) \rangle, \tag{7.1.25}$$

where $\mathbf{q}_v = \mathbf{k}_v - m_v/m_p \mathbf{K}_0$, with $\mathbf{R}_v = \beta_c \mathbf{R}_c + \mathbf{r}$ and $\beta_c = m_t/(m_t + m_c)$.

The breakup transition matrix of Eq. (7.1.25) separates nicely the structure part from the reaction part. The integrals $\langle \chi_{\mathbf{k}_c}^{(-)}(\mathbf{R}_c) e^{i\beta_c \mathbf{k}_v \cdot \mathbf{R}_c} | \chi_{\mathbf{K}_0}(\mathbf{R}_c) \rangle$ for Coulomb breakup are called the *bremsstrahlung integrals* and have been evaluated in closed form by Baur [9]. The structure part, which is simply the Fourier transform of the vertex function $V_{vc}(\mathbf{r})\phi_0(\mathbf{r})$, can be evaluated from some good structure model. Note that although there are some similarities in the form of Eq. (7.1.25) and a finite-range DWBA transition amplitude, the adiabatic model treats V_{vc} to all orders, and not just to first order. Applications to deuteron breakup show that this adiabatic model has good predictive power [8].

7.1.5 The adiabatic wave function for transfers

The few-body adiabatic model has also been applied to transfer reactions [10]. We consider in this section a cluster model with $t = c + v$, so the reaction $p + t \rightarrow (p + v) + c$ is called a pickup reaction.[2] Using now the prior form of the transfer matrix element, we have

$$\mathbf{T}_{prior}^{exact} = \langle \Psi^{(-)exact} | V_{pv} + U_{pc} - U_i | \phi_0(\mathbf{r}_{vc}) \chi_{\mathbf{k}_i}(\mathbf{R}_i) \rangle, \tag{7.1.26}$$

where $\chi_{\mathbf{k}_i}(\mathbf{R}_i)$ is the distorted wave in the entrance channel for distorting potential U_i. As the amplitude in Eq. (7.1.26) is exact, the result is independent of U_i. If we now choose U_i to be U_{pc}, and introduce on the right side the approximate adiabatic wave function of the Johnson special model satisfying

$$(\hat{T}_R + U_{pc}(\mathbf{R}_{pc}) - (E - \epsilon_0)) \Psi_{\mathbf{K}}^{sp}(\mathbf{r}_i, \mathbf{R}_i) = 0, \tag{7.1.27}$$

the prior form for the transfer amplitude reduces to

$$\mathbf{T}_{prior}^{sp} = \langle \Psi^{(-)exact} | V_{pv} | \Psi_{\mathbf{K}}^{sp}(\mathbf{r}_i, \mathbf{R}_i) \rangle. \tag{7.1.28}$$

---

[2] This is a time reversal of the (d,p) stripping reaction discussed in subsection 7.1.2 where a nucleon is transferred from the projectile onto the target, and $p = c + v$ is the initial bound state.

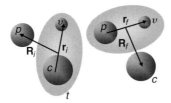

Fig. 7.3. Coordinates for adiabatic model of pickup reactions: prior and post form.

Expanding $\Psi_{\mathbf{K}}^{\text{sp}}$ by Eq. (7.1.17), we obtain a transition matrix element

$$\mathbf{T}_{\text{prior}}^{\text{sp}} = \langle \Psi^{(-)\text{exact}} | V_{pv} | \phi_0(\mathbf{r}_{vc}) \chi_{\mathbf{K}_0}(\mathbf{R}_{pc}) e^{im_v/m_t \mathbf{K}_0 \cdot \mathbf{r}} \rangle. \qquad (7.1.29)$$

One big advantage of the form of Eq. (7.1.29) is that the adiabatic wave function is only used for small r_{pv}, where it works best. The next possible step, to simplify Eq. (7.1.29) further, is to take the Johnson and Soper approach [2]: $\Psi^{(-)\text{exact}} \approx \chi_{\mathbf{k}_f}^{(-)}(\mathbf{R}_f) \phi_{pv}(\mathbf{r}_{pv})$ and the zero-range approximation for the interaction: $V_{vp}(\mathbf{r}_{pv}) \phi_{pv}(\mathbf{r}_{pv}) = D_0 \delta(\mathbf{R}_p - \mathbf{R}_v)$. We then obtain the transfer amplitude

$$\mathbf{T}_{\text{prior}}^{\text{ZR,sp}} = D_0 \langle \chi_{\mathbf{k}_f}^{(-)}(\mathbf{R}_f) | \phi_0(\mathbf{r}_{vc}) \chi_{\mathbf{K}_0}(\mathbf{r}) e^{im_v/m_t \mathbf{K}_0 \cdot \mathbf{r}} \rangle. \qquad (7.1.30)$$

One can see directly from this expression that the model contains recoil effects as well as breakup effect, and so goes well beyond DWBA.

7.2 Eikonal methods

7.2.1 The eikonal wave function

Eikonal methods are based on the high-energy approximation which assumes the projectile follows a straight-line trajectory. As the theory was originally developed by Glauber [11], it is often referred to as the *Glauber model*. We write the wave function in cylindrical coordinates $\mathbf{R} = (r, \theta, z) = (\mathbf{b}, z)$ (see Fig. 7.4) for a plane wave along the beam direction chosen as $\hat{z}$:

$$\Psi(\mathbf{R}) = e^{ikz} \phi(\mathbf{b}, z). \qquad (7.2.1)$$

The beam energy is $E = \hbar^2 k^2 / 2\mu$, and for high energies E the modulating factor $\phi(\mathbf{b}, z)$ should be a slowly varying function of both variables. The full Schrödinger equation written in cylindrical coordinates has a kinetic energy operator

$$\hat{T}_R = -\frac{\hbar^2}{2\mu} \left[\nabla_b^2 + \frac{\partial^2}{\partial z^2} \right]. \qquad (7.2.2)$$

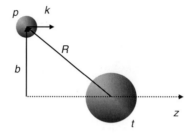

Fig. 7.4. Coordinates for eikonal model.

Substituting the Eq. (7.2.1) ansatz into $[\hat{T}_R + V(\mathbf{R})]\Psi = E\Psi$, we obtain

$$2ike^{ikz}\frac{\partial\phi}{\partial z} + e^{ikz}\frac{\partial^2\phi}{\partial z^2} + e^{ikz}\nabla_b^2\phi - \frac{2\mu}{\hbar^2}V(\mathbf{R})e^{ikz}\phi = 0. \tag{7.2.3}$$

Neglecting second-order derivatives of ϕ, we obtain what is called the eikonal equation:

$$\frac{\partial\phi}{\partial z} = -\frac{i}{\hbar v_p}V(\mathbf{R})\phi, \tag{7.2.4}$$

with relative velocity $v_p = \hbar k/\mu$. This first-order differential equation has solutions of the form

$$\phi(\mathbf{b}, z) = \exp\left[-\frac{i}{\hbar v_p}\int_{-\infty}^{z}V(\mathbf{b}, z')dz'\right]. \tag{7.2.5}$$

that satisfy the boundary condition $\phi(\mathbf{b}, -\infty) = 1$. The standard procedure is to define a *eikonal phase*

$$\chi(\mathbf{b}, z) = -\frac{1}{\hbar v_p}\int_{-\infty}^{z}V(\mathbf{b}, z')dz', \tag{7.2.6}$$

so that the wave function can be written as

$$\Psi(\mathbf{R}) = e^{i(kz+\chi(\mathbf{b},z))}. \tag{7.2.7}$$

The actual phase of the scattering wave function is thus accumulated along a classical path, in this case a straight-line trajectory. If after the reaction these phases are different along different parallel paths, then the wave front is tilted, and scattering is obtained to non-zero angles.

We have assumed the convergence in Eq. (7.2.6) of the integral over the potential $V(\mathbf{b}, z)$. While this is the case for short-range interactions, the long-range behavior of the Coulomb force needs special attention, and we obtain results [11] by screening at some large radius a_s. Then $\chi = \chi_n + \chi_c$, where χ_n is given by the integral over the nuclear potential, and the Coulomb contribution χ_c is calculated using a screening radius a_s of atomic dimensions, so that results can still be accurate at very forward angles. This can be expanded in powers of b_c/a_s and the leading term for $b_c > R_{Coul}$ is $\chi_c(b_c) = 2\eta_i \ln k_i b_c - 2\eta_i \ln(2k_i a_s)$. Here $\eta_i = Z_t Z_c e^2 \mu_i/\hbar^2 k_i$ is the elastic Sommerfeld parameter, where Z_t and Z_c are the target and core charges, and R_{Coul} the Coulomb radius as used in Eq. (4.1.2). The last term in χ_c only contributes a constant phase factor to all scattering amplitudes, and is not observable, so we need only use $\chi_c(b_c) = 2\eta_i \ln k_i b_c$.

7.2.2 Eikonal elastic scattering

The scattering amplitude can be obtained by replacing the exact wave function with the eikonal wave function in the plane-wave T-matrix integral

$$f(\theta) = -\frac{\mu}{2\pi\hbar^2} \int d\mathbf{R} \; e^{-i\mathbf{K}'\cdot\mathbf{R}} V(\mathbf{R}) \Psi_{\mathbf{K}_0}(\mathbf{R}). \tag{7.2.8}$$

We now decompose the position vector $\mathbf{R}$ into a component in the direction of the incident beam $\hat{\mathbf{n}}$ and another perpendicular to it, as $\mathbf{R} = z\hat{\mathbf{n}} + \mathbf{b}$. Defining the transferred momentum as $\mathbf{q} = \mathbf{K}_0 - \mathbf{K}'$ we find

$$f(\theta) = -\frac{\mu}{2\pi\hbar^2} \int d^2b \, e^{i\mathbf{q}\cdot\mathbf{b}} \int dz \, e^{i\mathbf{q}\cdot z\hat{\mathbf{n}}} V(\mathbf{b}, z) \exp\left[-\frac{i}{\hbar v_p} \int_{-\infty}^{z} V(\mathbf{b}, z')dz'\right], \tag{7.2.9}$$

where θ is the angle between $\mathbf{K}_0$ and $\mathbf{K}'$.

If we assume that $q \ll K$ (typically valid for forward angle scattering), so $\mathbf{q}\cdot\hat{\mathbf{n}} \approx 0$, then

$$\int dz \, V(\mathbf{b}, z) \exp\left[-\frac{i}{\hbar v_p} \int_{-\infty}^{z} V(\mathbf{b}, z')dz'\right] = i\hbar v_p (e^{i\chi(\mathbf{b})} - 1), \tag{7.2.10}$$

where $\chi(\mathbf{b}) = \chi(\mathbf{b}, \infty)$ corresponds to the (total) eikonal phase. The scattering amplitude can then be simplified to

$$f(\theta) = -\frac{iK_0}{2\pi} \int d^2b \, e^{i\mathbf{q}\cdot\mathbf{b}} (e^{i\chi(\mathbf{b})} - 1). \tag{7.2.11}$$

This plane integral can be converted to a one-dimensional integral over a cylindrical Bessel function:

$$f(\theta) = \frac{K_0}{i} \int_0^{\infty} b \, db \, J_0(2K_0 b \sin(\theta/2))(e^{i\chi(\mathbf{b})} - 1). \tag{7.2.12}$$

This expression for the scattering amplitude is valid for small angle scattering. The procedure then involves only two integrations, the first to obtain the eikonal phases, and the second to obtain the scattering amplitude. Further approximations can be made for these phases, for example the black-disk approximation where $e^{i\chi(\mathbf{b})} = 0$ for $b < R_n$ and $e^{i\chi(\mathbf{b})} = 1$ for $b \geq R_n$, with R_n the radius of the black disk.

Overall, the eikonal method requires three conditions for its validity. The first is that the characteristic wavelength of the projectile is small compared to the range of the interaction R_0, which translates into $K_0 R_0 \gg 1$. Thus, it is mostly useful for nuclear-dominated processes. This small wavelength condition is also seen in the WKB approximations presented in Section 7.4. The second is the high-energy approximation, which means that the beam energy should be much larger than the depth of the interaction: $E \gg |V_0|$. This condition is somewhat relaxed if absorptive optical potentials are used, and often one finds that $E \approx |V_0|$ still provides good results. It has been successfully applied for beam energies as low as 70 MeV/u. The third condition is the forward angle approximation, up to $10°$–$20°$ at most.

Finally we should note that

$$S^{\text{eik}}(\mathbf{b}) = e^{i\chi(b)} \tag{7.2.13}$$

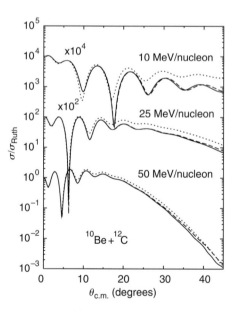

Fig. 7.5. Application [12] of the eikonal model to elastic scattering of ^{10}Be on ^{12}C at several beam energies (dotted lines). Comparison with the optical model (solid lines). Reprinted with permission from J. M. Brooke, J. S. Al-Khalili and J. A. Tostevin, *Phys. Rev. C* **59** (1999) 1560. Copyright (1999) by the American Physical Society.

are S-matrix type functions, like those introduced in Chapter 3, if we map $b = L/k$ semiclassically. We can also calculate required observables from these functions. An example is shown in Fig. 7.5 for the elastic scattering of ^{10}Be on ^{12}C at several beam energies.

7.2.3 Composite-body scattering and the optical limit

In order to calculate elastic and inelastic processes of composite objects with wave functions $\Phi(\mathbf{r})$ over some internal coordinates $\mathbf{r}$, the eikonal method above must be combined with the previous adiabatic treatment for the internal motion of the projectile. If we may assume that the excitation energies of the projectile are small compared with the beam energy, then scattering can be calculated separately for each value of $\mathbf{r}$ as in Eq. (7.1.5), and the elastic composite scattering then found by averaging each of these over the ground-state density $|\Phi(\mathbf{r})|^2$ as in Eq. (7.1.10). If the $\Phi_i(\mathbf{r})$ and $\Phi_f(\mathbf{r})$ are the initial and final projectile internal wave functions, then the inelastic scattering of the composite system would use the transition density $\Phi_f^*(\mathbf{r})\Phi_i(\mathbf{r})$. The scattering amplitude for each configuration $\mathbf{r}$ is a direct generalization from Eq. (7.2.11), so the general transition amplitude in the few-body eikonal method is

$$f_{fi}(\theta) = -\frac{iK_0}{2\pi} \int d^2b \, e^{i\mathbf{q}\cdot\mathbf{b}} \langle \Phi_f(\mathbf{r})|e^{i\chi(\mathbf{b}-\mathbf{b}_r)} - 1|\Phi_i(\mathbf{r})\rangle, \qquad (7.2.14)$$

with $\mathbf{b}$ being the impact parameter vector of the c.m. of the projectile and $\mathbf{b}_r$ being the projection of the internal coordinates $\mathbf{r}$ of the projectile. The coordinates for a two-body projectile are shown in Fig. 7.6. The equation reduces to the elastic amplitude when $f = i$.

If the projectile is composed of n particles, there are $n - 1$ internal coordinates, so $\Phi_i(\mathbf{r}) \equiv \Phi_i(\mathbf{r}_1, \mathbf{r}_2, \ldots, \mathbf{r}_{n-1})$. In the same way, the eikonal phases of Eq. (7.2.6) become $\chi(\mathbf{b} - \mathbf{b}_r) \equiv \sum_{j=1}^{n} \chi(\mathbf{b} - \mathbf{b}_{r_j})$. Putting these together and defining the

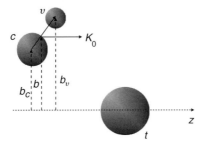

Fig. 7.6. Coordinates for eikonal model of reactions involving two-body projectiles, including the definition of various impact parameters.

S matrices for each single particle as

$$S_i(\mathbf{b}_i) = e^{i\chi_i(\mathbf{b}_i)}, \qquad (7.2.15)$$

where $\mathbf{b}_i = \mathbf{b} - \mathbf{b}_{r_i}$, we have the Glauber formula for composite projectiles,

$$f_{fi}(\theta) = -\frac{iK_0}{2\pi} \int d^2b \, e^{i\mathbf{q}\cdot\mathbf{b}} \int \cdots \int d\mathbf{r}_1 d\mathbf{r}_2 \ldots d\mathbf{r}_{n-1}$$
$$\Phi_f(\mathbf{r}_1, \ldots, \mathbf{r}_{n-1})^*[S_1(\mathbf{b}_1) \ldots S_n(\mathbf{b}_n) - 1]\Phi_i(\mathbf{r}_1, \ldots, \mathbf{r}_{n-1}). \qquad (7.2.16)$$

This corresponds to a truncated multiple scattering expansion to the order of the number of clusters [13]. Other multiple scattering frameworks can be used [14].

We can then define the n-body scattering S matrix from an initial state i to a final state f as

$$S_{fi}^n(\mathbf{b}) = \langle\Phi_f|\Pi_{j=1}^n S_j(\mathbf{b}_j)|\Phi_i\rangle. \qquad (7.2.17)$$

This scattering matrix contains correlations between the various particles, since the averaging over internal coordinates of the projectile occurs after the exponentiation. By expanding the exponential of Eq. (7.2.15) for $S_j(\mathbf{b}_j)$ into a power series, we see that it includes all higher orders of the potential V *before* averaging over the projectile's wave function. An application of this for ^{11}Be on ^{12}C and a comparison with the more exact adiabatic model can be found in [12] (see Fig. 7.7). Note that this figure is very similar to Fig. 7.5, as the extra neutron does not modify the scattering pattern significantly: it only introduces the form factor Eq. (7.1.21) to progressively reduce the cross section for larger angles.

The optical limit of the Glauber theory consists of neglecting all these correlations, so that instead of the full n-body $S_{fi}^n(\mathbf{b})$, we average the eikonal phases over the projectile wave function *before* taking the exponential:

$$S_{fi}^{OL}(\mathbf{b}) = \exp\left(\langle\Phi_f|i\sum_{j=1}^n \chi_j(\mathbf{b} - \mathbf{b}_j)|\Phi_i\rangle\right). \qquad (7.2.18)$$

Note that for elastic scattering, this equation can be expressed in terms of the ground-state density as

$$S_{fi}^{OL}(\mathbf{b}) = \exp\left\langle i\sum_{j=1}^n \chi_j(\mathbf{b} - \mathbf{b}_j)|\Phi_i|^2\right\rangle. \qquad (7.2.19)$$

As χ_j relates to the fragment T matrix, this is often referred to as the $t\rho$ method, since it picks up only the first order of the interaction multiplying the ground-state density.

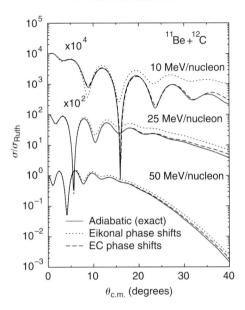

Fig. 7.7. ^{11}Be + ^{12}C elastic scattering at several beam energies [12]: adiabatic calculations (solid lines) versus the eikonal (dotted lines). Reprinted with permission from J. M. Brooke, J. S. Al-Khalili and J. A. Tostevin, *Phys. Rev. C* **59** (1999) 1560. Copyright (1999) by the American Physical Society.

7.2.4 Eikonal cross sections

Cross sections in the Glauber model can be obtained directly from the scattering amplitude $\sigma(\theta) = |f(\theta)|^2$. Based on the relationships introduced in Chapter 3, the angle-integrated elastic cross section is:

$$\sigma_{\text{el}} = \int d^2b \, |1 - S_{fi}^n(\mathbf{b})|^2$$
$$= \int d^2b \, |1 - \langle \Phi_0 | \Pi_{j=1}^n S_j(\mathbf{b} - \mathbf{b}_j) | \Phi_0 \rangle|^2. \qquad (7.2.20)$$

Similarly, the reaction cross section is

$$\sigma_R = \int d^2b \left(1 - |\langle \Phi_0 | \Pi_{j=1}^n S_j(\mathbf{b} - \mathbf{b}_j) | \Phi_0 \rangle|^2\right), \qquad (7.2.21)$$

where, if the projectile has spin I_p, we should average over its m-states.

For the breakup cross section we consider the two-body projectile $p = v + c$, for which the eikonal S matrix is factored as $S_{2b} = S_c(b_c)S_v(b_v)$. If $\mathbf{k}$ is the momentum of the two fragments in the projectile frame, the amplitude to the breakup continuum at **k** is the inelastic matrix element

$$f_{\text{bu}}(\mathbf{k}, \theta) = -\frac{iK_0}{2\pi} \int d^2b \, e^{i\mathbf{q}\cdot\mathbf{b}} \langle \Phi_{\mathbf{k},m'} | S_{2b} | \Phi_{0m} \rangle. \tag{7.2.22}$$

Here, Φ_{0m} represents the ground state of the two-body projectile, and $\Phi_{\mathbf{k},m'}$ represents the final breakup state, where m and m' are the angular momentum projections of the initial and final state respectively. If for example there is only one bound state, we can use a completeness relation to obtain a closed form for the total breakup cross section. Specifically, we use

$$\sum_{m'} \int d\mathbf{k} \, |\Phi_{k,m'}\rangle\langle\Phi_{k,m'}| = 1 - \sum_m |\Phi_{0m}\rangle\langle\Phi_{0m}| \tag{7.2.23}$$

to obtain

$$\sigma_{\text{bu}} = \frac{1}{2I_p+1} \int d^2b \sum_{mm'} \left(\langle\Phi_{0m'}| \, |S_{2b}|^2 |\Phi_{0m}\rangle\delta_{mm'} - |\langle\Phi_{0m'}|S_{2b}|\Phi_{0m}\rangle|^2 \right).$$

$$\tag{7.2.24}$$

The closure relation provides a simple form for the energy- and angular-integrated breakup cross section, but does not give angular or energy distributions.

7.2.5 Stripping reactions

Stripping reactions are those where a valence part of the projectile is removed by a non-elastic reaction with the target, leaving the core of the projectile intact. The construction of stripping cross sections is based on the probability concepts associated with $|S|^2$, with the probability of non-elastic reactions being $1 - |S|^2$.

Let us consider a reaction where the two-body projectile $p = c + v$ interacts with the target so only c survives, and v is absorbed by the target. Then $|S_c|^2$ gives the probability of the core surviving after the reaction, whereas $1 - |S_v|^2$ gives the probability of v being absorbed by the target non-elastically. The stripping cross section can then be directly written as

$$\sigma_{\text{str}} = \frac{1}{2I_p+1} \int d\mathbf{b} \sum_m \langle\Phi_{0m}| \left[|S_c|^2(1 - |S_v|^2) \right] |\Phi_{0m}\rangle. \tag{7.2.25}$$

The cross section for stripping processes of greater complexity can be constructed based on the same ideas. These eikonal formulae for *inclusive* cross sections are very useful for knock-out reactions [15, 16] (see Section 14.2).

7.3 First-order semiclassical approximation

If the charge distributions of the two nuclei involved in a reaction do not overlap at any time during the collision, and the reaction is Coulomb dominated, then we can assume that the projectile-target relative motion takes place on a classical Rutherford trajectory while treating the electromagnetic excitation of the projectile quantum mechanically. These are the basic ideas behind the Alder and Winther theory of Coulomb excitation [17, 18].

The cross section for the inelastic process from an initial state to a final state is written as a product of the Rutherford elastic cross section and the probability for making the transition to a specific final state:

$$\sigma_{fi}(\theta) = \sigma_{\text{Ruth}}(\theta) \, P_{fi}(\theta), \tag{7.3.1}$$

where P_{fi} is the probability for the transition $i \to f$ between the two projectile states. If $\mathbf{R}(t)$ is the Coulomb trajectory, let $V(\mathbf{R}(t))$ be the potential responsible for the excitation. In first order, the probability is

$$P_{fi} = |a_{fi}|^2, \tag{7.3.2}$$

$$\text{where} \qquad a_{fi} = \frac{1}{i\hbar} \int_{-\infty}^{\infty} dt \, e^{i\omega_{fi}t} \langle f | V(\mathbf{R}(t)) | i \rangle, \tag{7.3.3}$$

and the transition frequency relates to the energy difference between the initial and final states: $E_f - E_i = \hbar\omega_{fi}$. Usually, the Coulomb interaction is expanded in multipoles, so the integration over the Coulomb orbit can be simplified to isolate all dependence on the Coulomb orbitals. The transition amplitude can thus be written as:

$$a_{fi} = i \sum_{\lambda} \hat{a}_{fi}^{\lambda} f_{\lambda}(\xi), \tag{7.3.4}$$

where $\hat{a}_{fi}^{\lambda}$ depends on the properties of the projectile, the charge of the target, and the trajectory, as

$$\hat{a}_{fi}^{\lambda} = \frac{Z_t e}{\hbar v_p d_c^{\lambda}} \langle f | E(\lambda, \mu) | i \rangle. \tag{7.3.5}$$

Here we use the electric multipole operator of the Coulomb field $E(\lambda, \mu)$ (Eq. (4.7.21)), the distance of closest approach d_c, which at high energies is equal to the impact parameter, and the adiabaticity parameter $\xi = \omega_{fi} d_c / v_p$ which governs the kinematic part of the amplitude $f_{\lambda}(\xi)$. Using the explicit expressions for the Coulomb fields, it is possible to obtain analytic forms for a_{fi} [17, 18]. The cross sections are calculated summing the contributions of all classical orbitals. There

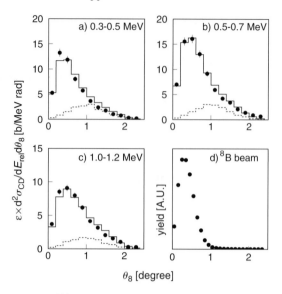

Fig. 7.8. ^{8}B breakup on ^{208}Pb at 254 MeV/u [20]: breakup yields as a function of c.m. scattering angle of the excited ^{8}B. The data are compared to first-order semiclassical predictions including E1+M1 (solid line) and estimates of E2 (dotted line). Reprinted with permission from N. Iwasa *et al.*, Phys. Rev. Lett. **83** (1999) 2910. Copyright (1999) by the American Physical Society.

is also a relativistic version of the theory, which uses straight-line trajectories but includes relativistic kinematics [19].

More recent formulations generalize Alder and Winther, to give excitation into the continuum, which is breakup [21]. It is possible to relate the Coulomb dissociation cross section to the capture cross section, and for this reason this theory is still in frequent use: the topic is discussed further in Chapter 8 and Chapter 14. An example of applying Coulomb dissociation to astrophysics is the study of the breakup of ^{8}B on ^{208}Pb for extracting the ^{7}Be(p,γ)^{8}B direct capture cross section. In Fig. 7.8 we show the comparison with the data of the first-order semiclassical predictions [20].

The Alder and Winther theory is a first-order semiclassical theory, and therefore should not be used in cases where multi-step effects are important. The quantum effects are restricted to the projectile excitation process, and in that sense 1-step DWBA theory in Chapter 8 is more complete. Alder and Winther assume the process is only Coulomb, yet, even in reactions where the nuclear component is small, we cannot eliminate the possibility of nuclear-Coulomb interference [22] as there may be nuclear diffraction to small scattering angles. It also assumes the projectile is point-like, which for nuclei on the dripline is clearly not adequate. Despite the many approximations, it is often used for Coulomb excitation and for Coulomb dissociation, as it provides a convenient factorization of structure and reaction, allowing for a simple, even if unreliable, extraction of $B(E\lambda)$.

Finally, it should be noted that semiclassical models which solve the coupled-channel time-dependent equations have been developed, and some include corrections to the trajectory due to the nuclear interaction (e.g. [23, 24, 25]). However, time-dependent solutions of the Schrödinger equations are outside the scope of this textbook.

7.4 WKB approximation

In many reactions close to the Coulomb barrier, neither the adiabatic nor the eikonal approximations can be made. Then it is common to apply the WKB approximation [26, 27], which is based on the assumption that the interaction is slowly varying over the typical wavelength of the projectile. This approximation was developed simultaneously by Wentzel, Kramers and Brillouin [28], thus the name WKB. Let us consider applying the method [26] to the radial equation

$$\left[\frac{d^2}{dr^2} - \frac{\ell(\ell+1)}{r^2} + k^2 - \check{U}(r) \right] u_\ell(r) = 0. \tag{7.4.1}$$

Here we use the scaled two-body central potential $\check{U}(r) = \frac{2\mu}{\hbar^2} V(r)$. We now change the variable of integration to $r = e^x/k$, and let the radial wave function take the form (omitting for simplicity the subscript ℓ)

$$u_\ell(r) = e^{x/2} w(x). \tag{7.4.2}$$

This form can only be assumed if the potential decreases faster than r^{-1} when $r \to \infty$ and is less singular than r^{-2} at the origin. Given the physical limits of $r \in [0, \infty]$, the new variable $x \in [-\infty, \infty]$. Substituting the variable and Eq. (7.4.2) into Eq. (7.4.1), we obtain

$$\frac{d^2 w}{dx^2} + Q^2(x) w = 0, \tag{7.4.3}$$

where $Q^2(x) = e^{2x}(1 - \check{U}/k^2) - (\ell + \frac{1}{2})^2$ and $w(-\infty) = 0$. For very large positive x, the function $Q^2(x)$ is positive, while for very large negative x, it is negative. There is thus at least one point x_0 where $Q^2(x_0) = 0$, the turning point.

As with the eikonal wave function, we consider further $w(x)$ in terms of another unknown function $q(x)$ appearing in the phase:

$$w(x) = \frac{1}{\sqrt{q(x)}} \exp\left[\pm i \int_{x_0}^x q(x') dx' \right]. \tag{7.4.4}$$

Inserting into Eq. (7.4.3), we have the radial WKB equation:

$$q^2(x) + R(x) - Q^2(x) = 0 \qquad (7.4.5)$$

$$\text{with} \qquad R(x) = \frac{1}{2q}\frac{d^2q}{dx^2} - \frac{3}{4q^2}\left(\frac{dq}{dx}\right)^2. \qquad (7.4.6)$$

The WKB approximation consists in neglecting the derivative terms $R(x)$. This is valid when $|R(x)| \ll |Q^2(x)|$, or when the potential changes slowly over the typical wavelength of the incident particle. As the potential increases, $Q^2(x)$ increases while $R(x)$ does not. Therefore, in the strong coupling limit ($\breve{U}(x) \gg k^2$), the WKB approximation becomes exact. For this reason it is useful in sub-barrier reactions, such as radioactive decay and fusion. It can also be used for scattering, although its condition of validity is not as strict as for the eikonal method. Note that in the WKB method, the trajectories are not restricted to straight lines.

The general solution of Eq. (7.4.5) under the approximation $R(x) \approx 0$ is

$$w(x) = \frac{A}{\sqrt{Q(x)}}e^{i\tilde{Q}(x)} + \frac{B}{\sqrt{Q(x)}}e^{-i\tilde{Q}(x)} \quad x > x_0,$$

$$w(x) = \frac{D}{\sqrt{|Q(x)|}}e^{\hat{Q}(x)} \quad x < x_0. \qquad (7.4.7)$$

Here we have introduced the phases $\tilde{Q}(x) = \int_{x_0}^x Q(x')dx'$ and $\hat{Q}(x) = \int_{x_0}^x |Q(x')|dx'$. For $x < x_0$ we already know that the wave function needs to be regular at the origin, so $w(-\infty) = 0$.

Around the turning point, none of the forms of Eq. (7.4.7) are valid. There are many procedures for treating the turning point. One such procedure [29] assumes a linear turning point: $Q^2(x) = a(x - x_0)$ for x near x_0, so Eq. (7.4.3) becomes

$$\frac{d^2w}{dx^2} - a(x - x_0)w = 0. \qquad (7.4.8)$$

This can be solved exactly to obtain coefficients A and B in terms of D:

$$A = -ie^{i\pi/4}D, \quad B = ie^{-i\pi/4}D. \qquad (7.4.9)$$

Then, for $x > x_0$ we have $w(x) = \frac{2D}{\sqrt{Q(x)}}\sin\left[\frac{\pi}{4} + \int_{x_0}^x Q(x')dx'\right]$. The coefficient D is obtained from the overall normalization condition.

Going back to the original radial wave function, the WKB solution is

$$u_\ell(r) = 2D\left(\frac{k^2}{\breve{U}(r)}\right)^{1/4}\sin\left[\frac{\pi}{4} + \int_{r_0}^r \sqrt{\breve{U}(r')}dr'\right], \qquad (7.4.10)$$

where $\tilde{U}(r) = k^2 - \check{U}(r) - (\ell+1/2)^2/r^2$, and $r_0 = e^{x_0}/k$ is the radius of the turning point.

In order to obtain the WKB phase shift, from which cross sections can be calculated, we note that $\tilde{U}(r) \rightarrow k^2$ for very large radii. Then the asymptotic form of Eq. (7.4.10) is given by:

$$u_\ell(r) \rightarrow 2D \sin\left[\frac{\pi}{4} + \int_{r_0}^\infty (\sqrt{\tilde{U}(r')} - k)\mathrm{d}r' + k(r-r_0)\right]. \qquad (7.4.11)$$

Comparing with the asymptotic form of Eq. (3.1.42), $\sin\left[kr - \ell\frac{\pi}{2} + \delta_\ell\right]$, we obtain the WKB phase shift

$$\delta_\ell^{\mathrm{WKB}} = (\ell + \tfrac{1}{2})\frac{\pi}{2} - kr_0 + \int_{r_0}^\infty (\sqrt{\tilde{U}(r)} - k)\mathrm{d}r. \qquad (7.4.12)$$

We can verify from this that the phase shift is zero for zero interaction. From these phase shifts, the WKB elastic scattering has amplitude

$$f^{\mathrm{WKB}}(\theta) = \frac{1}{k}\sum_{\ell=0}^\infty (2\ell+1)P_\ell(\cos\theta)e^{i\delta_\ell^{\mathrm{WKB}}}\sin\delta_\ell^{\mathrm{WKB}}. \qquad (7.4.13)$$

7.4.1 Coulomb penetration factors

It is very common to use the WKB method to estimate the probability of radioactive decay as well as the fusion penetration factor. If we consider the case of proton emission, the Coulomb interaction produces a barrier such that $\tilde{U}(r)$ contains two turning points corresponding to x_{in} and x_{out} as portrayed by r_n and r_0 in Fig. 7.9. The proton can tunnel through the barrier but the transmission probability is proportional to the square of the attenuation of the wave function through the classically forbidden region:

$$\frac{u_\ell(x_{\mathrm{out}})}{u_\ell(x_{\mathrm{in}})} = \exp\left[-\int_{x_{\mathrm{in}}}^{x_{\mathrm{out}}}|Q(x')|\mathrm{d}x'\right]. \qquad (7.4.14)$$

The penetrability normally includes the squared wave-function ratio as

$$P_\ell^{\mathrm{esc}}(E, r_n) = kr_n\left|\frac{u_\ell(\infty)}{u_\ell(r_n)}\right|^2, \qquad (7.4.15)$$

where k is the asymptotic wave number corresponding to energy E. The kr_n is a flux factor, included so that the resulting penetrability can be used directly in equations like Eq. (4.5.24).

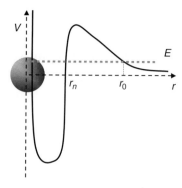

Fig. 7.9. Barrier associated with penetrability. Between the outside turning point at r_0 and the barrier at r_n, only Coulomb and centrifugal barriers are considered.

For fusion reactions starting at large radii, we need the other exponential solution that decays from x_{out} towards x_{in}. The fusion penetrability is now

$$P_\ell^{fus}(E, r_n) = k r_n \left| \frac{u_\ell(r_n)}{u_\ell(\infty)} \right|^2. \tag{7.4.16}$$

Because of time-reversal invariance, this (small) penetrability should be the same as that from Eq. (7.4.15).

If the Coulomb interaction dominates the process, the tunneling probability given by these equations is called the Coulomb penetrability. In this case, the potential barrier is the sum of Coulomb and centrifugal potentials: $V_\ell(r) = Z_A Z_v e^2/r + \ell(\ell+1)\hbar^2/(2\mu r^2)$. If r_n is located at or inside the barrier (see Fig. 7.9), the barrier there will be $E_B = E_C + E_\ell$: composed of the Coulomb barrier $E_C = Z_A Z_v e^2/r_n$ and the centrifugal barrier $E_\ell = \ell(\ell+1)\hbar^2/(2\mu r_n^2)$.

The penetrability can also be defined in terms of Coulomb wave functions as

$$P_\ell(E, r_n) = \frac{k r_n}{F_\ell^2(k r_n) + G_\ell^2(k r_n)}, \tag{7.4.17}$$

an expression which we will meet again in Chapter 10. A WKB approximation for the penetrability is still very useful, however, and may in fact be used to find the Coulomb functions themselves when far inside the classical turning point r_0.

We consider the escape process in detail, and therefore use the WKB forms Eq. (7.4.7) to write the wave at large separations that is purely outgoing:

$$u_\ell(r \gg r_0) = A[E - V_\ell(r)]^{-\frac{1}{4}} \exp\left\{ i\sqrt{\frac{2\mu}{\hbar^2}} \int_{r_0}^r \sqrt{E - V_\ell(r)}\, dr \right\}. \tag{7.4.18}$$

Inside the barrier, the wave function increases with decreasing r to

$$u_\ell(r \approx r_n) = A e^{i\pi/4} [V_\ell(r) - E]^{-\frac{1}{4}} \exp\left\{\sqrt{\frac{2\mu}{\hbar^2}} \int_r^{r_0} \sqrt{V_\ell(r) - E}\, dr\right\}. \quad (7.4.19)$$

Introducing Eqs. (7.4.18) and (7.4.19) into Eq. (7.4.15) and noting that at infinity the potential goes to zero, we obtain

$$P_\ell(E, r_n) = k r_n \left[\frac{V_\ell(r_n) - E}{E}\right]^{\frac{1}{2}} \exp\left\{-2\sqrt{\frac{2\mu}{\hbar^2}} \int_{r_n}^{r_0} \sqrt{V_\ell(r) - E}\, dr\right\}. \quad (7.4.20)$$

In terms of the height of the barriers, this is

$$P_\ell(E) = k r_n \sqrt{\frac{E_B - E}{E}} e^{-\mathcal{G}_\ell}$$

$$= \sqrt{\frac{2\mu(E_B - E) r_n^2}{\hbar^2}} e^{-\mathcal{G}_\ell}, \quad (7.4.21)$$

where the Gamow factor is introduced as

$$\mathcal{G}_\ell = 2\sqrt{\frac{2\mu}{\hbar^2}} \int_{r_n}^{r_0} dr \sqrt{\frac{E_C r_n}{r} + \frac{E_\ell r_n^2}{r^2} - E}. \quad (7.4.22)$$

We first consider the $\ell = 0$ case because it is the simplest and gives the largest penetrabilities. The integrations can be performed to give

$$\int_{r_n}^{r_0} \sqrt{E_C r_n / r - E}\, dr \propto \frac{\pi}{2} - \arcsin\sqrt{E/E_C} - \sqrt{E/E_C(1 - E/E_C)}.$$

We now expand in powers of E/E_C and keep only the first three terms. The Gamow factor becomes

$$\mathcal{G}_0 = -4\sqrt{\frac{2\mu E_C r_n^2}{\hbar^2}} + \pi\sqrt{\frac{2\mu E_C r_n^2}{\hbar^2}} \sqrt{\frac{E_C}{E}} \left[1 + \frac{2}{3\pi}\left(\frac{E}{E_C}\right)^{3/2}\right]. \quad (7.4.23)$$

When $\ell \neq 0$ we can make another low-energy approximation, and assume that in most of the range of the integration r_n/r is small, so the centrifugal barrier never dominates. Then

$$\mathcal{G}_\ell \approx 2\sqrt{\frac{2\mu}{\hbar^2}} \int_{r_n}^{r_0} \sqrt{\frac{E_C r_n}{r} - E}\, dr + \sqrt{\frac{2\mu}{\hbar^2}} \int_{r_n}^{r_0} \frac{E_\ell r_n^{3/2}}{E_C^{1/2}} \frac{dr}{r^{3/2}}. \quad (7.4.24)$$

The first term corresponds to $\mathcal{G}_0$. We retain only the leading order part of the second integral and rewrite the penetration factor for $E \ll E_B$. Using the Sommerfeld parameters $\eta = Z_A Z_v e^2/(\hbar v_p)$ for scattering energy E, and η_C for energy E_C, we have:

$$P_\ell(E) = 2\eta_C \exp\left\{-2\pi\eta + 8\eta_C - \frac{\ell(\ell+1)}{\eta_C}\right\}. \tag{7.4.25}$$

From this we can find the decay rate λ of an excited state of the compound nucleus. The decay rate relates to the lifetime τ and the partial width Γ through $\lambda = 1/\tau = \Gamma/\hbar$, and is the probability per unit time that particle x is found at large distances from A. From the WKB results,

$$\lambda = v|u_\ell(\infty)|^2 = \frac{\hbar}{\mu r_n}P_\ell(E, r_n)|u_\ell(r_n)|^2, \tag{7.4.26}$$

with the penetrability from Eq. (7.4.25). The asymptotic-velocity factors v are needed here to give the flux in terms of the particle densities from the wave function square moduli. These v factors have the same energy dependence as kr_n, so the second expression above gives decay rates with an energy dependence from tunneling exactly that of $P_\ell(E, r_n)$. Similar expressions for widths Γ will appear in Chapter 10.

The same penetrabilities can be used for fusion processes $A + x \rightarrow B + \gamma$, to find the energy behavior of the capture cross section at astrophysical energies ($\sim$ keV), much below the Coulomb barrier, so $\eta \gg \eta_C$. In this case the factor $e^{-2\pi\eta}$ in Eq. (7.4.25) gives the greatest energy dependence in the penetrability. When inserted in formulae for cross sections such as Eq. (4.5.24) which have an incoming-flux factor $k^{-2} \propto E^{-1}$, we can justify using the formula $\sigma = S(E)e^{-2\pi\eta}/E$ to define the astrophysical S-factor $S(E)$, which is then expected to be relatively constant at low energy (see Chapter 1 and Chapter 12).

This WKB estimate has been traditionally applied to one-particle emission (either proton or α). It has also been applied to three-body decay in a three-body model where an effective three-body potential creates the barrier that needs to be penetrated [30].

Exercises

7.1 Consider the elastic scattering of ^{11}Be on ^{12}C at 49.3 MeV/u within the Johnson special model. Determine the form factor of Eq. (7.1.21) for ^{11}Be and compare with the results shown in Fig. 7.2.

7.2 What is the penetrability for the reaction ^{3}He(^{4}He,γ)^{7}Be assuming you can neglect the centrifugal barrier? Estimate the error due to neglecting the centrifugal barrier.

7.3 Prove that the eikonal approximation satisfies the optical theorem in the limit of high energies.

7.4 Coulomb phase shifts χ_c can be obtained using a screening radius a_s of atomic dimensions in the Coulomb interaction which becomes $V_{\mathrm{Coul}}(R) = e^{-R/a_s}/R$. This ensures that results are still accurate at very forward angles. Expanding the screened Coulomb potential in powers of b/a_s, prove that the leading term for $b > R_{\mathrm{Coul}}$ is $\chi_c(b) = 2\eta \ln kb - 2\eta \ln(2ka_s)$ with R_{Coul} the Coulomb radius.

References

[1] J. S. Al-Khalili and J. A. Tostevin, *Scattering, Vol. 2*, ed. by Roy Pike and Pierre Sabatier. Boston: Academic Press, Chapter 3.1.3, p. 1375.

[2] R. C. Johnson and P. J. R. Soper, *Phys. Rev. C* **1** (1970) 055807.

[3] R. C. Johnson and P. C. Tandy, *Nucl. Phys.* **A235** (1974) 56.

[4] G. L. Wales and R. C. Johnson, *Nucl. Phys.* **A274** (1976) 168.

[5] R. C. Johnson, J. S. Al-Khalili and J. A. Tostevin, *Phys. Rev. Lett.* **79** (1997) 2771.

[6] M. L. Goldberger and K. M. Watson 1964, *Collision Theory*, New York: John Wiley and Sons.

[7] R. Crespo, A. M. Moro, I. J. Thompson, *Nucl. Phys. A* **771** (2006) 26.

[8] J. A. Tostevin, S. Rugmai and R. C. Johnson, *Phys. Rev. C* **57** (1998) 3225.

[9] G. Baur and D. Trautmann, *Nucl. Phys.* **A191** (1972) 321.

[10] N. K. Timofeyuk and R. C. Johnson, *Phys. Rev. C* **59** (1999) 1545.

[11] R. J. Glauber 1959, "High-energy collision theory", in *Lectures in Theoretical Physics Vol. 1*, ed. by W. E. Brittin. New York: Interscience, p. 315.

[12] J. M. Brooke, J. S. Al-Khalili and J. A. Tostevin, *Phys. Rev. C* **59** (1999) 1560.

[13] J. A. Al-Khalili, R. Crespo, A. M. Moro and I. J. Thompson, *Phys. Rev. C* **75** (2007) 024608.

[14] R. Crespo *et al.*, *Phys. Rev. C* **76** (2007) 014620.

[15] P. G. Hansen and J. A. Tostevin, *Annu. Rev. Nucl. Part. Sci.* **53** (2003) 219.

[16] A. Gade *et al.*, *Phys. Rev. C* **74** 021302 (2006); 047302 (2006).

[17] K. Alder and A. Winther 1975, *Electromagnetic Excitation*, Amsterdam: North Holland.

[18] K. Alder *et al.*, *Rev. Mod. Phys.* **28** (1956) 432.

[19] A. Winther and K. Alder, *Nucl. Phys.* **A319** (1979) 518.

[20] N. Iwasa *et al.*, *Phys. Rev. Lett.* **83** (1999) 2910.

[21] G. Baur, C. Bertulani and H. Rebel, *Nucl. Phys.* **A458** (1986) 188.

[22] M. Hussein, R. Lichtenthäler, F. M. Nunes and I. J. Thompson, *Phys. Lett. B* **640** (2006) 91.

[23] H. Esbensen and G. F. Bertsch, *Nucl. Phys.* **A600** (1996) 37.

[24] C. H. Dasso, S. M. Lenzi and A. Vitturi, *Nucl. Phys. A* **639** (1998) 635.

[25] H. D. Marta, L. F. Canto, R. Donangelo and P. Lotti, *Phys. Rev. C* **66** (2002) 024605.

[26] C. Joachain 1975, *Quantum Collision Theory*, New York: Elsevier Science Publishers B. V.

[27] C. A. Bertulani and P. Danielewicz 2004, *Introduction to Nuclear Reactions*, Bristol: IoP Publishing.

[28] G. Wentzel, *Z. Phys.* **38** (1926) 518; H. A. Kramers *Z. Phys.* **39** (1926) 828; L. Brillouin, *Comptes Rendues* **183** (1926) 24.

[29] M. L. Goldberger and K. M. Watson 1996, *Collision Theory*, New York: Wiley, Chapter 6.

[30] E. Garrido, D. V. Fedorov, A. S. Jensen and H. O. U. Fynbo, *Nucl. Phys. A* **748** (2005) 27.

8

Breakup

In science one tries to tell people, in such a way to be understood by
everyone, something that no one ever knew before. But in poetry, it's the
exact opposite.

Paul Dirac

Nuclei close to the driplines have large breakup cross sections. The **breakup
mechanism** connects a bound state of the projectile with its **continuum states**,
so the process is rather similar to **inelastic excitation**, except that the final state is
unbound. If the target has a strong Coulomb field, then breakup reactions measure
the electromagnetic response that populates scattering states. These can be relevant
for astrophysical processes either directly, in **photo-disintegration**, or in reverse,
for **capture reactions**. The excitation into the continuum also allows the study of
resonances, and this is important for structure studies because the understanding of
the spectra of unstable nuclei depends strongly on resonant properties. Because of
these many interests, it is important to establish reliable models for the description
of breakup reactions. For simplicity we shall neglect target spin in this chapter. The
connections to astrophysics and nuclear structure will be considered in Chapter 14.

In the simplest case, breakup will leave the target nucleus in its ground state. This
is called **elastic breakup** (also called **diffraction dissociation**). However, there
are experiments that measure the inclusive reaction, and then **inelastic breakup**
(also referred to as **stripping**) can be an important contribution. Elastic breakup
of a two-body projectile can be modeled as a three-body problem, and this is the
approach presented in Sections 8.1 and 8.2. Later in subsection 8.3.2 we discuss
methods for inelastic breakup.

8.1 Three-body wave equations

8.1.1 Wave function components

The breakup of a projectile p into a core + fragment ($c + v$) due to an interaction
with a target t can be described in a three-body model. We now must deal with

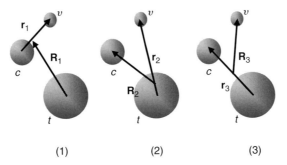

Fig. 8.1. Jacobi coordinates, typically used in the three-body models.

three-body systems which have two-body bound states between (at least) one of the pairs of bodies, if only because such a bound state is the initial state of a projectile whose breakup cross sections we wish to calculate. Three-body projectiles will be examined in more detail in Chapter 9, where the advantages of using Jacobi coordinates will be explained. We describe our three-body $(c + v) + t$ system using one or more of the coordinates illustrated in Fig. 8.1, with conjugate momenta and corresponding angular-momenta for each coordinate in the normal manner.

Let the initial projectile $p = c+v$ bound state be $\phi_0^{(1)}(\mathbf{r}_1)$ in terms of the coordinate $\mathbf{r}_1$ in Fig. 8.1. The incident projectile in the beam will therefore be described by a component $\phi_0^{(1)}(\mathbf{r}_1) \exp(i\mathbf{K}_1 \cdot \mathbf{R}_1)$ in the wave function $\Psi^{(1)}(\mathbf{r}_1, \mathbf{R}_1)$. However, if there are bound states of the $v + t$ or $c + t$ pairs, then the wave function $\Psi^{(1)}$ is not suitable to describe them, because they depend on the $\mathbf{r}_2$ and $\mathbf{r}_3$ coordinates respectively, and $\Psi^{(1)}(\mathbf{r}_1, \mathbf{R}_1)$ cannot be suitably factorized into products of those bound states and spherical outgoing waves. We therefore need to write the full system wave function as a sum of components, so there can be a component expressed in each of the Jacobi coordinate pairs,

$$\Psi = \sum_{n=1}^{3} \Psi^{(n)}(\mathbf{r}_n, \mathbf{R}_n). \tag{8.1.1}$$

Each component n is thus a function of the vector connecting two of the particles $\mathbf{r}_n$ and the vector connecting the c.m. of this pair to the third body $\mathbf{R}_n$. The three-body Hamiltonian is

$$H_{3b} = \hat{T} + V_{vc} + V_{vt} + V_{ct}, \tag{8.1.2}$$

where the kinetic energy operators are $\hat{T} = \hat{T}_n \equiv \hat{T}_{r_n} + \hat{T}_{R_n}$ for any choice of Jacobi coordinate set n.

We may consider writing the three-body wave equation in the *Schrödinger form* as

$$[H_{3b} - E]\Psi = 0, \tag{8.1.3}$$

but if there are partitions other than $n = 1$ with bound states, then the transfer channels still give an awkward complication. This is because each of the $\Psi^{(n)}(\mathbf{r}_n, \mathbf{R}_n)$ contains not just the bound states, but also all the continuum states, and can hence by itself completely represent the whole Ψ: even the bound states in other partitions $n \neq 1$! The problem with Eq. (8.1.3) is that we have not decoupled the bound-state components in the asymptotic region: in an integral reformulation of the differential equations, the problem can be delineated as the Green's function not having a *compact* or *connected kernel* [1]. Admittedly, the other bound states $\phi^{(2)}$ and $\phi^{(3)}$ can only be partly described by any finite partial-wave sum of spherical harmonics of the $(\mathbf{r}_1, \mathbf{R}_1)$ coordinates, but since matching is always performed at a finite radius, even a small part means that the outgoing amplitude for $\phi_{n'}$ in component $\Psi^{(n')}$ will have an admixture from component $\Psi^{(n)}$.

8.1.2 *Three-component Faddeev equations*

In order to have well-defined asymptotic behaviors, and to make sure that bound states $\phi^{(n)}$ only appear in the boundary conditions for wave-function component $\Psi^{(n)}(\mathbf{r}_n, \mathbf{R}_n)$, Faddeev [1] proposed a variant of Eq. (8.1.3) that *does* have unambiguous separations of the bound states at large distances.

The Faddeev equations consist of three equations coupling the three components in such a way that their sum is the same as before,

$$(E - T_1 - V_{vc})\Psi^{(1)} = V_{vc}(\Psi^{(2)} + \Psi^{(3)})$$

$$(E - T_2 - V_{ct})\Psi^{(2)} = V_{ct}(\Psi^{(3)} + \Psi^{(1)})$$

$$(E - T_3 - V_{tv})\Psi^{(3)} = V_{tv}(\Psi^{(1)} + \Psi^{(2)}). \qquad (8.1.4)$$

Components defined in this way are called *Faddeev components*. For finite-range potentials, these equations become uncoupled as $R_n \to \infty$ with r_n finite. In this limit the right-hand sides for $n' \neq n$ become zero, and hence each binding potential only influences *one* of the components $\Psi^{(n)}$. This gives a much better defined and numerically stable set of boundary conditions for the bound-state outgoing waves.

Asymptotically, each Faddeev component n thus contains contributions from the bound states associated to that pair and from three-body breakup. Transfer channels are confined to specific Faddeev components, but breakup is distributed among the three components. Each Faddeev component behaves asymptotically as

$$\Psi^{(n)} \to \sum_p \phi_p^{(n)}(\mathbf{r}_n)\psi_p^{(n)}(\mathbf{R}_n) + \text{breakup}; \quad \text{when } R_n \to \infty, \qquad (8.1.5)$$

where $\phi_p^{(n)}$ are bound states p of the corresponding subsystem (for $n = 1$ this corresponds to the bound states of $c + v$). If n is the initial channel then $\psi_p^{(n)}(\mathbf{R}_n)$ contains an incident plane wave and the outgoing spherical waves, otherwise it only contains the latter.

The three-body problem can be therefore solved exactly within the Faddeev formalism. This can be done in coordinate space with the appropriate boundary conditions [1, 2], or in momentum space [3]. In both cases, specific techniques must be used to regularize the Coulomb divergences.

8.1.3 Reduction to one Jacobi set

Work by Johnson and Soper [4] showed that deuteron breakup was very important to understand reactions involving the deuteron. In that work, a two-channel problem was solved where the deuteron continuum was represented by a single discrete s-state. Later, developments by Rawitscher [5] and Austern [6, 7, 8] helped to introduce a more realistic representation of the continuum. This approach used only the $n = 1$ wave-function component in which there is a bound state $(v + c)$, and so avoided solving the Faddeev equations. Their method solved the wave equation

$$[H_{3b} - E]\Psi^{(1)}(\mathbf{r}_1, \mathbf{R}_1) = 0. \tag{8.1.6}$$

How is it possible to reduce the number of components in this way?

When only one pair of particles has bound states, there is indeed no longer the original motivation for having multiple components. It should then be possible to simplify the set of Faddeev Eqs. (8.1.4) to an equation for a component written in terms of only one Jacobi coordinate pair. There may be only one partition with a bound state by accident, for example in $d + \alpha$ scattering, but there is another very important kind of breakup reaction where this also holds.

In projectile scattering at high incident energies, the potentials V_{vt} and V_{ct} should be appropriate for $v + t$ and $c + t$ scattering at (on average) the beam velocity, and most often these will be optical potentials, and so will have imaginary components. In this case, neither V_{vt} nor V_{ct} will support bound states, because their imaginary parts will simply 'absorb' such states in a short time, according to Eq. (3.1.109). When V_{ct} and V_{tv} are realistic optical potentials containing absorption, the wave functions in the rearrangement channels decay exponentially in $\mathbf{r}_{vt}$ and $\mathbf{r}_{ct}$, and should be describable with only low or moderate ℓ_{vc} components. The V_{vc} potential must be kept real, in order to support the initial projectile ground state.

In cases where the transfer channels are not important, or are inhibited by the imaginary potentials, the three-body wave equations may therefore be successfully simplified to Eq. (8.1.6). This component can be expanded in terms of the complete set of eigenstates of the internal Hamiltonian of the projectile $c + v$ as

$$\Psi_{\mathbf{K}_0}^{(1)}(\mathbf{r}_1, \mathbf{R}_1) = \sum_{p=1}^{n_b} \phi_p(\mathbf{r}_1)\psi_p(\mathbf{R}_1) + \int d\mathbf{k}\, \phi_{\mathbf{k}}(\mathbf{r}_1)\psi_{\mathbf{K}}^{\mathbf{k}}(\mathbf{R}_1). \tag{8.1.7}$$

We will from now on use $\mathbf{r}_1 \equiv \mathbf{r}$ and $\mathbf{R}_1 \equiv \mathbf{R}$, as only the first Jacobi component is necessary. The momentum $\mathbf{k}$ between the internal motion of $c + v$ is related to the momentum $\mathbf{K}$ between the projectile and the target through energy conservation:

$$E_{\text{cm}} + \epsilon_0 = E = \frac{\hbar^2 k^2}{2\mu_{vc}} + \frac{\hbar^2 K^2}{2\mu_{(vc)t}}. \tag{8.1.8}$$

This E is the center-of-mass energy of the three-body system: the sum of the projectile-beam center-of-mass energy E_{cm} and the (negative) projectile binding energy ϵ_0. The μ_{vc} is the reduced mass for $v + c$ relative motion, and $\mu_{(vc)t}$ is the reduced mass of the projectile-target motion.

The projectile eigenstates $\phi_p(\mathbf{r})$ are eigensolutions of the Hamiltonian $H_{\text{int}} = T_{vc}(\mathbf{r}) + V_{vc}(\mathbf{r})$ for energies $\epsilon_p < 0$, and the continuum states $\phi_{\mathbf{k}}(\mathbf{r})$ are the eigensolutions of the same Hamiltonian, but for scattering energies $\epsilon_k = \hbar^2 k^2/2\mu_{vc} > 0$. Both kinds of eigenstates (p, k) have the standard angular momentum decomposition

$$\phi_{(p,k)}^M(\mathbf{r}) = \frac{u_{(p,k)}(r)}{r}\left[[Y_\ell(\hat{\mathbf{r}}) \otimes \mathcal{X}_s]_j \otimes \mathcal{X}_{I_c}\right]_{I_p M}, \tag{8.1.9}$$

where ℓ is the orbital angular momentum of v relative to c, $s(I_c)$ is the spin of the fragment $v(c)$, and the total angular momentum of the projectile is I_p with projection M. The continuum integral in Eq. (8.1.7) in partial-wave form is

$$\int d\mathbf{k} \ldots \rightarrow \int_0^\infty dk \sum_{\ell s j I_c I_p} \ldots \tag{8.1.10}$$

Assuming that the effective interaction between the core c and fragment v is central, the radial wave functions $u_{(p,k)}(r)$ are solutions of the radial equation

$$\left[-\frac{\hbar^2}{2\mu_{vc}}\left(\frac{d^2}{dr^2} - \frac{\ell(\ell+1)}{r^2}\right) + V_{vc}(r) - \epsilon\right]u_{(p,k)}(r) = 0, \tag{8.1.11}$$

for both the bound states with $\epsilon_p < 0$ that are exponentially decaying at large distances, and also the continuum states with energy $\epsilon_k > 0$ that oscillate to infinity according to Eq. (3.1.85).

8.2 Continuum Discretized Coupled Channel method

In Chapter 7 we presented three different methods that could be used to compute breakup observables. The first involved using the adiabatic wave function containing the breakup channel in the exact T matrix with only a short-range interaction in the operator (subsection 7.1.4), mostly applicable at intermediate to high energies. The eikonal approximation can also be used to calculate breakup, as described in subsection 7.2.4. Finally, the first-order semiclassical theory for Coulomb excitation can also be used for Coulomb breakup, as discussed in Section 7.3. All these three methods rely on different but very significant approximations. In this chapter we lead up to a method that does not rely on such approximations.

8.2.1 Continuum bins

For a practical method, the three-body wave equation of Eq. (8.1.6) is too difficult to solve if it uses the expansions of Eqs. (8.1.7) and (8.1.10), because these are over a *continuous* variable k. It is more tractable to have a finite representation for the continuum, and there are three standard ways of generating this finite description:

(i) The first is the *mid-point method*, which consists of taking directly a scattering state $u_{k_p}(r)$ for a discrete set $p = p_0 \ldots p_n$ of scattering energies such as $\epsilon_p = \epsilon_{\min} + (p-p_0)\Delta\epsilon$, where $\epsilon_{\min}$ and $\epsilon_{\min} + (p_n-p_0)\Delta\epsilon$ are respectively the minimum and maximum excitation energies to be included in the model space.

(ii) One can also use *pseudostates*, which are simply the eigenstates of the internal Hamiltonian H_{int} on some convenient square-integrable basis. The basis could be harmonic oscillator states (as in the shell model), transformed harmonic oscillators [9], or a large set of Gaussians [10], amongst others. These pseudostates decay to zero at large distances, and have no simple relation to the $v + c$ scattering solutions $u_k(r)$, but we may hope that they form a sufficiently complete set in the reaction region.

(iii) Finally there is the *continuum bin* or *average* method, where the scattering wave functions $u_k(r)$ are averaged over k to be made square-integrable. The remainder of this section explains this average method and its applications.

The radial functions for the continuum bins in the average method, $\tilde{u}_p(r)$, are a superposition of the scattering eigenstates as

$$\tilde{u}_p(r) = \sqrt{\frac{2}{\pi N_p}} \int_{k_{p-1}}^{k_p} g_p(k) u_k(r) \, dk \qquad (8.2.1)$$

for some weight function $g_p(k)$. The normalization constant is chosen as $N_p = \int_{k_{p-1}}^{k_p} |g_p(k)|^2 \, dk$ to make the $\tilde{u}_p(r)$ form an orthonormal set when all the (k_{p-1}, k_p) are non-overlapping continuum intervals. We now generalize the meaning of the index p to extend beyond bound states to also include the bin wave functions $\tilde{u}_p(r)$ in given partial-wave channels and given continuum intervals, so $p = \{lsjI_cI_p; (k_{p-1}, k_p)\}$.

The weight function deserves particular attention. Suppose we want an amplitude for a cross section $a(k) = \langle u_k(r)|\Omega(r)\rangle$, for some source term $\Omega(r)$. If we insert a set of (approximately complete) bin functions

$$a(k) = \langle u_k(r)|\Omega(r)\rangle \approx \sum_p \langle u_k(r)|\tilde{u}_p(r)\rangle \, \langle \tilde{u}_p(r)|\Omega(r)\rangle \,, \qquad (8.2.2)$$

we now need $\langle u_k(r)|\tilde{u}_p(r)\rangle$ the overlap of a bin wave function and a true scattering wave function $u_k(r)$. From Eq. (8.2.1) we see that the energy dependence of this overlap is strongly influenced by the weight function $g_p(k)$. We should therefore choose $g_p(k)$ to reproduce some form of $a(k)$ that we can predict in advance.

If the bin is not near a resonance, then the large-r behavior will be important. In this case, $u_k(r) \propto e^{i\delta_\ell(k)} \sin(kr - \ell\pi/2 + \delta_\ell(k))$, where $\delta_\ell(k)$ are the phase shifts of the scattering states within the bin, so we would expect $a(k)$ to behave as $a(k) = e^{-i\delta_\ell(k)}\langle \sin(kr - \ell\pi/2 + \delta_\ell(k))|\Omega(r)\rangle$. It is easy to reproduce the most important feature of this dependence, the phase $e^{-i\delta_\ell(k)}$, by setting $g_p(k) = e^{-i\delta_\ell(k)}$. For resonant states, by contrast, the most important feature is the large enhancement of the small-r wave functions in proportion to $\sin\delta_\ell(k)$. This resonant behavior of $a(k)$ can be reproduced by including an additional factor in $g_p(k) = e^{-i\delta_\ell(k)} \sin\delta_\ell(k)$.

If the phase factor $e^{-i\delta_\ell(k)}$ is always included in $g_p(k)$, an advantage is that the bin wave functions can be made real-valued. This is because the boundary conditions of Eq. (3.1.85) for $u_k(r)$ are

$$u_k(r) \to e^{i\delta} \left[\cos\delta(k)F_\ell(kr) + \sin\delta(k)G_\ell(kr)\right], \qquad (8.2.3)$$

where F_ℓ and G_ℓ are the real regular and irregular partial-wave Coulomb functions, and the exponential phase factors cancel.

At large distances the bin wave functions must decay to zero. The bin method has the advantage that the extent and detail of the discrete representation of the continuum is adjustable: the convergence with respect to these adjustments will be discussed later. To gain an idea of the spatial size of the bin wave functions, we may calculate them analytically if the potential is zero in the $\ell = 0$ partial wave. In this case, we find

$$\tilde{u}_p(r) \propto \sin(k_p r)\frac{\sin((k_p - k_{p-1})r)}{r}, \qquad (8.2.4)$$

as opposed to the pure oscillatory behavior $\sin(k_p r)$ of $u_k(r)$. The $\tilde{u}_p(r)$ behave asymptotically with an extra factor $\sin((k_p - k_{p-1})r)/r$, which makes them square-integrable. Just as in this average method, the pseudostate methods are also based on square-integrable functions, however the mid-point method does not give square-integrable basis states.

We replace $u_p(r)$ in Eq. (8.1.9) by $\tilde{u}_p(r)$ to form a discrete set of continuum bin states $\tilde{\phi}_p(\mathbf{r})$, where now $p = 0, \ldots N$ ranges over all the continuum bins as well as the bound states. These bin states have average energies $\tilde{\epsilon}_p = \langle \tilde{\phi}_p(\mathbf{r}) | H_{int} | \tilde{\phi}_p(\mathbf{r}) \rangle$ that correspond to wave numbers $\tilde{k}_p$ near the mid-point momentum $(k_{p-1}+k_p)/2$.

8.2.2 CDCC equations and couplings

This Continuum Discretized Coupled Channel (CDCC) method expands the projectile continuum in terms of the bins defined above. The method solves the Schrödinger equation (8.1.6) for the first Jacobi component

$$(H_{3b} - E)\Psi^{CDCC}(\mathbf{r}, \mathbf{R}) = 0 \qquad (8.2.5)$$

in the model space defined by a discretized expansion

$$\Psi^{CDCC}(\mathbf{r}, \mathbf{R}) = \sum_{p=0}^{N} \tilde{\phi}_p(\mathbf{r})\psi_p(\mathbf{R}), \qquad (8.2.6)$$

for a given choice of N bins defined by all the $p = \{lsjI_cI_p; (k_{p-1}, k_p)\}$, including also projectile bound states. The CDCC wave function, which uses only one Jacobi set, therefore approximates the three-body asymptotic behavior (see Chapter 9) by a product of projectile-target two-body asymptotic forms. The uncertainties associated with this approximation have been studied in [11].

The three-body Hamiltonian Eq. (8.1.2) can be separated into the internal Hamiltonian of the projectile and the relative motion between the projectile and the target: $H_{3b} = \hat{T}_R + H_{int} + U_{vt} + U_{ct}$. We need a real potential for the projectile, whereas the fragment-target interactions will usually be optical potentials and contain absorption from channels not included explicitly in the model space, typically excitations of the fragments and the target (so are written U instead of V).

Multiplying Eq. (8.2.5) on the left by the conjugate projectile wave functions, and using the expansion of Eq. (8.2.6), we obtain coupled-channel equations

$$\sum_p \langle \tilde{\phi}_q(\mathbf{r}) | H - E | \tilde{\phi}_p(\mathbf{r}) \rangle \psi_p(\mathbf{R}) = 0. \qquad (8.2.7)$$

Using the above definition of $\tilde{\epsilon}_p$, this can be rewritten as

$$[\hat{T}_R + V_{pp}(R) - E_p]\psi_p(\mathbf{R}) + \sum_{p' \neq p} V_{pp'}(R)\psi_{p'}(\mathbf{R}) = 0, \qquad (8.2.8)$$

where $E_p = E - \tilde{\epsilon}_p$, and $V_{pp'}(R) = \langle \tilde{\phi}_p(r)|U_{vt}+U_{ct}|\tilde{\phi}_{p'}(r)\rangle$. These equations couple the projectile ground state to its continuum states by the $V_{p0}(\mathbf{R})$, and also couple projectile states within the continuum, by what are called *continuum-continuum couplings*.

In terms of the bin wave functions, the energy conservation of Eq. (8.1.8) requires that

$$\frac{\hbar^2 K_p^2}{2\mu_{(vc)t}} + \tilde{\epsilon}_p = \frac{\hbar^2 K_0^2}{2\mu_{(vc)t}} - \epsilon_0 = E, \qquad (8.2.9)$$

where K_p is the c.m. wave number of the projectile and $\tilde{\epsilon}_p$ is the average excitation energy in the corresponding bin state.

Let us consider the angular momenta involved in this three-body problem. A projectile wave function $\tilde{\phi}_p$ has the form of Eq. (8.1.9):

$$\tilde{\phi}_p(\mathbf{r}) = \tilde{u}_p(r)/r \left[\left[Y_\ell(\hat{\mathbf{r}}) \otimes \mathcal{X}_s\right]_j \otimes \mathcal{X}_{I_c}\right]_{I_p M}, \qquad (8.2.10)$$

where $\tilde{u}_p(r)$ are the radial bound-state wave functions or the continuum bins defined by Eq. (8.2.1). We use a standard multipole decomposition of $\psi_p(\mathbf{R})$:

$$\psi_p(\mathbf{R}) = \sum_L i^L \chi_{pL}(R) Y_L(\hat{\mathbf{R}})/R, \qquad (8.2.11)$$

where L is the orbital angular momentum of the relative motion between the projectile and the target (we neglect the target spin). The final three-body wave function carries total angular momentum and projection J_{tot}, M_{tot} resulting from the angular momentum coupling $L \otimes I_p$. We define a set of all quantum numbers $\alpha = \{p, L\} \equiv \{lsjI_cI_p; (k_{p-1}, k_p), L\}$, so that the CDCC partial-wave channels for each J_{tot} are indexed by α:

$$\left[-\frac{\hbar^2}{2\mu_{(vc)t}}\left(\frac{d^2}{dR^2} - \frac{L(L+1)}{R^2}\right) + V_{\alpha\alpha}^{J_{tot}}(R) + \tilde{\epsilon}_p - E\right]\chi_\alpha^{J_{tot}}(R)$$
$$+ \sum_{\alpha' \neq \alpha} i^{L'-L} V_{\alpha\alpha'}^{J_{tot}}(R)\chi_{\alpha'}^{J_{tot}}(R) = 0. \qquad (8.2.12)$$

The coupling potentials $V_{\alpha\alpha'}(R)$ describe the coupling between the different projectile relative-motion states:

$$V_{\alpha\alpha'}^{J_{tot}}(R) = \langle[\phi_p(\mathbf{r})Y_L(\hat{\mathbf{R}})]_{J_{tot}}|U_{ct}(\mathbf{R}_c) + U_{vt}(\mathbf{R}_v)|[\phi_{p'}(\mathbf{r})Y_{L'}(\hat{\mathbf{R}})]_{J_{tot}}\rangle, \qquad (8.2.13)$$

where $U_{ct}(\mathbf{R}_c)$ and $U_{pt}(\mathbf{R}_v)$ are again the total (nuclear and Coulomb) interactions between c, t and v, t respectively.

We need to consider carefully the properties of these coupling potentials. If we had used the mid-point method for the bins, the fact that the $v+t$ and $c+t$ interactions depend on $(\mathbf{R}_c, \mathbf{R}_v)$ means the continuum-continuum couplings would be between two true scattering states, and the integrals in Eq. (8.2.13) would diverge. Of course this is not the case for ground state to continuum couplings, as then the ground-state exponential decay provides a natural cutoff. The bin wave functions $\tilde{u}$ defined by the average method, however, as opposed to those defined by the mid-point method, are square-integrable and thus avoid this divergence problem. In this way, for CDCC breakup using the average method in the discretization of the continuum, the coupling potentials can be treated similarly to standard inelastic couplings.

Using the average method for the discretization, the long-range behavior of these coupling potentials is determined by any fragment-target Coulomb interactions. Using the multipole expansion of the Coulomb field it can be shown that, for couplings between different bins, the asymptotic behavior of the dipole component is $\sim 1/R^2$ but all higher-order multipoles will fall off as $\sim 1/R^3$. The diagonal interaction for a bin is of $\sim 1/R^2$ for all multipolarities: see [12] for more detail.

8.2.3 Calculating differential cross sections

The procedure in the CDCC method involves several steps:

(i) calculating the projectile wave functions (bound and scattering states) and the continuum bins;
(ii) calculating the coupling potentials $V_{\alpha\alpha'}(R)$;
(iii) solving the coupled equations Eq. (8.2.12) to obtain the S matrices **S**; and finally
(iv) constructing the observables, namely the cross sections.

From the coupled-equations solution we can calculate the (two-body) scattering amplitudes for populating each bin state $I_{p'}M'$ from initial state $I_p M$. These amplitudes contain a sum over all partial waves, and are a function of the angle $\theta(\mathbf{K}_{p'})$ of the c.m. of the excited projectile in the three-body c.m. frame:

$$\tilde{\mathcal{F}}_{M'M}(\mathbf{K}_{p'}) = \frac{4\pi}{K_0}\sqrt{\frac{K_{p'}}{K_0}}\sum_{LL'J}\langle L0, I_pM|JM\rangle\,\langle L'M-M', I_{p'}M'|JM\rangle$$

$$\times \exp(i[\sigma_L+\sigma_{L'}])\frac{1}{2i}\,\mathbf{S}^J_{\alpha'\alpha}(p')\,Y^0_L(\widehat{\mathbf{K}}_0)\,Y^{M-M'}_{L'}(\widehat{\mathbf{K}}_{p'}). \quad (8.2.14)$$

Here σ_L and $\sigma_{L'}$ are the initial and final Coulomb phase shifts, and the $\mathbf{S}^J_{\alpha'\alpha}(p')$ are the partial-wave S matrices for exciting bin state p' with c.m. wave number $K_{p'}$. The differential cross sections of the c.m. of the bin breakup states p of the projectile, averaged over all orientations, are

$$\frac{d\sigma(p')}{d\Omega_K} = \frac{1}{2I_p+1} \sum_{MM'} \left| \tilde{\mathcal{F}}_{M'M}(\mathbf{K}_{p'}) \right|^2 . \tag{8.2.15}$$

8.2.4 Model space and convergence of the CDCC equations

Convergence is the important consideration when calculating breakup observables with the CDCC method. The expansion of the three-body wave function is truncated in angular momentum, both in the internal motion of the projectile ℓ_{max} and the relative motion of the projectile and the target L_{max}. The coupling potentials Eq. (8.2.13) are integrated in r up to some maximum distance between fragments r_{bin}. We include only bins up to some maximum relative energy ϵ_{max}, and the CDCC equations are integrated up to R_{max}. The interactions $U_{ct}(\mathbf{R}_c) + U_{vt}(\mathbf{R}_t)$ are expanded to a maximum specified multipole order λ, so we also have to consider the multipole limit λ_{max}. A detailed study on convergence of the CDCC method for deuteron breakup on ^{58}Ni at 80 MeV, including nuclear couplings only, can be found in [13]. For Coulomb-dominated processes it may be necessary to go out to much larger distances, typically at least several hundred fm.

An optimal discretization of the continuum requires a consideration of the bin number, the boundaries k_i, the widths Δk_i and the weights g_p in forming the bins, which may depend on the I_p configuration. In Fig. 8.2 we show an example of the continuum discretization used to study the nuclear and Coulomb breakup of ^{7}Be [14]. The discretization is uniform in momentum, and partial waves $\ell \le 3$ are included. The maximum included excitation energy of the projectile is 20 MeV, and particular attention was given to the resonant bin in the $f_{7/2}$ channel. This reaction can serve as an indirect measurement for ^{3}He$(\alpha, \gamma)^7$Be, part of the pp

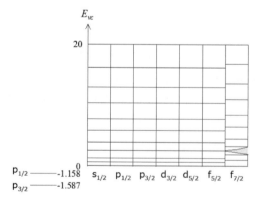

Fig. 8.2. The model space used in [14] for CDCC calculations of the ^{7}Be breakup into ^{3}He$+^4$He. Figure courtesy of Neil Summers.

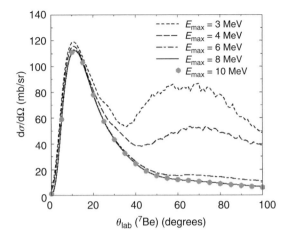

Fig. 8.3. Convergence for the angular distribution of ^{7}Be in the laboratory following the breakup of ^{8}B on ^{58}Ni at 26 MeV, by the method of subsection 8.2.6.

chain (see Chapter 1). Details of using indirect methods for astrophysics are given in Chapter 14.

Last but not the least, the size of the model space needs to be checked for each given observable. Experience has shown that the less integrated the cross section, the larger the model space needs to be to describe the same reaction. In Fig. 8.3, we show the progressive convergence of the angular distribution in the laboratory of the ^{7}Be fragment after the breakup of ^{8}B on ^{58}Ni at 26 MeV, as calculated by the method of subsection 8.2.6. While a maximum excitation of the ^{8}B of 3 MeV is sufficient to obtain converged angular distributions of ^{8}B in the center-of-mass system [15], this is no longer true for the three-body observables, where $\epsilon_{max} \lesssim 8$ MeV is necessary [16].

8.2.5 Relation to DWBA

The CDCC coupled equations of Eq. (8.2.12) may be solved either exactly using the methods of subsection 6.3.2, or else approximately by iteration. Starting with $\psi^{(-1)}(\mathbf{R}) = 0$, higher order can be iteratively determined by solving[1]

$$[\hat{T}_R + V_{\alpha\alpha}(R) - E_\alpha]\psi_\alpha^{(n)}(\mathbf{R}) = -\sum_{\alpha' \neq \alpha} V_{\alpha\alpha'} \psi_{\alpha'}^{(n-1)}(\mathbf{R}), \qquad (8.2.16)$$

---

[1] Note that the superscript (n) now refers to the order of the iteration and not the Faddeev component.

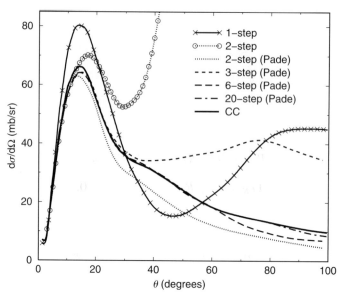

Fig. 8.4. Angular distribution for the breakup of ^{8}B on ^{58}Ni at 26 MeV in the c.m. [15]: the full CDCC result (solid line), and successive multi-step DWBA. Reprinted with permission from F. M. Nunes and I. J. Thompson, *Phys. Rev. C* **59** (1999) 2652. Copyright (1999) by the American Physical Society.

for $n = 0, 1, \ldots$ The function $\psi_\alpha^{(0)}(\mathbf{R})$ is thus the elastic channel resulting from the solution of

$$[\hat{T}_R + V_{\alpha\alpha}(R) - E_\alpha]\psi_\alpha^{(0)}(\mathbf{R}) = 0. \tag{8.2.17}$$

The potential $V_{\alpha\alpha}(R)$ is called the *bare potential*. Subsections 6.3.3 and 6.3.4 discussed the convergence of such iterative solutions.

The S matrix $\mathbf{S}^{(n)}$ of the wave functions $\psi^{(n)}(R)$ gives the cross section for an n'th-order DWBA that uses bare potentials (rather than optical potentials that reproduce elastic scattering). These multi-step DWBA results will converge to the coupled-channels solution if the off-diagonal couplings are sufficiently small. If they are large, the DWBA series diverges, but then the infinite series may be resumed by the method of Padé approximants (see page 214, or [15]). In Fig. 8.4 we show the cross section resulting from successive iterations of Eq. (8.2.16) for the breakup of ^{8}B on ^{58}Ni at 26 MeV. This is a case where off-diagonal couplings are strong and therefore Padé acceleration was required for convergence.

The usual DWBA uses optical potentials that reproduce elastic scattering in the entrance and exit channels. If these are $U_{\text{opt}}^0(R)$ and $U_{\text{opt}}^\alpha(R)$ respectively, then the

DWBA T matrix for the breakup process in prior form is

$$\mathbf{T}_{\alpha 0}^{\text{prior DW}} = \langle \tilde{\psi}_\alpha^{(-)}(\mathbf{R})\phi_\alpha(\mathbf{r})|U_{vt} + U_{ct} - U_{\text{opt}}^\alpha(R)|\phi_0(\mathbf{r})\tilde{\psi}_0(\mathbf{R})\rangle$$

$$\equiv \langle \tilde{\psi}_\alpha^{(-)}(\mathbf{R})|\mathcal{V}_{\alpha 0}(R)|\tilde{\psi}_0(\mathbf{R})\rangle \tag{8.2.18}$$

where we abbreviate $\mathcal{V}_{\alpha 0}(R) = \langle\phi_p(\mathbf{r})|U_{vt} + U_{ct} - U_{\text{opt}}^\alpha|\phi_0(\mathbf{r})\rangle$, and where the initial and final distorted waves are found by solving

$$[\hat{T}_R + U_{\text{opt}}^\alpha(R) - E_\alpha]\tilde{\psi}_\alpha(\mathbf{R}) = 0,$$

$$[\hat{T}_R + U_{\text{opt}}^0(R) - E_0]\tilde{\psi}_0(\mathbf{R}) = 0. \tag{8.2.19}$$

We usually approximate $U_{\text{opt}}^\alpha(R) = U_{\text{opt}}^0(R) = U_{\text{opt}}(R)$ when only the elastic channel with the projectile in the ground state can be measured.

The Eq. (8.2.18) has another form when using only the bare potential for the exit channel:

$$\mathbf{T}_{\alpha 0}^{\text{prior BP–DW}} = \langle \tilde{\psi}_{\alpha,\text{BP}}^{(-)}(\mathbf{R})|\mathcal{V}_{\alpha 0}(R)|\tilde{\psi}_0^{(+)}(\mathbf{R})\rangle, \tag{8.2.20}$$

where the bare wave functions $\tilde{\psi}_{\alpha,\text{BP}}^{(-)}(\mathbf{R})$ are solutions of a decoupled equation from the CDCC set:

$$[\hat{T}_R + V_{\alpha\alpha}(R) - E_\alpha]\tilde{\psi}_{\alpha,\text{BP}}^{(-)}(\mathbf{R}) = 0. \tag{8.2.21}$$

The one-step solution of the CDCC equation (Eq. (8.2.16) for $n = 1$) corresponds to taking the bare interaction as the distorting potential in both the entrance and exit channels:

$$\mathbf{T}_{\alpha 0}^{1\text{step}} = \langle \tilde{\psi}_{\alpha,\text{BP}}^{(-)}(\mathbf{R})|\mathcal{V}_{\alpha 0}(R)|\tilde{\psi}_{0,\text{BP}}^{(+)}(\mathbf{R})\rangle. \tag{8.2.22}$$

The exact solution is given by

$$\mathbf{T}_{\alpha 0}^{\text{exact}} = \langle \tilde{\psi}_{\alpha,\text{BP}}^{(-)}(\mathbf{R})\phi_\alpha(\mathbf{r})|U_{vt} + U_{ct} - V_{\alpha\alpha}(R)|\Psi^{\text{exact}}(\mathbf{r},\mathbf{R})\rangle. \tag{8.2.23}$$

The CDCC method replaces the exact wave function by that of Eq. (8.2.6) so

$$\mathbf{T}_{\alpha 0}^{\text{CDCC}} = \langle \tilde{\psi}_{\alpha,\text{BP}}^{(-)}(\mathbf{R})\phi_\alpha(\mathbf{r})|U_{vt} + U_{ct} - V_{\alpha\alpha}(R)|\sum_{\alpha'}\phi_{\alpha'}(\mathbf{r})\psi_{\alpha'}(\mathbf{R})\rangle. \tag{8.2.24}$$

The first term of Eq. (8.2.24) for $\alpha' = 0$ is $\mathbf{T}_{\alpha 0}^{\text{priorBP–DW}}$. Then the various differences between $T_{\alpha 0}^{\text{priorDW}}$ and $T_{\alpha 0}^{\text{CDCC}}$ can be separated:

$$\mathbf{T}_{\alpha 0}^{\text{CDCC}} - \mathbf{T}_{\alpha 0}^{\text{priorDW}} = (\mathbf{T}_{\alpha 0}^{\text{CDCC}} - \mathbf{T}_{\alpha 0}^{1\text{step}}) + (\mathbf{T}_{\alpha 0}^{1\text{step}} - \mathbf{T}_{\alpha 0}^{\text{BP–DW}})$$

$$+ (\mathbf{T}_{\alpha 0}^{\text{BP–DW}} - \mathbf{T}_{\alpha 0}^{\text{priorDWBA}}). \tag{8.2.25}$$

The first term in brackets represents multi-step effects, the second the effect of distortion in the entrance channel and finally, in the third, the distortion in the exit channel.

The effect of breakup on the elastic channel can be understood as a dynamic polarization potential: that potential which, when combined with the bare potential, reproduces the elastic scattering amplitudes. Polarization potentials provide a very approximate way of understanding the effect of additional channels on the elastic channel, and are discussed further in subsection 11.5.2.

8.2.6 Three-body observables

To predict the angular and energy distributions of fragments after breakup, such as shown in Fig. 8.3, we need to undo the averaging that generated the bins. To calculate three-body observables for final state momenta $\mathbf{k}$ and $\mathbf{K}$, therefore, we rewrite the T matrix from the CDCC method of Eq. (8.2.24) as

$$\mathbf{T}_{\mu\sigma:M}(\mathbf{k}, \mathbf{K}) = \langle \phi_{\mathbf{k}\mu\sigma}^{(-)}(\mathbf{r}) \, e^{i\mathbf{K}\cdot\mathbf{R}} | U(\mathbf{r}, \mathbf{R}) | \Psi_{\mathbf{K}_0 M}^{\text{CDCC}}(\mathbf{r}, \mathbf{R}) \rangle. \qquad (8.2.26)$$

Here $\phi_{\mathbf{k}\mu\sigma}$ is the $c + v$ exact final state and $U = U_{vt} + U_{ct}$. Next we insert the complete set of bin states to obtain

$$\mathbf{T}_{\mu\sigma:M}(\mathbf{k}, \mathbf{K}) = \sum_{p,M'} \langle \phi_{\mathbf{k}\mu\sigma}^{(-)} | \tilde{\phi}_p^{M'} \rangle \langle \tilde{\phi}_p^{M'} \, e^{i\mathbf{K}\cdot\mathbf{R}} | U(\mathbf{r}, \mathbf{R}) | \Psi_M(\mathbf{r}, \mathbf{R}) \rangle, \qquad (8.2.27)$$

where the sum is over all bins p which contain wave number k so the overlap is non-zero. The first factor in Eq. (8.2.27) is

$$\langle \hat{\phi}_{\mathbf{k}\mu\sigma}^{(-)} | \phi_p^{M'} \rangle = \frac{(2\pi)^{3/2}}{k\sqrt{N_\alpha}} \sum_v (-i)^\ell \langle \ell v, s\sigma | jm \rangle \langle jm, I_c \mu | I'_p M' \rangle e^{i\bar{\delta}_p(k)} \tilde{u}_p(k) Y_{\ell v}(\hat{\mathbf{k}}). $$

$$(8.2.28)$$

Here $\bar{\delta}_p(k) = \delta_p(k) + \sigma_p(k)$ is the sum of the nuclear and Coulomb phase shifts for $c + v$ scattering at relative wave number k.

The transition matrix elements appearing in the second term in Eq. (8.2.27),

$$\widehat{\mathcal{T}}_{M'M}^p(\mathbf{K}_p) = \langle \hat{\phi}_p^{M'} \, e^{i\mathbf{K}_p\cdot\mathbf{R}} | U(\mathbf{r}, \mathbf{R}) | \Psi_{\mathbf{K}_0 M}^{\text{CDCC}}(\mathbf{r}, \mathbf{R}) \rangle, \qquad (8.2.29)$$

relate to the matrix elements $\tilde{\mathcal{F}}_{M'M}(\mathbf{K}_p)$ obtained from the coupled-channels solution through:

$$\widehat{\mathcal{T}}_{M'M}^p(\mathbf{K}_p) = -\frac{2\pi\hbar^2}{\mu_{(vc)t}} \sqrt{\frac{K_0}{K_p}} \, \tilde{\mathcal{F}}_{M'M}(\mathbf{K}_p). \qquad (8.2.30)$$

Through Eq. (8.2.30), the inelastic amplitudes are converted to those of the T matrix by removal of their two-body phase space factors. In this way, we find the asymptotic amplitudes for plane-wave final states $\exp(i\mathbf{k}\cdot\mathbf{r})$.

Equation (8.2.15) can therefore be written in terms of $\widehat{\mathcal{T}}^{\alpha}_{M'M}(\mathbf{K}_{\alpha})$, so the angular distribution in the center-of-mass frame for excitation of each bin state p, averaged over all projectile orientations, is

$$\frac{d\sigma(p)}{d\Omega_K} = \frac{1}{2I_p+1}\left[\frac{\mu_{(vc)t}}{2\pi\hbar^2}\right]^2 \frac{K_p}{K_0}\sum_{MM'}|\widehat{\mathcal{T}}^{p}_{M'M}(\mathbf{K}_p)|^2. \tag{8.2.31}$$

Inserting Eq. (8.2.28) and Eq. (8.2.29) into Eq. (8.2.27), we get for the three-body breakup T matrix

$$\mathbf{T}_{\mu\sigma:M}(\mathbf{k},\mathbf{K}) = \frac{(2\pi)^{3/2}}{k}\sum_{p\nu}(-i)^{\ell}\langle \ell\nu, s\sigma|jm\rangle\langle jm, I_c\mu|I'_pM'\rangle$$
$$\times \exp[i\bar{\delta}_p(k)]Y^{\nu}_{\ell}(\hat{\mathbf{k}})\, g_p(k)\,\widehat{\mathcal{T}}_{M'M}(p,\mathbf{K}). \tag{8.2.32}$$

We usually express scattering amplitudes in a coordinate system with the x-axis in the plane of $\mathbf{K}_0$ and $\mathbf{K}_p$, such that the azimuthal angle ϕ_{K_p} of $\mathbf{K}_p$ is zero. For calculating three-body observables, however, the x-coordinate axis should rather be defined such that the coordinate system is fixed relative to the detectors in the laboratory. In that case, the amplitudes must be multiplied by $\exp(i[M-M']\phi_K)$, where ϕ_K is the azimuthal angle of $\mathbf{K}_p$ with respect to the new x-axis.

The triple-differential cross sections for breakup can be obtained from the three-body amplitudes of Eq. (8.2.32). For example, let us consider a case where both fragment angles are detected, and the energy of the core c is measured. Then the triple-differential cross section is

$$\frac{d^3\sigma}{d\Omega_c d\Omega_v dE_c} = \frac{2\pi\mu_{(vc)t}}{\hbar^2 K_0}\frac{1}{(2I_p+1)}\sum_{\mu\sigma M}\left|\mathbf{T}_{\mu\sigma:M}(\mathbf{k},\mathbf{K})\right|^2 \rho_{\mathrm{ps}}(E_c,\Omega_c,\Omega_v),$$
$$\tag{8.2.33}$$

where the three-body phase-space factor $\rho_{\mathrm{ps}}(E_c,\Omega_c,\Omega_v)$ [17] is the density of states per unit core-particle energy, for detection at solid angles Ω_v and Ω_c. It can be derived by evaluating the integral

$$\rho_{\mathrm{ps}}(E_c,\Omega_c,\Omega_v)dE_c d\Omega_c d\Omega_v =$$
$$\int d^3\mathbf{k}_c d^3\mathbf{k}_v d^3\mathbf{k}_t\, \delta(\mathbf{P}-\hbar\mathbf{k}_c-\hbar\mathbf{k}_v-\hbar\mathbf{k}_t)\, \delta(E-E_c-E_v-E_t), \tag{8.2.34}$$

where E is the available kinetic energy. The core, valence particle and the target momenta in the final state are $\hbar\mathbf{k}_c$, $\hbar\mathbf{k}_v$ and $\hbar\mathbf{k}_t$, and $\mathbf{P} = \hbar\mathbf{K}_{\mathrm{tot}}$ is the total momentum of the system. The momenta can be evaluated either in the c.m. frame or the laboratory frame, and relate to $\mathbf{k}$ and $\mathbf{K}$ through

$$\mathbf{K} = \mathbf{k}_v + \mathbf{k}_c - \frac{m_c + m_v}{m_c + m_v + m_t}\mathbf{K}_{\text{tot}}, \quad \mathbf{k} = \frac{m_c}{m_c + m_v}\mathbf{k}_v - \frac{m_v}{m_c + m_v}\mathbf{k}_c. \quad (8.2.35)$$

The resulting phase-space density is

$$\rho_{\text{ps}}(E_c, \Omega_c, \Omega_v) = \frac{m_c m_v \hbar k_c \hbar k_v}{(2\pi\hbar)^6} \frac{m_t}{m_v + m_t + m_v(\mathbf{k}_c - \mathbf{K}_{\text{tot}}) \cdot \mathbf{k}_v / k_x^2}. \quad (8.2.36)$$

If an experiment only detects the core, for example, we need to integrate over Ω_v. Furthermore, as detectors always have a solid-angle coverage $\Delta\Omega_c \neq 0$ with an efficiency profile $e(\Omega_c)$, the prediction to be compared with experiment would be

$$\left\langle \frac{d^2\sigma}{d\Omega_c dE_c} \right\rangle = \frac{1}{\Delta\Omega_c} \int_{\Delta\Omega_c} d\Omega_c \left\{ e(\Omega_c) \int d\Omega_v \frac{d^3\sigma}{d\Omega_c d\Omega_v dE_c} \right\}. \quad (8.2.37)$$

Here, it would be most convenient to choose the x–z plane to be that defined by the beam and the core-particle detector.

8.3 Other breakup measures and methods

8.3.1 Momentum distributions

When a nucleus is broken into two or more pieces in a breakup reaction, it is common to measure the *momentum distributions* of the fragments. Usually, these distributions are transformed into the rest frame of the initial nucleus, whether it be a target or a projectile, and either the longitudinal or transverse components are plotted.

The most accurate way of calculating these momentum distributions is to calculate the triple-differential cross sections according to Eq. (8.2.33), and then integrate over the angle of the unobserved particle, as well as over the unobserved components of the observed particle. This gives the elastic breakup (diffraction dissociation) part of the cross section, as would be appropriate when the unobserved particle is definitely produced, even if not registered. The lack of γ-rays from compound-nucleus decay may be taken to indicate elastic breakup.

8.3.2 Inclusive measurements

An important component of observed inclusive momentum distributions is from inelastic breakup (stripping, or breakup-fusion) as described in subsection 7.2.5. There, the other particle may well still be fused with the target in some complicated inelastic process. This component cannot be calculated by the CDCC breakup method. There is an approximate spectator expression suitable for high-energy beams, based on the eikonal approach of subsection 7.2.5, given by Hussein and McVoy [18]. At the highest energies, the results become similar to that of the Serber model [19], which calculates the momentum distribution directly as a Fourier transform of the initial bound-state wave function.

The post-form DWBA approach [20, 21] can also approximate the stripping component for light-nucleus breakup using a surface-closure approximation, as does the method of Udagawa *et al.* [22].

8.3.3 Semiclassical and time-dependent methods

The breakup process can be modeled in some cases as an inelastic excitation of the projectile while moving in a classical orbit, such that the excitation into the continuum is treated as a perturbation. This is the well-known semiclassical model introduced by Alder and Winther that was briefly discussed in Section 7.3. Its popularity is mainly due to the simple way in which it relates the structure of the projectile (namely the $dB(E\lambda)/dE$) and the Coulomb dissociation cross section. We leave a discussion of its application to Chapter 14. Non-perturbative time-dependent methods, such as the time-dependent Schrödinger equation (TDSE) [23] and the time-dependent Hartree-Fock (TDHF) [24] are beyond the scope of this book.

8.3.4 Transfer to the continuum

The breakup process can also be modeled as transfer to the continuum: this is particularly useful to populate unbound states. An example are the measurements of the states in ^{10}Li through ^{9}Li(d,p) performed at REX-ISOLDE [25]. Several earlier theories for transfer to the continuum are based in the first-order DWBA for transfer processes, such as the semiclassical method of Brink and Bonaccorso [26], or those with specific approximations for the final unbound state of interest (e.g. [27, 28, 29]). An obvious improvement of these approaches [30] is to calculate the transfer T matrix by replacing the exit-channel wave function by the CDCC three-body wave function expanded using the second Jacobi set of coordinates shown in Fig. 8.1. The convergence of this method as well as its applicability is still under study.

Exercises

8.1 The breakup of ^{8}B has been studied in a number of experiments in order to determine the ^{7}Be(p,γ) reaction rate, of importance in the solar neutrino problem. Consider the breakup of ^{8}B on Pb at 82 MeV/u. The CDCC method provides a theoretical framework in which to study this reaction. In this method, the ^{7}Be+p continuum is sliced into energy bins (there is an example input in Appendix B).

(a) For $E = 0.2$ MeV, determine the bin wave function and compare with the corresponding pure scattering wave function.

(b) Perform a CDCC calculation of ^{8}B→^{7}Be+p on Pb at 82 MeV/u and obtain the angular distributions.

(c) Estimate the uncertainty from the ^{7}Be optical potential and the proton optical potential for this reaction.

(d) Determine the model space needed for convergence for calculating the total cross section for ^{8}B→^{7}Be+p on Pb at 82 MeV/u if charged particle detectors were placed from $\theta_{cm} = 0.1$-1.0 degrees.

(e) In the conditions above, what would be the percentage of the nuclear contribution to the total cross section?

(f) In many cases, nuclear/Coulomb interference is important. If one were to subtract a nuclear contribution from the total cross section, what would be the error in the 'Coulomb only' cross section due to interference?

(g) How would the energy distribution E_{Be-p} for the breakup ^{8}B→^{7}Be+p on Pb at 82 MeV/u change if the $p-$wave resonance were at 0.1 MeV?

References

[1] L. D. Faddeev, *JETP* **39** (1960) 1459.
[2] C. J. Joachain 1987, *Quantum Collision Theory,* Amsterdam: North-Holland, section 19.3.
[3] A. Deltuva, A. C. Fonseca and P. U. Sauer, *Phys. Rev. C* **72** (2005) 054004.
[4] R. C. Johnson and P. J. R. Soper, *Phys. Rev. C* **1** (1970) 976.
[5] G. H. Rawitscher, *Phys. Rev. C* **9** (1974) 2210.
[6] N. Austern, Y. Iseri, M. Kamimura, G. Rawitscher and M. Yahiro, *Phys. Rep.* **154** (1987) 125.
[7] M. Yahiro, N. Nakano, Y. Iseri and M. Kamimura, *Prog. Theo. Phys.* **67** (1982) 1464; *Prog. Theo. Phys. Suppl.* **89** (1986) 32.
[8] N. Austern, M. Kawai and M. Yahiro, *Phys. Rev. Lett.* **63** (1989) 2649; *Phys. Rev. C* **53** (1996) 314.
[9] A. M. Moro, F. Pérez-Bernal, J. M. Arias and J. Gómez-Camacho, *Phys. Rev. C* **73** (2006) 044612.
[10] T. Egami, K. Ogata, T. Matsumoto, Y. Iseri, M. Kamimura and M. Yahiro, *Phys. Rev. C* **70** (2004) 047604.
[11] E. O. Alt, B. F. Irgaziev and A. M. Mukhamedzhanov, *Phys. Rev. C* **71** (2005) 024605.
[12] F. M. Nunes and A. M. Mukhamedzhanov, *Nucl. Phys. A* **736** (2004) 255.
[13] R. A. D. Piyadasa, M. Kawai, M. Kamimura and M. Yahiro, *Phys. Rev. C* **60** (1999) 044611.
[14] N. C. Summers and F. M. Nunes, *Phys. Rev. C* **70** (2004) 011602.
[15] F. M. Nunes and I. J. Thompson, *Phys. Rev. C* **59** (1999) 2652.
[16] J. A. Tostevin, F. M. Nunes and I. J. Thompson, *Phys. Rev. C* **63** (2001) 024617.
[17] H. Fuchs, *Nucl. Inst. and Meth.* **200** (1982) 361.

[18] M. S. Hussein and K. W. McVoy, *Nucl. Phys.* **A445** (1985) 124.

[19] R. Serber, *Phys. Rev.* **72** (1947) 1008.

[20] G. Baur and D. Trautmann, *Phys. Rep.* **25** (1976) 293.

[21] R. Chatterjee, P. Banerjee and R. Shyam, *Nucl. Phys. A* **675** (2000) 477.

[22] T. Udagawa, T. Tamura and R. C. Mastroleo, *Phys. Rev. C* **37** (1988) 2261.

[23] V. S. Melezhik and D. Baye, *Phys. Rev. C* **59** (1999) 3232.

[24] P. Bonche, S. E. Koonin and J. N. Negele, *Phys. Rev. C* **13** (1976) 1226.

[25] H. B. Jeppesen *et al.*, *Phys. Lett. B* **642** (2006) 449.

[26] A. Bonaccorso and D. M. Brink, *Phys. Rev. C* **38** (1988) 1786; *Phys. Rev. C* **43** (1991) 299; *Phys. Rev. C* **44** (1991) 1559; *Phys. Rev. C* **46** (1992) 700; *Phys. Rev. C* **63** (2001) 044604.

[27] R. Shyam and S. Mukherjee, *Phys. Rev. C* **13** (1976) 2099.

[28] R. Huby and D. Kelvin, *J. Phys. G* **1** (1975) 203.

[29] C. M. Vincent and H. T. Fortune, *Phys. Rev. C* **2** (1970) 782.

[30] A. M. Moro and F. M. Nunes, *Nucl. Phys. A* **767** (2006) 138.

9

Three-body nuclei

One never notices what has been done; one can only see what remains to be done.

Marie Curie

The direct reaction of a two-body projectile with a target constitutes a three-body problem, as discussed in the previous chapter. The next most complicated group of processes involve a three-body projectile, which with a target make a four-body reaction problem. This group includes reactions where the projectile is a two-nucleon halo nucleus, and in this chapter we present theories for the structure and reactions of such nuclei. First we introduce the topic of halo nuclei, then describe three-body models for bound and scattering states. Finally we discuss four-body reaction models within DWBA, the adiabatic approximation and the eikonal approximation, and conclude by looking at four-body CDCC.

9.1 Definitions of halo and deeply bound states

Stable nuclei are characterized by large binding energies and extremely long lifetimes. Figure 9.1 shows the nuclear chart for light nuclei. As we add protons or neutrons to the system, and move away from the valley of stability (black squares in Fig. 9.1), the binding energy of the valence nucleons becomes smaller and smaller until eventually the system can no longer bind. Around the nuclear dripline, we frequently find exotic structures called *halo nuclei* [1]. The halo phenomenon comes from a significant decoupling of the valence nucleon (or nucleons) from the remaining nucleons, which form a core. The valence nucleon(s) reside mostly in the classically forbidden region ($E < V$) and contribute to a much-increased rms radius of the halo nucleus. The phenomenon is hindered by Coulomb and centrifugal barriers, so one-neutron halos appear typically when filling the s- or p-wave shells. Examples of one-neutron halos are ^{11}Be and ^{19}C. The first evidence for the existence of halo nuclei was the very large increase in radii extracted from

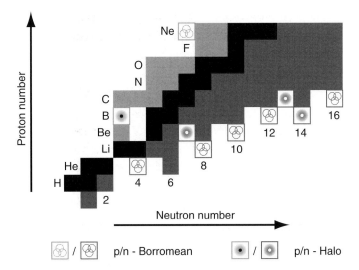

Fig. 9.1. The chart of light nuclides up to $N = 16$, indicating halo and Borromean phenomena at the driplines. Figure courtesy of Marc Hausmann.

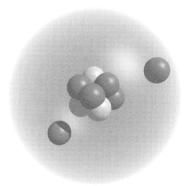

Fig. 9.2. Three-body model for ^{11}Li based on realistic relative sizes of the ^{9}Li core and the two neutrons of the halo.

reaction cross sections [2], when for example adding only one neutron from ^{10}Be to ^{11}Be. Later experiments revealed momentum distributions following the breakup of halo systems that are very narrow compared to the stable nuclear neighbors. This, through Heisenberg's uncertainty principle, implies a strong delocalization in space.

Borromean nuclei are a particular type of two-nucleon halo systems. If two valence nucleons are loosely bound and not strongly coupled to the core, and have an rms radius much larger than that of the core, then the nucleus is a good three-body halo nucleus. If, in addition, none of its pairs form bound states, the system is said to be Borromean [1]. In Fig. 9.2 we show one of the prime examples of a Borromean nucleus: ^{11}Li is bound by only 0.3 MeV relative to the three-body threshold, but neither the dineutron n–n nor ^{10}Li are bound. We need all three bodies to tie such a system together. Although less common, Borromean two-proton halo nuclei such as ^{17}Ne can also exist. The Borromean nuclei are represented by squares with the Borromean rings in Fig. 9.1.

This chapter deals with the theory of Borromean three-body systems: those where there are no two-body bound states. The previous Chapter 8 examined breakup theory when there *are* two-body bound states present. We saw that the presence of bound states in one or more partitions led to complications, and often the need to solve Faddeev equations, but these complications do not arise in the Borromean case. Here we discuss three-body bound states and three-body continuum states, as both are important in reactions.

9.2 Three-body models for bound states

We now formulate the three-body problem for bound states of core+n+n systems using the Schrödinger equation with the hyperspherical harmonics method. The hyperspherical method was first introduced in atomic physics [3], and often used both in atomic and molecular physics [4]. It was brought into nuclear physics in 1959 by Delves [5] for developing a general nuclear reaction theory. The formalism is also introduced for halo nuclei in the review [1].

A 'good' halo nucleus will have its valence nucleons far from the core, and is hence unlikely to excite the core to higher-energy states. Strictly speaking, however, we should allow for this possibility, as in some nuclei this is an important effect [6, 7, 8]. We therefore carry forward the sum over core states I, while assuming that one, or at most two, core states are necessary in the summation in practice. We will not include isospin dependence since the interactions to be used have a fixed isospin.

The kinetic energy operator for the relative motion in a three-body system, after extracting the center-of-mass motion, will depend on just *two* coordinates. One possible pair are the 'physical' neutron-core coordinates $\mathbf{r}_{31}$ and $\mathbf{r}_{32}$ in Fig. 9.3, where bodies 1 and 2 are the neutrons n, and body 3 the core c. In that coordinate system the kinetic energy is

$$\hat{T} = -\frac{\hbar^2}{2\mu_{31}}\nabla^2_{r_{31}} - \frac{\hbar^2}{2\mu_{32}}\nabla^2_{r_{32}} - \frac{\hbar^2}{m_3}\nabla_{r_{31}} \cdot \nabla_{r_{32}}, \qquad (9.2.1)$$

where the reduced masses are defined in the usual way $\mu_{ij} = m_i m_j/(m_i+m_j)$. However, we would much prefer *Jacobi coordinates* in which the coordinates can be uncoupled, without any dot-product, so the free-field solution can be factorized

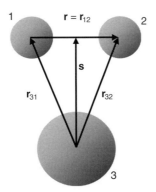

Fig. 9.3. Coordinates in which the physical two-body interactions are defined, for nucleons 1 and 2, and core 3.

into a product of functions of the two cordinates. Because of the dot-product term above, the $(\mathbf{r}_{31}, \mathbf{r}_{32})$ are not Jacobi coordinates. A better choice is $(\mathbf{r}, \mathbf{s})$ represented in Fig. 9.3 because now the kinetic energy is simply

$$\hat{T} = \hat{T}_{\mathbf{r}} + \hat{T}_{\mathbf{s}} \equiv -\frac{\hbar^2}{2\mu_{12}}\nabla_r^2 - \frac{\hbar^2}{2\mu_{(12)3}}\nabla_s^2, \tag{9.2.2}$$

where μ_{12} is the reduced mass in the $1+2$ system, and $\mu_{(12)3}$ is the reduced mass in the $(12)+3$ system.

The full Hamiltonian contains these two kinetic energy operators, the intrinsic Hamiltonians of the three bodies, and the three possible two-body interactions. In the system shown in Fig. 9.3, the Hamiltonian is

$$H_{3b} = \hat{T}_{\mathbf{r}} + \hat{T}_{\mathbf{s}} + H_{\text{core}}(\boldsymbol{\xi}) + V_{nc}(\mathbf{r}_{13}) + V_{nc}(\mathbf{r}_{23}) + V_{nn}(\mathbf{r}). \tag{9.2.3}$$

The distances between each neutron and the core, r_{31} and r_{32}, can be expressed in terms of the Jacobian coordinates $(\mathbf{r}, \mathbf{s})$ where $\mathbf{r} \equiv \mathbf{r}_{12}$ is the distance between the two neutrons and $\mathbf{s} \equiv \mathbf{r}_{(12)3}$ is the distance between the core and the neutrons' center of mass:

$$\mathbf{r}_{31} = \mathbf{s} + \tfrac{1}{2}\mathbf{r} \quad \text{and} \quad \mathbf{r}_{32} = \mathbf{s} - \tfrac{1}{2}\mathbf{r}. \tag{9.2.4}$$

The intrinsic Hamiltonian of the core determines a set of eigenstates ϕ_I and eigenenergies ϵ_I according to

$$H_{\text{core}}(\boldsymbol{\xi})\,\phi_I(\boldsymbol{\xi}) = \epsilon_I\,\phi_I(\boldsymbol{\xi}), \tag{9.2.5}$$

where $\boldsymbol{\xi}$ represents the internal core degrees of freedom. The procedure consists then of summing a product of a few core wave functions with their corresponding valence wave functions, $\Psi^{JM}(\boldsymbol{\xi}, \mathbf{r}, \mathbf{s}) = \sum_I [\phi_I(\boldsymbol{\xi}) \otimes \psi_I^J(\mathbf{r}, \mathbf{s})]_{JM}$, and, with this ansatz, solving the Schrödinger equation:

$$H_{3b}\,\Psi^{JM} = E\,\Psi^{JM}. \qquad (9.2.6)$$

The valence wave function $\psi_I^J(\mathbf{r}, \mathbf{s})$ contains the radial, angular and spin dependence for the valence neutrons, and thus will contain further sums over additional quantum numbers. The meaning of E is the energy of the three-body system relative to the breakup threshold for all three bodies.

9.2.1 The hyperspherical coordinates

In order to facilitate the hyperspherical method, we introduce scaled Jacobi coordinates. If the three bodies have masses m_i, and therefore ratios $A_i = m_i/m$ to some unit mass m, we devise *scaled Jacobi coordinates* $(\mathbf{x}, \mathbf{y})$ in terms of which the two kinetic energy terms have the same coefficient of ∇^2. From Eq. (9.2.2), we see that to get

$$\hat{T} = -\frac{\hbar^2}{2\mu_{12}}\nabla_r^2 - \frac{\hbar^2}{2\mu_{(12)3}}\nabla_s^2 = -\frac{\hbar^2}{2m}\nabla_x^2 - \frac{\hbar^2}{2m}\nabla_y^2, \qquad (9.2.7)$$

we need

$$\mathbf{x} = \sqrt{\mu_{12}/m}\,\mathbf{r} \quad \text{and} \quad \mathbf{y} = \sqrt{\mu_{(12)3}/m}\,\mathbf{s}. \qquad (9.2.8)$$

We now generalize this to any cyclic permutation of the labels 1, 2 and 3 of the three bodies, to define $(\mathbf{x}_k, \mathbf{y}_k)$ for $k = 1, 2, 3$. The relative coordinates can be generally defined in terms of the position of each body $\mathbf{r}_1, \mathbf{r}_2, \mathbf{r}_3$ as $\mathbf{r}_{ij} = \mathbf{r}_j - \mathbf{r}_i$. We also introduce the vector connecting the c.m. of a given pair to the third body as $\mathbf{r}_{(ij)k} = \mathbf{r}_{ij}^{cm} - \mathbf{r}_k$. We use an 'odd man out' notation, so $\mathbf{x}_1$ refers to the separation of particles 2 and 3, and have therefore the general definitions

$$\mathbf{x}_k = \sqrt{\frac{A_i A_j}{A_i + A_j}}\,\mathbf{r}_{ij} \quad \text{and} \quad \mathbf{y}_k = \sqrt{\frac{(A_i + A_j)A_k}{A_i + A_j + A_k}}\,\mathbf{r}_{(ij)k}, \qquad (9.2.9)$$

where (ijk) are cyclic permutations of (123). The A_i are dimensionless ratios, so $(\mathbf{x}_k, \mathbf{y}_k)$ are still lengths, while their scale is set by the unit mass m, commonly taken as the nucleon mass, or otherwise 1 amu. The coordinates are all illustrated in Fig. 9.4 for the two-neutron + core case, where each separation vector is labeled by its rescaled Jacobi coordinate.

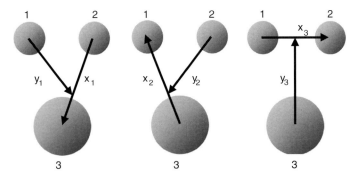

Fig. 9.4. Jacobi coordinates for the core $+ n + n$ system: the first two are usually referred to as **Y** basis, and the third as **T** basis.

If particles 1 and 2 are both nucleons of equal unit mass, the coordinate systems $(\mathbf{x}_1, \mathbf{y}_1)$ and $(\mathbf{x}_2, \mathbf{y}_2)$ are essentially equivalent. The scaled coordinates are then related to the previous $(\mathbf{r}, \mathbf{s})$ Jacobi coordinates through

$$\mathbf{x}_3 = \frac{\mathbf{r}}{\sqrt{2}} \qquad \mathbf{y}_3 = \sqrt{\frac{2A_3}{A_3 + 2}}\, \mathbf{s},$$

$$\mathbf{x}_1 = \sqrt{\frac{A_3}{A_3 + 1}}\, \mathbf{r}_{23} \qquad \mathbf{y}_1 = \sqrt{\frac{A_3 + 1}{A_3 + 2}}\, \mathbf{r}_{(23)1}. \qquad (9.2.10)$$

The $(\mathbf{x}_3, \mathbf{y}_3)$ is called the **T**-basis set, and both $(\mathbf{x}_1, \mathbf{y}_1)$ and $(\mathbf{x}_2, \mathbf{y}_2)$ are called **Y**-basis sets, because of the shapes in Fig. 9.4.

The hyperspherical radius ρ (also called 'hyper-radius') and hyperangles θ_i are then defined by

$$\rho^2 = x_i^2 + y_i^2 \quad \text{and} \quad \theta_i = \arctan\left(\frac{x_i}{y_i}\right). \qquad (9.2.11)$$

The hyper-radius, $\rho^2 = x_i^2 + y_i^2$, is the same for all $i = 1, 2, 3$, and this invariance is a basic advantage offered by the hyperspherical coordinate system. The hyperangles $\theta_i = \arctan(x_i/y_i)$, however, are different for the **T** and the two **Y** bases. The hyperspherical transformation is essentially a transformation from Cartesian to polar coordinates according to $(x, y) = (\rho \sin \theta, \rho \cos \theta)$ plane,[1] for θ in the quadrant $[0, \pi/2]$. The transformation is useful because the kinetic energy of

[1] The definition $\tan \theta = x/y$ is used rather than the more natural y/x for purely historical reasons.

Eq. (9.2.7) now separates into a sum of ρ, θ, $\hat{\mathbf{x}}$ and $\hat{\mathbf{y}}$ differential operators as

$$\hat{T} = -\frac{\hbar^2}{2m}\left[\frac{1}{\rho^5}\frac{\partial}{\partial\rho}\left(\rho^5\frac{\partial}{\partial\rho}\right) + \frac{1}{\rho^2\sin^2 2\theta}\frac{\partial}{\partial\theta}\left(\sin^2 2\theta\frac{\partial}{\partial\theta}\right)\right.$$

$$\left. - \frac{\hat{L}_x^2}{\rho^2\sin^2\theta} - \frac{\hat{L}_y^2}{\rho^2\cos^2\theta}\right]. \tag{9.2.12}$$

Here the operators $\hat{L}_x$ and $\hat{L}_y$ are the angular momenta associated with coordinates $\hat{\mathbf{x}}$ and $\hat{\mathbf{y}}$ respectively. Although the absolute values of x and y no longer coincide with the physical distances between the bodies of the system under study, the hyper-radius ρ is invariant under translations, rotations and all permutations, and is directly related to the overall size of the nucleus. It is proportional to the three-body moment of inertia as

$$\rho^2 = \sum_{i=1}^{3} A_i d_i^2, \tag{9.2.13}$$

where d_i are the distances of the three bodies i from the center of mass of the whole three-body system. For a c+n+n system, the rms radius of the three-body nucleus $r_{\mathrm{rms}}(c+n+n)$ relates to the rms radius of the core $r_{\mathrm{rms}}(c)$ and the mean square of the hyper-radius $\langle\rho^2\rangle$ through

$$r_{\mathrm{rms}}^2(c+n+n) = \frac{A_3}{A_3+2}r_{\mathrm{rms}}^2(c) + \frac{1}{A_3+2}\langle\rho^2\rangle, \tag{9.2.14}$$

where the mass of the core is $m_c = A_3 m$.

On the other hand, the *hyperangles* contain radial correlations, and are related to the relative magnitudes of the respective pair of Jacobi coordinates. As an example, and thinking in the T basis, $\theta \approx 0$ means that the two neutrons are much closer to each other than to the core. The other extreme happens if the two neutrons are far from each other but the core is sitting in between, as then $\theta \approx \pi/2$. Note that the value of θ has in itself *no* implications for either of the $\hat{\mathbf{x}}$ or $\hat{\mathbf{y}}$ angles.

Any one of the scaled Jacobi sets $(\mathbf{x}_k, \mathbf{y}_k)$ is sufficient to describe the system, so the development of formalism for two-neutron halo nuclei will preferentially use the Jacobian coordinates in the **T**-basis set, where $k = 3$. This is because of the antisymmetrization of the wave function between the two neutrons: when constructing the channels for the **T** basis the neutron antisymmetrization can be accommodated by selecting only partial waves with $l_x + S + T = $ odd, where l_x is the relative orbital angular momentum between the two neutrons, and S and T are the total spin and isospin of the two-neutron subsystem (see subsection 3.4.1). As the isospin for the two-neutron subsystem is $T = 1$, the condition is equivalent to $l_x + S = $ even.

9.2.2 Hyperspherical expansions

The transformation from a Jacobian coordinate system to the hyperspherical coordinate system is

$$(\mathbf{x}, \mathbf{y}, \sigma_1, \sigma_2) \equiv (x, y, \hat{\mathbf{x}}, \hat{\mathbf{y}}, \sigma_1, \sigma_2) \rightarrow (\rho, \theta, \hat{\mathbf{x}}, \hat{\mathbf{y}}, \sigma_1, \sigma_2) \equiv (\rho, \Omega_5, \sigma_1, \sigma_2),$$
(9.2.15)

and does not affect the angular and spin variables of the two neutrons. Note that $\hat{\mathbf{x}} = \hat{\mathbf{r}}$ and $\hat{\mathbf{y}} = \hat{\mathbf{s}}$.[2] We therefore expand on spherical harmonics for $\hat{\mathbf{r}}$ and $\hat{\mathbf{s}}$ as usual, using sums over angular-momentum quantum numbers (l_x, l_y) respectively.

Given orbital angular momenta (l_x, l_y), the spin of the core I, and the spin of the neutrons (σ_1, σ_2), we can write down the partial-wave expansion for the total wave function, specifying a coupling order such as

$$\Psi^{JM} = \sum_{l_x l_y \ell SjI} \psi_{l_x l_y}^{\ell SjIJ}(x, y) \left\{ ([Y_{l_x} \otimes Y_{l_y}]_\ell \otimes [X_{\sigma_1} \otimes X_{\sigma_2}]_S)_j \otimes \phi_I \right\}_{JM}.$$
(9.2.16)

Here we have assumed, as anticipated, that we are using the **T** basis.

The *two-dimensional radial* wave function $\psi(x, y)$ can now be expanded in the hyperspherical variables. We factorize the hyperangle and hyper-radial dependence of the wave function by means of a discrete sum over a new index K:

$$\psi_{l_x l_y}^{\ell SjIJ}(r, s) = \rho^{-\frac{5}{2}} \sum_K \chi_{Kl_x l_y}^{\ell SjIJ}(\rho) \; \varphi_K^{l_x l_y}(\theta).$$
(9.2.17)

The $\rho^{-5/2}$ is chosen here to cancel the ρ^5 factors in the first derivative term in Eq. (9.2.12). We further require that the new hyperangular functions $\varphi_K^{l_x l_y}(\theta)$ are eigensolutions for some eigenvalue B of the hyperangular equation

$$\left[\frac{1}{\sin^2 2\theta} \frac{\partial}{\partial \theta} \left(\sin^2 2\theta \frac{\partial}{\partial \theta} \right) - \frac{l_x(l_x+1)}{\sin^2 \theta} - \frac{l_y(l_y+1)}{\cos^2 \theta} \right] \varphi_K^{l_x l_y}(\theta) = -B \varphi_K^{l_x l_y}(\theta)$$
(9.2.18)

that are regular at $\theta = 0$ and $\theta = \pi/2$. The eigenvalues of this equation are $B = K(K+4)$ for $K = l_x + l_y + 2n$, with integer $n = 0, 1, 2, \ldots$, and the corresponding eigenfunctions are

$$\varphi_K^{l_x l_y}(\theta) = N_K^{l_x l_y} (\sin \theta)^{l_x} (\cos \theta)^{l_y} P_n^{l_x + \frac{1}{2}, l_y + \frac{1}{2}}(\cos 2\theta),$$
(9.2.19)

---

[2] Note that in this chapter, due to the need for additional coordinates to describe the reaction, s is an internal coordinate of the projectile, and spins are denoted by σ. This differs from previous chapters where s was used for spin.

where $P_n^{l_x+1/2,l_y+1/2}(\cos 2\theta)$ are Jacobi polynomials. The $N_K^{l_x l_y}$ are normalization factors so that the functions $\varphi_K^{l_x l_y}(\theta)$ form an orthonormal set with weight factor $\sin^2\theta\cos^2\theta$:

$$\int_0^{\pi/2} \varphi_K^{l_x l_y}(\theta)\varphi_{K'}^{l_x l_y}(\theta)\,\sin^2\theta\cos^2\theta\,\mathrm{d}\theta = \delta_{KK'}. \qquad (9.2.20)$$

This discrete expansion therefore introduces a new quantum number – the *hyperangular momentum K* – directly related to the order of the corresponding Jacobi polynomial. The integer n gives the number of nodes in $\varphi_K^{l_x l_y}(\theta)$ as a function of θ (excluding $\theta = 0$ and $\pi/2$).

We can now define the hyperharmonic basis functions, functions of all the $(\Omega_5,\sigma_1,\sigma_2,\boldsymbol{\xi})$ excluding only the dependence on ρ, as

$$\mathcal{Y}_{Kl_x l_y}^{\ell S j I,JM}(\Omega_5,\sigma_1,\sigma_2,\boldsymbol{\xi}) = \varphi_K^{l_x l_y}(\theta)\,\left\{([Y_{l_x}\otimes Y_{l_y}]_\ell \otimes [X_{\sigma_1}\otimes X_{\sigma_2}]_S)_j \otimes \phi_I\right\}_{JM} \qquad (9.2.21)$$

allowing the total wave function to be summarized as

$$\Psi^{JM} = \rho^{-\frac{5}{2}}\sum_{l_x l_y \ell S j I,K} \chi_{Kl_x l_y}^{\ell S I j,J}(\rho)\,\mathcal{Y}_{Kl_x l_y}^{\ell S I j,JM}(\Omega_5,\sigma_1,\sigma_2,\boldsymbol{\xi}). \qquad (9.2.22)$$

9.2.3 Coupled hyper-radial equations

When the expansions (9.2.16, 9.2.17, 9.2.19) of the wave functions are substituted in the three-body Schrödinger equation (9.2.6), we obtain a set of coupled equations, similar to those seen in Chapter 3:

$$\left(-\frac{\hbar^2}{2m}\left[\frac{\mathrm{d}^2}{\mathrm{d}\rho^2} - \frac{(K+\frac{3}{2})(K+\frac{5}{2})}{\rho^2}\right] - E\right)\chi_\gamma^J(\rho) + \sum_{\gamma'} V_{\gamma\gamma'}(\rho)\chi_{\gamma'}^J(\rho) = 0, \qquad (9.2.23)$$

upon rewriting using the multi-index $\gamma = \{l_x l_y \ell S j I K\}$. Here, the coupling potentials depend on the sum of the three pairwise potentials V_{ij} as

$$V_{\gamma'\gamma}(\rho) = \langle \mathcal{Y}_{\gamma'}(\Omega_5,\sigma_1,\sigma_2,\boldsymbol{\xi}) \mid \sum_{j>i=1}^{3} V_{ij}(\rho,\Omega_5,\sigma_1,\sigma_2,\boldsymbol{\xi}) \mid \mathcal{Y}_\gamma(\Omega_5,\sigma_1,\sigma_2,\boldsymbol{\xi})\rangle. \qquad (9.2.24)$$

In these coupled hyper-radial equations, the standard two-body centrifugal barrier is seen to be replaced by an effective centrifugal repulsive potential that depends on the hyperangular momentum K with equivalent L-value of $\mathcal{L} = K + 3/2$. In contrast to the two-body case, the barrier does not now vanish when the two valence neutrons are in the $s_{1/2}$ orbital and $K = 0$. This centrifugal barrier contains not only the *single-particle centrifugal barriers* associated with each variable, but an added repulsion term reflecting the difficulty of finding both neutrons close to the core simultaneously.

For three-body bound states in Borromean systems with finite-range two-body interactions, the three-body asymptotic forms are easily specified [9] as an exponential decay similar to that of the two-body single particle case:

$$\chi_\gamma(\rho) \overset{\rho \to \infty}{\to} \exp(-\kappa\rho), \quad \text{where} \quad \kappa = \sqrt{\frac{2m|E|}{\hbar^2}}. \tag{9.2.25}$$

However, if any of the two-body subsystems is bound when the third is removed, the three-body asymptotics in one part of configuration space will be ruled by the two-body asymptotics for the bound state moving relative to the third body, and no simple representation for the boundary conditions can be found in the hyperspherical coordinate system. The area of configuration space where Eq. (9.2.25) is valid is very close to the total space if [9] the ratio of the two-body to three-body binding energies E_{2b}/E_{3b} is close to zero, which makes it specially suitable for Borromean systems. Since three-body halo systems are weakly bound, the tail of the wave function offers a large contribution to most physical observables, contrary to standard nuclei. Therefore, if one is to calculate good estimates for observables in three-body halo systems, it is vital to expand not only where the potentials are non-zero, but also to treat the asymptotic forms correctly.

The coupled equations (9.2.23) may be solved by the iterative method of subsection 6.4.1 to find the eigensolution which exists only for specific eigen-energies E, or they may be solved by expanding on some convenient square-integrable basis, as long as the basis is sufficient to describe the long-range behavior with ρ. Both methods are capable of solving Eq. (9.2.23) to find three-body wave functions $\Phi_0(\mathbf{x}, \mathbf{y})$ for a ground state with $E < 0$.

For the numerical solutions it is useful to know the asymptotic behavior of the three-body couplings. For short-range two-body interactions, the three-body potential behaves as $V_{\gamma\gamma'}(\rho \to \infty) \sim \rho^{-n}$ with $n \geq 3$, with $n = 3$ for the diagonal terms. If the valence particles are charged, we also need to include the Coulomb forces. The slow rate of decay reflects the peculiar feature of three-body systems, where two particles can still interact when at large distances from the third. The long tail for the coupling terms has numerical implications, namely that a larger radial range will be needed for the calculations.

The three-body Schrödinger equation is usually solved, as here, in the $\mathbf{T}$ coordinate system. However, the n-core interaction matrix elements depending on

r_{32}, for example, are most simply calculated in the **Y** coordinate system since the scaled Jacobi coordinate $\mathbf{x}_1$ is proportional to the $\mathbf{r}_{32}$ vector. One therefore needs to determine the coefficients relating the components of the wave function in one Jacobi coordinate system with the components in the other, and this relation is the linear combination

$$\mathcal{Y}^{LSjI}_{Kl^T_x l^T_y}(T) = \sum_{l^Y_x l^Y_y} \langle l^T_x l^T_y | l^Y_x l^Y_y \rangle_L \; \mathcal{Y}^{LSjI}_{Kl^Y_x l^Y_y}(Y). \tag{9.2.26}$$

The coefficients $\langle l^T_x l^T_y | l^Y_x l^Y_y \rangle_L$, called the Raynal-Revai coefficients [10], have analytic forms. Their calculation is based on seeing the transformation between the two coordinate systems as a six-dimensional rotation. The relation between the **Y** coordinates $(\mathbf{x}^Y, \mathbf{y}^Y)$, labelled (1) in Fig. 9.3, and the **T** coordinates $(\mathbf{x}^T, \mathbf{y}^T)$, labeled (3) in Fig. 9.3, can be expressed as

$$\mathbf{x}^T = -\cos\phi \, \mathbf{x}^Y + \sin\phi \, \mathbf{y}^Y \tag{9.2.27}$$

$$\mathbf{y}^T = -\sin\phi \, \mathbf{x}^Y - \cos\phi \, \mathbf{y}^Y. \tag{9.2.28}$$

The rotation angle is related to the mass of the constituents of a two-neutron three-body system by

$$\cos\phi = \sqrt{\frac{A_3}{2 + 2A_3}} \quad \text{and} \quad \sin\phi = \sqrt{\frac{2 + A_3}{2 + 2A_3}}, \tag{9.2.29}$$

for core mass A_3. Thus, when calculating the matrix elements for the n-core interaction, there are three steps involved: transforming the wave function from the **T** basis to the **Y** basis, performing the calculation of the matrix elements in the **Y**-coordinate system and rotating back from the **Y** basis to the **T** basis.

9.2.4 Pauli principle

Since the microscopic detail of the core is not specified in this three-body model, it is difficult to fully antisymmetrize the three-body wave function. Antisymmetrization can be approximately taken into account through Pauli blocking, to guarantee that the valence nucleons do not occupy the lowest core+n states that are already filled by the core nucleons. In the simple case of ^{6}He, the valence neutrons should not be in the lowest $s_{1/2}$ orbital, since the four nucleons in the α core already fill it up. A crude way of taking into account the Pauli principle is to add by hand a repulsive short-range interaction in the s-wave channel to keep the valence neutrons out. A more systematic way of introducing this same effect is by generating the supersymmetric potential [11]. Alternatively, we can project the forbidden states out from the existing model space before diagonalization [12]. These methods are not exactly equivalent, and a comparison of various Pauli blocking methods is given in [13].

9.3 Three-body continuum

One can solve the Eq. (9.2.23) for positive energies, with the appropriate scattering boundary conditions, to obtain three-body scattering states $\Psi(\mathbf{x}, \mathbf{y}; E)$ for $E > 0$, as well as the bound states $\Phi(\mathbf{x}, \mathbf{y})$ for $E < 0$ as outlined in Section 9.2. This section provides a summary of three-body scattering states needed for reactions. More detail can be found in [14, 15]. Here we consider the full dynamics of a core+n+n system, but ignore the spins σ or I of the three bodies, so here L is the total angular momentum. Unless otherwise stated, all expressions refer to the **T** basis.

We first define the Jacobi momenta for the system considering core $= 3$. If $\mathbf{k}_1, \mathbf{k}_2, \mathbf{k}_3$ are the momenta of the two nucleons and the core in the center-of-mass frame of the whole nucleus (so $\mathbf{k}_1 + \mathbf{k}_2 + \mathbf{k}_3 = 0$), then the momenta $(\mathbf{k}_x, \mathbf{k}_y)$ conjugate to the Jacobi coordinates $(\mathbf{x}, \mathbf{y})$ are

$$\mathbf{k}_x = \sqrt{\mu_{12}/m}(\mathbf{k}_2 - \mathbf{k}_1),$$
$$\mathbf{k}_y = \sqrt{\mu_{(12)3}/m}\left(\frac{\mathbf{k}_3}{A_3} - \frac{\mathbf{k}_1 + \mathbf{k}_2}{2}\right), \tag{9.3.1}$$

so the *hyper-radial momentum* (or 'hypermomentum') defined by

$$\kappa^2 = k_x^2 + k_y^2 \tag{9.3.2}$$

is the same from all three sets of Jacobi coordinates. The total energy of the system is given by $E = \epsilon_x^2 + \epsilon_y^2 \equiv \frac{\hbar^2}{2m}(k_x^2 + k_y^2) = \hbar^2\kappa^2/2m$.

The three-body incoming plane wave can be expressed in terms of the scaled Jacobi coordinates $(\mathbf{x}, \mathbf{y})$ and their conjugate momenta $(\mathbf{k}_x, \mathbf{k}_y)$ by

$$\frac{1}{(2\pi)^3} \exp \mathrm{i}(\mathbf{k}_x\cdot\mathbf{x} + \mathbf{k}_y\cdot\mathbf{y}) = \frac{1}{(\kappa\rho)^2} \sum_{KLMl_xl_y} \mathrm{i}^K J_{K+2}(\kappa\rho)\mathcal{G}_\gamma^{LM}(\Omega_5^\rho)\mathcal{G}_\gamma^{LM}(\Omega_5^\kappa)^*, \tag{9.3.3}$$

where the functions $\mathcal{G}_\gamma^{LM}(\Omega_5) = \varphi_K^{l_xl_y}(\theta)\,[Y_{l_x} \otimes Y_{l_y}]_{LM}$ may use either coordinate or momentum angles for the two spherical harmonics with $\tan\theta = y/x$ or k_y/k_x, respectively.

Under the influence of interactions, the incoming three-body wave function can be written as

$$\Psi^{(+)}(\mathbf{x}, \mathbf{y}, \hat{\mathbf{k}}_x, \hat{\mathbf{k}}_y, \theta_\kappa) = \frac{1}{(\kappa\rho)^{5/2}} \sum_{L\gamma\gamma'} i^K \chi^L_{\gamma\gamma'}(\kappa\rho) \mathcal{G}^{LM}_\gamma(\Omega^\rho_5) [\mathcal{G}^{LM}_\gamma(\Omega^\kappa_5)]^* ,$$

(9.3.4)

for wave functions $\chi^J_{\gamma\gamma'}(\kappa\rho)$ that are no longer Bessel functions. For uncharged bodies, these hyper-radial continuum wave functions behave asymptotically in the standard form

$$\chi^L_{\gamma\gamma_i}(\kappa\rho) \rightarrow \frac{i}{2} [\delta_{\gamma\gamma_i} H^-_{K+3/2}(0, \kappa\rho) - \mathbf{S}^L_{\gamma\gamma_i} H^+_{K+3/2}(0, \kappa\rho)],$$

(9.3.5)

when there is a plane-wave component in channel γ_i. Since the Coulomb functions H^+ and H^- as defined in Box 3.1 behave at large distances as $e^{\pm i\kappa\rho}$, they describe out- and in-going three-body spherical waves. The $\mathbf{S}^L_{\gamma\gamma_i}$ is the three-body to three-body scattering matrix. Phase shifts can be determined through the diagonal elements as $\mathbf{S}^L_{\gamma\gamma} = e^{i\delta^L_\gamma}$, or as eigenphases of the whole S matrix. The above scattering wave functions are normalized as

$$\int_0^\infty \chi^L_{\gamma\gamma_i}(\kappa\rho) \chi^L_{\gamma\gamma_i}(\kappa'\rho) \mathrm{d}\rho = \frac{\pi}{2} \delta(\kappa - \kappa').$$

(9.3.6)

The constituent energies depend on the three-body continuum wave functions through the Jacobi polynomials and powers of the relative energies as $\epsilon_x^{l_x/2} \epsilon_y^{l_y/2} P^{l_x+1/2,l_y+1/2}_{(K-l_x-l_y)/2}((\epsilon_y - \epsilon_x)/E)$.

The wave function for three-body elastic scattering has the general form

$$\hat{\mathbf{A}}_{nn} \frac{1}{(2\pi)^2} \exp i(\mathbf{k}_x \cdot \mathbf{x} + \mathbf{k}_y \cdot \mathbf{y}) + f(E, \Omega^{\kappa_i}_5, \Omega^{\kappa_f}_5) \frac{e^{i\kappa\rho}}{\rho^{5/2}},$$

(9.3.7)

where the scattering amplitude can be found from the S matrix by

$$f(E, \Omega^{\kappa_i}_5, \Omega^{\kappa_f}_5) = \frac{i}{2} \frac{e^{-\frac{3\pi i}{4}}}{\kappa^{5/2}} \sum_{\gamma\gamma_i LM} (\delta_{\gamma\gamma_i} - \mathbf{S}^L_{\gamma\gamma_i}) [\mathcal{G}^{LM}_{\gamma_i}(\Omega^{\kappa_i}_5)]^* \mathcal{G}^{LM}_\gamma(\Omega^{\kappa_f}_5),$$

(9.3.8)

and $\Omega^{\kappa_i}_5$ and $\Omega^{\kappa_f}_5$ are initial and final angles and hyperangles, associated with the momenta $(\mathbf{k}^i_x, \mathbf{k}^i_y)$ and $(\mathbf{k}^f_x, \mathbf{k}^f_y)$ respectively. Note that in Eq. (9.3.7) we have introduced the antisymmetrization operator $\hat{\mathbf{A}}_{nn}$ which picks out channels that are allowed after antisymmetrization.

Finally, the elastic differential cross section for scattering in the three-body continuum, if such an experiment could be performed, would be

$$\frac{d^5\sigma(3)}{d\Omega_5^{K_f}} = |f(E, \Omega_5^{K_i}, \Omega_5^{K_f})|^2. \tag{9.3.9}$$

As usual, if spins are included we will need to perform an incoherent sum over all final spin projections and an average over all initial spin projections.

9.4 Reactions with three-body projectiles

Reactions with three-body projectiles are particularly challenging as they consist of a four-body system and thus exact solutions would require solving the Faddeev-Yakubowski equations [16]. As in the two-body projectile case, it is usual to factorize the four-body wave function into a part describing the relative motion between the projectile and the target, and another consisting of the three-body projectile wave function. Then, the three-body projectile wave functions introduced in this chapter can be used in calculating the T matrix and the cross sections for various processes. We will briefly mention a few models that have been developed specifically to handle three-body projectiles under different approximations. In this section we assume that subscripts (1) and (2) refer to the two halo nucleons while (3) refers to the core.

The simplest approximation to calculating the elastic cross section for a three-body projectile p incident on a target t could be obtained by solving the single-channel Schrödinger equation with the Watanabe potential, consisting of a single-folding over the three-body wave function

$$U_{pt}(R) = \langle \Phi_0(\mathbf{x}, \mathbf{y}) | U_{pt} | \Phi_0(\mathbf{x}, \mathbf{y}) \rangle, \tag{9.4.1}$$

where U_{pt} is the sum of the optical interactions of each of the three fragments in the projectile with the target:

$$U_{pt} = U_{1t} + U_{2t} + U_{3t}. \tag{9.4.2}$$

Typically, this is the starting point for further improvements, such as by adiabatic models in subsection 9.4.2, or by CDCC calculations in subsection 9.4.4.

9.4.1 Born approximations

One can calculate the elastic breakup as a 1-step inelastic excitation of the projectile into the continuum. For a Borromean system, where there are no bound two-body states, at high energies this excitation into the continuum would most probably happen in one step. In this case the transition matrix element for the process $p + t \rightarrow 1 + 2 + 3 + t$ can be written in the DWBA as:

$$\mathbf{T}_{fi} = \langle \Psi^{(-)}(\mathbf{x}, \mathbf{y}, \hat{\mathbf{k}}_x, \hat{\mathbf{k}}_y, \theta_\kappa) \chi_f^{(-)}(\mathbf{R}, \mathbf{k}_f) | U_{pt} | \Phi_0(\mathbf{x}, \mathbf{y}) \chi_i^{(+)}(\mathbf{R}, \mathbf{k}_i) \rangle. \qquad (9.4.3)$$

Studies of correlations in the three-body continuum is an ongoing research topic and more detail can be found for example in [14, 15, 17]. Cross sections for the excitation of ^{6}He on ^{12}C at 240 MeV/u, studied in detail in [17], are shown in Fig. 9.5. An explicit analysis is given for (a) the various continuum final states of ^{6}He, (b) the comparison of nuclear and Coulomb contributions, as well as for (c) elastic and inelastic processes.

9.4.2 Adiabatic models

A four-body adiabatic model has been developed for reactions at relatively high energy [18], based on the approximations described in subsection 7.1.1, whereby the internal coordinates of the projectile (x, y), or (r, s), are essentially frozen during the reaction: the coordinates illustrated in Fig. 9.6. The full four-body Schrödinger equation is

$$[\hat{T}_R + H_{3b}(\mathbf{r}, \mathbf{s}) + U_{pt}(\mathbf{r}, \mathbf{s}, \mathbf{R}) - E_{pt}]\Psi(\mathbf{r}, \mathbf{s}, \mathbf{R}) = 0, \qquad (9.4.4)$$

where $H_{3b}(\mathbf{r}, \mathbf{s})$ is the three-body Hamiltonian of the projectile described in Section 9.2, the $U_{pt}(\mathbf{r}, \mathbf{s}, \mathbf{R})$ is again the sum of the two-body potentials between each projectile fragment and the target, and E_{pt} is the center-of-mass energy in the p+t incident relative motion. As in subsection 7.1.1, we expand the wave function in terms of the three-body ground state Φ_0 and all inelastic and breakup states Φ_i:

$$\Psi = \Phi_0(\mathbf{r}, \mathbf{s})\psi_0(\mathbf{R}) + \sum_{i=1}^{\infty} \Phi_i(\mathbf{r}, \mathbf{s})\psi_i(\mathbf{R}), \qquad (9.4.5)$$

where the Φ_i are orthonormal eigenstates of the projectile satisfying

$$H_{3b}(\mathbf{r}, \mathbf{s})\Phi_i(\mathbf{r}, \mathbf{s}) = \epsilon_i \Phi_i(\mathbf{r}, \mathbf{s}). \qquad (9.4.6)$$

Here, i is a generic index covering both discrete and continuous parts of the spectrum. Using this form for the wave function in Eq. (9.4.4), we obtain

$$\sum_{i=0}^{\infty} [\hat{T}_R + U_{pt}(\mathbf{r}, \mathbf{s}, \mathbf{R}) - (E - \epsilon_i)]\Phi_i(\mathbf{r}, \mathbf{s})\psi_i(\mathbf{R}) = 0. \qquad (9.4.7)$$

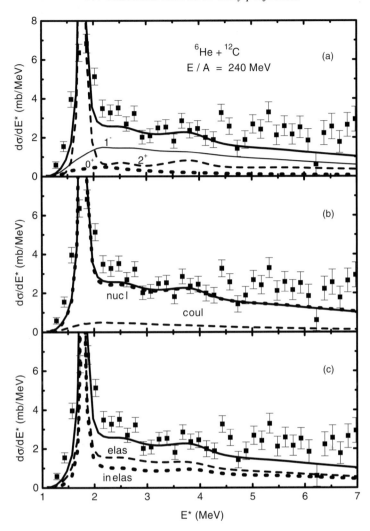

Fig. 9.5. Comparison of DWBA predictions for ^{6}He inelastic scattering on ^{12}C at 240 MeV/u with experimental data, as a function of excitation energy [17]: (a) contribution of different final states; (b) contribution of Coulomb versus nuclear; (c) contribution of elastic and inelastic. Reprinted with permission from S. N. Ershov, *et al.*, *Phys. Rev. C* **64** (2001) 064609. Copyright (2001) by the American Physical Society.

The adiabatic approximation consists of replacing all $\epsilon_i \to \epsilon_0$ in Eq. (9.4.7), which then simplifies to

$$[\hat{T}_R - (E - \epsilon_0)]\Psi^{\mathrm{ad}}(\mathbf{r}, \mathbf{s}, \mathbf{R}) = U_{pt}(\mathbf{r}, \mathbf{s}, \mathbf{R})\Psi^{\mathrm{ad}}(\mathbf{r}, \mathbf{s}, \mathbf{R}), \qquad (9.4.8)$$

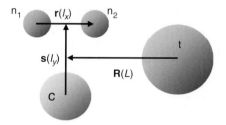

Fig. 9.6. Coordinates for a four-body adiabatic model.

where the adiabatic wave function is a superposition of many projectile excited states. We represent this as

$$\Psi^{\text{ad}} = \Phi_0(\mathbf{r}, \mathbf{s})\psi_0^{\text{ad}}(\mathbf{R}) + \sum_i \Phi_i(\mathbf{r}, \mathbf{s})\Psi_i^{\text{ad}}(\mathbf{R}). \tag{9.4.9}$$

for some new functions $\Psi_i^{\text{ad}}(\mathbf{R})$.

The advantage of the method is that in Eq. (9.4.8) the internal coordinates of the projectile appear only parametrically. The equation can be therefore solved as a two-body dynamical problem for each fixed $(\mathbf{r}, \mathbf{s})$ combination that appears in the projectile initial bound state.

We use a partial-wave decomposition of the adiabatic wave function:

$$\Psi^{\text{ad}} = \sum_{\gamma \gamma_i JM} A_{\gamma_i}^{JM}(r, s) \frac{\psi_{\gamma \gamma_i}^J(R; r, s)}{R} \left[\left[Y_{l_x}(\hat{\mathbf{r}}) \otimes Y_{l_y}(\hat{\mathbf{s}}) \right]_{\ell} \otimes Y_L(\hat{\mathbf{R}}) \right]_{JM}.$$
$$\tag{9.4.10}$$

Here γ_i, γ denote quantum number sets $(l_x, l_y)\ell, L$ for the incoming and outgoing channels respectively. The coefficients $A_{\gamma_i}^{JM}(r, s)$ are chosen to reproduce the boundary condition of an incident plane wave. The resulting coupled radial equations, obtained after substituting Eq. (9.4.10) into Eq. (9.4.8), are

$$\left[-\frac{\hbar^2}{2\mu_{pt}} \left(\frac{d^2}{dR^2} - \frac{L(L+1)}{R^2} \right) - (E - \epsilon_0) \right] \psi_{\gamma \gamma_i}^J(R; r, s)$$
$$= \sum_{\gamma'} \langle \gamma J | U_{pt}(\mathbf{r}, \mathbf{s}, \mathbf{R}) | \gamma' J \rangle \, \psi_{\gamma' \gamma_i}^J(R; r, s), \tag{9.4.11}$$

with the radial wave functions satisfying standard boundary conditions

$$\psi_{\gamma \gamma_i}^J(R; r, s) \rightarrow \delta_{\gamma \gamma_i} F_L(k_i R) + \mathbf{T}_{\gamma \gamma_i}^J(r, s) H_L^+(kR), \tag{9.4.12}$$

giving therefore T-matrix elements depending on the r, s parameters. The final scattering amplitude, and consequently the cross sections, can be constructed in terms of these $\mathbf{T}^J_{\gamma \gamma_i}(r, s)$ by averaging over the internal coordinates $(\mathbf{r}, \mathbf{s})$. See [18] for more detail.

9.4.3 Three-body eikonal models

As we have seen in Chapter 7, for the high-energy limit and for forward angle scattering, a straight line (eikonal) approximation is adequate. An eikonal model has been developed specifically for projectiles with three-body structure [19].

For the description of this model it is convenient to represent all positions/ distances in cylindrical coordinates, since the projection onto the impact parameter plane plays an important role. These are represented in Fig. 9.7: $\mathbf{R}, \mathbf{r}$ and $\mathbf{s}$ are the coordinates of the projectile's center of mass, the n–n separation and the core–nn separation, respectively. When the incident beam is in the $+z$ direction, the corresponding projections onto the impact-parameter (xy) plane are $\mathbf{b}, \mathbf{b}_r$ and $\mathbf{b}_s$.

In the Glauber approximation the elastic scattering amplitude is

$$f_{\mathrm{el}}(\mathbf{q}) = \frac{-iK_0}{2\pi} \int d\mathbf{b}\, e^{i\mathbf{q}\cdot\mathbf{b}} \int d\mathbf{s} \int d\mathbf{r} |\Phi_0|^2 \left(e^{i\chi(\mathbf{b},\mathbf{b}_r,\mathbf{b}_s)} - 1 \right), \qquad (9.4.13)$$

where K_0 is the incident wave number for the relative motion, and $\mathbf{q}$ is the momentum transfer, by a derivation exactly analogous to that in Section 7.2.

The Glauber phase shift is the sum of the phase shift of each fragment,

$$\chi(\mathbf{b}, \mathbf{b}_r, \mathbf{b}_s) = \chi_c(|\mathbf{b}_c|) + \chi_n(|\mathbf{b}_1|) + \chi_n(|\mathbf{b}_2|). \qquad (9.4.14)$$

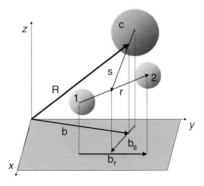

Fig. 9.7. Coordinates of the projectile-target four-body system, and their projections on the impact parameter xy-plane.

Each phase is defined in terms of the corresponding fragment-target optical potential, with the appropriate coordinate including recoil:

$$\chi_c(|\mathbf{b}_c|) = -\frac{\mu}{\hbar^2 K_0} \int_{-\infty}^{\infty} U_{ct}(|\mathbf{R} - \frac{2}{A_c+2}\mathbf{s}|) \, dR_z$$

$$\chi_n(|\mathbf{b}_1|) = -\frac{\mu}{\hbar^2 K_0} \int_{-\infty}^{\infty} U_{nt}(|\mathbf{R} + \frac{A_c}{A_c+2}\mathbf{s} + \frac{1}{2}\mathbf{r}|) \, dR_z$$

$$\chi_n(|\mathbf{b}_2|) = -\frac{\mu}{\hbar^2 K_0} \int_{-\infty}^{\infty} U_{nt}(|\mathbf{R} + \frac{A_c}{A_c+2}\mathbf{s} - \frac{1}{2}\mathbf{r}|) \, dR_z, \qquad (9.4.15)$$

where we consider only central potentials between the fragments and the target. The integration is made over the z-projection of $\mathbf{R}$. Here $\mathbf{b}_c$, $\mathbf{b}_1$ and $\mathbf{b}_2$ are the projections of the core and neutron coordinates (relative to the target) on the impact parameter plane. Both the reduced mass μ and wave number K_0 may be calculated using relativistic kinematics as in subsection 2.3.3, and in [20].

The Glauber phase shift $\chi(\mathbf{b}, \mathbf{b}_r, \mathbf{b}_s)$ depends on vectors which lie in the impact parameter plane, consequently the integration in the z components of Eq. (9.4.13) involves only the projectile density. It is convenient then to evaluate the projectile density projected onto the impact parameter plane, which is usually denoted by the *Glauber thickness function*,

$$\xi(\mathbf{b}_r, \mathbf{b}_s) = \int_{-\infty}^{\infty} ds_z \int_{-\infty}^{\infty} dr_3 \, \langle \, |\Phi_0(\mathbf{x}, \mathbf{y})|^2 \rangle_{\text{spin}}, \qquad (9.4.16)$$

where $\langle \ \rangle_{\text{spin}}$ denotes an integration over spin coordinates, and s_z and r_z correspond to the z-projection of $\mathbf{s}$ and $\mathbf{r}$ respectively.

In terms of the thickness function, the elastic amplitude is:

$$f_{\text{el}}(\theta) = \frac{-iK_0}{2\pi} \int d\mathbf{b} \, e^{i\mathbf{q}\cdot\mathbf{b}} \int d\mathbf{b}_r \int d\mathbf{b}_s \, \xi(\mathbf{b}_r, \mathbf{b}_s) \left(e^{i\chi(\mathbf{b},\mathbf{b}_r,\mathbf{b}_s)} - 1 \right), \quad (9.4.17)$$

where the momentum transfer and the scattering angle are related through $q = 2K_0 \sin(\theta/2)$. So far, the phase shift χ is due to the strong interaction only. The Coulomb phase χ_{Coul}^a due to the Coulomb force between the core and the target can be determined from a screened Coulomb potential of screening radius a_s, as described in Section 7.2. Upon adding the point-charge Coulomb amplitude to the Glauber amplitude, with elastic Sommerfeld parameter η, we have [21]

$$f_{\text{el}}(\theta) = e^{i\chi_s^a} \left\{ f_c(\theta) - \frac{iK_0}{2\pi} \int d\mathbf{b} \, e^{i\mathbf{q}\cdot\mathbf{b} + 2i\eta_0 \ln K_0 b} \left(e^{i\chi_{\text{opt}}(\mathbf{b})} - 1 \right) \right\}, \qquad (9.4.18)$$

where the new elastic Glauber phase is

$$e^{i\chi_{\text{el}}(\mathbf{b})} = \int d\mathbf{b}_s \int d\mathbf{b}_r \, \xi(\mathbf{b}_s, \mathbf{b}_r) \left(e^{i(\chi(\mathbf{b},\mathbf{b}_s,\mathbf{b}_r) + \chi_\rho(\mathbf{b},\mathbf{b}_s) - 2\eta_0 \ln K_0 b)} \right). \qquad (9.4.19)$$

The overall phase factor χ_s^a (the only effect of the screening radius) has no effect on the cross sections. In this model, the only necessary ingredients to calculate the elastic scattering cross section are the projectile three-body density and the neutron-target and core-target optical potentials. This model has been applied to the elastic scattering of ^{11}Li on ^{12}C [19].

Another high-energy theory, of application for reactions with three-body projectiles, is that of the multiple scattering expansion of the total transition amplitude (MST) [22, 23]. Although this framework is beyond the scope of this book, we note several applications of MST to halo nuclei [23].

9.4.4 Four-body CDCC

For lower-energy regimes, the above DWBA or eikonal methods become inaccurate. There are several other quantum methods being developed specifically to handle three-body projectiles without high-energy approximations. All of these are based on some form of discretization of the continuum, so we include them in this section on four-body CDCC.

As discussed in Chapter 8, it is convenient to have a square-integrable representation of the continuum. One important difference in the wave function of three-body projectiles compared to the definition in Chapter 8 is that here the scattering wave function is always a coupled-channel wave function $\chi_{\gamma\gamma_i}^J(\rho)$. This hyper-radial function describes a system initially in channel γ_i being scattered to a final channel γ, where the label γ_i is the channel containing the plane-wave component in addition to the usual outgoing components. The asymptotic properties of this wave function are defined in Eq. (9.3.5). The corresponding bin wave functions are

$$\tilde{\chi}_{\gamma\gamma_i}^{J(n)}(\rho) = \sqrt{\frac{2}{\pi N_{\gamma_i}}} \int_{\kappa_{n-1}}^{\kappa_n} g_{\gamma_i}(\kappa) \chi_{\gamma\gamma_i}^J(\rho) \, d\kappa \qquad (9.4.20)$$

for some weight factor $g_{\gamma_i}(\kappa)$ whose square norm integrates over the bin to N_{γ_i}. Here $n = 1, \ldots, N_{bin}$ is the bin number, starting at $\kappa_0 = 0$ as the three-body breakup threshold. The full bin wave function describing the projectile for momentum bin n and plane wave in channel γ_i is the sum over all γ of the coupled-channels bins defined in Eq. (9.4.20). Coupled-channel bins were first introduced for modeling core excitation [24], and are still under development for three-body projectiles. In particular, no choice of $g_{\gamma_i}(\kappa)$ makes the bin wave functions $\tilde{\chi}_{\gamma\gamma_i}^{J(n)}(\rho)$ all real-valued, in contrast to two-body continuum states.

Another way of discretizing the three-body continuum is to use pseudostates. In particular, Gaussian expansions have been proposed as an efficient method for calculating reaction observables involving Borromean systems [25]. In that method, no hyperspherical expansion is used, but rather it is the product of Gaussians in the several Jacobi coordinates that provide the basis.

Similar to the Gaussian method, the transformed harmonic oscillator (THO) method uses a discrete square-integrable representation for the three-body continuum [26]. As in the standard THO, a local scale transformation is applied to each channel component of the bound-state wave function of the three-body system $\chi_{B\beta}(\rho)$ in order to obtain a seed for the THO basis $\chi_{0K}^{\mathrm{THO}}(s_\rho)$, from which the corresponding harmonic oscillator set is generated $\chi_{i\beta}^{\mathrm{THO}}(\rho)$. Recent applications [27] to the inelastic scattering of ^{6}He by protons show a comparative study of two pseudostate methods.

Exercises

9.1 Derive the rms radius expression of Eq. (9.2.14).

9.2 The K_{min} approximation reduces the coupled-channel hyper-radial Eq. (9.2.23) to a single equation for a dominant K-channel. Consider a simplified model of ^{6}He as two valence neutrons in p-waves outside the α core in the K_{min} approximation.

   (a) Work out the dominant K-channel, K_{min}.
   (b) Take a square well potential in hyper-radius, of size 3.5 fm and adjust the depth V_0 so that you reproduce the experimental binding energy of ^{6}He.
   (c) Determine the hyper-radial wave function and use it to calculate the average distance between the α and the two neutrons.

9.3 One model often used in reactions with ^{6}He is the dineutron model. Assume a *dineutron* as a particle with no internal energy but a finite size of 1 fm. Assume the interaction between the α core and the dineutron can be modeled by a square well potential of (physical) radius 1.8 fm. Adjust the depth to reproduce the binding energy and use the resulting wave function to estimate the radius. Compare your result with Ex. 9.2(c).

9.4 Using global potentials for the fragments and the ^{6}He wave function in the K_{min} approximation, derive the single-folded (Watanabe) potential for three-body ^{6}He on ^{209}Bi at $E = 21.4$ MeV.

9.5 Compare the potential derived in Ex. 9.4 with that coming from a direct fit to the data [28] (use e.g. SFRESCO to fit the optical potential).

9.6 Work out the eikonal phase shift for the three-body ^{6}He on ^{12}C and compare with that obtained with the dineutron model of Ex. 9.3.

9.7 ^{22}C is believed to be a halo nucleus with two valence neutrons. These neutrons could be in d-waves or in s-waves. Determine the relevant K-channel for having d^2 and s^2 configuration, in the K_{min} approximation. Take a square well in hyper-radius of size 4.0 fm.

   (a) For each configuration, d^2 and s^2, adjust the depth of the interaction to reproduce the binding energy of ^{22}C.

(b) Work out the wave function obtained in each case and study the consequences of the different structure on the rms radius of ^{22}C.

(c) For each configuration, determine the probability that the valence neutrons are found in the classically forbidden region. Which of the two cases is more likely to produce a halo nucleus?

References

[1] M. V. Zhukov, B. V. Danilin, D. V. Fedorov, J. M. Bang, I. J. Thompson and J. S. Vaagen, *Phys. Rep.* **231** (1993) 151.

[2] I. Tanihata, *J. Phys. G* **22** (1996) 157.

[3] T. H. Gronwall, *Phys. Rev.* **51** (1937) 655.

[4] C. D. Lin, *Phys. Rep.* **257** (1995) 1.

[5] L. M. Delves, *Nucl. Phys.* **9** (1959) 391; **20** (1962) 268.

[6] F. M. Nunes, J. A. Christley, I. J. Thompson, R. C. Johnson and V. D. Efros, *Nucl. Phys.* **A609** (1996) 43.

[7] F. M. Nunes, I. J. Thompson and J. A. Tostevin, *Nucl. Phys. A* **703** (2002) 593.

[8] I. J. Thompson, F. M. Nunes and B. V. Danilin, *Comp. Phys. Comm.* **161** (2004) 87.

[9] S. P. Merkuriev, C. Gignoux and A. Laverne, *Ann. Phys. (NY)* **99** (1976) 1.

[10] J. Raynal and J. Revai, *Nouvo Cimento* **A68** (1970) 612.

[11] D. Baye, *Phys. Rev. Lett.* **58** (1997) 2738; *J. Phys. A* **20** (1987) 5529.

[12] J. M. Bang and I. J. Thompson, *Phys. Lett.* **B279** (1992) 201.

[13] I. J. Thompson, B. V. Danilin, V. D. Efros, J. S. Vaagen, J. M. Bang and M. V. Zhukov, *Phys. Rev. C* **61** (2000) 24318.

[14] B. V. Danilin, T. Rodge, J. S. Vaagen, I. J. Thompson and M. V. Zhukov, *Phys. Rev. C* **69** (2004) 024609.

[15] B. V. Danilin, J. S. Vaagen, T. Rodge, E. N. Ershov, I. J. Thompson and M. V. Zhukov, *Phys. Rev. C* **73** (2006) 054002.

[16] L. D. Faddeev, *Zh. Eksp. Teor. Fiz.* **39**, (1960) [*Sov. Phys. JETP* **12**, (1961) 1014]; O. A. Yakubowsky, *Sov. J. Nucl. Phys.* **5** (1967) 937.

[17] S. N. Ershov, B. V. Danilin and J. S. Vaagen, *Phys. Rev. C* **62** (2000) 041001; *Phys. Rev. C* **64** (2001) 064609; *Phys. Rev. C* **74** (2006) 014603.

[18] J. A. Christley, J. S. Al-Khalili, J. A. Tostevin and R. C. Johnson, *Nucl. Phys. A* **624** (1997) 275.

[19] I. J. Thompson, J. S. Al-Khalili, J. A. Tostevin and J. M. Bang, *Phys. Rev. C* **47** (1993) R1364.

[20] W. R. Coker, L. Ray and G. W. Hoffman, *Phys. Lett.* **64B** (1976) 403.

[21] J. S. Al-Khalili and R. C. Johnson, *Nucl. Phys.* **A546** (1992) 622.

[22] M. L. Goldberger and K. M. Watson 1964, *Collision Theory*, New York: John Wiley and Sons.

[23] R. Crespo and R. C. Johnson, *Phys. Rev. C* **60** (1999) 034007.

[24] N. C. Summers, F. M. Nunes and I. J. Thompson, *Phys. Rev. C* **74** (2006) 014606.

[25] T. Matsumoto *et al.*, *Phys. Rev. C* **70** (2004) 061601.

[26] M. Rodríguez-Gallardo *et al.*, *Phys. Rev. C* **72** (2005) 024007.

[27] A. M. Moro, M. Rodríguez-Gallardo, R. Crespo and I. J. Thompson, *Phys. Rev. C* **75** (2007) 017603.

[28] A. R. Garcia *et al.*, *Phys. Rev. C* **76** (2007) 067603.

10

R-matrix phenomenology

The simplicities of natural laws arise through the complexities
of the language we use for their expression.

Eugene Wigner

10.1 R-matrix parameters

In the previous Section 6.5, we saw how the R matrix could be constructed from
a Hamiltonian and its potentials, using the wave functions of eigenstates in the
interior region $R \leq a$, the R-matrix radius. For scattering, the only properties
of the eigenstates used are their energies, and the values of their wave functions
at the R-matrix radius. The energies are the R-matrix *pole energies* e_p, and the
surface values of the wave functions $g_\alpha^p(R)$ give the *reduced width amplitudes*
$\gamma_{p\alpha} = \sqrt{t_\alpha/a}\, g_\alpha^p(a)$ where $t_\alpha = \hbar^2/2\mu_\alpha$. The R matrix is then easily constructed
at any desired scattering energy E by Eq. (6.5.25):

$$\mathbf{R}_{\alpha'\alpha}(E) = \sum_{p=1}^{P} \frac{\gamma_{p\alpha}\gamma_{p\alpha'}}{e_p - E}. \tag{10.1.1}$$

From this R matrix, the scattering S matrix can be found by Eq. (6.5.33):

$$\mathbf{S} = \frac{t^{\frac{1}{2}}\mathbf{H}^- - a\mathbf{R}t^{\frac{1}{2}}(\mathbf{H}^{-\prime} - \beta\mathbf{H}^-)}{t^{\frac{1}{2}}\mathbf{H}^+ - a\mathbf{R}t^{\frac{1}{2}}(\mathbf{H}^{+\prime} - \beta\mathbf{H}^+)} \quad \text{most generally,} \tag{10.1.2a}$$

$$= \frac{\mathbf{H}^- - a\mathbf{R}(\mathbf{H}^{-\prime} - \beta\mathbf{H}^-)}{\mathbf{H}^+ - a\mathbf{R}(\mathbf{H}^{+\prime} - \beta\mathbf{H}^+)} \quad \text{if } t_\alpha \text{ are all equal,} \tag{10.1.2b}$$

$$= \frac{\mathbf{H}^- - a\mathbf{R}\mathbf{H}^{-\prime}}{\mathbf{H}^+ - a\mathbf{R}\mathbf{H}^{+\prime}} \quad \text{when also } \beta = 0, \tag{10.1.2c}$$

and hence all the channel cross sections.

This suggests that, if our aim is not so much to start with a Hamiltonian,
but to fit a set of reaction measurements over a range of channels, energies and

angles, then all we need to do is to find the reduced width amplitudes $\gamma_{p\alpha}$ and the corresponding pole energies e_p. These parameters are sufficient to describe all asymptotic properties of the scattering wave functions, and hence all cross sections that may be measured. After fitting these parameters, it is comparatively easy to interpolate or extrapolate as desired in energies and angles. This fitting programme is called *R-matrix phenomenology* [1, 2, 3].

The *methods* of fitting data to find reduced widths and pole positions will be described in Chapter 15. For now, we presume that this fitting has been done, and in the present chapter discuss the interpretation of the results.

To facilitate the understanding of an R-matrix fit, it is useful to modify the derivation in Eq. (10.1.2) of the S matrix so that approximations may be made to give simpler formulae that can be more easily understood. These reformulations will refer in particular to **widths**, either **reduced**, **partial** or **total**, as well as to **shift functions** and **penetrabilities**. We will clarify the connection between R-matrix poles and **resonances** (which are S-matrix poles). To begin with, we examine the one-channel R-matrix theory in more detail, beginning with a simpler derivation of elastic phase shifts δ. Later we will derive a **level-matrix** formulation more suitable for multi-channel modeling.

10.2 Single-channel R matrix

10.2.1 Phase shifts from the one-channel R matrix

In the one-channel case, all the reduced masses in the t_α in Eq. (10.1.2) are equal, and matrices are only 1×1, so we find the S-matrix element according to Eq. (6.5.12):

$$
\begin{aligned}
\mathbf{S} &= \frac{H^- - a\mathbf{R}(H^{-\prime} - \beta H^-)}{H^+ - a\mathbf{R}(H^{+\prime} - \beta H^+)} \\
&= \frac{1 - a\mathbf{R}(H^{-\prime}/H^- - \beta)}{1 - a\mathbf{R}(H^{+\prime}/H^+ - \beta)} \frac{H^-}{H^+}.
\end{aligned} \tag{10.2.1}
$$

At a given scattering energy E, using the Wronskian $\dot{F}G - \dot{G}F = 1$,

$$
\frac{H^{+\prime}}{H^+} = \frac{G' + iF'}{G + iF} = k\frac{\dot{G} + i\dot{F}}{G + iF} = k\frac{\dot{F}F + \dot{G}G}{F^2 + G^2} + ik\frac{1}{F^2 + G^2}. \tag{10.2.2}
$$

Thus we can write

$$
a\frac{H^{+\prime}}{H^+} = S + iP \quad \text{and} \quad a\frac{H^{-\prime}}{H^-} = S - iP, \tag{10.2.3}
$$

by defining a *shift function*

$$S(E) = ka \frac{\dot{F}F + \dot{G}G}{F^2 + G^2} \tag{10.2.4}$$

and a *penetrability*

$$P(E) = \frac{ka}{F^2 + G^2}, \tag{10.2.5}$$

where the Coulomb functions F and G are all evaluated for argument $\rho = ka$. The wave number k and Sommerfeld parameter η have their usual relation to the c.m. scattering energy E.

The final factor in Eq. (10.2.1) can be rewritten

$$\frac{H^-}{H^+} = \frac{G - iF}{G + iF} = e^{2i\phi}, \tag{10.2.6}$$

$$\text{where} \quad \phi = -\arctan\frac{F}{G} \tag{10.2.7}$$

is called the *hard-sphere phase shift* since this would be the scattering phase shift if the wave function were forced to go to zero at $R = a$, as can be seen from Eq. (3.1.41).

In terms of the shift function S, penetrability P and hard-sphere phase shift ϕ, the **S** matrix is

$$\mathbf{S} = \frac{1 - \mathbf{R}(S - iP - a\beta)}{1 - \mathbf{R}(S + iP - a\beta)} \, e^{2i\phi} \tag{10.2.8}$$

$$= \frac{1 - \mathbf{R}(S - a\beta) + i\mathbf{R}P}{1 - \mathbf{R}(S - a\beta) - i\mathbf{R}P} \, e^{2i\phi}. \tag{10.2.9}$$

From $\mathbf{S} = e^{2i\delta}$, we see the scattering phase shift is $\delta = \phi + \delta_R$, where the *R-matrix phase* δ_R is

$$\delta_R = \arctan \frac{\mathbf{R}P}{1 - \mathbf{R}(S - a\beta)}. \tag{10.2.10}$$

For single-channel scattering, this is a practical formula to calculate the phase shifts. Often we will abbreviate $S^0 \equiv S - a\beta$, but not forget that S, S^0 and P are all functions of the scattering energy E.

In some special cases:

- For bound states at negative energies E, the penetrability is zero, but the shift function can still be defined as the logarithmic derivative of the Whittaker function.
- For neutrons and photons and $ka \ll L$, the penetrability may be found using Eq. (3.1.18), namely $P \propto k^{2L+1}$. For *s*-wave neutrons, this simplifies to $P_{L=0} = ka$, with the shift function $S_0 = 0$, and the hard-sphere phase shift is $\phi_0 = -\arctan ka$ in this partial wave.

- For charged particles with $ka \ll L$, the penetrability is dominated by the large value of the irregular function G for small ka. By Eq. (3.1.65), we find $P_L \approx (2L + 1)^2$ $(ka)^{2L+1}C_L(\eta)^2$. For s-wave scattering, this simplifies to $P_0 = 2\pi \eta ka/(\exp(2\pi \eta) - 1)$ $\approx 2\pi \eta ka \exp(-2\pi \eta)$ (noting that ηk is energy independent).

10.2.2 Isolated poles in single-channel scattering

The simplest R-matrix fit is for a single pole in a one-channel problem. Suppose that we have found a reduced width amplitude γ and a pole energy e_p such that

$$\mathbf{R} = \gamma^2/(e_p - E) \tag{10.2.11}$$

gives a suitable fit to some small range of experimental data. Can we tell if there is a resonance or a bound state? What would be the width of the resonance? Why is γ^2 called the reduced width?

From Eq. (10.2.8), the S matrix is now

$$\mathbf{S} = \frac{1 - \gamma^2(S^0 - iP)/(e_p - E)}{1 - \gamma^2(S^0 + iP)/(e_p - E)} \, e^{2i\phi}$$

$$= \left[1 - \frac{2i\gamma^2 P}{E - (e_p - \gamma^2 S^0 - i\gamma^2 P)} \right] e^{2i\phi} \tag{10.2.12}$$

$$= \frac{E - [e_p - \gamma^2 S^0 + i\gamma^2 P]}{E - [e_p - \gamma^2 S^0 - i\gamma^2 P]} \, e^{2i\phi} \tag{10.2.13}$$

which, at first glance, appears to be a resonance of Breit-Wigner form of Eq. (3.1.94):

$$\mathbf{S}(E) = e^{2i\delta_{\mathrm{bg}}(E)} \frac{E - E_r - i\Gamma/2}{E - E_r + i\Gamma/2} \tag{10.2.14}$$

for a pole at $E = E_r - i\Gamma/2$ having parameters

$$E_r^f = e_p - \gamma^2 S^0 = e_p - \gamma^2(S - a\beta)$$

$$\Gamma^f = 2\gamma^2 P. \tag{10.2.15}$$

These values are called the *formal* resonance position and width, hence the superscripts f. We can see why γ^2 is called the 'reduced width', since it is the result of removing (twice) the penetrability factor from Γ. We can see why $S(E)$ is called a 'shift function', as it contributes to the shift from the R-matrix pole at e_p towards the S-matrix pole at E_r^f. (As $S^0 = S - a\beta$, we will see on page 302 it might be useful to choose $\beta = S(E)/a$ in advance in order to fix that $S^0(E) = 0$ for some predetermined energy E of interest. This is called the *natural boundary condition* for β.)

These 'formal' values would be exact if the shifts S and penetrabilities P were constants independent of energy, but they are not, so the true position is more complicated. The *true* resonance position is taken as the complex energy where the S matrix has a pole, as discussed in subsection 3.1.3. That is, we should solve for complex energy E the equation

$$E = e_p - \gamma^2 S^0(E) - i\gamma^2 P(E)$$

$$\text{or} \quad E_r - i\tfrac{\Gamma}{2} = e_p - \gamma^2 S^0(E_r - i\Gamma/2) - i\gamma^2 P(E_r - i\tfrac{\Gamma}{2}). \tag{10.2.16}$$

If we are unwilling to directly evaluate these right-hand sides for complex E, we may use Taylor series to extrapolate off the real axis as $S^0(E_r - i\Gamma/2) = S^0(E_r) - iS'(E_r)\Gamma/2$, etc. The *Thomas approximation* [4] assumes that the shift function is locally linear, and this appears to be generally quite accurate. That is, we solve instead the linearized form

$$E_r - i\tfrac{\Gamma}{2} = e_p - \gamma^2[S^0(E_r) - iS'(E_r)\tfrac{\Gamma}{2}] - i\gamma^2[P(E_r) - iP'(E_r)\tfrac{\Gamma}{2}]. \tag{10.2.17}$$

Equating the real and imaginary parts gives

$$E_r = e_p - \gamma^2 S^0(E_r) - \tfrac{1}{2}\gamma^2\Gamma P'(E_r) \tag{10.2.18}$$

$$\text{and} \quad -\Gamma = \gamma^2\Gamma S'(E_r) - 2\gamma^2 P(E_r),$$

$$\text{so} \quad \Gamma = \frac{2\gamma^2 P(E_r)}{1 + \gamma^2 S'(E_r)} \tag{10.2.19}$$

$$\text{and} \quad E_r = e_p - \gamma^2 S^0(E_r) - \frac{\gamma^4 P(E_r)P'(E_r)}{1 + \gamma^2 S'(E_r)}. \tag{10.2.20}$$

Usually the third term in the E_r expression is neglected, along with all higher-order terms, and the solutions are now called the *observed* resonance energy and width. These satisfy

$$E_r^{\text{obs}} = e_p - \gamma^2 S^0(E_r^{\text{obs}}) \tag{10.2.21}$$

$$\Gamma^{\text{obs}} = \frac{2\gamma^2 P(E_r^{\text{obs}})}{1 + \gamma^2 S'(E_r^{\text{obs}})}. \tag{10.2.22}$$

A further simplification, still in the spirit of the Thomas approximation, is to evaluate $S^0(E)$ not at E_r^{obs} but at the R-matrix pole energy e_p. The Eq. (10.2.21) becomes now an explicit rather than an implicit equation for E_r^{obs}. Because the penetrability is extremely dependent on energy, Eq. (10.2.22) must still use E_r^{obs} rather than e_p.

The 'observed' resonance energy $E = E_r^{\text{obs}}$ on the real axis has the great virtue that in Eq. (10.2.13) the first fraction $\mathbf{S}e^{-2i\phi} = -1$, showing that the R-matrix phase shift $\delta_R = \pi/2$. Examples of δ_R plots are shown in Fig. 10.1.

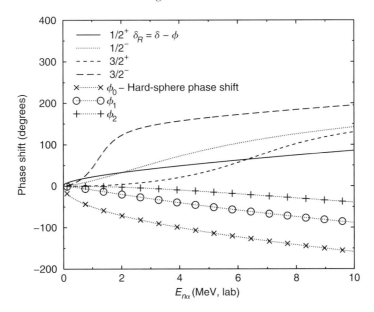

Fig. 10.1. Examples of R-matrix phase shifts δ_R for low-energy n–α scattering, along with the hard-sphere phase shifts ϕ_L in the s, p and d-waves that have been subtracted. The resonance in the $p_{1/2}$ channel is now much more visible than in Fig. 3.2.

For these reasons, the 'observed' resonance position is often *defined* as the energy where $\delta_R = \pi/2$. This implies from Eq. (10.2.10) that the 'observed' energy is in general where

$$R(E_r^{\text{obs}})S^0(E_r^{\text{obs}}) = 1, \qquad (10.2.23)$$

an expression which can be used with any number of R-matrix poles to find one-channel resonance positions. The 'observed' width Γ^{obs} can in the general case be obtained from

$$\Gamma^{\text{obs}} = 2 \left[\frac{\mathrm{d}\delta_R}{\mathrm{d}E} \right]_{E=E_r}^{-1}, \qquad (10.2.24)$$

using again the Thomas approximation. As definitions of a resonance energy, however, the equations (10.2.21, 10.2.23) have residual dependence on the R-matrix radius a which enters into the evaluation of the hard-sphere phase shift ϕ.

Note that neither the 'observed' resonance position, nor the real part of complex pole energy, is in general the energy where the full scattering phase shift satisfies $\delta = \pi/2$. A somewhat better estimate is where the derivative $\mathrm{d}\delta/\mathrm{d}E$ has its maximum, because the hard-sphere phase shifts vary only moderately with energy. If the poles are wide and a long way from the real axis, however, then the Thomas approximation is not so reliable. In general, the Coulomb functions F and G are then needed at complex energies to find the *true* S-matrix pole by explicit searching in the complex plane.

It is possible for the R-matrix pole energy e_p to be negative like a bound state, but for it to still describe a resonance at $E_r > 0$ because of the shift function in $E_r = e_p - \gamma^2 S^0(E_r)$. To find the width of this resonance, at least in this case we cannot approximate the penetrability $P(E_r)$ by $P(e_p)$ since the latter value is zero.

It is most likely that negative e_p values correspond to genuine bound states for $E_r < 0$, also called *sub-threshold states*. The shift $-\gamma^2(S(E_r) - a\beta)$ may still be found, showing the difference between the R-matrix pole energy and the actual bound state. If the boundary-condition parameter β is chosen as the logarithmic derivative of the bound-state wave function at $R = a$, then it can be prearranged that $S^0 = 0$ at a specific energy. The application of this natural boundary condition to a specific sub-threshold state is good practice if the energy of that state is already known from other experiments; we do not want to vary it in a fitting procedure. For each spin and parity J_{tot}^π we can choose β so that $S^0(E_b) = 0$ at some preferred energy E_b.

Magnitude of the reduced width

The magnitude of the reduced width amplitude γ has physical meaning. In a potential model, it is proportional to the magnitude of an interior eigenfunction $w(a)$ at the surface $R = a$. If this corresponds to a physical resonance then γ will be small, since the resonant wave function will be large and oscillatory in the interior region of the potential, and its overall normalization to unity along with tunneling through the barrier will mean that $w(a)$ will be small outside the potential. The larger the interior resonant wave function, the smaller $w(a)$ and γ will be. It is physically reasonable that a resonance with a large interior amplification of the wave function will be 'narrow' or 'sharp' with energy, with a small reduced width γ^2 and hence small width $\Gamma = 2\gamma^2 P$.

Although both γ^2 and penetrability P depend on the matching radius a, the resulting Γ should be almost independent of radius. This is because the value of $w(a)$ in γ reflects the tunneling at internal radii $R < a$, whereas the penetrability P reflects the tunneling from $R = a$ out to infinity. The combination of these two factors should not depend on where the R-matrix theory changes its description of tunneling.

If we were to make a list of reduced widths of all the poles p, some will be large and the others small. The small widths generally correspond to resonances or bound states, whereas if not a resonance, the reduced widths will be larger.

In the limit of weak or uniform potentials, non-resonant γ^2 may be estimated by considering an eigenstate of uniform probability density inside a radius a. The value of a in this case should be as close as possible to the edge of the potential, in which case we derive what is called the *Wigner limit* γ_W^2 for the reduced widths. A constant probability density indicates constant $w(R)/R$, which implies $w(R) = 3^{1/2}a^{-3/2}R$ for unit normalization over $R < a$. The reduced width in this limit is thus

$$\gamma_W^2 = \frac{\hbar^2}{2\mu a}|w(a)|^2 = \frac{3\hbar^2}{2\mu a^2}. \qquad (10.2.25)$$

The Wigner limit can sometimes be used in the absence of any specific structural information. It is strongly dependent on the radius a.

Another estimate of reduced widths, slightly more realistic, is to use the single-particle states in some potential. Let us choose some standard Woods-Saxon geometry and spin-orbit potential, and adjust its depth so that it has a single-particle eigenstate whose energy ε_n agrees with the resonance state of interest, and whose wave function $w_n(R)$ has the number of interior nodes expected from a simple shell model. The one-channel R-matrix theory of subsection 6.5.1 then gives us the reduced width amplitudes of Eq. (6.5.14), allowing us to define a *single-particle reduced width* of

$$\gamma_{sp}^2 = \frac{\hbar^2}{2\mu a} w_n(a)^2. \qquad (10.2.26)$$

These single-particle widths will vary with energy and partial wave L in a more realistic manner than γ_W^2.

Given actual reduced widths γ^2, the Wigner or single-particle *ratios*

$$\theta_W^2 = \gamma^2/\gamma_W^2 \quad \text{or} \quad \theta_{sp}^2 = \gamma^2/\gamma_{sp}^2 \qquad (10.2.27)$$

are dimensionless, and can sometimes be used *like* spectroscopic factors to express a ratio between the actual reduced width and the width in some simple limit. They are however *not the same* as the spectroscopic factor, as γ^2 (and hence the θ^2) depend on the outer surface properties of an eigenstate, not on its volume integral as does the spectroscopic factor. The reduced width for a bound state is more closely related to the asymptotic normalization coefficient (ANC) defined in Eq. (4.5.24), as both are measures of the outer asymptotic properties.

If $\phi(r)$ is a true single-channel bound state, and β chosen as $\beta = \phi'/\phi$ at $r = a$ from the 'natural boundary condition,' then $S^0 = 0$ at the bound state. The R-matrix eigenfunction $w(r)$ will be strictly proportional to $\phi(r)$ where it exists inside $r = a$: $\phi(r) = Aw(r)$, where $A^2 = 1 - \int_a^\infty |\phi(r)|^2 dr$ reflects the different normalization requirements. The ANC value C is the coefficient in $\phi(r) = CW(-2kr)$ for the Whittaker function $W(-2kr)$ specified in Eq. (4.5.24). The reduced width is therefore

$$
\begin{aligned}
\gamma^2 &= \frac{\hbar^2}{2\mu a} |w(a)|^2 \\
&= \frac{\hbar^2}{2\mu a} \frac{C^2 \, W(-2ka)^2}{A^2} \\
&= \frac{\hbar^2}{2\mu a} \frac{C^2 \, W(-2ka)^2}{1 - C^2 \int_a^\infty |W(-2kr)|^2 \mathrm{d}r}.
\end{aligned}
\qquad (10.2.28)
$$

For deeply bound states $A \approx 1$, and then we will have the proportionality $\gamma^2 \propto C^2$, but the integral in the denominator may become important for weakly bound or halo states.

10.2.3 Multiple poles in one channel

More general behavior of the energy dependence of $\mathbf{R}(E)$ may be obtained by expanding it using more than one pole:

$$
\mathbf{R}(E) = \sum_p \gamma_p^2 / (e_p - E). \qquad (10.2.29)
$$

Resonances in this case, probably several, may be found by solving the Eq. (10.2.23), namely $\mathbf{R}(E_r)S(E_r) = 1$, and their widths by differentiation of the R-matrix phase shift δ_R by Eq. (10.2.24) at the various 'observed' resonance positions E_r. Note however that this definition of resonances implies that a resonance exists between every successive pair of R-matrix poles, but these may be too broad to be noticeable, especially if they are higher in energy than the Coulomb and centrifugal barriers.

If the energy of interest is away from any poles or resonances e_p, then the $\mathbf{R}(E)$ is approximately constant. The 'distant pole' approach is to replace the R matrix by a constant

$$
\mathbf{R}(E) = \mathbf{R}_0, \qquad (10.2.30)
$$

so one simple approximation around a single resonance is to fit the parameters of one pole in addition to some constant:

$$
\mathbf{R} = \gamma^2 / (e_p - E) + \mathbf{R}_0. \qquad (10.2.31)
$$

Adding such term allows us, for example, to describe the slow energy variation of some non-resonant contribution, as the value of $\mathbf{R}_0$ can often be fitted to simple forms of phase shifts.

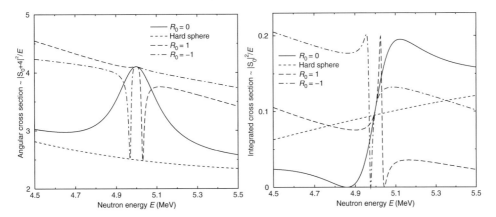

Fig. 10.2. Possible kinds of interference patterns in the 'differential' (left) and integrated cross sections (right) for *s*-wave scattering on a nucleus. A single R-matrix pole at 5 MeV was used with a reduced width of 30 keV at $a = 8$ fm, for varying additive contributions $\mathbf{R}_0$. The 'differential' cross section is obtained by coherent addition of the *s*-wave S matrix at an amplitude set at 4 to represent scattering from other partial waves.

Later in Section 10.4 we will describe an alternative *hybrid model*, which uses potentials to describe the non-resonant processes. The R matrix from the potentials will be combined with phenomenological pole terms for specific resonances beyond the scope of a potential model.

10.3 Coupled-channels R matrix

To facilitate the understanding of a multi-channel R-matrix fit, it is again useful to modify the derivation of Eq. (10.1.2) of the S matrix from the R matrix, so various approximations allow simple formulae to be derived that may be easier to understand.

10.3.1 Revised derivation of the scattering S matrix

We start as before with the matrix expression of Eq. (6.5.33). We follow the pattern of the derivation giving Eq. (10.2.1), but remember that these terms are all matrices:

$$\mathbf{S} = [t^{\frac{1}{2}}\mathbf{H}^{+} - a\mathbf{R}\,t^{\frac{1}{2}}(\mathbf{H}^{+\prime} - \beta\mathbf{H}^{+})]^{-1}[t^{\frac{1}{2}}\mathbf{H}^{-} - a\mathbf{R}\,t^{\frac{1}{2}}(\mathbf{H}^{-\prime} - \beta\mathbf{H}^{-})]$$

$$= (t^{\frac{1}{2}}\mathbf{H}^{+})^{-1}[1 - a\mathbf{R}(\mathbf{H}^{+\prime}/\mathbf{H}^{+} - \beta)]^{-1}[1 - a\mathbf{R}(\mathbf{H}^{-\prime}/\mathbf{H}^{-} - \beta)](t^{\frac{1}{2}}\mathbf{H}^{-})$$

$$(10.3.1)$$

where for matrices $A/B \equiv AB^{-1}$. We have assumed that the asymptotic Coulomb wave functions $\mathbf{H}$ have no off-diagonal terms when the diagonal elements of

$t = \hbar^2/2\mu$ vary for different mass partitions, which is normally the case. Following now the previous pattern of subsection 10.2.1, we now define a 'logarithmic' matrix

$$L = \mathbf{H}^{+\prime}/\mathbf{H}^+ - \beta = \frac{1}{a}(S + iP - a\beta) \tag{10.3.2}$$

where P, S are taken as matrices with diagonal elements

$$P_\alpha = k_\alpha a/(F_\alpha^2 + G_\alpha^2) \tag{10.3.3}$$

$$\text{and} \ \ S_\alpha = (\dot{F}_\alpha F_\alpha + \dot{G}_\alpha G_\alpha)P_\alpha. \tag{10.3.4}$$

Since $\mathbf{H}^{-\prime}/\mathbf{H}^- = L^*$ also, we have

$$\mathbf{S} = (t^{\frac{1}{2}}\mathbf{H}^+)^{-1}[1 - a\mathbf{R}L]^{-1}[1 - a\mathbf{R}L^*](t^{\frac{1}{2}}\mathbf{H}^-)$$

$$= \sqrt{\frac{\mathbf{H}^-}{\mathbf{H}^+}}\frac{1}{\sqrt{t\mathbf{H}^-\mathbf{H}^+}}[1 - a\mathbf{R}L]^{-1}[1 - a\mathbf{R}L^*]\sqrt{t\mathbf{H}^-\mathbf{H}^+}\sqrt{\frac{\mathbf{H}^-}{\mathbf{H}^+}}$$

$$= \Omega\frac{1}{\sqrt{t\mathbf{H}^-\mathbf{H}^+}}[1 - a\mathbf{R}L]^{-1}[1 - a\mathbf{R}L^*]\sqrt{t\mathbf{H}^-\mathbf{H}^+}\ \Omega \tag{10.3.5}$$

where Ω is the matrix with diagonal elements $e^{i\phi_\alpha}$ for hard-sphere phase shifts $\tan \phi_\alpha = -F_\alpha/G_\alpha$. The matrix product $(t\mathbf{H}^-\mathbf{H}^+)$ is diagonal, with elements

$$\mathbf{H}_\alpha^-\mathbf{H}_\alpha^+ t_\alpha = (F_\alpha^2 + G_\alpha^2)\frac{\hbar^2}{2\mu_\alpha} = \frac{k_\alpha a}{P_\alpha}\frac{\hbar^2}{2\mu_\alpha} = \frac{\hbar v_\alpha a}{2P_\alpha} \tag{10.3.6}$$

in terms of the penetrability defined above in Eq. (10.3.3), and channel velocities $v = \hbar k/\mu$.

The symmetric $\tilde{\mathbf{S}}$ matrix is constructed from the above $\mathbf{S}$ by Eq. (3.2.12): $\tilde{\mathbf{S}} = \mathbf{v}^{\frac{1}{2}}\mathbf{S}\mathbf{v}^{-\frac{1}{2}}$, using the same velocity factors put into a diagonal matrix $\mathbf{v}$. This means that we have the simpler form for

$$\tilde{\mathbf{S}} = \mathbf{v}^{\frac{1}{2}}\Omega\frac{1}{\sqrt{\frac{1}{2}\hbar\mathbf{v}a/P}}[1 - a\mathbf{R}L]^{-1}[1 - a\mathbf{R}L^*]\sqrt{\frac{1}{2}\hbar\mathbf{v}a/P}\ \Omega\mathbf{v}^{-\frac{1}{2}}$$

$$= \Omega P^{\frac{1}{2}}[1 - a\mathbf{R}\,L]^{-1}[1 - a\mathbf{R}\,L^*]P^{-\frac{1}{2}}\Omega. \tag{10.3.7}$$

We now use the matrix identity

$$[1 - RA]^{-1}[1 - RB] = 1 + [1 - RA]^{-1}R(A - B) \tag{10.3.8}$$

to find

$$\tilde{\mathbf{S}} = \Omega P^{\frac{1}{2}}[1 + (1 - a\mathbf{RL})^{-1}a\mathbf{R}\,(L - L^*)]P^{-\frac{1}{2}}\Omega \qquad (10.3.9)$$

$$= \Omega P^{\frac{1}{2}}[1 + (1 - a\mathbf{RL})^{-1}a\mathbf{R}\,2iP/a]P^{-\frac{1}{2}}\Omega \qquad (10.3.10)$$

$$= \Omega[1 + 2iP^{\frac{1}{2}}(1 - a\mathbf{RL})^{-1}\mathbf{R}P^{\frac{1}{2}}]\Omega. \qquad (10.3.11)$$

This is the main result: a derivation of the S matrix from the R matrix, using the penetrabilities and shifts, with $aL = S^0 + iP$.

In the *one*-channel case, the Eq. (10.3.11) reduces to the equations of subsection 10.2.1. In the *two*-channel case, we can perform the matrix inversion by hand, yielding the 2×2 $\tilde{\mathbf{S}}$ matrix

$$\tilde{\mathbf{S}}_{11} = e^{2i\phi_1}[1 + 2iP_1[\mathbf{R}_{11} - aL_2(\mathbf{R}_{11}\mathbf{R}_{22} - \mathbf{R}_{12}^2)]d^{-1}] \qquad (10.3.12a)$$

$$\tilde{\mathbf{S}}_{22} = e^{2i\phi_2}[1 + 2iP_2[\mathbf{R}_{22} - aL_1(\mathbf{R}_{11}\mathbf{R}_{22} - \mathbf{R}_{12}^2)]d^{-1}] \qquad (10.3.12b)$$

$$\tilde{\mathbf{S}}_{12} = \tilde{\mathbf{S}}_{21} = e^{i(\phi_1+\phi_2)}2iP_1^{1/2}\mathbf{R}_{12}P_2^{1/2}d^{-1}, \qquad (10.3.12c)$$

where the determinant is

$$d = (1 - a\mathbf{R}_{11}L_1)(1 - a\mathbf{R}_{22}L_2) - a^2L_1\mathbf{R}_{12}^2L_2. \qquad (10.3.12d)$$

These equations have a very simple form if there is one isolated level e_p, as here $\mathbf{R}_{\alpha'\alpha} = \gamma_\alpha\gamma_{\alpha'}/(e_p - E)$ and hence $\mathbf{R}_{11}\mathbf{R}_{22} = \mathbf{R}_{12}^2$. In this one-pole case all the elements have the general form

$$\tilde{\mathbf{S}}_{\alpha'\alpha} = e^{i\phi_\alpha}\left[\delta_{\alpha'\alpha} + \frac{2iP_\alpha^{1/2}\gamma_\alpha\gamma_{\alpha'}P_{\alpha'}^{1/2}}{(e_p - E)(1 - a\mathbf{R}_{11}L_1 - a\mathbf{R}_{22}L_2)}\right]e^{i\phi_{\alpha'}}$$

$$= e^{i\phi_\alpha}\left[\delta_{\alpha'\alpha} + \frac{i\Gamma_\alpha^{1/2}\Gamma_{\alpha'}^{1/2}}{e_p - E - \gamma_1^2S_1^0 - i\gamma_1^2P_1 - \gamma_2^2S_2^0 - i\gamma_2^2P_2}\right]e^{i\phi_{\alpha'}}, \qquad (10.3.13)$$

since the formal widths of Eq. (10.2.15) are $\Gamma_\alpha = 2\gamma_\alpha^2 P_\alpha$. We have used the shifts and penetrabilities defined in Eq. (10.3.2), with $S^0 = S - a\beta$ as before.

This formula suggests we define a *total formal width*

$$\Gamma_{\text{tot}} = 2\gamma_1^2 P_1 + 2\gamma_2^2 P_2 = \Gamma_1 + \Gamma_2, \qquad (10.3.14)$$

in terms of which the two-channel one-pole $\tilde{\mathbf{S}}$ matrix begins to look like an isolated Breit-Wigner resonance:

$$\tilde{\mathbf{S}}_{\alpha'\alpha} = e^{i\phi_\alpha}\left[\delta_{\alpha'\alpha} - \frac{i\Gamma_\alpha^{1/2}\Gamma_{\alpha'}^{1/2}}{E - (e_p - \gamma_1^2S_1^0 - \gamma_2^2S_2^0) + i\Gamma_{\text{tot}}/2}\right]e^{i\phi_{\alpha'}} \qquad (10.3.15)$$

with formal resonance energy $E_r^f = e_p - \gamma_1^2 S_1^0 - \gamma_2^2 S_2^0$. We see that the shift contributions from the individual channels are added together to produce the shift for the coupled-channels resonance.

The non-elastic cross section contribution from a specific coupled-channels set J_{tot}^π is given by Eq. (3.2.29). Using the spin weighting factor g_{tot} defined by Eq. (3.2.31), the total cross section for scattering to channel α from α_i is

$$\sigma_{\alpha\alpha_i}(J_{\text{tot}}^\pi) = \frac{\pi}{k_i^2}\, g_{J_{\text{tot}}}\, |\tilde{\mathbf{S}}_{\alpha\alpha_i}|^2$$

$$= \frac{\pi}{k_i^2}\, g_{J_{\text{tot}}}\, \frac{\Gamma_\alpha \Gamma_{\alpha_i}}{|E - E_r^f - i\Gamma_{\text{tot}}/2|^2}$$

$$= \frac{\pi}{k_i^2}\, g_{J_{\text{tot}}}\, \frac{\Gamma_\alpha \Gamma_{\alpha_i}}{(E - E_r^f)^2 + \Gamma_{\text{tot}}^2/4}, \qquad (10.3.16)$$

which is exactly the form of an isolated Breit-Wigner resonance with a strong peak at $E \approx E_r^f$ and fwhm of Γ_{tot}.

The individual Γ_α are called the *partial widths*, because they are ($\hbar$ times) the decay rates of a resonance through specific exit channels. The total (formal) width $\Gamma_{\text{tot}} = \sum_\alpha \Gamma_\alpha$ is the sum of all the partial widths, and describes the resonance's overall decay rate. Note that, by time-reversal invariance, the same width applies to the *entrance* channel as to the exit channels, and hence Γ_{α_i} measures also the rate at which a resonance could be populated from a given initial scattering configuration α_i.

This summation of partial widths and shifts has been proved for the two-channel one-pole case. To see how partial widths add together in the multi-channel case, we first need to present the level-matrix formulation of R-matrix theory.

10.3.2 Level-matrix formulation

All the matrix operations so far in this chapter involve inversion of a matrix with the dimensionality M of the number of *partial-wave channels* in a coupled-channels set for a specific overall spin and parity J_{tot}^π. Sometimes, however, it is more convenient to reformulate the theory so that the inversion is only needed of a matrix with dimensions P of the number of *levels* e_p. In R-matrix phenomenology, this is often a smaller number. So we now show how to construct a $P{\times}P$ symmetric level matrix A with elements A_{pq} for level indices p, q, such that the $\tilde{\mathbf{S}}$ matrix depends on calculating the inverse A^{-1}.

To simplify the derivation of A, we write the previous theory as much as possible in matrix form. We write the initial $M \times M$ R matrix of Eq. (10.1.1) as

$$\mathbf{R} = \gamma^T F \gamma \tag{10.3.17}$$

where γ is the $P \times M$ rectangular matrix with elements $\gamma_{p\alpha}$, and F is the matrix of diagonal reciprocals $1/(e_p - E)$. From the previous matrix form Eq. (10.3.11), we have

$$\tilde{\mathbf{S}} = \Omega[1 + 2iP^{\frac{1}{2}}(1 - \gamma^T F \gamma (S^0 + iP))^{-1} \gamma^T F \gamma] P^{\frac{1}{2}} \Omega. \tag{10.3.18}$$

We now try to find a matrix $\mathbf{A}$ so that this can be rewritten

$$\tilde{\mathbf{S}} = \Omega[1 + 2iP^{\frac{1}{2}} \gamma^T \mathbf{A} \gamma P^{\frac{1}{2}}] \Omega, \tag{10.3.19}$$

for which we need to satisfy the identity

$$(1 - \gamma^T F \gamma (S^0 + iP))^{-1} \gamma^T F \gamma = \gamma^T \mathbf{A} \gamma, \tag{10.3.20}$$

which is

$$\gamma^T F \gamma = (1 - \gamma^T F \gamma (S^0 + iP)) \gamma^T \mathbf{A} \gamma$$
$$= \gamma^T \mathbf{A} \gamma - \gamma^T F \gamma (S^0 + iP) \gamma^T \mathbf{A} \gamma,$$

$$\text{or} \quad \gamma^T \left[F - \mathbf{A} - F \gamma (S^0 + iP) \gamma^T \mathbf{A} \right] \gamma = 0. \tag{10.3.21}$$

This condition will always be satisfied if we can choose $\mathbf{A}$ such that

$$F - \mathbf{A} - F \gamma (S^0 + iP) \gamma^T \mathbf{A} = 0 \tag{10.3.22}$$

$$\text{or} \quad \mathbf{A}^{-1} = F^{-1} - \gamma S^0 \gamma^T - i\gamma P \gamma^T. \tag{10.3.23}$$

This last equation is the defining equation we are looking for. If we construct diagonal and symmetric off-diagonal shift and width *level matrices* as

$$\hat{\Delta}_{pq} = (\gamma S^0 \gamma^T)_{pq} = \sum_\alpha \gamma_{p\alpha} S^0_\alpha \gamma_{q\alpha} \tag{10.3.24}$$

$$\text{and} \quad \hat{\Gamma}_{pq} = 2(\gamma P \gamma^T)_{pq} = 2 \sum_\alpha \gamma_{p\alpha} P_\alpha \gamma_{q\alpha}, \tag{10.3.25}$$

then the defining equation for the symmetric level matrix $\mathbf{A}$ is

$$(\mathbf{A}^{-1})_{pq} = \delta_{pq}(e_p - E) - \hat{\Delta}_{pq} - \tfrac{i}{2}\hat{\Gamma}_{pq} \tag{10.3.26}$$

$$\text{or} \quad -(\mathbf{A}^{-1})_{pq} = E\delta_{pq} - (e_p \delta_{pq} - \hat{\Delta}_{pq} - \tfrac{i}{2}\hat{\Gamma}_{pq}). \tag{10.3.27}$$

The $\hat{\Delta}_{pq}$ and $\hat{\Gamma}_{pq}$ are generalized shifts and widths that now have off-diagonal as well as diagonal effects on the levels. Their signs are now important, and reflect the interference between levels, as they enter into the matrix inversion

$$
\mathbf{A} = - \begin{pmatrix} E - e_1 + \hat{\Delta}_{11} + \frac{i}{2}\hat{\Gamma}_{11} & \hat{\Delta}_{12} + \frac{i}{2}\hat{\Gamma}_{12} & \cdots \\ \hat{\Delta}_{21} + \frac{i}{2}\hat{\Gamma}_{21} & E - e_2 + \hat{\Delta}_{22} + \frac{i}{2}\hat{\Gamma}_{22} & \cdots \\ & \cdots & \end{pmatrix}^{-1},
$$

(10.3.28)

remembering the symmetry of the $\hat{\Gamma}$ and $\hat{\Delta}$ matrices. From Eq. (10.3.19), the full multi-channel multi-level $\tilde{\mathbf{S}}$ matrix is constructed in terms of this matrix inverse $\mathbf{A}$ as

$$
\tilde{\mathbf{S}}_{\alpha'\alpha} = \Omega_\alpha \left[\delta_{\alpha'\alpha} + i \sum_{\lambda\lambda'} \Gamma_{\alpha\lambda}^{1/2} \mathbf{A}_{\lambda\lambda'} \Gamma_{\alpha'\lambda'}^{1/2} \right] \Omega_{\alpha'},
$$

(10.3.29)

where we define $\Gamma_{\alpha\lambda}^{1/2} = \gamma_{\alpha\lambda}(2P_\alpha)^{1/2}$ to preserve the signs of the $\gamma_{\alpha\lambda}$. This level-matrix construction is also a proof that the $\tilde{\mathbf{S}}$ matrix is *symmetric* for symmetric potential interaction matrices, as originally claimed in subsection 3.2.4, as in that case the R matrix is also symmetric.

One level: an isolated resonance

The coupled-channels case with just one level $p = q = 1$ is now particularly easy, as the level matrix $\mathbf{A}$ is then a number. Using

$$
\hat{\Delta}_{11} = \sum_\alpha \gamma_{1\alpha}^2 S_\alpha^0
$$

(10.3.30)

$$
\text{and} \quad \Gamma_{\text{tot}} \equiv \hat{\Gamma}_{11} = \sum_\alpha 2\gamma_{1\alpha}^2 P_\alpha \equiv \sum_\alpha \Gamma_\alpha,
$$

(10.3.31)

the $\mathbf{A}$ is now

$$
-\mathbf{A}^{-1} = E - (e_1 - \hat{\Delta}_{11} - \tfrac{i}{2}\Gamma_{\text{tot}}) \equiv E - E_p,
$$

(10.3.32)

where $E_p = e_1 - \hat{\Delta}_{11} - \frac{i}{2}\Gamma_{\text{tot}}$ is the complex resonance pole position. Thus the matrix of scattering from this isolated pole is

$$
\tilde{\mathbf{S}}_{\alpha'\alpha} = \Omega_\alpha \left[\delta_{\alpha'\alpha} - \frac{i\Gamma_\alpha^{1/2}\Gamma_{\alpha'}^{1/2}}{E - E_p} \right] \Omega_{\alpha'}.
$$

(10.3.33)

Elastic scattering: The contributing part to an *elastic* cross section ($\alpha = \alpha'$) is therefore identical to Eq. (10.2.12),

$$\tilde{\mathbf{S}}_{\alpha\alpha} = \Omega_\alpha^2 \left[1 - \frac{i\Gamma_\alpha}{E - E_p} \right]. \tag{10.3.34}$$

A function of $\tilde{\mathbf{S}}$ which appears in the elastic amplitude $f(\theta)$ (see Eq. (3.1.34)) is

$$\tilde{\mathbf{S}}_{\alpha\alpha} - 1 = 2ie^{i\phi} \left[\sin\phi - \frac{i}{2} \frac{e^{i\phi}\Gamma_\alpha}{E - E_p} \right] \tag{10.3.35}$$

using $\Omega_\alpha = e^{2i\phi}$. The scattering arising from the first term $\sin\phi$ is sometimes called in R-matrix theory the 'potential scattering', and that from the $\Gamma/(E - E_p)$ term the 'resonance scattering', and there will also be an interference term. However, the 'potential' under discussion here is just the 'hard sphere' at radius $R = a$, and should not be mistaken for the attractive optical-potential well, more commonly called the scattering potential. We discuss below in Section 10.4 a better physical way to combine potential ('shape') and resonant scattering.

Non-elastic reactions: The contributing part to an *inelastic* cross section ($\alpha \neq \alpha'$) is analogously

$$\sigma_{\alpha\alpha_i}(J_{\text{tot}}^\pi) = \frac{\pi}{k_i^2} g_{J_{\text{tot}}} \frac{\Gamma_\alpha \Gamma_{\alpha_i}}{(E - e_1 + \hat{\Delta}_{11})^2 + \Gamma_{\text{tot}}^2/4} \tag{10.3.36}$$

which is a Breit-Wigner form that is similar to Eq. (10.3.16), but now established for any number of partial-wave channels. The total (formal) width $\Gamma_{\text{tot}} = \sum_\alpha \Gamma_\alpha$ is again the sum over all the partial widths for each coupled channel. Because, strictly, the shifts and widths are energy dependent:

$$\sigma_{\alpha\alpha_i}(J_{\text{tot}}^\pi; E) = \frac{\pi}{k_i^2} g_{J_{\text{tot}}} \frac{\Gamma_\alpha(E)\Gamma_{\alpha_i}(E)}{(E - e_1 + \hat{\Delta}_{11}(E))^2 + \Gamma_{\text{tot}}(E)^2/4}, \tag{10.3.37}$$

the observed and true resonance widths will be as usual slightly different from the formal value Γ_{tot}.

10.4 Combining direct and resonant contributions

In most reactions there is a combination of a smooth background cross section and large peaks at the several compound-nucleus resonance energies. From the potential or coupled-channels method of Chapters 3 and 4, it is usually possible to describe the smooth background as a direct reaction, and perhaps also a few of the resonances as arising from particular channel couplings. Rarely is a coupled-channels method able to describe *all* the resonances that are seen in experiments, whereas the R-matrix method can simply add further poles with partial widths at each level chosen to reproduce the experimental data. The R-matrix method, however, does not describe the smooth direct contributions so well: this is usually done by specifying some distant poles at fixed energies outside the energy range of interest, but these *background poles* do not have specific physical significance.

In order to get the best of both worlds – the use of channel couplings to describe the direct parts, and the use of R-matrix poles to describe the compound nuclear resonances – we could reasonably consider a *hybrid model* which combines both sorts of contributions. The combination is most naturally done in the R-matrix theory, as there it is explicit that all the pole contributions should simply be added together. We therefore define hybrid models in which the R matrix $\mathbf{R}^{cc}$ from the coupled channels – Eq. (6.5.25) – is combined with additional phenomenological terms $\mathbf{R}^{ph}$ as previously in this chapter. We construct

$$\mathbf{R}^{hyb} = \mathbf{R}^{cc} + \mathbf{R}^{ph} \tag{10.4.1}$$

$$\text{or} \quad \mathbf{R}^{hyb}_{\alpha'\alpha} = \sum_{p=1}^{P} \frac{\gamma_{p\alpha}\gamma_{p\alpha'}}{e_p - E} + \sum_{q=1}^{Q} \frac{g_{q\alpha}g_{q\alpha'}}{f_q - E}, \tag{10.4.2}$$

where the phenomenological poles at energies f_q for $q = 1, \ldots, Q$ have reduced width amplitudes $g_{q\alpha}$ for channel α. Each total spin and parity J^{π}_{tot} of a coupled-channels set has its own set of phenomenological poles.

Two analyses shown in Chapter 1 for the reactions of protons on ^{8}B and ^{14}N used this hybrid method. Figure 1.9 shows the S-factor for the ^{7}Be(p,γ)^{8}B reaction, where the final proton state is very weakly bound at –0.137 MeV. This calculation includes a direct-capture mechanism for the smooth contribution, combined with a hybrid R-matrix pole for the 1^+ resonance at 640 keV. An R-matrix radius of 20 fm was used, along with asymptotic couplings out to 300 fm using the methods of Section 6.6 in order to obtain correct S-factors at the lowest energies shown here. The calculation shown in Fig. 1.13 was principally a phenomenological R-matrix fit with $a = 6.5$ fm. The level at 6.79 MeV, however, is weakly bound at -0.502 MeV, so, in order to get the flat S-factor line around 1 mb MeV, a direct coupling was added for the transition between 6.5 and 50 fm.

Exercises

10.1 Derive penetrability and shift analytic expressions for s-, p- and d-wave neutrons from the formulae in Box 3.1.

10.2 For given R-matrix pole parameters for s, p or d-wave neutrons, use results of Ex. (10.1), when analytically continued to complex energies, to find the exact pole positions. Compare with the 'formal' and 'observed' positions to assess their accuracy.

10.3 Use results of Ex. (10.1), for neutrons in a square well, to find positions and reduced widths of R-matrix poles, then positions and formal widths of S-matrix poles, using the radius of the well as the R-matrix radius. Are there any poles in s-wave scattering?

10.4 Determine the accuracy of one-pole expansions for neutron scattering on a square well, for a range of R-matrix radii outside the radius of the potential. Repeat numerically for a Woods-Saxon binding potential. What do you conclude about the optimal R-matrix radius?

10.5 Isolated resonances may be represented by an R-matrix expansion in the energy region E around the pole at ε. Consider the one-channel case, so Eq. (6.5.6) only uses $w_1(R)$.

(a) Determine the interior wave function $\chi(R; E)$ for energies E around the resonance, using Eqs. (6.5.16) and (10.3.34).

(b) One measure of the intensity of a resonance is the interior norm integral $\rho(E) = \int_0^a |\chi(R; E)|^2 dR$. Show, after [5], that

$$\rho(E) = \frac{E}{k} \frac{\Gamma(E)}{(E - \varepsilon - \Delta(E))^2 + \Gamma(E)^2/4}. \tag{E10.1}$$

References

[1] A. M. Lane and R. G. Thomas, *Rev. Mod. Phys.* **30** (1958) 257.
[2] E. W. Vogt, *Rev. Mod. Phys.* **34** (1962) 723.
[3] P. Descouvemont 2004, *Theoretical Models for Nuclear Astrophysics*, Hauppauge NY: Nova Science Publishers.
[4] R. G. Thomas, *Phys. Rev.* **81** (1951) 148.
[5] F. C. Barker, *Phys. Rev. C* **59** (1999) 535.

11

Compound-nucleus averaging

Science makes people reach selflessly for truth and objectivity;
it teaches people to accept reality, with wonder and admiration,
not to mention the deep joy and awe that the natural order of things
brings to the true scientist.

Lise Meitner

11.1 Compound-nucleus phenomena

In Chapter 2 we saw how nuclear reactions are broadly dominated by two kinds
of timescales: the fast direct reactions and the slower compound-nucleus (CN)
reactions. The direct reactions are typically described in R-matrix theory by a few
poles with large widths, whereas there are usually very many compound-nucleus
resonances, each of which has a narrow width.

Direct reactions are generally foward-peaked with respect to the incident
direction, whereas the CN process has less 'memory' about that direction and
gives products which are typically symmetric about $90°$. Usually it is possible to
experimentally separate the symmetric contributions to a given outgoing channel,
and theoretically the direct and CN cross sections are calculated by quite different
methods. We will sometimes try to model the detailed resonance structure of the CN
process, but usually calculate the reaction rates averaging over many CN levels, and
use only statistical features of these levels, such as their average spacings and widths.

To describe these CN processes we will use the R-matrix phenomenology of
Chapter 10, in particular the level-matrix formalism of subsection 10.3.2 when
we consider the case of extremely many levels at energies e_p and with widths Γ_p
for states of given spin and parity J_{tot}^{π}. The simplest statistical properties will be
the **average level spacing** D (the inverse of the **level density** $\rho_{\text{ld}}(E)$) and the
average width $\langle \Gamma \rangle$, for each J_{tot}^{π} value.

The exact effects of all the CN states on scattering are expressed by the width
level matrices defined in Eq. (10.3.25): $\hat{\Gamma}_{pq} = 2 \sum_{\alpha} \gamma_{p\alpha} P_{\alpha} \gamma_{q\alpha}$, which depend on

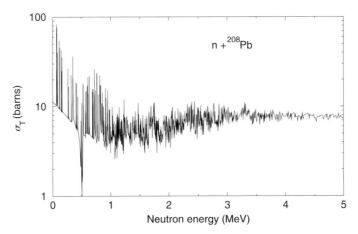

Fig. 11.1. Numerous compound-nucleus resonances can be seen in this plot of the total cross section $\sigma_{\text{tot}}(E)$ for neutrons incident on ^{208}Pb. At low energies, individual resonances can be distinguished ($\Gamma \ll D$). In the 0.5–2 MeV range they overlap more, and above 3 MeV they are not resolvable ($\Gamma \gg D$). Note that ^{208}Pb is a closed shell nucleus: most heavy nuclei have resonances too numerous to be seen on a plot with this scale.

the individual reduced width amplitudes $\gamma_{p\alpha}$ for state p in partial decay channel α, and on the penetrabilities P_α. The average behavior of the $\hat{\Gamma}_{pq}$ will depend first of all on the statistical properties of the $\gamma_{p\alpha}$. Since by Eq. (6.5.24) these are $\gamma_{p\alpha} = \sqrt{\frac{\hbar^2}{2\mu_\alpha a}}\, g_\alpha^p(a)$, they depend on the asymptotic amplitudes $g_\alpha^p(a)$ of the resonant states in channel α at radius a.

11.1.1 Porter-Thomas statistics

If the resonant states $|g_p\rangle$ are very complicated configurations over the channels α, then we should expect the reduced width amplitudes to be statistically distributed with a mean of zero. Because these amplitudes arise from many random influences in the Hamiltonian, we expect to be able to use the central-limit theorem, which says that the overall distribution of the $\gamma_{p\alpha}$ should be a normal distribution, centered here about zero. This was suggested by Porter and Thomas [1], who proposed that the normal distribution

$$\mathcal{P}_\alpha^{\text{PT}}(\gamma) = \frac{1}{\sqrt{2\pi}}\frac{1}{\langle\gamma^2\rangle_\alpha}\exp\left(-\frac{\gamma^2}{\langle\gamma^2\rangle_\alpha}\right) \qquad (11.1.1)$$

describe the probability density function for an individual reduced-width amplitude γ in channel α, where $\langle\gamma^2\rangle_\alpha$ is the variance (mean square value) of $\gamma_{p\alpha}$ in the energy region being considered.

The partial widths $\Gamma_{\alpha p} = 2\gamma_{p\alpha}^2 P_\alpha$ would then have the statistical distribution of a square of a normal variate. If we define a *mean partial width* $\langle\Gamma\rangle_\alpha = 2\langle\gamma_{p\alpha}^2\rangle P_\alpha$, and a normalized variate $x = \Gamma/\langle\Gamma\rangle_\alpha$, then x should follow the probability density function

$$\mathcal{P}_1(x) = \frac{1}{\sqrt{2\pi}} x^{-\frac{1}{2}} e^{-x/2}. \tag{11.1.2}$$

The penetrability factor P_α will vary with energy, but only slowly over the range of many narrow resonances in medium and heavy nuclei.

The total widths $\Gamma_p = \sum_\alpha \Gamma_{\alpha p} = 2\sum_\alpha \gamma_{p\alpha}^2 P_\alpha$ are composed of the sums of squares of normally distributed variables $\gamma_{p\alpha}$. Such a sum, for n terms called 'degrees of freedom', follows the χ^2 distribution $f(\mathcal{X}^2)$, namely

$$\mathcal{P}_n(x) = nf(nx) = \frac{n}{2\Gamma(\frac{n}{2})} \left(\frac{nx}{2}\right)^{\frac{n}{2}-1} e^{-\frac{nx}{2}}, \tag{11.1.3}$$

where $x = \Gamma/\langle\Gamma\rangle$, and $\Gamma(z)$ is the mathematical gamma function. The distribution (11.1.3) has a mean at $x = 1$ and a variance of $2/n$ about this mean.

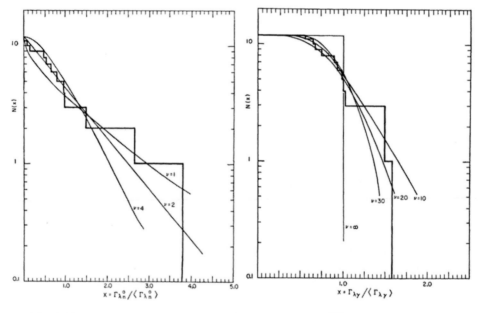

Fig. 11.2. An early analysis of distributions of widths, in ^{234}U* from analysis of 12 low-lying resonances in n + ^{233}U scattering, from [2]. The $N(x)$ is the number of levels having a value of $\Gamma/\langle\Gamma\rangle$ greater than x. The curves are the Porter-Thomas distributions for the indicated number of degrees of freedom (ν, our n). We see (left) that neutron widths show 1 or 2 open channels (degrees of freedom), while the photon widths (right) have $\sim$20. Reprinted with permission from M. S. Moore and C. W. Reich, *Phys. Rev.* **118** (1960) 718. Copyright (1960) by the American Physical Society.

For low-energy scattering, the number of degrees of freedom n will be the number of partial widths which add together to give the total width. At low energies and without photo-capture reactions, the distribution of total widths will be $\mathcal{P}_n$ for some small n, and hence will have a large variance about the mean (Fig. 11.2, left). When gamma-decay processes dominate, by contrast, because there are so many of these in heavy nuclei – to all the allowed states lower in energy – the value of n will be quite large (Fig. 11.2, right). Since the variance of $\Gamma / \langle \Gamma \rangle$ is $2/n$, the fractional fluctuations of Γ_γ around $\langle \Gamma_\gamma \rangle$ will be smaller for gamma decays than for particle decays.

11.2 Approximations neglecting interference

11.2.1 Reich-Moore approximation

In this section we exhibit a series of approximations that neglect progressively more interference terms, when compared with the exact R-matrix formulae of Chapter 10. The first is due to Reich and Moore [3], who considered the statistics in the sum for the width level matrices of Eq. (10.3.25), $\hat{\Gamma}_{pq} = 2 \sum_\alpha^n \gamma_{p\alpha} P_\alpha \gamma_{q\alpha}$. When n is large, they argue that the many terms with $p \neq q$ should average to zero, since $\gamma_{p\alpha}$ and $\gamma_{q\alpha}$ should have random signs and exhibit random size variations for the n different decay channels α. The off-diagonal terms of $\hat{\Gamma}_{pq}$ should at least be rather small compared with the diagonal terms $\hat{\Gamma}_{pp}$.

The *Reich-Moore approximation* is to neglect the off-diagonal terms $\hat{\Gamma}_{pq}$ just in the case of γ-channels, since in medium and heavy nuclei, as discussed above, there are so many partial γ-decay channels. The number of *particle* channels is much smaller, and all their off-diagonal elements should be preserved.

The Reich-Moore approximation, which has proved to be enormously successful in fitting low and medium energy nucleon-nucleus scattering, therefore approximates

$$
\hat{\Gamma}_{pq} = \begin{cases} \Gamma_p & p = q \\ 0 & p \text{ or } q \text{ are gamma channels} \\ 2 \sum_\alpha \gamma_{p\alpha} P_\alpha \gamma_{q\alpha} & p \text{ and } q \text{ are particle channels.} \end{cases} \tag{11.2.1}
$$

Reich and Moore show how to then divide the level matrix **A** of Eq. (10.3.26) into blocks so that it is easily inverted.

11.2.2 Multi-level Breit-Wigner approximation

The next simplification is to neglect *all* the off-diagonal terms of $\hat{\Gamma}_{pq}$, even for particle channels, and hence to neglect some more of the interference terms between different resonant levels for the particles. In this more extreme approximation, we have from Eq. (10.3.26)

$$
\mathbf{A}_{pq} = \delta_{pq} \frac{1}{e_p - E - \hat{\Delta}_{pp} - \frac{i}{2} \Gamma_{pp}}, \tag{11.2.2}
$$

from which the scattering $\tilde{\mathbf{S}}$ matrix by Eq. (10.3.29) is

$$\tilde{\mathbf{S}}_{\alpha'\alpha} = \Omega_\alpha \left[\delta_{\alpha\alpha'} + i \sum_p \frac{\Gamma_{\alpha p}^{1/2} \Gamma_{\alpha' p}^{1/2}}{E_p - E - \frac{i}{2}\Gamma_{pp}} \right] \Omega_{\alpha'}, \qquad (11.2.3)$$

where we have abbreviated $E_p \equiv e_p - \hat{\Delta}_{pp}$ as the energy of the resonant state p. If these are compound-nucleus resonances, then the compound-nucleus part of the angle-integrated cross section to channel α' from α is

$$\sigma_{\alpha'\alpha}(J_{\text{tot}}^\pi; E) = \frac{\pi}{k^2} \, g_{J_{\text{tot}}} \sum_p \frac{\Gamma_{\alpha p}\Gamma_{\alpha' p}}{(E - E_p)^2 + \Gamma_p^2/4}, \qquad (11.2.4)$$

where we remind the reader that $\Gamma_{\alpha p}$ is energy-dependent via the penetrability, and moreover that the sum over p terms is only for the levels of specific spin and parity J_{tot}^π. Each level p has a total width of $\Gamma_p = \sum_\alpha \Gamma_{\alpha p}$, an energy-integrated partial cross section of

$$\int_0^\infty dE \, \sigma_{\alpha'\alpha}^{(p)}(J_{\text{tot}}^\pi; E) = \frac{2\pi^2}{k^2} \, g_{J_{\text{tot}}} \frac{\Gamma_{\alpha p}\Gamma_{\alpha' p}}{\Gamma_p}, \qquad (11.2.5)$$

and an integrated non-elastic cross section, for incoming channel α, of

$$\int_0^\infty dE \, \sigma_\alpha^{(p)}(J_{\text{tot}}^\pi; E) = \frac{2\pi^2}{k^2} \, g_{J_{\text{tot}}} \frac{\Gamma_{\alpha p}(\Gamma_p - \Gamma_{\alpha p})}{\Gamma_p}. \qquad (11.2.6)$$

These values, for narrow resonances, will be used directly in energy integrals such as Eq. (12.1.17).

This is the *multi-level Breit-Wigner formula*, whereby the angle-integrated cross section is an incoherent sum of contributions from all the contributing resonances, each peaking near its position E_p with full width at half maximum of Γ_p. We expect that this expression would be most accurate if all the widths were much smaller than the mean spacing D between the levels, that is for mean widths $\langle\Gamma\rangle \ll D$. This is the case for non-overlapping and well-separated resonances.

11.3 Hauser-Feshbach models

If we add the cross sections of the multi-level Breit-Wigner formula (11.2.4) over all final states, the total should equal the summed reaction cross section σ_R defined by Eq. (3.2.32) for the loss of flux from the entrance channel. This should also match the total absorption cross section σ_A given by a one-channel optical potential for elastic scattering in channel α. Conversely, if we already know the optical potential and especially its imaginary component, we may use this knowledge to normalize the total widths for the decay of compound nuclear states. This is the basis for the very successful **Hauser-Feshbach approximation** [4] for calculating the production and all the decay channels of compound nuclear states. We now derive the Hauser-Feshbach formula in the case of well-separated resonances, using the above multi-level Breit-Wigner expression (11.2.4), and in subsection 11.3.5 show how to use this to calculate the decay chains of a compound nucleus.

To find the cross sections averaged over many resonances, let us average over an interval I that contains many resonances: $I \gg D$ where D is the mean level spacing for CN resonances of given total spin and parity J_{tot}^π. We define an *energy average cross section* for a sum of narrow peaks $\sigma_p(E)$ as appear in Eq. (11.2.4) by

$$
\begin{aligned}
\langle \sigma(E) \rangle &= \frac{1}{I} \int_{E-I/2}^{E+I/2} \sum_p \sigma_p(E) \mathrm{d}E \\
&= \frac{1}{I} \sum_{p,\, E_p \in [E \pm I/2]} \int_0^\infty \sigma_p(E) \mathrm{d}E \\
&= \frac{1}{I} \frac{I}{D} \int_0^\infty \sigma_p(E) \mathrm{d}E = \frac{1}{D} \int_0^\infty \sigma_p(E) \mathrm{d}E
\end{aligned}
\tag{11.3.1}
$$

since there are I/D peaks within the averaging interval that are similar on average.

The integral over a single resonance of a Breit-Wigner resonance peak in Eq. (11.2.4) gives

$$
\int_0^\infty \mathrm{d}E \, \frac{\Gamma_{\alpha p} \Gamma_{\alpha' p}}{(E - E_p)^2 + \Gamma_p^2/4} = \frac{2\pi \Gamma_{\alpha p} \Gamma_{\alpha' p}}{\Gamma_p}
\tag{11.3.2}
$$

(as was used to derive Eq. (11.2.5)), so the energy-averaged cross section is

$$
\langle \sigma_{\alpha'\alpha}(J_{\text{tot}}^\pi; E) \rangle = \frac{\pi}{k^2} g_{J_{\text{tot}}} \left\langle \frac{\Gamma_{\alpha p} \Gamma_{\alpha' p}}{\Gamma_p} \right\rangle \frac{2\pi}{D}.
\tag{11.3.3}
$$

11.3.1 Width fluctuation corrections

Unfortunately the average *ratio* $\langle \Gamma_{\alpha p} \Gamma_{\alpha' p}/\Gamma_p \rangle$ is *not* simply given in terms of the average values of the numerator factors and the denominator. Because of possible correlations in the averaging, we define a *width fluctuation factor* $W_{\alpha\alpha'}$ by

$$
\left\langle \frac{\Gamma_{\alpha p} \Gamma_{\alpha' p}}{\Gamma_p} \right\rangle = W_{\alpha\alpha'} \frac{\langle \Gamma_\alpha \rangle \langle \Gamma_{\alpha'} \rangle}{\langle \Gamma \rangle},
\tag{11.3.4}
$$

so the energy average cross section can be written

$$
\langle \sigma_{\alpha'\alpha}(J_{\text{tot}}^\pi; E) \rangle = \frac{\pi}{k_i^2} g_{J_{\text{tot}}} W_{\alpha\alpha'} \frac{2\pi}{D} \frac{\langle \Gamma_\alpha \rangle \langle \Gamma_{\alpha'} \rangle}{\langle \Gamma \rangle},
\tag{11.3.5}
$$

where the average sum $\langle \Gamma \rangle = \sum_\alpha \langle \Gamma_\alpha \rangle$.

To estimate the width fluctuation corrections $W_{\alpha\alpha'}$ in our $\Gamma \ll D$ limit, we focus on the numerators of Eq. (11.3.4). Factorizing out the penetrability factors, the $W_{\alpha\alpha'}$ must satisfy $\langle \gamma_{p\alpha}^2 \gamma_{p\alpha'}^2 \rangle = W_{\alpha\alpha'} \langle \gamma_{p\alpha}^2 \rangle \langle \gamma_{p\alpha'}^2 \rangle$. For inelastic reactions $\alpha \neq \alpha'$, so the $\gamma_{p\alpha}^2$ and $\gamma_{p\alpha'}^2$ should be statistically independent, giving $\langle \gamma_{p\alpha}^2 \gamma_{p\alpha'}^2 \rangle \sim \langle \gamma_{p\alpha}^2 \rangle \langle \gamma_{p\alpha'}^2 \rangle$ and $W_{\alpha\alpha'} \approx 1$. For elastic channels with $\alpha = \alpha'$, however, we have $\langle \gamma_{p\alpha}^4 \rangle = W_{\alpha\alpha} \langle \gamma_{p\alpha}^2 \rangle^2$, so, defining $x = \gamma_{p\alpha}/\langle \gamma_{p\alpha}^2 \rangle^{1/2}$, the elastic $W_{\alpha\alpha}$ is the fourth moment $\langle x^4 \rangle$. If the $\gamma_{p\alpha}$ follow the Porter-Thomas distribution, then this is the fourth moment of a normal distribution, namely $W_{\alpha\alpha} = 3$. To conserve the total cross section, however, this enhancement of the elastic channel should be compensated by a proportional reduction of all the other channels. This will only be significant if there are not too many of those, such as at low energies.

11.3.2 Transmission coefficients

For an incoming channel α, the reaction cross section is the sum of the $\sigma_{\alpha'\alpha}(J_{\text{tot}}^\pi; E)$ over the outgoing channels α'. We now also include the elastic channel $\alpha' = \alpha$ in this sum, as this is the *compound elastic* contribution to be discussed in subsection 11.5.1. Since $\sum_{\alpha'} \Gamma_{\alpha'p} = \Gamma_p$, we thus have

$$\int_0^\infty \sigma_\alpha^R(J_{\text{tot}}^\pi; E) \mathrm{d}E = \frac{\pi}{k^2} g_{J_{\text{tot}}} 2\pi \Gamma_\alpha,$$

so the average is $\langle \sigma_\alpha^R(J_{\text{tot}}^\pi; E) \rangle = \frac{\pi}{k^2} g_{J_{\text{tot}}} \frac{2\pi \langle \Gamma_\alpha \rangle}{D}.$ (11.3.6)

We now establish the absolute scale of the $\langle \Gamma_\alpha \rangle$ by connection with the reaction cross sections predicted by the optical model. From Eq. (3.2.32), the optical model gives a reaction cross section of

$$\sigma_\alpha^R(J_{\text{tot}}^\pi; E) = \frac{\pi}{k^2} g_{J_{\text{tot}}} (1 - |\mathbf{S}_{\alpha\alpha}^{J_{\text{tot}}\pi,\text{opt}}|^2),$$ (11.3.7)

where $\mathbf{S}_{\alpha\alpha}^{\text{opt}}$ is the elastic optical S-matrix element, and comparison of these two expressions gives

$$1 - |\mathbf{S}_{\alpha\alpha}^{\text{opt}}|^2 = \frac{2\pi \langle \Gamma_\alpha \rangle}{D}.$$ (11.3.8)

We now define the so-called *transmission coefficients*:[1]

$$\mathcal{T}_\alpha = 1 - |\mathbf{S}_{\alpha\alpha}^{\text{opt}}|^2,$$ (11.3.9)

---

[1] They should be distinguished from e.g. barrier tunneling coefficients because these $\mathcal{T}_\alpha = 0$ for all real optical potentials, and $\mathcal{T}_\alpha = 1$ for strongly absorbed partial waves.

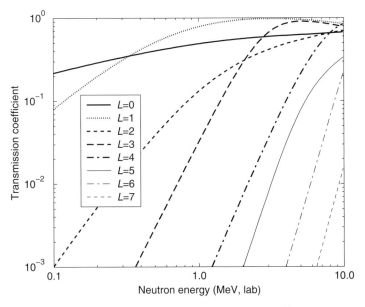

Fig. 11.3. Transmission coefficients for neutrons incident on ^{90}Zr in various partial waves L, using the optical potential of Koning and Delaroche [5] without the spin-orbit force.

which measure the 'coupling' described by the imaginary potentials between the external scattering and the internal compound-nucleus production. Figure 11.3 shows these for neutrons on ^{90}Zr in various partial waves.

In terms of the transmission coefficients, Eq. (11.3.5) becomes

$$\langle \sigma_{\alpha'\alpha}(J_{\text{tot}}^{\pi}; E) \rangle = \frac{\pi}{k^2}\, g_{J_{\text{tot}}} W_{\alpha\alpha'} \frac{\mathcal{T}_\alpha \mathcal{T}_{\alpha'}}{\sum_{\alpha''} \mathcal{T}_{\alpha''}}. \tag{11.3.10}$$

We see in this expression what are called the *Hauser-Feshbach branching ratios*:

$$\mathcal{B}_{\alpha'} = \frac{\mathcal{T}_{\alpha'}}{\sum_{\alpha''} \mathcal{T}_{\alpha''}}. \tag{11.3.11}$$

These Hauser-Feshbach formulae have simple physical interpretations. The branching ratio for the decay $\mathcal{B}_{\alpha'}$ of a given compound nuclear state to final channel α' is proportional to the transmission coefficient $\mathcal{T}_{\alpha'}$, normalized in the denominator by the summed coefficients so the total decay probability is unity. The *production* of the CN state is assumed to be a time-reversal of the decay mechanism, and hence is proportional to the same $\mathcal{T}_\alpha$. In subsection 11.3.5 we will see how to use the Hauser-Feshbach formula to calculate all the decay chains that occur after compound-nucleus production.

11.3.3 Weisskopf-Ewing approximation

The cross section summed over all total spins and parities is

$$\langle \sigma_{\alpha'\alpha}(E) \rangle = \frac{\pi}{k^2} \sum_{J_{\text{tot}}^\pi} g_{J_{\text{tot}}} W_{\alpha\alpha'} \frac{\mathcal{T}_\alpha \mathcal{T}_{\alpha'}}{\sum_{\alpha''} \mathcal{T}_{\alpha''}}. \tag{11.3.12}$$

This expression does *not* factorize into a product of production and decay probabilities, because J_{tot} and parity π are conserved quantum numbers that are not affected by averaging. It only factorizes if $W_{\alpha\alpha} \approx 1$ *and* the J_{tot}^π sum can be ignored. This latter occurs for example if only one incoming partial wave α is significant (e.g. $1/2^+$ for thermal neutrons on a light spin-zero target), or if all the $\mathcal{T}_\alpha$ have the same J_{tot}^π dependence. The earlier Weisskopf-Ewing theory [6] can be obtained as a limit of the Hauser-Feshbach theory by assuming a fixed distribution of spins. This enables a complete factorization as

$$\langle \sigma_{\alpha'\alpha}(E) \rangle = \sigma_\alpha^R(E) \frac{\mathcal{T}_{\alpha'}}{\sum_{\alpha''} \mathcal{T}_{\alpha''}} \equiv \sigma_\alpha^R(E) \, \mathcal{B}_{\alpha'}. \tag{11.3.13}$$

This theory simply states that the reaction cross section $\sigma_\alpha^R(E)$ of Eq. (11.3.7) decays according to branching ratios $\mathcal{B}_{\alpha'}$ obtained by normalizing the $\mathcal{T}_{\alpha'}$ to unit probability.

11.3.4 Strong couplings and overlapping resonances

The above derivation of the Hauser-Feshbach formula for the branching ratios proceeded in the limit of isolated resonances, which holds for example in neutron scattering on heavy nuclei up to 100 or 200 keV. Above that energy, as the resonances become both wider and more closely spaced, the condition of $\Gamma \ll D$ is no longer satisfied. However, the Hauser-Feshbach method still works rather well, and the above theory can be improved in three ways to work in the 'strong absorption' limit. The first change is that Γ/D may become large, but $\mathcal{T} \leq 1$ always, so the relation $\mathcal{T}_\alpha = 2\pi \langle \Gamma_\alpha \rangle / D$ can no longer hold. Simonius [7] gives reasons for using

$$\mathcal{T}_\alpha = 1 - \exp(-2\pi \langle \Gamma_\alpha \rangle / D), \tag{11.3.14}$$

an expression which behaves properly with both weak and strong couplings.

Secondly, the width-fluctuation correction $W_{\alpha\alpha'}$ has to be recalculated. Investigations by Moldauer [8] and others shows that $W_{\alpha\alpha'}$ is generally near unity except for the elastic channel. The elastic value was $W_{\alpha\alpha} \approx 3$ for small $\mathcal{T}_\alpha$, but is found to be near 2 for larger $\mathcal{T}_\alpha \lesssim 1$ in the strong absorption limit. Numerical analysis of statistical models suggests that it varies smoothly between these two limits, and various parametric forms have been found to work well in practice. A fit to results of Monte Carlo calculations by Tepel and co-workers [9], for example, gave

$$W_{\alpha\alpha} = 1 + \frac{2}{1 + \mathcal{T}_\alpha^{1/2}} \qquad (11.3.15)$$

for the elastic width-fluctuation correction, which appears to be quite adequate [10]. See [11] for more general calculations of $W_{\alpha\alpha'}$.

Finally, at higher energies there will be more open channels from target excitations, and hence direct-reaction transitions during the transition from scattering to compound-nucleus absorption. There will more likely be rotational or vibrational excitation of the target by the incoming nucleon while it is in the surface region. Several theories have been developed for this case. Satchler [12] proposes generalizing the transmission coefficients to a *matrix*, $\mathcal{T} = 1 - \mathbf{S}^\dagger \mathbf{S}$, which has the same diagonal elements as before:

$$\mathcal{T}_{\alpha'\alpha} = \delta_{\alpha\alpha'} - \sum_{\alpha''} (\mathbf{S}^{\text{opt}}_{\alpha''\alpha'})^* \, \mathbf{S}^{\text{opt}}_{\alpha''\alpha} \qquad (11.3.16)$$

in terms of the optical S-matrix elements. The diagonalizing transformation of this matrix may be used to transform a coupled-channels scattering S matrix into an uncoupled set of diagonal matrices [13], and the standard theory [14] reused. Some more general theories such as that of Kawai, Kerman and McVoy [15] attempt to include direct and compound reactions on an equal footing, but have not yet been systematically tested.

11.3.5 Decay models

The simplest kind of decay network is that of the irreversible decay chain of an excited compound nucleus. These form a Markov chain with different branching probabilities at each stage, and constitute a *Hauser-Feshbach model* for the irreversible stages in the decay of that compound nucleus. These statistical decay models can be used whenever there is a large number of resonances in the energy region of interest, so that the cross section can be calculated by averaging over resonances. This can occur for light nuclei at higher excitation energies, for medium-mass nuclei away from closed shells and from the dripline, and for most heavy nuclei.

In subsection 11.3.1, the theory of one-stage cross sections from isolated resonances was used to derive the key equation (11.3.10), for the energy-averaged cross section. We now present this scattering in two stages: first the production of compound-nucleus resonances, and secondly the decay of such resonances according to branching ratios $\mathcal{B}_\alpha$. The Hauser-Feshbach decay model then assumes that *further* decays by particle emissions from the residual nuclear states can be calculated by using similar branching ratios. Successive particle emissions result in 3- and 4-body final states and more. These cannot be coherently described by our set of quantum numbers α for two-body channels, but if each decay is regarded as statistically independent (decoherent), then the probabilistic branching ratios for the possible two-body decays of any residual state may still be calculated.

We first find the production of a compound-nucleus state when the incoming projectile p_i fuses with the target t_i. Any compound-nucleus state may be labeled by multi-index $\beta = \{x, t\}$ such that x labels the mass and charge of the nucleus, and t its excited state, so spin I_t, parity π_t, and excitation energy ϵ_t depend on t. With this notation, the cross section for fusing with a target initially in its ground state $\beta_i = \{x_i, 1\}$, to produce a CN state β in nucleus $x = x_i + p_i$ with spin and parity $I_t^{\pi t}$, is

$$\sigma_\beta^{(0)} = \sum_{L_i J_{p_i}} \frac{\pi}{k_i^2} \frac{2I_t + 1}{(2I_{p_i} + 1)(2I_{t_i} + 1)} T_{\beta_i : \beta}^{L_i J_{p_i}}, \tag{11.3.17}$$

for incoming partial wave $L_i J_{p_i}$.

The transmission coefficients $T_{\beta' : \beta}^{L J_p}$ are found by using the previous formula for optical-model scattering on a nuclear state β' when absorption produces a state β:

$$T_{\beta' : \beta}^{L J_p} = 1 - |\mathbf{S}_{\alpha\alpha}^{L J_p, \text{opt}} (\epsilon_t - \epsilon_{t'} - Q_x^{x'})|^2. \tag{11.3.18}$$

This S matrix is calculated for optical-model scattering in partial wave $L J_p$ with kinetic energy $E = \epsilon_t - \epsilon_{t'} - Q_x^{x'}$, total angular momentum I_t and parity π_t. In our previous notation, this is the two-body partial wave $\alpha = \{y + x', 1, t', L, I_y, J_p, I_t\}$ for a particle y incident on a target x' (neglecting excited states of y). We find $y = x - x'$ from the differences of the charges and masses of the initial and final CN states. The $Q_x^{x'}$ is the Q-value, namely the energy released on capturing the particle on the x' ground state to form the ground state of x. We set $T_{\beta' : \beta}^{L J_p} = 0$ if the constituting angular momenta cannot be coupled to the required total I_t, or if parity is not conserved.

We can now rewrite Eq. (11.3.10) as the $n = 1$ case of a general formula for the decay of a $\sigma_\beta^{(n-1)}$ CN-population cross section, with n the number of particles that will have been emitted after this decay:

$$\sigma_{\beta'}^{(n)} = \sum_\beta \mathcal{B}_{\beta' : \beta} \sigma_\beta^{(n-1)}, \tag{11.3.19}$$

where the branching ratio from state β to β' is

$$\mathcal{B}_{\beta' : \beta} = \sum_{L J_p} \frac{T_{\beta' : \beta}^{L J_p}}{\sum_{\beta'' L'' J_p''} T_{\beta'' : \beta}^{L'' J_p''}} \tag{11.3.20}$$

in terms of the transmission coefficients of Eq. (11.3.18). The sum over β'' in the denominator is over the set of all open channels that can be reached from the initial state β at energy ϵ_t, which is the same as the set of all the allowed outgoing β' values.

The decays will in general include emission of all kinds of particles and clusters, including deuterons, tritons and α particles if the energies are sufficiently high that such channels are open. Knowledge is therefore needed of the optical potentials of all the incident and emitted particles and clusters, with particular attention to the imaginary parts of these potentials. This is because the imaginary part describes the 'transmission' or 'coupling' between exterior scattering and the interior population of compound-nucleus states. Photons can also be produced, allowing a nuclear state to γ-decay within the cascade, but the transmission coefficients for photons depend on giant-resonance parameters, and discussion of these is beyond the scope of this book.

The formula (11.3.19) ignores the width-fluctuation corrections $W_{\alpha\alpha_i}^{J_{\text{tot}}^{\pi}E_i}$ discussed in subsection 11.3.1. To include them would require modification of the structure of the equations, since the first decay cross sections $\sigma_{\beta'}^{(1)}$ to state β' now strictly depend in part on the incoming β_i in Eq. (11.3.17).

In practical calculations, the discrete spectrum of energy levels is only comprehensively known at low excitation energies. Above those energies, we have to rely on some estimates of the average *level densities* for eigenstates of each compound nucleus. In the spectral region $\epsilon \geqslant \epsilon_s$ above where discrete levels are known, we approximate the level density by a continuous function. In this region, sums over open channels α may be estimated by a sum over the discrete quantum numbers $\{I_t \pi_t\}$ of integrals over the excitation energy using the level density. This level density will depend on excitation energy ϵ, spin I_t and parity π_t as, say, $D^{-1} = \rho_{\text{ld}}(\epsilon, I_t, \pi_t)$, in units of MeV^{-1}. The denominator in Eq. (11.3.20) is thus rewritten to include integrals

$$\sum_{\beta''} \mathcal{T}_{\beta'':\beta} \sim \sum_{\beta''}^{\epsilon_{t''} < \epsilon_s} \mathcal{T}_{\beta'':\beta} + \sum_{I_{t''}, \pi_{t''}} \int_{\epsilon_s}^{\epsilon - Q_t^{t''}} \mathcal{T}_{\{x''I_{t''}\pi_{t''}\epsilon_{t''}\}:\beta} \, \rho_{\text{ld}}(\epsilon_{t''}, I_{t''}, \pi_{t''}) d\epsilon_{t''}$$

(11.3.21)

up to the highest possible excitation energy of an open channel. We discuss in Section 11.4 how to estimate the level densities.

The final distribution of residues will consist of nuclei in their ground states, or perhaps in isometric states. For specified final nuclear states $\{\beta_f\}$, the cumulative cross section is

$$\sigma_f(\{\beta_f\}) = \sum_{n=0}^{\infty} \sum_{\beta \in \{\beta_f\}} \sigma_{\beta}^{(n)},$$

(11.3.22)

which includes a sum over all emitted particles during the decay sequences.

The energy spectrum of specific emitted particles y may be calculated. As a function of energy E_y, it is a cumulative sum of terms that select the specific out-going kinetic energy:

$$\sigma_y(E_f) = \sum_{n=1}^{\infty} \sum_{\beta'} \sum_{\substack{\beta:\, \epsilon_t = \epsilon_{t'} + E_y \\ x = x' + y}} \mathcal{B}_{\beta':\beta} \sigma_{\beta}^{(n-1)}. \tag{11.3.23}$$

By using Hauser-Feshbach models we therefore have, as illustrated in Fig. 11.4, a theory of the initial production rate given by $\sigma^{(0)}$, as well as of all the successive decay cross sections $\sigma^{(n)}$ of a compound nucleus involving the emission of n particles or photons. The theory may include known discrete levels at low excitation energies, but at higher excitations it must average over energy intervals containing many resonances.

In practice, the dependence of the transmission coefficients on the projectile spin is usually ignored, as polarized projectiles are not under consideration. In this simplified case, we should average Eq. (11.3.18) over the J_p quantum number, and keep only the L dependence.

The Hauser-Feshbach method is implemented systematically in the codes STAPRE [16], EMPIRE [17], and TALYS [18], all of which are publically available.

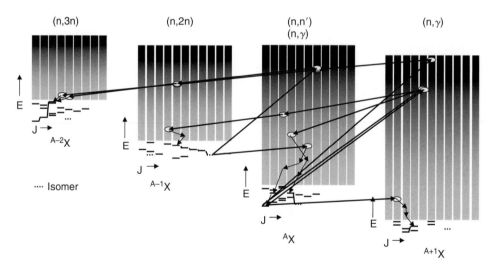

Fig. 11.4. Multi-stage decay processes may be modeled with the Hauser-Feshbach theory, following an initial reaction of neutrons incident on ^{A}X. The initial compound nucleus is $^{A+1}$X on the right, which may decay by emission of 1, 2 or 3 neutrons to the nuclei to the left. Figure courtesy of Erich Ormand.

11.4 Level densities

As well as the optical potentials, a key ingredient of Hauser-Feshbach models are the level densities, as these are needed at high energies, including above particle emission (breakup) thresholds. Just above the threshold, measurements of low-energy scattering can be used to determine the spacing D_0 of individual *s*-wave resonances up to several hundred keV, until the resonances overlap too much. For neutrons, such measurements can only be accomplished with stable target nuclei in, for example, time-of-flight experiments. For higher energies, and for all unstable nuclei, discrimination of individual neutron resonances is almost impossible, and theoretical estimates must be used.

Level densities may be calculated microscopically by the shell model in light nuclei for excitation energies up to several tens of MeV, or by the more general Monte Carlo shell model [19] for somewhat heavier nuclei. More simple estimates may be obtained by combinatorial analysis of single-particle coupling combinations [20], and the most simple method is to use Fermi gas approximations as first suggested by Bethe [21].

A simple ansatz is that of the *back-shifted Fermi gas* [22], which introduces an empirical energy shift into the Fermi gas model to include the effect of the pairing correlation energy. The level density in this approach is given by

$$\rho_{bsfg}(\epsilon) = \frac{\sqrt{\pi}}{12 a_{ld}^{1/4}} \frac{\exp 2\sqrt{a_{ld} U}}{U^{5/4}}, \qquad U = \epsilon - \Delta_{bs}, \qquad (11.4.1)$$

where Δ_{bs} is an adjustable *back-shift* independent of excitation energy ϵ. The level-density parameter a_{ld} is standardly taken near $a_{ld} = A/8 \text{ MeV}^{-1}$ for nuclei of mass number A, and the back-shift parameter is most simply near $\Delta_{bs} = \pm 12/\sqrt{A}$, taking the plus sign for even-even nuclei, the minus sign for odd-odd nuclei, and $\Delta_{bs} = 0$ for odd-A nuclei.

In order to describe the dependence on compound-nucleus spin I, and in particular the lack of states for large I, it is common to include a Gaussian cutoff factor of width σ_{spc}:

$$\mathcal{F}(\epsilon, I) = \frac{2I+1}{2\sigma_{spc}^2} \frac{1}{\sqrt{2\pi}\,\sigma_{spc}} \exp\left(-\frac{I(I+1)}{2\sigma_{spc}^2}\right). \qquad (11.4.2)$$

The spin cutoff parameter σ_{spc} can be chosen in proportion to the average angular momentum of a rigid body:

$$\sigma_{spc}^2 = \frac{\mathcal{M}_{rigid}}{\hbar^2}\sqrt{\frac{U}{a_{ld}}}, \qquad (11.4.3)$$

using a moment of inertia $\mathcal{M}_{\text{rigid}} = \frac{2}{5}m_u A R_0^2$. Here the surface radius is R_0, and m_u is the atomic mass unit (1 amu).

Including also a factor 1/2 for an equiprobable distribution over parities, we have

$$\rho_{\text{ld}}(\epsilon, J_{\text{tot}}^\pi) = \frac{1}{2}\mathcal{F}(\epsilon, J_{\text{tot}})\rho_{\text{bsfg}}(\epsilon). \qquad (11.4.4)$$

Improvements on this back-shifted Fermi gas model are necessary at low excitation energies, because the above formulae diverge when $\epsilon \lesssim \Delta_{bs}$ as then $U \lesssim 0$. The simplest solution is that of Gilbert and Cameron [23], who at low energies match Eq. (11.4.1) onto a level density for constant temperature

$$\rho_{\text{ld}}(\epsilon; T) \propto \exp(\epsilon/T)/T, \qquad (11.4.5)$$

by requiring continuity of the level density and its derivative at the matching energy to fix the temperature T and the matching energy. The resulting level density curves for ^{67}Zn and ^{68}Zn are shown in Fig. 11.5.

All the level density parameters may be fine-tuned to reproduce known excitation spectra and s-wave resonance spacings D_0 where these are known. For unstable

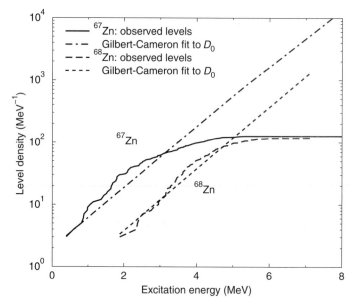

Fig. 11.5. Level densities in ^{67}Zn and ^{68}Zn. The curves becoming flat at high energies are the cumulative count of the observed levels, of which there are just over 100 in each nucleus. The rising lines are plots using the Gilbert and Cameron model fitted to the D_0, the s-wave resonance spacing at the breakup threshold, and fitted (approximately) to the experimental curves at low energies. We see that the back-shift Δ_{bs} differs by $\sim$1.5 MeV for these two nuclei.

nuclei, where default parameters have so far been used [24, 25], we expect that these Fermi gas models will eventually be superseded by the results of microscopic methods, either from the shell model directly, or from Monte Carlo simulations of the shell-model partition functions at finite temperatures. The more microscopic models will especially correct the assumption of equal distributions for each parity, an assumption that becomes less accurate towards the driplines.

11.5 Average amplitudes and the optical model

11.5.1 Sources of the optical potential

The larger part of the optical potential is the real part from folding the nucleon-nucleon forces over the density distributions of the interacting bodies. This is the folding procedure described in Section 5.2. The smaller part of the potential, but no less significant, is the *dynamic polarization potential* (DPP) that comes from all other processes. The main component of the DPP is the imaginary part arising from the absorptive effect of channels not explicitly included in the direct-reaction model. For a one-channel model, that refers to *all* the non-elastic channels.

These omitted channels are of two kinds – they may be further direct-reaction channels, or they may involve compound-nucleus production. Other direct-reaction processes obviously remove flux from the elastic channel, but so does the excitation of CN states. This is because then a fraction of the initial wave packet is temporarily captured for the long lifetimes of the CN states, and then released incoherently at a later time. Even if it is released back to the elastic channel, the effect on a short-duration wave packet is the same as absorption. The optical potential is defined as that which reproduces the prompt scattering in a given model space.

One way of delineating prompt scattering is to consider a wave packet with a wide energy spread, according to $\Delta t \cdot \Delta E \sim \hbar$. The scattering amplitude of a coherent wave packet is the average *amplitude* over the energy range, *not* the average cross section. Since the optical potential is defined as that which reproduces the prompt scattering, the scattering amplitudes from the optical potential will be the *energy-averaged amplitude* of exact scattering. This average amplitude for scattering will involve cancellations over any narrow resonances, as the S-matrix element (the asymptotic amplitude) changes in phase as $e^{2i\delta}$, and the average will be reduced in amplitude as if absorption were present. The flux that does eventually return incoherently to the elastic channel is called the *compound elastic cross section*, and has been evaluated by the Hauser-Feshbach theory of the previous section.

Thus we now come to the question of how to derive from first principles the optical potential itself, as distinct from fitting to data, when we have neglected direct-reaction channels, neglected (or averaged over) compound-nucleus resonances, or both.

11.5.2 Effects of neglected direct-reaction channels

Whenever the Hamiltonian couples elastic to non-elastic channels, the elastic scattering will be changed. To get a one-channel model to reproduce the new elastic scattering (affected by couplings), we should have to modify some of the potentials in the elastic channel. The Green's function method can be used to show that the effect of specific reaction channels can be equivalent to adding a dynamic polarization potential to the elastic channel.

Suppose we take a schematic pair of coupled channels

$$[T_1 + U_1 - E_1]\psi_1(R) + V_{12}\psi_2(R) = 0$$
$$[T_2 + U_2 - E_2]\psi_2(R) + V_{21}\psi_1(R) = 0, \qquad (11.5.1)$$

then we can formally solve these equations using Green's function methods to find the effective equation for ψ_1 alone. We find

$$[T_1 + U_1 + V_{12}\hat{G}_2^+ V_{21} - E_1]\psi_1(R) = 0. \qquad (11.5.2)$$

Here we use the distorted-wave Green's function for channel 2, $\hat{G}_2^+ = [E_2 - T_2 - U_2]^{-1}$, with the notation of Eq. (3.3.10). This implies that the scattering in channel 1 is not governed purely by U_1, but by an additional effective potential $V_{12}\hat{G}_2^+ V_{21}$ that depends on energy (E_2) and is non-local. This additional part is the dynamic polarization potential (DPP) :

$$V_{\text{DPP}} = V_{12}\hat{G}_2^+ V_{21}, \qquad (11.5.3)$$

and its effect is to change the elastic scattering from that caused by U_1 alone. The dynamic polarization potential will have an imaginary part that reflects the loss of flux from channel 1 into channel 2, and will also have a real part that may modify potential barriers and hence any low-energy tunneling probabilities.

This illustrates how coupled-channels calculations may be reduced to a one-channel problem, by building all the coupling effects into an extra term in the single-channel Hamiltonian. This extra term appears in Eq. (11.5.2) for the channel 1, because we have neglected to explicitly couple to the direct-reaction channel 2, and the operation of this extra term on the elastic-channel wave function $\psi_1(R_1)$ can be written with explicit coordinates as

$$V_{\text{DPP}}\psi_1 = V_{12}\hat{G}_2 V_{21}\psi_1$$
$$= V_{12}(R) \int_0^\infty G_2(R, R'; E_2)V_{21}(R')\psi(R')dR', \qquad (11.5.4)$$

where $G_2(R, R')$ is the Green's function kernel for propagation in channel 2 at energy E_2. We see that V_{DPP} is not simple. First, it is non-local as $R \neq R'$ in the

Green's function. Second, it is energy dependent as G_2 depends explicitly on the channel 2 energy E_2, and, third, it will almost certainly depend on the elastic partial wave $|\alpha\rangle = |LI_pJ_pI_tJ_{\text{tot}}\pi\rangle$. This last dependence may be an L-dependence or a parity-dependence, or something more complicated.

The effects of non-locality may be estimated by the Perey-Buck procedure [26], or by finding what is called the *trivially-equivalent local potential* (telp) if the coupled-channels solutions ψ_α are all known:

$$V_{\text{telp}}(R) = \frac{V_{12}\psi_2(R)}{\psi_1(R)}. \tag{11.5.5}$$

This $V_{\text{telp}}(R)$ will have imaginary parts that are both positive (emissive) and negative (absorptive) at different radii, but the total change of elastic flux should be a removal. Since the V_{telp} will have poles where the elastic channel has a node $\psi_1(R) \sim 0$, a *weighted-equivalent local potential* (welp) is sometimes calculated [27] by averaging $V_{\text{telp}}(R)$ over all partial waves with weight factor $\psi_1(R)^*$, giving

$$V_{\text{welp}}(R) = \frac{\sum_{J_{\text{tot}},\alpha'\neq\alpha} \psi_\alpha^{J_{\text{tot}}}(R)^* V_{\alpha\alpha'} \psi_{\alpha'}^{J_{\text{tot}}}(R)}{\sum_{J_{\text{tot}}} |\psi_\alpha^{J_{\text{tot}}}(R)|^2}. \tag{11.5.6}$$

if α is the elastic channel. This expression does avoid zero denominators as it is extremely improbable that all partial waves pass through zero at the same radius. The real test of any of these potentials averaged over partial waves, however, is how well they reproduce all the partial-wave S-matrix elements, and hence the angular distribution for elastic scattering. For this reason, the optical potentials are often best found by fitting to the elastic S-matrix elements or angular distribution. Figure 11.6 shows recent results for the effect of proton- and neutron-pickup channels on the optical potential for deuterons incident on ^{40}Ca at 52 MeV. Obtaining the optical potential from the S matrix is often referred to as the inverse method and a recent review can be found in [29].

11.5.3 Effects of neglected compound-nucleus channels

We will now find the contributions to the optical potential that arise when we neglect (or average over) compound-nucleus resonances. This is done by assuming that we can average S-matrix amplitudes over an energy range that includes many resonances. The resulting averaged S-matrix element is that which should be given by the optical potential, so

$$\mathbf{S}_{\alpha'\alpha}^{\text{opt}} = \overline{\mathbf{S}_{\alpha'\alpha}}, \tag{11.5.7}$$

averaging over some defined energy interval $\mathcal{I}$. We hence talk about **optical average** S-matrix elements.

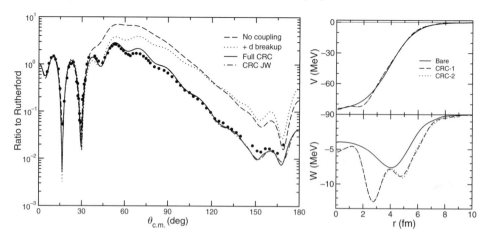

Fig. 11.6. Left: Data for the ^{40}Ca(d,d) elastic scattering compared with calculations including no coupling (dashed curve), deuteron breakup only (dotted curve), and deuteron breakup plus (d,t) and (d,^{3}He) pickup (solid curve) [28]. The dot-dashed curve gives the result for the J-weighted S matrix from the full CRC calculation. Right: Comparing the bare potential (solid line) and the two inverted potentials, with the real part in the upper panel and the imaginary part below. Reprinted with permission from N. Keeley and R. S. Mackintosh, *Phys. Rev. C* **77** (2008) 054603. Copyright (2008) by the American Physical Society.

Consider the case of low-energy scattering in which there are many prominent CN resonances, typically at energies below 1 MeV for neutrons, as above this energy they become broader and overlap more. Now we do not average the cross sections as in Section 11.3, but average the scattering *amplitudes*: either the whole wave function or just its asymptotic coefficient the S-matrix element. Let us average these complex functions over an interval $\mathcal{I}$ that is much larger than the average level spacing D, which in turn is assumed larger than the average width $\langle \Gamma \rangle$. That is, we deal with non-overlapping resonances, and assume that the Coulomb functions (the shift functions and penetrabilities) do not vary significantly over the energy interval $\mathcal{I}$.

First we define an *optical average* S-matrix element as $\overline{\mathbf{S}} \equiv \langle \mathbf{S} \rangle$ by

$$\overline{\mathbf{S}}(E) = \int dE' \, f(E - E') \, \mathbf{S}(E') \qquad (11.5.8)$$

for some smoothing function $f(\Delta E)$. Originally in Eq. (11.3.1) we used a square step function, but this now gives insufficiently smooth effects as particular resonances enter or leave the interval. An alternative suggested for this purpose by Brown [30] is to use a Lorentzian smoothing function

$$f(\Delta E) = \frac{\mathcal{I}}{\pi} \frac{1}{\Delta E^2 + \mathcal{I}^2} \tag{11.5.9}$$

where now $2\mathcal{I}$ is the full width at half maximum. This shape will be seen to yield a mathematically simpler conclusion than for the square step.

Averaged S-matrix elements

Any smoothed optical $\overline{\mathbf{S}}$ does not itself give an average elastic or average reaction cross section after integrating over many resonances, since these are quadratic in $\mathbf{S}$. But it does give the average total cross section σ_T of Eq. (3.1.50), since this is linear in $\mathbf{S}$. More specifically, since

$$\sigma_T = \frac{2\pi}{k_i^2} \sum_{J_{\text{tot}}\pi} g_{J_{\text{tot}}} (1 - \text{Re}\mathbf{S}), \tag{11.5.10}$$

we have $$\overline{\sigma}_T = \frac{2\pi}{k_i^2} \sum_{J_{\text{tot}}\pi} g_{J_{\text{tot}}} (1 - \text{Re}\overline{\mathbf{S}}) = \sigma_T(\text{opt}). \tag{11.5.11}$$

The reaction and angle-integrated elastic cross sections are however quadratic in $\mathbf{S}$:

$$\sigma_R = \frac{\pi}{k_i^2} \sum_{J_{\text{tot}}\pi} g_{J_{\text{tot}}} (1 - |\mathbf{S}|^2) \text{ and } \sigma_{\text{el}} = \frac{2\pi}{k_i^2} \sum_{J_{\text{tot}}\pi} g_{J_{\text{tot}}} |1 - \mathbf{S}|^2. \tag{11.5.12}$$

Since $1 - |\mathbf{S}|^2 = 1 - |\mathbf{S} - \overline{\mathbf{S}}|^2 - 2\text{Re}(\mathbf{S} - \overline{\mathbf{S}})\overline{\mathbf{S}} - |\overline{\mathbf{S}}|^2$, the average reaction cross section contains the factor

$$\overline{1 - |\mathbf{S}|^2} = 1 - |\overline{\mathbf{S}}|^2 - \overline{|\mathbf{S} - \overline{\mathbf{S}}|^2}. \tag{11.5.13}$$

From this, we find that the energy-average reaction cross section differs from that calculated from the optical potential according to

$$\overline{\sigma}_R = \sigma_R(\text{opt}) - \sigma_{\text{fl}}, \tag{11.5.14}$$

where

$$\sigma_{\text{fl}} = \frac{\pi}{k_i^2} \sum_{J_{\text{tot}}\pi} g_{J_{\text{tot}}} \overline{|\mathbf{S} - \overline{\mathbf{S}}|^2} \tag{11.5.15}$$

is the *fluctuation cross section* that depends on the variance of the S matrix about its mean as we vary the energy across the resonances. Similarly, we find that the average elastic cross section differs from that of the optical model as

$$\overline{\sigma}_{\text{el}} = \sigma_{\text{el}}(\text{opt}) + \sigma_{\text{fl}}. \tag{11.5.16}$$

Physically, σ_{fl} is the compound elastic cross section mentioned previously: that flux which comes back to the elastic channel after the decay of CN states. It enhances the elastic cross section over that predicted by the optical model, but this does not invalidate the optical model because it is produced after a time delay and is quantum mechanically incoherent, and hence simply adds a symmetric background to the elastic cross section. It reduces the reaction cross section from the optical prediction, since it describes flux that does not, after all, leave the elastic channel.

The Hauser-Feshbach approach may be regarded as producing an approximation for the fluctuation component for any S-matrix element, namely

$$\overline{|\mathbf{S}_{\alpha'\alpha} - \overline{\mathbf{S}_{\alpha'\alpha}}|^2} = W_{\alpha\alpha'} \frac{\mathcal{T}_\alpha \mathcal{T}_{\alpha'}}{\sum_{\alpha''} \mathcal{T}_{\alpha''}}. \tag{11.5.17}$$

Averaging and complex energies

Suppose that we had a microscopic theory or multi-channel **R**-matrix fit which describes all of the CN resonances in the energy interval $\mathcal{I}$ of interest, and then we averaged the S matrix over the resonances using the Lorentzian smoothing kernel of Eq. (11.5.9). We start from the multi-level Breit-Wigner formula (11.2.3) for the **S** matrix:

$$\mathbf{S}_{\alpha'\alpha}(E) = \Omega_\alpha \left[\delta_{\alpha\alpha'} + \mathrm{i} \sum_p \frac{\sqrt{\Gamma_{\alpha p}}\sqrt{\Gamma_{p\alpha'}}}{E_p - E - \frac{\mathrm{i}}{2}\Gamma_{pp}} \right] \Omega_{\alpha'}. \tag{11.5.18}$$

If we average over a suitably small interval $\mathcal{I}$, we may assume that the Ω_α and $\Gamma_{\alpha p}$ are constant, as discussed earlier, so the essential averaging integral is

$$\overline{\mathcal{S}} = \int_0^\infty \frac{\mathcal{I}}{\pi} \frac{1}{(E - E') + \mathcal{I}^2} \frac{c_p}{E_p - \mathrm{i}\Gamma_p/2 - E'} \, \mathrm{d}E', \tag{11.5.19}$$

for some numerator c_p. The poles of the integrand are at $E' = E_p - \mathrm{i}\Gamma_p/2$ and $E' = E \pm \mathrm{i}\mathcal{I}$. We may complete the integral in the complex E' plane by continuing it with a semicircular loop in the upper half plane which encloses the pole at $E' = E + \mathrm{i}\mathcal{I}$. The residue at this pole is $c_p/[2\pi\mathrm{i}(E_p - \mathrm{i}\Gamma_p/2 - E - \mathrm{i}\mathcal{I})]$, so

$$\overline{\mathcal{S}} = \frac{c_p}{E_p - \mathrm{i}\Gamma_p/2 - E - \mathrm{i}\mathcal{I}} \approx \frac{c_p}{E_p - E - \mathrm{i}\mathcal{I}} \tag{11.5.20}$$

as $\mathcal{I} \gg \Gamma_p/2$.

The Lorentz-smoothed S-matrix element has in general the simple form

$$\overline{\mathbf{S}_{\alpha'\alpha}(E)} = \mathbf{S}_{\alpha'\alpha}(E + \mathrm{i}\mathcal{I}), \tag{11.5.21}$$

showing that this smoothed S matrix can be very conveniently found by an analytic continuation of the scattering to a complex energy. This applies to all scattering functions $\mathbf{S}(E)$ which have poles in the third quadrant, which correspond to CN resonances with positive widths. It is therefore expected to be true, even for overlapping or dense resonance spectra, that the averaging interval $\mathcal{I}$ adds $2\mathcal{I}$ to the CN width Γ_p, and spreads out all the resonance strengths over the whole averaging interval. This tells us that we can find the S matrix to be obtained from the optical potential by evaluating the S matrix from a microscopic model at some complex energy $E + \mathrm{i}\mathcal{I}$.

Substituting Eq. (11.5.18) into Eq. (11.5.21), and approximating $\sum_p$ by $\frac{1}{D}\int dE_p$ where D is the level spacing assumed to be small and the $\Gamma_{\alpha p}$ are assumed to be constant, we find

$$
\overline{\mathbf{S}_{\alpha'\alpha}(E)} = \Omega_\alpha \left[\delta_{\alpha\alpha'} + \mathrm{i}\frac{1}{D}\int dE_p \frac{\sqrt{\Gamma_{\alpha p}\Gamma_{p\alpha'}}}{E_p - E - \mathrm{i}\mathcal{I}} \right] \Omega_{\alpha'}
$$

$$
= \Omega_\alpha \left[\delta_{\alpha\alpha'} + \mathrm{i}\sqrt{\Gamma_{\alpha p}\Gamma_{p\alpha'}}\frac{1}{D}\int dE_p \frac{(E_p - E) + \mathrm{i}\mathcal{I}}{(E_p - E)^2 + \mathcal{I}^2} \right] \Omega_{\alpha'}. \quad (11.5.22)
$$

The real part of the E_p integrand is odd about E and its integral should approximately cancel, whereas the imaginary integral has a value of π. This gives the energy-averaged

$$
\overline{\mathbf{S}_{\alpha'\alpha}(E)} \approx \Omega_\alpha \left[\delta_{\alpha\alpha'} - (\Gamma_{\alpha p}\Gamma_{p\alpha'})^{\frac{1}{2}}\frac{\pi}{D} \right] \Omega_{\alpha'}. \quad (11.5.23)
$$

At low energies $\Omega_\alpha \approx 1$ as $ka \ll 1$, so the average elastic S-matrix element is

$$
\overline{\mathbf{S}_{\alpha\alpha}(E)} \approx 1 - \frac{\pi\langle\Gamma_\alpha\rangle}{D} \quad (11.5.24)
$$

which agrees with the result of Eq. (11.3.8), to first order in $\pi\langle\Gamma_\alpha\rangle/D$.

Averaged CN polarization potentials

As well as averaging the S-matrix elements by $\overline{\mathbf{S}_{\alpha'\alpha}(E)} = \mathbf{S}_{\alpha'\alpha}(E + \mathrm{i}\mathcal{I})$, we could also similarly average the dynamic polarization potential (DPP) that comes from the couplings to the compound-nucleus states. Let us generalize the definition of this polarization potential from Eq. (11.5.3) to use a Feshbach projection operator P for the direct-reaction channels and Q for the compound-nucleus channels, with $P+Q = 1$. From the full Hamiltonian H, the CN sector is governed by $H_{QQ} = QHQ$ and the couplings between CN states and direct-reaction channels use $H_{PQ} = PHQ$.

We now calculate the average DPP by evaluating it for scattering at a complex energy $E + \mathrm{i}\mathcal{I}$. A complete set of CN states is taken to be $|E_p\rangle$ satisfying $H_{QQ}|E_p\rangle = E_p|E_p\rangle$, and the sum over these is (as above) approximated by $\sum_p \sim \frac{1}{D}\int dE_p$. On this basis,

$$\overline{V_{\mathrm{DPP}}(E)} = H_{PQ}\frac{1}{E + \mathrm{i}\mathcal{I} - H_{QQ}}H_{QP}$$

$$= \sum_{p} H_{PQ}|E_p\rangle \frac{1}{E + \mathrm{i}\mathcal{I} - E_p}\langle E_p|H_{QP}$$

$$= \frac{1}{D}\int \mathrm{d}E_p H_{PQ}|E_p\rangle \frac{1}{E + \mathrm{i}\mathcal{I} - E_p}\langle E_p|H_{QP}$$

$$= \frac{1}{D}\int \mathrm{d}E_p H_{PQ}|E_p\rangle \frac{(E - E_p) - \mathrm{i}\mathcal{I}}{(E - E_p)^2 + \mathcal{I}^2}\langle E_p|H_{QP}. \qquad (11.5.25)$$

We might now assume that the *coupling operators* H_{PQ} to the CN states do not *themselves* vary significantly over the energy range $\mathcal{I}$, so that all the resonant peaks arise from the Breit-Wigner denominators $(E - E_p)^2 + \mathcal{I}^2$. In this case, we could approximate the imaginary part of $\overline{V_{\mathrm{DPP}}(E)}$ by

$$\mathrm{Im}(\overline{V_{\mathrm{DPP}}(E)}) \approx \frac{1}{D}\int \mathrm{d}E_p H_{PQ}|E\rangle \frac{-\mathcal{I}}{(E - E_p)^2 + \mathcal{I}^2}\langle E|H_{QP}$$

$$= \frac{1}{D}H_{PQ}|E\rangle\langle E|H_{QP}\int \mathrm{d}E_p \frac{-\mathcal{I}}{(E - E_p)^2 + \mathcal{I}^2}$$

$$= -\frac{\pi}{D}H_{PQ}|E\rangle\langle E|H_{QP}. \qquad (11.5.26)$$

From this expression we see that the contribution of the CN averaging to the direct-reaction optical potentials is to give a negative-definite imaginary component.

We cannot make such an approximation for the real part since the $E - E_p$ factor means it depends more on distant than on nearby resonances. We might therefore expect the real part to average near zero when the level density for compound nuclear states is approximately equal above and below E, except perhaps at the lower end of an energy spectrum where eigenstates are no longer equally dense above and below.

Fitting averaged S-matrix elements

Sometimes we do not have a microscopic theory or detailed R-matrix fit to the resonances in neutron-nucleus scattering, but do have an accurate experimental spectrum for the total neutron cross sections $\sigma_T(E)$. This does constrain the optical S matrix at least for the elastic channel. From the neutron data we can calculate $\overline{\sigma_T(E)}$ by (for example) Lorentzian averaging with width $\mathcal{I}$. This gives the smooth function of energy shown by the dashed line in Fig. 11.7, which at low energies is directly related by Eq. (11.5.11) to the real part of the $\mathbf{S}$(opt) from the optical potential.

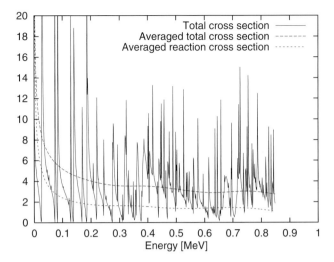

Fig. 11.7. Comparison [31] of the actual total cross section σ_{tot} (with resonances) in units of barns, with the Lorentz-averaged total and reaction cross sections that can be reproduced by the optical potential. Reprinted with permission from T. Kawano and F. H. Fröhner, *Nucl. Sci. Eng.* **127** (1997) 130. Copyright (1997) by the American Nuclear Society.

Exercises

11.1 Find the experimental spacing D_0 of low-energy *s*-wave resonances for a nucleus of your choice listed in the RIPL-2 database (URL:**www-nds.iaea.org/ripl-2**). How well is this D_0 predicted by the global level-density expression in Section 11.4? What is the best refit of the parameters to reproduce the observed D_0? Plot the resulting level density against the cumulative integral of discrete levels from the RIPL-2 database. Fit a constant-temperature level-density expression at low energies following Gilbert and Cameron. Does this improve the agreement with the discrete level integral?

11.2 The *strength function* for neutron resonances is the function $\langle\Gamma\rangle/D$ as a function of energy. Usually we plot $S_{L=0} = \langle\Gamma\rangle/D\sqrt{E_0/E}$ to remove most of the energy dependence, where $E_0 = 1$ eV is an arbitrary energy. Show how measurements of the *s*-wave strength function S_0 constrain the optical potential up to 1 MeV. For a chosen nucleus, test the accuracy of some neutron global potentials.

11.3 Use the global level-density expressions and a random-number generator to construct a hypothetical (n,γ) spectrum of a new radioactive nucleus from 0 to 1 keV. Assume that the photon width Γ_γ is 30 meV, typical of heavy nuclei, and the neutron reduced width has a mean value of 1 eV and follows the Porter-Thomas distribution for the one open channel. Consider neutron channels only for $L = 0$. Use a Wigner probability distribution for the level spacing: $p(s) = \pi s/(2D)\exp(-\pi(s/D)^2/4)$. Compare with the shapes of experimental (n,γ) results on nearby stable nuclei that have been measured in time-of-flight experiments.

11.4 Isomers have been proposed as a possible means of storing energy. Consider as a possible neutron trigger of a ^{178m2}Hf isomer the 16^+ state at 2.447 MeV. The inelastic outgoing channel for a neutron may have more energy than the elastic channel: this is called 'superelastic scattering,' or 'inelastic neutron acceleration.' It has been proposed that 100 keV neutrons could trigger the release of inelastic neutrons with more than 1.5 MeV kinetic energy. Examine the spectrum of Hf to determine possible final states in such a reaction, and the necessary exit partial waves. Use a global neutron optical potential to determine the transmission coefficients for exit neutrons from an initial 16^+ compound state of ^{179}Hf*. By using a photon width of $\Gamma_\gamma = 30$ meV (standard value for heavy nuclei), and a global level density for ^{179}Hf*, determine the branching ratio for superelastic neutron emission in competition with (n,γ) decays. Use Eq. (3.1.114) if necessary. Assess the feasibility of the Hf isomer for this kind of energy storage.

References

[1] C. E. Porter and R. G. Thomas, *Phys. Rev.* **104** (1956) 483.
[2] M. S. Moore and C. W. Reich, *Phys. Rev.* **118** (1960) 718.
[3] C. W. Reich and M. S. Moore, *Phys. Rev.* **111** (1958) 929.
[4] W. Hauser and H. Feshbach, *Phys. Rev.* **87** (1952) 366.
[5] A. J. Koning and J. P. Delaroche, *Nucl. Phys. A* **713** (2003) 231.
[6] V. F. Weisskopf and D. H. Ewing, *Phys. Rev.* **57** (1940) 472.
[7] M. Simonius, *Phys. Lett. B* **52** (1974) 279.
[8] P. Moldauer, *Phys. Rev.* **135** (1964) B642; *Nucl. Phys. A* **344** (1980) 185.
[9] J. W. Tepel, H. Hoffmann and H. A. Weidenmüller, *Phys. Lett. B* **49** (1974) 1.
[10] J. Thomas, M. Zirnbauer and K.-H. Langanke, *Phys. Rev. C* **33** (1986) 2197.
[11] S. Hilaire, Ch. Lagrange and A. J. Koning, *Ann. Phys. (NY)* **306** (2003) 209.
[12] G. R. Satchler, *Phys. Lett.* **7** (1963) 55.
[13] H. M. Hofmann, J. Richert, J. W. Tepel and H. A. Weidenmüller, *Ann. Phys. (NY)* **90** (1975) 403.
[14] C. Mahaux and H. A. Wiedenmüller, *Ann. Rev. Nucl. Part. Sci.* **29** (1979) 1.
[15] M. Kawai, A. K. Kerman and K. W. McVoy, *Ann. Phys. (NY)* **75** (1973) 156.
[16] M. Uhl and B. Strohmaier, 'Computer code for particle induced activation cross sections and related quantities', IRK report No. 76/01, Vienna (1976).
[17] M. Herman, in Workshop on Nuclear Reaction Data and Nuclear Reactors: 'Physics, Design and Safety,' N. Paver, M. Herman and A. Gandini (eds.), Trieste, Italy, (2001) 137. URL: www.nndc.bnl.gov/empire.
[18] A. J. Koning, S. Hilaire and M. C. Duijvestijn, *TALYS-1.0*, Proceedings of the International Conference on Nuclear Data for Science and Technology - ND2007, April 22–27 2007, Nice, France (2008). URL:www.talys.eu.
[19] K.-H. Langanke, *Nucl. Phys. A* **778** (2006) 233.
[20] V. Paar and R. Pezer, *Phys. Rev. C* **55** (1997) R1637.
[21] H. A. Bethe, *Phys. Rev.* **50** (1936) 332.
[22] W. Dilg, W. Schantl, H. Vonach and M. Uhl, *Nucl. Phys. A* **217** (1973) 269.
[23] A. Gilbert and A. G. W. Cameron, *Canadian J. Phys.* **43** (1965) 1446.
[24] J. J. Cowan, F.-K. Thielemann and J. W. Truran, *Phys. Rep.* **208** (1991) 267.
[25] T. Rauscher, F.-K. Thielemann and K.-L. Kratz, *Phys. Rev. C* **56** (1997) 1613.
[26] F. Perey and B. Buck, *Nucl. Phys.* **32** (1962) 353.

[27] I. J. Thompson, M. A. Nagarajan, J. S. Lilley and M. J. Smithson, *Nucl. Phys. A* **505** (1989) 84.
[28] N. Keeley and R. S. Mackintosh, *Phys. Rev. C* **77** (2008) 054603.
[29] V. I. Kukulin and R. S. Mackintosh, *J. Phys. G* **30** (2004) 1.
[30] G. E. Brown, *Rev. Mod. Phys.* **31** (1959) 893.
[31] T. Kawano and F. H. Fröhner, *Nucl. Sci. Eng.* **127** (1997) 130.

12

Stellar reaction rates and networks

I found that the best ideas usually came, not when one was actively striving for them, but when one was in a more relaxed state.

Paul Dirac

12.1 Thermal averaging

12.1.1 Reaction rates $\langle \sigma v \rangle$ and lifetimes

The reactions that we have discussed so far all have cross sections $\sigma(E)$ that depend strongly on center-of-mass energy E. This dependence may be because of a repulsive Coulomb barrier, so that $\sigma(E) = E^{-1} \exp(-2\pi\eta)S(E)$ for an 'astrophysical S-factor' $S(E)$ that is relatively less variable with energy. It may be because neutrons have a $\sigma(E) \propto 1/v$ behavior for projectile-target relative velocity v near zero. Or it may be because of resonances giving sharply peaked cross sections, like $\Gamma/[(E - E_R)^2 + \Gamma^2/4]$ for a single resonance centered at E_R with a full width at half maximum of Γ.

In a stellar plasma of a mixture of two or more nuclear species, there will be a considerable range of relative energies E (or, velocities v) because of the statistical distribution of thermal energy among all the particles in the plasma. The actual rate of nuclear reactions will therefore require an averaging of the cross sections $\sigma(E)$ over the thermal distribution of relative energies. We will therefore define an average *reaction rate* as the number of reactions per second per unit volume, and find an expression for this in terms of $\sigma(E)$ and the distribution $\phi(v)$ of the relative velocities of the interacting particles.

Reaction rates

Let us consider two nuclear species 1 and 2 in a plasma, with number densities n_1 and n_2 particles per unit volume, respectively. For a specific relative velocity v, the flux of particles 1 (considered as the projectile) will be

$$j = n_1 v \text{ particles/unit area/unit time.} \tag{12.1.1}$$

Now the cross section σ is the frequency of scattering per unit incident flux, for a single target. So the scatterings per unit time will be $\sigma j = \sigma n_1 v$ for one target nucleus. So r_{12}, the total *reaction rate* per second per unit volume, will be this expression times n_2, the number of particles 2 (considered as the target) per unit volume:

$$r_{12} = n_1 n_2 \sigma v. \tag{12.1.2}$$

In a plasma, if the distribution of relative velocities is given by $\phi(v)$ normalized to unity as $\int_0^\infty \phi(v) dv = 1$, then the *average* scattering rate will be proportional to the integral

$$\langle \sigma v \rangle = \int_0^\infty \sigma(E) \phi(v) v dv \tag{12.1.3}$$

called the *reaction rate per particle pair*. In a given plasma, the net reaction rate will then be

$$r_{12} = n_1 n_2 \langle \sigma v \rangle.$$

This formula needs one modification, should particles 1 and 2 be identical. In this case, the number of distinct particle pairs is not n_1^2, but rather $\frac{1}{2} n_1^2$, and so a factor of $\frac{1}{2}$ must be introduced to avoid counting each pair of particles twice. This factor may be introduced formally into the above formula as

$$r_{12} = \frac{1}{1 + \delta_{12}} n_1 n_2 \langle \sigma v \rangle, \tag{12.1.4}$$

where $\delta_{12} = 1$ if the particles 1 and 2 are identical, otherwise zero.

In the integral (12.1.3) defining $\langle \sigma v \rangle$ the integration starts from zero for exothermic reactions, but for endothermic reactions that absorb energy the integrand is zero for velocities too small to reach the threshold in relative energy.

Decay lifetimes

Resonances and unstable nuclei will decay spontaneously, with some lifetime. If a resonance pole is at a complex energy $E = E_R - i\Gamma/2$, for example, then the time evolution of the resonant probability follows the square norm of the wave function

$$|\psi|^2 \propto |e^{-iEt/\hbar}|^2 = |e^{-iE_R t/\hbar}|^2 |e^{-\Gamma t/2\hbar}|^2 = e^{-\Gamma t/\hbar}, \tag{12.1.5}$$

which corresponds to

$$\frac{dn}{dt} = -\lambda n(t) \text{ for a decay constant of } \lambda = \frac{\Gamma}{\hbar}. \tag{12.1.6}$$

Its *half-life* is

$$t_{\frac{1}{2}} = \frac{\ln 2}{\lambda} = \frac{\hbar}{\Gamma} \ln 2. \tag{12.1.7}$$

12.1.2 *Maxwell-Boltzmann distributions*

In the environment of a stellar plasma, the particles can be described by non-relativistic kinematics except at the very highest temperatures. We consider first the case of non-relativistic thermal equilibrium, according to which the statistical distribution of velocities is governed by a temperature T according to the Maxwell-Boltzmann distribution. According to this theory, the distribution of absolute velocities v_i of species i in a thermal gas is given by

$$\Phi_i(\mathbf{v}_i) = \left(\frac{m_i}{2\pi k_B T}\right)^{\frac{3}{2}} \exp\left(-\frac{m_i v_i^2}{2k_B T}\right), \tag{12.1.8}$$

where k_B is Boltzmann's constant, $v_i = |\mathbf{v}_i|$, m_i is the mass of particle i, and this distribution is normalized for the three dimensions of velocity according to

$$\int \Phi(\mathbf{v})d\mathbf{v} = 1. \tag{12.1.9}$$

The mean particle energy is $k_B T$, and the number density of particles i per unit volume of 'velocity space' is thus $n_i \Phi_i(\mathbf{v}_i)$.

Remember that v in Eq. (12.1.2) is the *relative* velocity $v = |\mathbf{v}_1 - \mathbf{v}_2|$, so the integral (12.1.3) for the reaction rate per pair of particles must be calculated as

$$\langle \sigma v \rangle = \int \int \sigma(E) |\mathbf{v}_1 - \mathbf{v}_2| \, \Phi_1(\mathbf{v}_1)\Phi_2(\mathbf{v}_2) \, d\mathbf{v}_1 d\mathbf{v}_2, \tag{12.1.10}$$

where the relative energy $E = \frac{1}{2}\mu_{12}|\mathbf{v}_1 - \mathbf{v}_2|^2$ using the reduced mass μ_{12} for collisions between particles 1 and 2. We now evaluate this expression using the Maxwell-Boltzmann distribution of Eq. (12.1.8).

In terms of the relative velocity $\mathbf{v}$ and the velocity of the center of mass $\mathbf{V}$ of particles 1 and 2, the individual velocities are

$$\mathbf{v}_1 = \mathbf{V} + \frac{m_2}{m_1 + m_2}\mathbf{v},$$
$$\mathbf{v}_2 = \mathbf{V} - \frac{m_1}{m_1 + m_2}\mathbf{v}. \tag{12.1.11}$$

When we multiply $\Phi_1(\mathbf{v}_1)\Phi_2(\mathbf{v}_2)$ in Eq. (12.1.10), the exponents have a factor which may be rewritten

$$\frac{1}{2}m_1v_1^2 + \frac{1}{2}m_2v_2^2 = \frac{1}{2}(m_1 + m_2)V^2 + \frac{1}{2}\mu_{12}v^2, \tag{12.1.12}$$

where the first term on the right is the energy of motion of the whole center of mass, and the second term is the relative energy E defined above.

Since the cross section $\sigma(E)$ only depends on E and not on the center-of-mass motion, we should transform the integral (12.1.10) to separate the relative and center-of-mass motions. Using the unit Jacobean of the transformation (12.1.11), the differential may be transformed using $d\mathbf{v}_1 d\mathbf{v}_2 = d\mathbf{V}d\mathbf{v}$ to give

$$\langle \sigma v \rangle = \int \int \sigma(E)v \left\{ \left(\frac{m_1 + m_2}{2\pi k_B T} \right)^{\frac{3}{2}} \exp\left(-\frac{(m_1 + m_2)V^2}{2k_B T} \right) \right\}$$

$$\times \left[\left(\frac{\mu_{12}}{2\pi k_B T} \right)^{\frac{3}{2}} \exp\left(-\frac{\mu_{12}v^2}{2k_B T} \right) \right] d\mathbf{v}d\mathbf{V}. \tag{12.1.13}$$

The first factor $\{\cdot\}$ describes the Maxwell-Boltzmann distribution of the whole center of mass of the two particles, and the second $[\cdot]$ factor describes the distribution over their relative velocity v. The integral over $\mathbf{V}$ may be done immediately, and gives unity by Eq. (12.1.9). The reaction-rate integral therefore simplifies to

$$\langle \sigma v \rangle = \int \sigma(E)v \left(\frac{\mu_{12}}{2\pi k_B T} \right)^{\frac{3}{2}} \exp\left(-\frac{\mu_{12}v^2}{2k_B T} \right) d\mathbf{v}. \tag{12.1.14}$$

We can also integrate analytically over the angle $\hat{\mathbf{v}}$, leaving

$$\langle \sigma v \rangle = 4\pi \int_0^\infty v^3 \sigma(E) \left(\frac{\mu_{12}}{2\pi k_B T} \right)^{\frac{3}{2}} \exp\left(-\frac{\mu_{12}v^2}{2k_B T} \right) dv. \tag{12.1.15}$$

Comparing this with (12.1.3) shows that $\phi(v)$, the distribution of relative velocities, is

$$\phi(v) = 4\pi v^2 \left(\frac{\mu_{12}}{2\pi k_B T} \right)^{\frac{3}{2}} \exp\left(-\frac{\mu_{12}v^2}{2k_B T} \right). \tag{12.1.16}$$

Transforming the variable of integration to the energy E, we have

$$\langle \sigma v \rangle = \sqrt{\frac{8}{\pi \mu_{12}(k_B T)^3}} \int_0^\infty E\sigma(E) \exp\left(-\frac{E}{k_B T} \right) dE. \tag{12.1.17}$$

The integrand $E\sigma(E)\exp(-E/k_B T)$ is sometimes called the *Gamow distribution function*, and shows how the energy dependence of the cross section is weighted by the thermal distribution of energies to give the reaction rate. If we replace the cross section by the astrophysical S-factor, the above integral becomes

$$\langle\sigma v\rangle = \sqrt{\frac{8}{\pi\mu_{12}(k_B T)^3}}\int_0^\infty S(E)\exp\left(-\frac{E}{k_B T}-\sqrt{\frac{E_G}{E}}\right)\,dE. \qquad (12.1.18)$$

where the energy constant is $E_G = 4\pi^2\eta^2 E = 2\mu_{12}(\pi Z_1 Z_2 e^2)^2/\hbar^2$.

12.1.3 The Gamow peak

Note that $\exp(\sqrt{E_G/E})$ is small at low energies, while $\exp(-E/k_B T)$ is small at higher energies. As Fig. 12.1 shows, the product of these two terms gives a value for the integral which has a peak, called the *Gamow peak*, at an energy E_0 where

$$E_0 = \left(\frac{E_G k_B^2 T^2}{4}\right)^{\frac{1}{3}}, \qquad (12.1.19)$$

which is known as the *effective burning energy*.

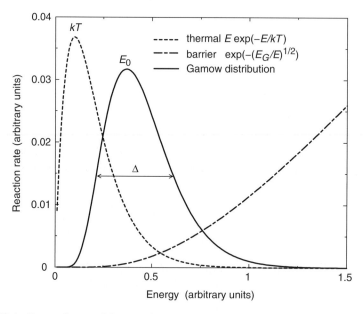

Fig. 12.1. Dependence of the reaction cross section for charged-particle reactions as a function of energy (arbitrary units).

The width Δ of the peak about this effective burning energy can be estimated, if $S(E)$ is assumed to be constant, by approximating the peak as a Gaussian function

$$\exp\left(-\frac{E}{k_B T} - \sqrt{\frac{E_G}{E}}\right) \approx I_{\max} \exp\left[-\left(\frac{E - E_0}{\Delta/2}\right)^2\right], \qquad (12.1.20)$$

where

$$I_{\max} = \exp\left(-\frac{3E_0}{k_B T}\right). \qquad (12.1.21)$$

Differentiating (12.1.20) twice, the width Δ is found to be

$$\Delta = \frac{4}{3^{\frac{1}{2}}} (E_0 k_B T)^{\frac{1}{2}} = \frac{4}{3^{\frac{1}{2}} 2^{\frac{1}{3}}} E_G^{\frac{1}{6}} (k_B T)^{\frac{5}{6}}. \qquad (12.1.22)$$

Using (12.1.18), an approximation to the integral is

$$\int_0^\infty \exp\left[-\frac{E}{k_B T} - \sqrt{\frac{E_G}{E}}\right] dE \approx I_{\max} \Delta = \frac{4}{3^{\frac{1}{2}}} (E_0 k_B T)^{\frac{1}{2}} \exp\left(-\frac{3E_0}{k_B T}\right). \qquad (12.1.23)$$

Nuclear reactions occur mainly in the energy region straddled by the *energy window* defined by

$$E = E_0 \pm \frac{\Delta}{2}. \qquad (12.1.24)$$

This energy may often be too low to measure the reaction cross section directly in the laboratory, so one typically finds $S(E)$ over a range of available lab energies and then extrapolates down to the region around E_0.

Using the approximation given in Eq. (12.1.23), if the $S(E)$ is approximately constant then the non-resonant reaction rate per particle pair can be estimated as

$$\langle \sigma v \rangle \approx \left(\frac{2}{\mu}\right)^{\frac{1}{2}} \frac{\Delta}{(k_B T)^{\frac{3}{2}}} S(E_0) \exp\left(-\frac{3E_0}{k_B T}\right). \qquad (12.1.25)$$

If the S-factor varies significantly around the Gamow peak, then we may expand it in a Taylor series about zero energy up to the quadratic term

$$S(E) \approx S(0) + E\, S'(0) + \frac{1}{2} E^2 S''(0). \qquad (12.1.26)$$

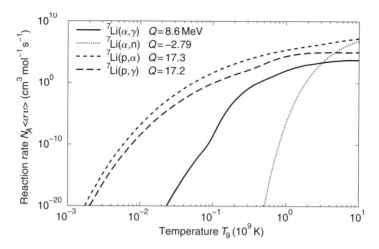

Fig. 12.2. Reaction rates for four processes starting with ^{7}Li. These are the adopted reaction rates from the NACRE compilation [2]. The exothermic reactions ($Q > 0$) have much larger rates at low temperatures, and endothermic reactions are only significant above a minimum temperature.

In terms of these derivatives, the $S(E_0)$ in Eq. (12.1.25) can be more accurately replaced at E_0 by an *effective S-factor* of

$$S_{\text{eff}} = \left(1 + \frac{5k_BT}{36E_0}\right)S(0) + \left(E_0 + \frac{35}{36}k_BT\right)S'(0) + \frac{1}{2}E_0\left(E_0 + \frac{89}{36}k_BT\right)S''(0),$$

(12.1.27)

and further approximate forms are discussed in [1]. Some of the reaction rates calculated from the cross sections adopted by the NACRE compilation [2] are shown in Fig. 12.2.

12.1.4 Averaging over resonances

The cross section for a single Breit-Wigner resonance cross section from Eq. (11.2.4), if substituted in the formula of Eq. (12.1.17) for the reaction rate per particle pair, gives

$$\langle \sigma v \rangle_{\alpha'\alpha} = \sqrt{\frac{2\pi\hbar^4}{(\mu_{12}k_BT)^3}}\, g_{J_{\text{tot}}} \int_0^\infty \frac{\Gamma_{\alpha p}\Gamma_{\alpha'p}}{(E - E_p)^2 + \Gamma_p^2/4}\, \exp\left(-\frac{E}{k_BT}\right) dE, \quad (12.1.28)$$

for the cross section from incoming channel α to outgoing channel α'. Remembering that the widths $\Gamma_{\alpha p} = 2\gamma_{p\alpha}^2 P_\alpha(E)$ include the Coulomb penetrability, we see that this integrand has two peaks: one near the resonance energy E_p, and the other near the Gamow peak. If these are well separated then their contributions may be found separately and simply added. The dominant contribution will typically be from the resonance energy, giving a reaction rate per particle pair of

$$\langle \sigma v \rangle_{\alpha'\alpha} = \left(\frac{2\pi}{\mu_{12} k_B T} \right)^{\frac{3}{2}} \hbar^2 g_{J_{\text{tot}}} \frac{\Gamma_{\alpha p} \Gamma_{\alpha' p}}{\Gamma_p} \exp\left(-\frac{E_p}{k_B T} \right). \qquad (12.1.29)$$

When a narrow resonance is near to the Gamow peak, then combined formulae may be useful, as suggested by Ueda *et al.* [3]. For wide resonances, however, there is no substitute for numerical integration of Eq. (12.1.28).

Sub-threshold resonances

Given enough energy, any nucleus C will eventually break up into a variety of pairs A, B of clusters. The threshold energies at which these fragmentations begin are called the various *breakup thresholds* E_{thr}, as at these energies the fragments A and B have zero relative kinetic energies. An excited state in the C nucleus that is *above* the threshold will show up as a clear resonance peak in A+B scattering. An excited state that is *below* that threshold may not show up as a peak, but may still have an effect on the A+B scattering. These latter states (mentioned in subsection 10.2.2) are called sub-threshold resonances, and may still have some effect on positive energy scattering if their widths Γ_p are of the same order as their threshold energies.

Every excited state of a nucleus will eventually decay by some means, and its width Γ_p is inversely proportional to this lifetime. If the nucleus only decays by γ emission, then the resonance lifetime will be long ($\gtrsim 10^{-12}$ s), its width will be rather narrow, and it will hardly have any effect above threshold. If, however, the sub-threshold resonance width is larger (because of proton or α emission perhaps) then the factor of $[(E - E_p) + \Gamma_p^2/4]^{-1}$ will describe a high-energy tail to the resonance that extends up above the threshold, into the continuum. It is then possible for a fusion to have a large cross section at very low continuum energies because of the long tail of a sub-threshold resonance. One example of this is the $^{14}\text{N}(\text{p},\gamma)^{15}\text{O}$ reaction, which proceeds strongly because of a sub-threshold resonance at -22 keV.

12.1.5 Neutron and photon reaction rates
Neutron reaction rates

Neutrons do not see any repulsive Coulomb barrier, so there is not a Gamow peak as there is for charged particles. Most low-energy neutron reactions are dominated by the *s*-wave ($L = 0$) entrance channel, so their cross sections at low energies are proportional to v^{-1}. This can be seen from the R-matrix formula, as this is the dependence of $\Gamma(E)/k_i^2$, since for *s*-waves $P(E) = k_i a$ and the velocity v is proportional to the entrance-channel wave number k_i. It can also be seen from phase shift analyses, as phase shifts go to zero linearly with k at low energies according to the scattering length expression of Eq. (3.1.95).

Instead of defining an 'astrophysical S-factor', for neutrons we define the alternate function

$$R(E) \equiv \sigma v. \tag{12.1.30}$$

The $R(E)$, when far from resonances, is thus nearly constant at low energies. The reaction rate per particle pair is hence practically constant in the non-resonant regions:

$$\langle \sigma v \rangle \simeq R(E), \tag{12.1.31}$$

evaluated for any energy E below the lowest resonance. For reactions in higher partial waves ($L > 0$) there will be some energy dependence, but much weaker than with the reactions of charged particles.

When neutron resonances can be be individually enumerated, we should integrate using Eq. (12.1.28). In heavy nuclei, where the level density of resonances is high, the statistical Hauser-Feshbach methods of Chapter 11 can be used. In that case, the average ratio of level widths to level spacings is found using the imaginary part of the neutron optical potential, as explained in subsection 11.3.2.

Photo-nuclear reaction rates

When one of the interacting particles is a photon, then the relative velocity $|\mathbf{v}_1 - \mathbf{v}_2|$ will always be c, and the energy for the evaluation of $\sigma(E)$ will be the photon energy E_γ. In the integral of Eq. (12.1.10), the integral over the particle velocity can be done immediately, and the other integration variable must now be the photon energy E_γ. For photo-disintegration we assume in stars that the number distribution of the photons is fixed in terms of the temperature by the black-body radiation law, so the reaction rate per particle pair is now only proportional to the target number density. The proportionality coefficient we write as $\lambda_{1\gamma}$, and has the same units as that of a spontaneous decay rate. Its value for a target nucleus called 1 is therefore

$$\lambda_{1\gamma} = \int_{-Q}^{\infty} c\, \sigma_{1\gamma}(E_\gamma)\, n_\gamma(E_\gamma)\, dE_\gamma, \tag{12.1.32}$$

where $-Q$ is the photon energy threshold, and $\sigma_{1\gamma}(E_\gamma)$ is the cross section for γ+target photo-disintegration, calculated using Eq. (3.5.23). This applies to the disintegration into any number $(2, 3,\ldots)$ of particles in the final state. In this integral we use the number density of photons $n_\gamma(E_\gamma)$ in a thermodynamic environment at temperature T. The number is not conserved as for particles, but depends on the stellar temperature. According to the Planck black-body law, the photon number density per unit energy is

$$n_\gamma(E_\gamma) = \frac{1}{\pi^2\hbar^3 c^3} \frac{E_\gamma^2}{\exp(E_\gamma/k_B T) - 1}. \qquad (12.1.33)$$

This distribution is *not* normalized to unity, because integrated black-body radiation energy increases in proportion to T^4.

12.1.6 Inverse reaction rates

For low temperatures, nuclear reactions with positive Q-values (those which *release energy*) are preferred, that is

$$1 + 2 \rightarrow 3 + 4 \quad Q > 0. \qquad (12.1.34)$$

However, as the temperature increases, so does the number of nuclei with kinetic energies greater than the Q-value, and for these particles, the *inverse* process can occur

$$3 + 4 \rightarrow 1 + 2 \quad Q < 0. \qquad (12.1.35)$$

In order to fully understand and model the evolution and burning of stars and the production of heavier nuclear species, the rates for these inverse processes must be known.

Using the detailed balance property of the $\tilde{\mathbf{S}}$ matrix discussed in subsection 3.2.4, the forward and reverse cross sections are related according to

$$\frac{\sigma_{12}}{\sigma_{34}} = \frac{k_{34}^2}{k_{12}^2} \frac{(2I_3+1)(2I_4+1)}{(2I_1+1)(2I_2+1)} \frac{(1+\delta_{12})}{(1+\delta_{34})}, \qquad (12.1.36)$$

where the δ-functions are unity if particles 1 and 2 (or 3 and 4) are identical. This is a very useful result since it is often easier experimentally to measure a nuclear reaction cross section in the *reverse direction*. By using the above equation, one can directly obtain the reaction cross section of interest.

Since the reaction rate per particle pair can be given by (12.1.17) for both the forward $\langle \sigma v \rangle_{12}$ and the reverse $\langle \sigma v \rangle_{34}$, by the conservation of energy $E_{34} = E_{12} + Q$, we can write the *ratio* of reaction rates per particle pair, for Q of either sign, as

$$\frac{\langle \sigma v \rangle_{34}}{\langle \sigma v \rangle_{12}} = \frac{(2I_1+1)(2I_2+1)}{(2I_4+1)(2I_3+1)} \frac{(1+\delta_{34})}{(1+\delta_{12})} \left(\frac{\mu_{12}}{\mu_{34}}\right)^{\frac{3}{2}} \exp\left(-\frac{Q}{k_B T}\right). \qquad (12.1.37)$$

This ratio is dominated by the exponential term, $\exp(-Q/k_B T)$, and is hence a rather small number except at very high temperatures. For example, if $Q = 8\,\text{MeV}$ and $T = 10^{10}\,\text{K}$, then $\langle\sigma v\rangle_{34}/\langle\sigma v\rangle_{12} \approx 10^{-4}$.

12.1.7 Electron screening

Whether in the laboratory or in the star, nuclear reaction rates are modified due to the presence of electrons. There are two limiting cases, referred to as *weak screening* and *strong screening*. Weak screening [4] is applicable when there is a low-density plasma of ions and electrons characterized by a Debye radius and when the average Coulomb energy is much smaller than the thermal energy. Nucleosynthesis reaction rates in main sequence stars are corrected within this picture.

In the Debye approximation, the potential between nuclei with nucleon numbers (A_1, Z_1) and (A_2, Z_2) becomes of Yukawa form:

$$V_{\text{screen}}(r) = \frac{Z_1 Z_2 e^2}{r} e^{(-r/r_D)}, \qquad (12.1.38)$$

with the Debye radius being $r_D = \sqrt{\frac{kT}{4\pi e^2 \rho_e N_A \zeta}}$, ρ_e the electronic density, N_A the Avogadro's number and $\zeta = \sum_i [Z_i(Z_i + 1)X_i/A_i]$ the rms charge of the plasma. If the fusion reaction happens at distances much smaller than the Debye radius, then the potential can be expanded to first order and the energy shift from the screened potential compared to the unscreened potential becomes $U_0 = -Z_1 Z_2 e^2/r_D$. If $U_0 \ll E$, it can be shown that in the weak screening case, the reaction rates become

$$\langle\sigma v\rangle_{\text{ws}} = f_{\text{ws}}\,\langle\sigma v\rangle_{\text{vacuum}} \qquad (12.1.39)$$

with a factor $f_{\text{ws}} = \exp[U_0/kT]$.

The other screening limit corresponds to a picture where the nucleus is surrounded by a cloud of uniformly distributed electrons of radius comparable to the inter-ionic distance. This strong screening is the typical laboratory scenario. In this case, the medium is no longer locally neutral and reactions take place under the influence of strong screening [5]. Many astrophysical reactions are somewhere in between and a number of models have been developed to account for these processes more accurately [6].

12.2 Reaction networks

In the previous section we have calculated the reaction rates per particle pair $\langle\sigma v\rangle$ by averaging the reaction cross sections over the thermal velocities in a gas. Spontaneous one-body decays are characterized by their decay constant λ. In terms of these quantities and the number densities n_i of the various nuclear species, the reaction rates r per unit volume per second are, respectively,

$$r_{12} = \frac{1}{1 + \delta_{12}} n_1 n_2 \langle \sigma v \rangle_{12}$$

$$r_1 = -\lambda_1 n_1. \tag{12.2.1}$$

There may also be three-body reactions, but in the expression r_{123} we should allow for a general number of identical pairs. For two-body reactions let us define the number of distinct particle pairs per $n_j n_k$ as

$$C_{jk} = \frac{1}{1 + \delta_{jk}}, \tag{12.2.2}$$

and for three-body reactions the number of distinct particle pairs per $n_j n_k n_l$ as

$$C_{jkl} = \frac{1}{2!} \quad \text{for one identical pair among } jkl$$

$$= \frac{1}{3!} \quad \text{for all } jkl \text{ identical.} \tag{12.2.3}$$

12.2.1 Coupled rate equations

We now find the *effects* of these reactions on the number densities of both the reactants n_1, n_2 and the products (say n_3 and n_4). To do this, we need to know the numbers of the different nuclei created or destroyed in each kind of reaction. Let us define

$$N_i(j) = \text{number of species } i \text{ produced by the decay of species } j, \text{ and}$$

$$N_i(jk) = \text{number } i \text{ produced by the reaction of species } j \text{ with } k,$$

and $N_i(jkl)$ similarly for three-body reactions.[1] We take these numbers as *negative* if the species i are *consumed* in the reaction. In terms of these parameters, the general equation for the rate of change of the number densities is

$$\frac{\partial n_i}{\partial t} = \sum_j N_i(j) r_j + \sum_{jk} N_i(jk) r_{jk} + \sum_{jkl} N_i(jkl) r_{jkl}. \tag{12.2.5}$$

---

[1] Using these $N_i(jkl)$ numbers, we may generalize the above C coefficients for an arbitrary number of reacting nuclei to

$$(C_{jkl\ldots})^{-1} = \prod_s^{n_s} |N_{i_s}(jkl\cdots)|!, \tag{12.2.4}$$

where $s = 1, \ldots, n_s$ is the index to the distinct species i_s among the $jkl, \ldots$.

The first term describes the decay of species j into i, the second term describes the reaction of j and k together where one of the products is i, and the third term describes the three-body reaction of j, k, and l together to produce i.

We substitute in Eq. (12.2.5) the reaction rates from Eq. (12.2.1), and, using the $C_{jk...}$ coefficients, obtain

$$\frac{\partial n_i}{\partial t} = \sum_j N_i(j)\lambda_j n_j + \sum_{jk} N_i(jk)C_{jk}\, n_j n_k\, \langle \sigma v \rangle_{jk}$$

$$+ \sum_{jkl} N_i(jkl)C_{jkl}\, n_j n_k n_l\, \langle \sigma v \rangle_{jkl}. \tag{12.2.6}$$

We now rewrite the number densities n_i in terms of the fractional *nuclear abundances* Y_i, in order to exclude changes to the n_i that come from simply the expansion or contraction of the plasma. We define, in a given volume of gas, the fractional nuclear number abundances by

$$Y_i = \text{number of nuclei } i \text{ per number of all nucleons.} \tag{12.2.7}$$

If each species i has atomic mass A_i, then these abundances are normalized by $\sum_i A_i Y_i = 1$. A gas of purely one species, for example, would have $Y = 1/A$ and, for number density n, would have mass density $\rho = A m_u n = m_u n / Y$. This implies that n is given in terms of Y by $n = Y\rho/m_u$, and for a mixture of several species we have $n_i = Y_i \rho/m_u$ for overall gas density ρ. We now define $N_u = 1/m_u$,[2] so $n_i = N_u Y_i \rho$ is the scale relation between number densities and nuclear abundances, and Eq. (12.2.6) may be rewritten as

$$\frac{\partial Y_i}{\partial t} = \sum_j N_i(j)\lambda_j Y_j + \rho N_u \sum_{jk} N_i(jk)C_{jk}\, Y_j Y_k\, \langle \sigma v \rangle_{jk}$$

$$+ \rho^2 N_u^2 \sum_{jkl} N_i(jkl)C_{jkl}\, Y_j Y_k Y_l\, \langle \sigma v \rangle_{jkl}, \tag{12.2.8}$$

which are coupled rate equations that form a reaction network. In Box 12.1 we show the coupled equations corresponding to the CNO cycle of Fig. 1.11.

12.2.2 *Explicit and implicit solution methods*

Chapter 6 discussed methods for solving the second-order scattering equations, whereas here the network Eqs. (12.2.8) are of first order, and in the form $dy/dt =$

---

[2] If masses are measured in grams, then $1/m_u = N_A$, Avogadro's number of 6.022×10^{23} mol^{-1}. For a convention-free number we write $N_u = 1/m_u$, which coincides with N_A if cgs units are adopted.

In order to calculate the flow around the CNO cycle of Fig. 1.11, we need

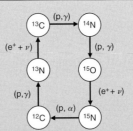

(i) the three (p,γ) reaction rates on $X = {}^{12}$C, ^{13}C and ^{14}N: $\langle\sigma v\rangle_{p,X}$,
(ii) the β-decay rates for ^{13}N and ^{15}O: λ_X, and
(iii) the (p,α) reaction rate on ^{15}N: $\langle\sigma v\rangle_{p,15n}$.

For total plasma density ρ, since there are no identical particles or three-body reactions, the coupled rate equations for the abundances Y_i are

$$\dot{Y}_p = -\rho N_u[Y_p Y_{12c}\langle\sigma v\rangle_{p,12c} + Y_p Y_{13c}\langle\sigma v\rangle_{p,13c}$$
$$+ Y_p Y_{14n}\langle\sigma v\rangle_{p,12c} + Y_p Y_{15n}\langle\sigma v\rangle_{p,15n}]$$
$$\dot{Y}_{12c} = +\rho N_u[-Y_p Y_{12c}\langle\sigma v\rangle_{p,12c} + Y_p Y_{15n}\langle\sigma v\rangle_{p,15n}]$$
$$\dot{Y}_{13n} = -\lambda_{13n}Y_{13n} + \rho N_u Y_p Y_{12c}\langle\sigma v\rangle_{p,12c}$$
$$\dot{Y}_{13c} = +\lambda_{13n}Y_{13n} - \rho N_u Y_p Y_{13c}\langle\sigma v\rangle_{p,13c}$$
$$\dot{Y}_{14n} = +\rho N_u[-Y_p Y_{14n}\langle\sigma v\rangle_{p,14n} + Y_p Y_{13c}\langle\sigma v\rangle_{p,13c}]$$
$$\dot{Y}_{15o} = -\lambda_{15o}Y_{15o} + \rho N_u Y_p Y_{14n}\langle\sigma v\rangle_{p,14n}$$
$$\dot{Y}_{15n} = +\lambda_{15o}Y_{15o} - \rho N_u Y_p Y_{15n}\langle\sigma v\rangle_{p,15n}$$
$$\dot{Y}_{\alpha} = +\rho N_u Y_p Y_{15n}\langle\sigma v\rangle_{p,15n}. \tag{12.2.9}$$

Here, we have not tracked the positrons or neutrinos, and the photons are quickly absorbed by the plasma electrons. The protons are only consumed by this set of reactions, and α particles only produced.

Box 12.1 **Coupled rate equations for the CNO cycle**

$f(y)$ for a vector y of abundances, where $f(y)$ is a non-linear function. The numerical solution of these equations y_n at times t_n may be accomplished by *explicit* or *predictor* integration algorithms such as

$$\frac{y_{n+1} - y_n}{t_{n+1} - t_n} = f(y_n), \tag{12.2.10}$$

or higher-order forms such as Runge-Kutta that are more accurate. Methods based on Eq. (12.2.10) are called explicit because they yield y_{n+1} directly in terms of the earlier y_n. However, if the reaction network contains a mixture of both fast and slow processes (large and small reaction rates), then the time step $t_{n+1} - t_n$ has to be smaller than the relaxation time of the fastest process. If the full dynamical time evolution is desired, then this might be satisfactory, but most stellar evolution is slow and there is no need to calculate the physical oscillations around the moving near-equilibria that are not of primary interest.

Practical solutions of stellar networks, therefore, use *implicit* numerical methods such as

$$\frac{y_{n+1} - y_n}{t_{n+1} - t_n} = \frac{1}{2}(f(y_{n+1}) + f(y_n)), \qquad (12.2.11)$$

or more accurate forms, such as those used in *corrector* integration algorithms. These methods are called implicit because the unknown y_{n+1} appears on both sides of the equation. Finding y_{n+1} therefore requires solving a multivariable non-linear equation, and this is traditionally done by Newton-Raphson schemes. The implicit Eq. (12.2.11) has very different behavior for large time steps, because if a physical equilibrium or static solution with $f(y) = 0$ exists, then this solution will be found even with one large time step.

12.3 Equilibria

12.3.1 Fixed points of the rate equations

When all the abundances are constant, $\partial Y_i/\partial t = 0$, a kind of balance holds. However, this might not be a *thermodynamic* equilibrium, if there is a loop or cycle of reactions with static abundances, but without individual reactions being balanced. The CNO cycle of Fig. 1.11 is like this: the reactions go only forwards, and rarely backwards at each step. *Thermodynamic equilibrium* is defined to be rather the case of *detailed balance*, the state of statistical equilibrium when *all* reaction steps in the network have forward and backward rates that are *equal*.[3] Let us consider the different kinds of equilibrium conditions in stars.

Consider first a reaction of particles 1 and 2 that produce particles 3 and 4. The *net reaction rate* is the *difference* between the forward and reverse reaction rates, so if for example r_{12} is the 'production' rate and r_{34} is the 'decay' rate,

$$r = r_{12} - r_{34} = \frac{n_1 n_2}{1 + \delta_{12}} \langle \sigma v \rangle_{12} - \frac{n_3 n_4}{1 + \delta_{34}} \langle \sigma v \rangle_{34}. \qquad (12.3.1)$$

By substitution of Eq. (12.1.37),

$$r = \frac{\langle \sigma v \rangle_{12}}{1 + \delta_{12}} \left[n_1 n_2 - \frac{n_3 n_4 (2I_1 + 1)(2I_2 + 1)}{(2I_3 + 1)(2I_4 + 1)} \left(\frac{\mu_{12}}{\mu_{34}} \right)^{\frac{3}{2}} \exp \left(-\frac{Q}{k_B T} \right) \right]. \qquad (12.3.2)$$

---

[3] Note that this kind of detailed balance is a property of a gas or plasma, and is distinct from that principle described in subsection 3.2.4, which was a property of a single reaction step.

In thermodynamic equilibrium conditions, the rate of production equals the rate of decay for each step including this one, so $r = 0$, which implies

$$\frac{n_1 n_2}{n_3 n_4} = \frac{(2I_1+1)(2I_2+1)}{(2I_3+1)(2I_4+1)} \left(\frac{\mu_{12}}{\mu_{34}}\right)^{\frac{3}{2}} \exp\left(-\frac{Q}{k_B T}\right). \tag{12.3.3}$$

This describes a statistical or thermodynamic equilibrium between two pairs of nuclei in their ground states $1 + 2 \rightleftharpoons 3 + 4$, the details of which are independent of the actual reaction rates $\langle \sigma v \rangle$, and vary with temperature T only according to the Q-value energy difference.

Secondly we may consider a reaction of two particles 1 and 2 to form a compound system 12 as a resonance that lives for some period of time. Consider therefore a resonance at energy E of total width $\Gamma = \Gamma_{el} + \Gamma_o$ composed of partial width for the elastic channel of Γ_{el} and Γ_o for a reaction channel. Let us assume that the Maxwell-Boltzmann distribution does not vary significantly across the width of the resonance, so we may use the energy integral of Eq. (11.2.5),

$$\int_0^\infty dE \, \sigma(E) = \frac{2\pi^2}{k_i^2} g_{J_{tot}} \frac{\Gamma_{el} \Gamma_o}{\Gamma_{el} + \Gamma_o}, \tag{12.3.4}$$

for the cross section for populating the resonance and decaying by the reaction channel. If $\Gamma_o \gg \Gamma_{el}$ then this non-elastic decay will happen almost every time the resonance is formed, whereas if $\Gamma_o \ll \Gamma_{el}$ then the reaction will be infrequent and most often the resonance will decay back to the elastic channel.

Substituting this in Eq. (12.1.17), with $\hbar^2 k_i^2 / 2\mu_{12} = E$, the energy of the resonance, we obtain the reaction rate per particle pair of

$$\langle \sigma v \rangle_{12} = \sqrt{\frac{8}{\pi \mu_{12}(k_B T)^3}} \frac{2\pi^2}{k_i^2} g_{J_{tot}} \frac{\Gamma_{el} \Gamma_o}{\Gamma_{el} + \Gamma_o} E \exp\left(-\frac{E}{k_B T}\right)$$

$$= \left(\frac{2\pi}{\mu_{12} k_B T}\right)^{\frac{3}{2}} \hbar^2 g_{J_{tot}} \frac{\Gamma_{el} \Gamma_o}{\Gamma_{el} + \Gamma_o} \exp\left(-\frac{E}{k_B T}\right). \tag{12.3.5}$$

In terms of $\langle \sigma v \rangle_{12}$, the volume production rate of the resonance is $r_{12} = n_1 n_2 \langle \sigma v \rangle_{12}$. The resonance however decays to the reaction channel with decay constant $\lambda = \Gamma_o/\hbar$, and the volume decay rate is $dn_{12}/dt = -\lambda n_{12}$, where n_{12} is the number density of nuclei pairs trapped in this resonance.

In a state of statistical equilibrium, static abundance of the resonance requires that $r_{12} = -dn_{12}/dt$, so that $\lambda n_{12} = n_1 n_2 \langle \sigma v \rangle_{12}$, or

$$\frac{\Gamma_o}{\hbar} n_{12} = n_1 n_2 \left(\frac{2\pi}{\mu_{12} k_B T}\right)^{\frac{3}{2}} \hbar^2 g_{J_{tot}} \frac{\Gamma_{el} \Gamma_o}{\Gamma_{el} + \Gamma_o} e^{-E/k_B T} \tag{12.3.6}$$

or

$$\frac{n_{12}}{n_1 n_2} = \left(\frac{2\pi\hbar^2}{\mu_{12}k_BT}\right)^{\frac{3}{2}} g_{J_{tot}} \frac{\Gamma_{el}}{\Gamma_{el}+\Gamma_o} e^{-E/k_BT}. \tag{12.3.7}$$

12.3.2 The Saha equation

The condition of static abundances is not thermodynamic unless the formation and decay of the resonance by each step are equal. This would require here that the rate of population of the resonance from the elastic channel is equal to the rate of decaying back to the elastic channel. This can only happen if $\Gamma_{el} \gg \Gamma_o$, in which case

$$\frac{n_{12}}{n_1 n_2} = \left(\frac{2\pi\hbar^2}{\mu_{12}k_BT}\right)^{\frac{3}{2}} g_{J_{tot}} e^{-E/k_BT}. \tag{12.3.8}$$

We call the number

$$n_Q = \left(\frac{\mu_{12}k_BT}{2\pi\hbar^2}\right)^{\frac{3}{2}} \tag{12.3.9}$$

the *quantum concentration*, in terms of which

$$\frac{n_{12}}{n_1 n_2} n_Q = g_{J_{tot}} e^{-E/k_BT}. \tag{12.3.10}$$

This condition on the number densities imposed here by thermodynamic equilibrium is a special case of the *Saha equation* obtained for the simple case of reacting nuclei 1 and 2 with no excited states separately. It can be derived more directly from thermodynamic arguments, but above we have presented a microscopic derivation for a simple case of a reaction $1 + 2 \leftrightharpoons 12$, where we know formulae for both the forward and reverse reaction rates. The final formula, as befits a thermodynamic result, makes no mention of the individual partial widths Γ_α.

There are more general forms of the Saha reaction, both for nuclei with excited states, and for reacting pairs that have bound states. We now derive the first of these for the above $1 + 2 \leftrightharpoons 12$ reactions.

Let the nuclei species 1, 2 and 12 have energy levels $E_1^{(a)}$, $E_2^{(b)}$ and $E_{12}^{(c)}$ respectively, for level indices $a, b, c = 1, 2, \ldots$, and corresponding spins $I_1^{(a)}$, $I_2^{(b)}$ and $I_{12}^{(c)}$. Let $a = b = c = 1$ be the ground states. The statistical g factors are then $g_{c:ab} = (2I_{12}^{(c)} + 1)/[(2I_1^{(a)}+1)(2I_2^{(b)}+1)]$, and the c.m. kinetic energy that remains on constituting state c out of ingredients in states a and b is $E_{c:ab} = E_{12}^{(c)} - E_1^{(a)} - E_2^{(b)}$. Let n_a, n_b and n_c be the volume number densities of the three kinds of species, and let Γ_c be the reaction decay width of resonance c which is assumed again to be much less than the elastic width for both producing the resonance and reverting it back to its a and b parts.

With these conditions on the partial widths, if we equate the reaction decay rate of the resonance c, ($\lambda_c = \Gamma_c/\hbar$), with its rate of production using Eq. (12.3.5), we have, just as with Eq. (12.3.8),

$$\frac{n_c}{n_a n_b} n_Q = g_{c:ab}\, e^{-E_{c:ab}/k_B T}$$

$$= \frac{2I_{12}^{(c)}+1}{(2I_1^{(a)}+1)(2I_2^{(b)}+1)} e^{-(E_{12}^{(c)}-E_1^{(a)}-E_2^{(b)})/k_B T}. \qquad (12.3.11)$$

If we sum the number density of the resonances c over some range of states $c = 1, \ldots, C$, the total number density will be $n_C^* = \sum_{c=1}^{C} n_c$, which gives

$$\frac{n_C^*}{n_a n_b} n_Q = \sum_{c=1}^{C} \frac{2I_{12}^{(c)}+1}{(2I_1^{(a)}+1)(2I_2^{(b)}+1)} e^{-(E_{12}^{(c)}-E_1^{(a)}-E_2^{(b)})/k_B T}$$

$$= \frac{e^{(E_1^{(a)}+E_2^{(b)})/k_B T}}{(2I_1^{(a)}+1)(2I_2^{(b)}+1)} \sum_{c=1}^{C} (2I_{12}^{(c)}+1) e^{-E_{12}^{(c)}/k_B T}. \qquad (12.3.12)$$

The sum of weighted nuclear spins in this last equation occurs so often that it is given a special name. We define the *partition function $G(T)$* of any nucleus at a given temperature T as the sum of $2I+1$ over its excited states $p = 1, \ldots, P$ according to

$$G_P(T) = \sum_{p=1}^{P} (2I_p+1)\, e^{-(E^{(p)}-E^{(1)})/k_B T}. \qquad (12.3.13)$$

The partition function is thus proportional the thermal average of the number of m-substates available to the nucleus.

The exponential on the right of Eq. (12.3.13) contains the excitation energies, so at zero temperature $G_P(0) = 2I_1+1$, and depends only on the ground-state spin. In terms of the partition function, the cumulative number density over the given range of excited states is

$$\frac{n_C^*}{n_a n_b} n_Q = \frac{e^{(E_1^{(a)}+E_2^{(b)})/k_B T}}{(2I_1^{(a)}+1)(2I_2^{(b)}+1)} G_C(T) e^{-E_{12}^{(1)}/k_B T}. \qquad (12.3.14)$$

Then, by taking the reciprocals of both sides of this equation, we can sum over the excited states of both the constituent nuclei 1 and 2. Using their similar partition functions $G_A(T)$ and $G_B(T)$ we have

$$\frac{n_A^* n_B^*}{n_C^*} = n_Q \frac{G_A(T) G_B(T)}{G_C(T)} e^{(E_{12}^{(1)} - E_1^{(1)} - E_2^{(1)})/k_B T}, \tag{12.3.15}$$

which is the more usual form of the Saha equation, with $n_Q = (\mu_{12} k_B T / 2\pi \hbar^2)^{3/2}$ as above. The exponential on the right contains the difference of the ground-state energies.

As stated above, the Saha equation is quite general, and applies also to the population of excited states of bound nuclei in a thermal plasma, as well as to bound atoms in an electrical plasma, the context where Saha originally derived this equation [7]. We cannot give a simple derivation of this application of the Saha equation, since we have not yet derived formulae for the excitation and de-excitation cross sections of a bound system on collision with another body. In fact, this is the Saha equation's most useful application in nuclear astrophysics, since the sums for the number densities and the partition functions can be limited just to the bound states. The equation can then be used at a given temperature to derive the number of nuclei summed over all the states stable to particle decay, states which are excited in a thermal equilibrium, with relative probabilities $(2J+1) \exp(-\epsilon/k_B T)$. We give two applications of this result.

12.3.3 Reactions with excited states

Laboratory experiments can only measure cross sections $\sigma_{\alpha\alpha_i}$ in which the target is initially in its ground state, $\epsilon_i = 0$. In a stellar plasma at temperature T, however, nuclear excited states are populated with relative probabilities $\exp(-\epsilon_i/k_B T)$. This means that the astrophysical cross section σ^*, summed over all final states, is hence given by the thermal average

$$\sigma^* = \frac{\sum_{\alpha_i} (2I_i+1) \exp(-\epsilon_i/k_B T) \sum_\alpha \sigma_{\alpha\alpha_i}}{\sum_i (2I_i+1) \exp(-\epsilon_i/k_B T)}, \tag{12.3.16}$$

where the I_i are the spins of the excited states.

12.3.4 Nuclear statistical equilibrium

In certain advanced stages of stellar evolution, such as in the 'silicon burning' stages, the collection of nuclear species is almost in thermodynamic equilibrium. On the short timescale of nuclear reactions they are in equilibrium, but there is a longer timescale of β-decay because of weak interactions, which slowly change the isotopes, typically by changing neutrons into protons. For practical analysis, therefore, it is very convenient to first establish a *nuclear statistical equilibrium* for all the nucleonic reactions, and regard the β-decays as a slow perturbation.

Consider therefore a chain of isotopes $^A Z$ of element Z in thermodynamic equilibrium at temperature T, in a plasma with a large number of neutrons. Define n_0 as the number density of the neutrons, n_A as the isotopic number densities, m_0 as the neutron mass, and m_A as the isotopic masses. The capture reaction $n + {}^A Z \rightleftharpoons {}^{A+1} Z + \gamma$ releases an energy of $Q = (m_A + m_0 - m_{A+1} - E_\gamma^{(A+1)})c^2$ for the ground state to ground state transition. If in thermodynamic equilibrium with a high-density photon bath, the number densities follow the Saha equation

$$\frac{n_0 n_A^*}{n_{A+1}^*} = \left(\frac{A}{A+1}\right)^{\frac{3}{2}} \frac{2 G_A(T)}{G_{A+1}(T)} \theta_u \, e^{(m_A + m_0 - m_{A+1})c^2 / k_B T}, \qquad (12.3.17)$$

where $\theta_u = (m_u k_B T / (2\pi \hbar^2))^{3/2}$. The partition functions here are summed over the states of the isotopes bound to neutron emission, neglecting any resonances as being too short-lived in this context. The neutron does not have any excited states, so we have used $G_0(T) = 2$.

By multiplying equations like (12.3.17) along the isotopic chain, we find that the densities of isotopes $^A Z$ and $^{A+s} Z$ are related by

$$\frac{n_0^s n_A^*}{n_{A+s}^*} = \left(\frac{A}{A+s}\right)^{\frac{3}{2}} \frac{2^s G_A(T)}{G_{A+s}(T)} \theta_u^s \, e^{(m_A + s m_0 - m_{A+s})c^2 / k_B T}. \qquad (12.3.18)$$

If the total number of bound and unbound neutrons is fixed to say n_{tot}, then the absolute normalization of the isotopic densities is determined by

$$\sum_s (A+s-Z) n_{A+s}^* + n_0 = n_{\text{tot}}, \qquad (12.3.19)$$

and Eq. (12.3.18) therefore enables us to calculate all the n_{A+s}^* in terms only of the nuclear binding energies (that is, masses). Figure 12.3 shows how the equilibrium distributions near the ^{56}Fe peak are very sensitive to slight variations in the fraction of excess neutrons.

These states of nuclear statistical equilibrium will slowly change with β-decays, which cannot be in equilibrium as the neutrinos escape irreversibly. The heavier elements $Z + 1$ are therefore slowly produced by decays of the more neutron-rich members of the isotopic chain, but, if this is sufficiently slow, the *nuclear* statistical equilibrium is adiabatically preserved.

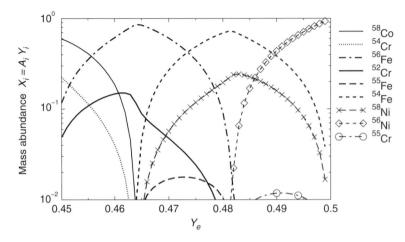

Fig. 12.3. Nuclear statistical equilibrium at various neutron excesses, at a density of 10^7 g/cm^3 and temperature of $T_9 = 3.5$. Y_e is the number fraction of electrons and protons, after Fig. 2 in [8].

12.3.5 Freeze-out

When the neutrons begin to disappear, the (n,γ)–(γ,n) equilibrium becomes more difficult to maintain and β-decays play a larger role in determining the most abundant isotopes of a given element. Eventually the neutron-capture and photo-disintegration reactions 'freeze out,' and the nuclei simply β-decay back to the stability line. In these final stages, according to Surman *et al.* [9], neutron capture and photo-disintegration continue to compete with β-decay, and their detailed rates have an influence on the final isotopic distributions.

12.4 Sensitivities to nuclear data

An important question is to what extent the astrophysical phenomena are dependent on nuclear properties, and hence what sort of experimental and theoretical research programmes should be carried out. The principal kinds of nuclear data that enter into astrophysical calculations are the masses that set the Q-values, the level densities for compound-nucleus decays, β-decay lifetimes, and the cross sections for the various capture and disintegration reactions (averaging over resonances as necessary).

In general, the reaction rates that are necessary to measure and/or calculate are those that are *not* in statistical equilibrium. Such *non-equilibrium* reactions are all those where the reaction products are not sufficiently numerous that the reverse reaction does not occur in detailed balance, and therefore the precise forward reaction rate is a 'bottleneck' that does influence the number density of the reaction products. Most reactions in young stars are in this class, if only for the reason that

Sensitivity analyses in network calculations

In order to see the effects of uncertainties of the cross sections, we need *sensitivity calculations*. These typically use Monte Carlo combinations of input data within their error bars, and analyze the fluctuations in the final results in order to determine their overall accuracy. Alternatively, a linearized analysis of the network calculations may be made, in order to construct a sensitivity matrix which shows how the results depend on specific input nuclear properties.

Sensitivity analyses are outlined by Langanke and Wiescher [10], and recent work includes

- Fiorentini *et al.* [11], Nollett and Burles [12] and Serpico *et al.* [1] for primordial nucleosynthesis,
- Haxton *et al.* [13] and Weiss *et al.* [14] for solar production,
- Hoffman *et al.* [15] for nucleosynthesis in massive stars, and
- Iliadis *et al.* [16] and Hoffman *et al.* [17], for nova and supernova nucleosynthesis.

Sensitivity in nuclear statistical equilibrium

In the nuclear statistical equilibrium discussed in subsection 12.3.4, the *nuclear* rates are in equilibrium, but not the β-decay processes. In this case, the important ingredients are the masses which enter the exponents of Eq. (12.3.18), along with the β-decay lifetimes. It is not necessary to know the exact neutron capture and photo-disintegration cross sections, since these occur much faster than the overall change of the number densities of the interacting and produced nuclei. Hauser-Feshbach predictions of decays, if used, always assume some kind of statistical decay process in which specific fluctuations (e.g. from individual resonances) are not significant.

During nuclear statistical equilibrium, the sensitivities of element production to masses and level densities have been examined by Rauscher *et al.* [18], Demetriou *et al.* [19], and Qian [20], for *r*-process sensitivities to nuclear masses.

Box 12.2 **Sensitivity analyses**

the reaction products have not yet accumulated. As each new cycle (3α, CNO, Si-burning) begins and later ends, there are always certain non-equilibrium reactions that set the overall speed of the processes, and which therefore need to be known with considerable accuracy.

In Box 12.2 we point to some of the sensitivity analyses that have been performed, both in direct network calculations, and also concerning nuclear statistical equilibrium.

We should remember that most of the nuclei used to construct the 'average theory' of nuclear physics are those near the valley of stability, and we may well expect changes in these theories to be necessary as we move towards the dripline, a movement that will be required in order to understand the *r*-process. We do not know, for example, the variation in the optical potentials needed for neutron-rich nuclei, and the imaginary part of these potentials is exactly that which determines the 'transmission coefficients' $\mathcal{T}_\alpha$ for the production and decay of compound-nucleus states. This imaginary part may well depend on the neutron excess (an *isovector part*, we call it), and determining this part will require a new range of experiments and theoretical modeling. At the dripline, the level densities become rather small, and it may not indeed be possible to use the common energy-averaging procedures of Chapter 11. If, for example, the ground state is near to threshold, then it will not be possible to find an averaging interval that includes the many excited state resonances, and in which there are simultaneously only small variations in the Coulomb scattering functions and the penetrabilities that depend on these. The theory of capture reactions on heavy nuclei near the dripline will again require analysis of individual direct-capture mechanisms, as is already the case for light nuclei.

Need for indirect experiments

Direct measurements with exotic or dripline nuclei are impossible if their half lives are too short for them to be produced and then used as targets in experiments with light-ion beams. Reactions involving radioactive nuclei with half lives shorter than a day or so will require experiments in which these nuclei are the beam, rather than the target. Even so, capture reactions are too weak to be measured directly, and thus must be inferred *indirectly* from *other* direct reactions that measure excitation energies, partial wave composition, spectroscopic factors, asymptotic normalization coefficients, level densities and other properties of the relevant states. Transfer reactions such as (d,p) will therefore continue to be an essential spectroscopic tool, but now with a deuteron target so the reaction is conducted in inverse kinematics. The beams of radioactive nuclei will have to be produced by some means, focused, and sometimes reaccelerated in order to make reaction with a deuterated target. The accelerators necessary to produce these radioactive beams are outlined in the next chapter. In Chapter 14 we discuss using reactions at higher energies (above the Coulomb barrier) to extract spectroscopic information about the participating nuclei, and hence indirectly predict capture rates.

Exercises

12.1 Calculate the reaction rates $\langle \sigma v \rangle$ for two proton and neutron reactions on ^{7}Li using cross sections from [2], and compare with Fig. 12.2. There is a resonance in p+^{7}Li scattering at 440 keV. Compare calculations with and without this resonance to find how it affects the reaction rates.

12.2 Using the WKB approximation, derive the screening factor $f_{ws} = \exp[U_0/kT]$ presented in subsection 12.1.7.

12.3 Integrate numerically the CNO network equations of Box 12.1. Compare explicit and implicit methods of solution for long-time evolution.

12.4 Write out the reaction network for the Big Bang reactions shown in Fig. 1.6. Import reaction rates from the compilation of [21].

12.5 Use nuclear statistical equilibrium to find the most stable isotopes, when varying the temperature, if there are equal proton and neutron numbers. At lower temperatures, we should see (as in Chapter 1) that nuclei around ^{56}Fe are the most stable, whereas at high temperatures they would be broken in favor of α particles, and then into separate nucleons. What is the effect of using the partition functions for non-zero temperatures? Compare with the results of [22].

References

[1] P. D. Serpico, S. Esposito, F. Iocco, G. Mangano, G. Miele and O. Pisanti, *J. of Cosm. and Astro. Phys.* **12** (2004) 010.

[2] C. Angulo *et al.*, *Nucl. Phys. A* **656** (1999) 3; URL:pntpm3.ulb.ac.be/Nacre.

[3] M. Ueda, A. J. Sargeant, M. P. Pato and M. S. Hussein, *Phys. Rev. C* **70** (2004) 025802.

[4] E. E. Salpeter, *Austr. J. Phys.* **7** (1954) 373.

[5] E. E. Salpeter and H. M. Van Horn, *Astro. J.* **155** (1969) 183.

[6] H. Dzitko, S. Turck-Chièze, P. Delbourgo-Salvador and C. Lagrange, *Astro. J.* **447** (1995) 428.

[7] M. N. Saha, *Phil. Mag.* **40** (1920) 472.

[8] D. Hartmann, S. E. Woosley and M. F. El Eid, *Astro. J.* **297** (1985) 837.

[9] R. Surman, J. Engel, J. R. Bennett and B. S. Meyer, *Phys. Rev. Lett.* **79** (1997) 1809; R. Surman, J. Beun, G. C. McLaughlin, W. R. Hix, nucl-th/0806.3753v1.

[10] K. Langanke and M. Wiescher, *Rep. Prog. Phys.* **64** (2001) 1657.

[11] G. Fiorentini, E. Lisi, S. Sarkar and F. L. Villante, *Phys. Rev. D* **58** (1998) 064506.

[12] K. M. Nollett and S. Burles, *Phys. Rev. D* **61** (2000) 123505.

[13] W. C. Haxton, P. D. Parker and C. E. Rolfs, *Nucl. Phys. A* **777** (2006) 226.

[14] A. Weiss, A. Serenelli, A. Kitsikis, H. Schlattl and J. Christensen-Dalsgaard, *Astron. Astrophys.* **441** (2005) 1129.

[15] R. Hoffman, T. Rauscher, A. Heger and S. Woosley, *J. Nucl. Sci. Techn.*, supplement 2 (2002) 512.

[16] C. Iliadis, A. Champagne, J. José, S. Starrfield and P. Tupper, *Astro. J. Suppl.* **142** (2002) 105.

[17] R. D. Hoffman, S. E. Woosley, T. A. Weaver, T. Rauscher and F.-K. Thielemann, *Astro. J.* **521** (1999) 735.

[18] T. Rauscher, F.-K. Thielemann and K.-L. Kratz, *Phys. Rev. C* **56** (1997) 1613.

[19] P. Demetriou and S. Goriely, *Nucl. Phys. A* **695** (2001) 95.

[20] Y.-Z. Qian, *Nucl. Phys. A* **752** (2005) 550.

[21] P. Descouvemont, A. Adahchoura, C. Angulob, A. Cocc and E. Vangioni-Flam, *At. Data and Nucl. Data Tables* **88** (2004) 203.

[22] F. E. Clifford and R. J. Taylor, *Mem. R.A.S.* **69** (1965) 21.

13

Connection to experiments

> If your experiment needs statistics, you ought to have done a better experiment.
>
> *Ernest Rutherford*

Making predictions for an experiment is much better achieved if the person performing the calculation is aware of the experimental details. In this chapter we therefore address some of the important issues to take into consideration when applying reaction theories. There is a wide variety of experiments related to astrophysics: the direct measurements (e.g. (p, γ), (n, γ), (α, γ)) performed at low energy are obviously important, but there are also many indirect measurements (e.g. Coulomb dissociation method), a large fraction of which make use of higher-energy accelerators. Many of the forefront research experiments involve radioactive beams, while there are still important rates to determine using stable beams.

Since a large part of the research activity is taking place in rare-isotope facilities, we will first focus (Section 13.1) on some specifics of these facilities, where the exotic nuclei are produced, including both low- and high-energy laboratories. Next, Section 13.2 focuses on different aspects of present-day detectors that need to be considered when comparing calculations with data. Finally, in Section 13.3, we briefly mention some of the direct measurements involving less-exotic nuclei which are stable enough to be made into targets. Included are reactions with light charged-particle beams (protons and alphas), neutron beams and photon beams.

13.1 New accelerators and their methods

Many of the recent leading studies in nuclear physics involve unstable nuclei. The first step in an experiment with radioactive beams is the production of the radio-isotope of interest. The standard setup for experiments with radioactive beams thus contains an ion source from which stable ions are extracted, the accelerator that gives energy to the primary beam, and the production target where the exotic species are produced and focused into a radioactive nuclear beam (RNB). In some cases, further manipulation of the RNB is required in order to obtain a beam with the desired properties. It is with this secondary beam that experimental studies are performed, requiring still a secondary target and finally the detectors. Beam properties are to be kept in mind when performing reaction calculations, and other important considerations are associated with the detection system.

13.1.1 Beam production

The first element in an accelerator facility is the ion source. Almost all ion sources consist of a gas kept at high voltage in a plasma phase from which, in a more or less selective way, ions are extracted. There are a large number of ways to confine the plasma which give rise to a variety of ion sources [1], but since these concern only the primary beam, their details are not so important for the reaction modeling.

Techniques to produce RNBs can be essentially divided into two broad categories, called *fast fragmentation* (FF) and *isotope separator on-line* (ISOL). In *fast fragmentation*, the primary beam is a heavy ion with high energy and the target is relatively thin ($\approx 100 \, \text{mg/cm}^2$), so that after the collision the desired fragments do not lose much energy (remaining at $\gtrsim 50 \, \text{MeV/u}$), and reacceleration is not needed. One of the first realizations of this technique was at Berkeley [2]. The beam produced is a cocktail beam (containing a mixture of isotopes with similar charge over mass ratio) that needs to be purified and focused with electromagnetic filters. Degraders can be used to reduce the energy of the beam if required for specific experiments, but there is a trade-off, as the use of degraders reduces the quality of the beam: the beam spot size increases as well as its energy and angular spreads. Wedges are often introduced in the beam line as a way to purify the beam, and in some cases they can reduce the energy spread. Examples of fast-fragmentation facilities include NSCL (USA), GSI (Germany), RIKEN (Japan), and GANIL (France).

In the *isotope separator on-line* (ISOL) technique, light particles (typically protons) are driven into a thick target, typically of a heavy element such as uranium. The thickness of the target and the beam energy are matched so that the beam is stopped by central collisions producing a wide range of isotopic fragments. The radioactive species diffuse out of the target and then need to be extracted and reaccelerated to energies up to $\approx 10 \, \text{MeV/u}$ [3]. In these facilities, the chemical reactions between the target and the radioactive atoms limit the rate of extraction, and thus only a number of species can effectively be produced. Examples of ISOL facilities include TRIUMF (Canada), REX-ISOLDE (Switzerland), SPIRAL/GANIL (France) and Oak Ridge (USA).

Other techniques exist. The Ion Guided Isotope Separation On-Line (IGISOL) consists on a variation of the ISOL technique where the beam is not completely stopped [4]. There are so far only two implementations of this technique: one at Louvain-la-Neuve (Belgium) and the other at Jyväskylä (Finland). Also one can extract radioisotopes from the core of a reactor as the planned Munich Accelerator for Fission Fragments (MAFF) in Germany [5]. Finally a few radioactive beams can be produced using specific direct reactions, as is the case at ATLAS, Texas A&M, and Notre-Dame, all in the USA.

More than the ion sources, the method of acceleration influences the quality of the secondary beam, namely its energy spread, beam spot size, momentum structure and of course intensity. Cyclotrons are frequently used to accelerate the primary beam in fast-fragmentation facilities, for example: GSI (> 200 MeV/u), RIKEN (≈ 100 MeV/u), NSCL (≈ 100 MeV/u). These typically produce pulsed secondary beams that have large energy spreads (up to 10%), beam spots of > 2 mm^2 and distorted momentum distributions, which make focusing harder. Also the energy range is constrained by the primary beam, since reducing the secondary-beam energy degrades its properties. Nevertheless, fast-fragmentation facilities can produce a very large variety of exotic species and usually have the largest intensities.

Tandems and LINACS are linear accelerators often used in low-energy facilities, such as those using the ISOL technique. They are characterized by a very good energy resolution, beam spot size less than 1 mm^2, as well as small angular spread. There is also a choice of energies for the ISOL beams, but they are typically of lower energy, as otherwise these would become too costly (for a linear accelerator, the higher the beam energy required, the larger the facility needs to be). Here, though, there is a limit to intensities due to space charge effects in the extraction from the target area [6]. Note that cyclotrons and LINACs produce pulsed beams (from the RF structure), whereas tandems can produce continuous beams, but both are equally suited for reaction studies. Beams at Notre-Dame and Oak Ridge are accelerated with tandems (among the older facilities), whereas LINACs are used for reacceleration at ATLAS, ISOLDE, and TRIUMF.

An important advantage of fast-fragmentation facilities is that they can study nuclei with much shorter lifetimes than the ISOL facilities. This is due to the fact that diffusion out of the production target takes time, and for nuclei with lifetimes shorter than seconds, by the time it is extracted and reaccelerated, the isotope has already decayed. In fast-fragmentation facilities, secondary beams with lifetimes down to microseconds can be used.

Finally, there is a correlation between the beam energy of a facility and the physics that can be done. Experimental programs at low-energy facilities focus on fusion (Chapter 7), transfer (Chapter 14) or breakup (Chapter 8) at the Coulomb barrier, whereas fast-fragmentation facilities measure Coulomb excitation, breakup (Chapter 8), knockout and charge-exchange reactions (Chapter 14) well above the barrier.

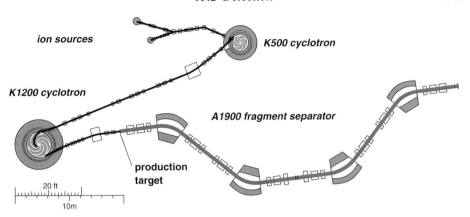

Fig. 13.1. Accelerators at a fast-fragmentation facility: the National Super-conducting Cyclotron Laboratory. Figure courtesy of Thomas Baumann.

13.1.2 An example of a fast-fragmentation facilty

The coupled-cyclotron facility at the National Superconducting Cyclotron Laboratory outlined in Fig. 13.1 is an example where exotic species are produced through fragmentation. A stable heavy isotope is extracted from an electron cyclotron resonance (ECR) ion source. This primary beam is fed into the first superconducting cyclotron (K500) and follows into the second one (K1200), resulting in a beam of 80–160 MeV/u. Finally the primary beam collides with the production target, generating a cocktail of nuclei, among which are the exotic nuclei of interest. The subsequent sequence of electromagnetic devices consists of the fragment separator A1900 [7], which provides the selectivity and focusing for the secondary beam to be guided to the final target.

13.1.3 An example of an ISOL facility

The ISAC facility at TRIUMF, pictured in Fig. 13.2, is an example of an ISOL machine. The primary beam consists of a very intense beam of 500 MeV protons accelerated in the largest non-superconducting cyclotron in the world. These collide with a thick uranium target, which functions as an ion source for the secondary beam. The radioactive nuclei are extracted and go first through a high-resolution mass separator. A beam of the nucleus of interest is then accelerated in a LINAC, which in its first construction phase can attain a few MeV/u.

13.2 Detection

A fast evolving area is that of detector development. The materials used and the electronics are both important components for the developer, but are not so relevant

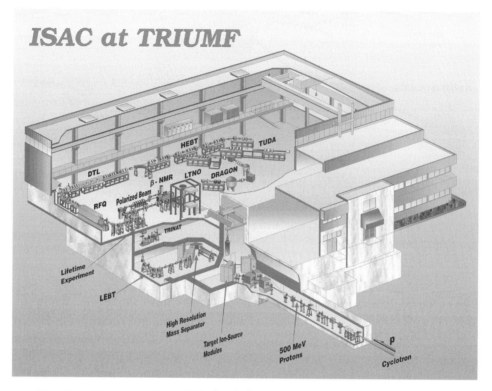

ISAC at TRIUMF

Fig. 13.2. Illustration of an ISOL facility: ISAC at TRIUMF. Figure courtesy of Barry Davids.

for reaction modeling. However, a few aspects need to be considered.[1] In reactions with RNBs one may be interested in measuring: (i) light or heavy charged particles, over a large energy range; (ii) neutrons, specially in reactions with nuclei on the neutron dripline; (iii) gamma rays, for identification of final states.

RNB experiments often start with cocktail beams. A spectrograph for particle identification which can also measure energies is an essential tool in the analysis of the fragments coming out from the reaction area. In Fig. 13.3 we show a picture of a high-resolution, large acceptance magnetic spectrograph in Michigan, the S800, that is able to separate the various reaction products and focus them into the detector plate.

As the secondary beams are cocktails and the targets are typically a mixture, there is usually a background from other radionuclides that needs to be subtracted. A simple example is the (d,p) reaction in inverse kinematics, where one has the carbon background from the hydrocarbon used to make solid targets with

---

[1] Note that, even though here we focus on detectors used in RNB experiments, many aspects are equally important for stable beam experiments.

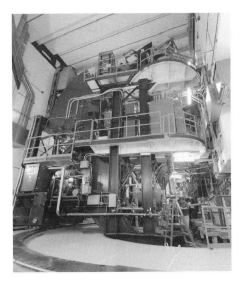

Fig. 13.3. The magnetic spectrograph at NSCL: the S800. Picture courtesy of Thomas Baumann.

deuterons. The subtraction is usually possible when the contaminant has well-known properties, but it is an additional complication. In the specific measurement [8] of ^{9}Li(d,p)^{10}Li, the carbon background is large, and angular ranges were chosen to clean the data. More elaborate methods, involving coincidence or tagging, are often chosen because then the backgrounds can be reduced unambiguously.

In many cases, the secondary beam nucleus to be studied is heavier than the target nucleus, and this inverse-kinematics arrangement imposes specific technical challenges. Inverse kinematics squeezes down the angular range for the reaction products moving in the forward direction. Detectors need to be put at very forward angles (where the elastic background may be large) and the angular resolution requirements may need to be as fine as 10^{-2} degrees, ≈ 100 times better than in normal kinematics. Light-particle detection is very dependent on the experimental setup and properties of the target. For example, in (d,p) reactions in inverse kinematics, protons need to be detected at forward angles in the laboratory but also at backward angles. In Fig. 13.4, an example of such a light charged-particle detector is shown (MUST). Each detector is finely striped to obtain the angular precision needed.

Another aspect that is equally challenging is the energy of the charged particles to be measured, as the energy of outgoing fragments may vary widely with angle (sometimes by orders of magnitudes). Since there is no one detector that can measure energies in such a large range, in some cases particular angular cuts are made.

Fig. 13.4. Picture of a segmented charged-particle array: MUST.

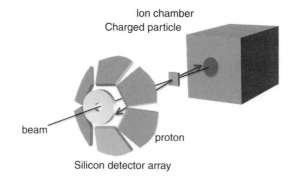

Fig. 13.5. Illustration of the usual setup for (d,p) reaction in inverse kinematics at Oak Ridge National Laboratory.

If the target is thick, there is an additional complication: one cannot determine accurately the energy of the outgoing fragment at the time of the reaction. Using again the (d,p) reactions as an example, whereas in the traditional normal kinematics, measuring the proton would be sufficient for a meaningful analysis, in some inverse kinematic experiments with thick targets, both outgoing particles need to be detected. An example is the setup used to measure (d,p) reactions at Oak Ridge, sketched in Fig. 13.5.

If the RNB intensity is low, as is typically the case for many of the leading studies, detector efficiency and acceptance become important issues. The efficiency of gamma array detection is far from ideal (varying from 1–30%) and is normally traded for resolution. Gamma efficiency is very energy dependent, and the efficiency curve should be measured for every experiment and folded with the

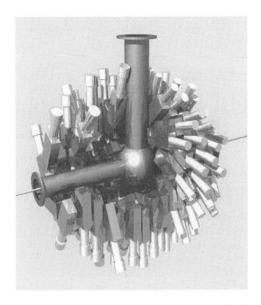

Fig. 13.6. Illustration of the DALI2 gamma array used at RIKEN.

theoretical prediction. A complication occurs when gamma rays are emitted in-flight at very high velocities, because then they show large Doppler shifts when detected. Gamma-ray spectra in the projectile reference frame should be Doppler reconstructed event-by-event, and high granularity of the gamma array is needed to accurately determine the gamma emission angles that enter into this reconstruction. An example of a gamma array is the DALI2 detector from Japan with 25% efficiency and high granularity (see Fig. 13.6).

Neutron detectors are usually placed downstream, and need to be physically extended to cover a significant angular range. Acceptance and efficiency require special attention. In experiments involving neutron charged-particle coincidences, theoretical predictions sometimes need to be processed through appropriate Monte Carlo simulations. An example is the large neutron array MoNA at the NSCL, shown in Fig. 13.7.

Due to limited production yield or detection efficiency, in many cases only inclusive measurements are performed. In such cases, reaction predictions become generally less reliable, especially if it is not possible to calculate all the various processes included in the data in the same theoretical framework. Even when there is just one dominant reaction channel, it is desirable that theorists make predictions of appropriate integrated observables in the laboratory system. If this is not done, experimentalists may try to reconstruct center-of-mass observables based on approximations which are not very accurate. For example, consider the breakup

Fig. 13.7. Picture of the modular neutron array MoNA at NSCL. Picture courtesy
of Thomas Baumann.

of $A \rightarrow B + x$ after collision with a target, where the angle of B is measured and
x is not detected. Theorists normally produce cross sections as a function of the
c.m. angle of the group $B + x$, but this is not the same as the B angle. A simple
relation like that of Eq. (2.3.14) might be tried to guess the $\theta_{\mathrm{exp}}(B) \rightarrow \theta_{\mathrm{cm}}(B)$
transformation, but this is not correct within a three-body final state. Predicting the
multiple-differential cross section from the theory, and then integrating over the
unobservable coordinates, is the best way to compare with the data, and is essential
for an accurate analysis.

The future points toward *complete kinematics*, which is where all outgoing
particles are detected, as well as an identification of possible excited states.
Examples of triple-coincidence measurements are no longer rare. In Fig. 13.8
we show a setup at GSI where neutrons, photons and charged particles are all
measured in coincidence. Such a setup allows, for instance, the measurements
of breakup of neutron-rich isotopes into specific states of the core (e.g. [9]).
The gamma detection array surrounds the target, while the outgoing particles go
through a powerful magnet that bends the charged particles into their detector
to provide particle identification and momentum analysis. Neutrons are not bent
and therefore go straight into the LAND detector. Throughout the beam lines
there are tracking devices to enable energy determination using the time-of-flight
technique. The drawback of this type of experiment is the level of complexity of
the devices. In many cases, a comparison with the data will require folding the

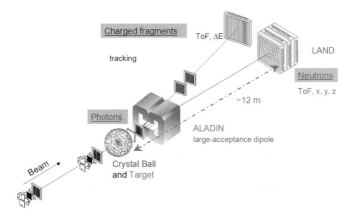

Fig. 13.8. Typical setup for complete kinematic experiments with neutron-rich nuclei at GSI. Figure courtesy of Thomas Aumann.

theoretical prediction with Monte Carlo simulations for the detectors. Again here it is necessary that the theory produces multiple-differential cross sections in the appropriate laboratory frame.

13.3 Direct measurements

There are several important programs that involve direct measurements for astrophysics. For reaction rates involving charged particles such as (p, γ) and (α, γ), usually small accelerators are used. We also need many (n, γ) rates and those require intense beams of neutrons for which there are dedicated facilities. Finally, sometimes the photo-nuclear reaction rate is the preferred measurement, and techniques to make photon beams are well developed. In this section we cover these three cases.

13.3.1 Charged-particle beams

At temperatures of stellar burning (see Section 1.3), the cross sections of charged-particle induced reactions are extremely small because of the low tunneling probability through the Coulomb barrier. Instead, high-energy data are often extrapolated down to stellar energies, but this extrapolation procedure introduces other uncertainties into the reaction rates. For example, resonances in the continuum can make these extrapolations unreliable, as would sub-threshold states (see, for example, Fig. 1.9).

At the astrophysical prohibitively low energies, it is crucial to eliminate any source of background. To eliminate cosmic ray and other background particular to the experimental setting, experiments are performed with complex coincidence schemes. Such is the option followed at the Laboratory for Experimental Nuclear

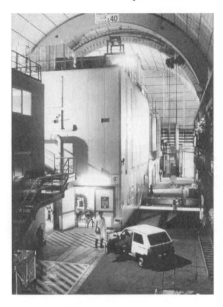

Fig. 13.9. Picture of the LUNA laboratory for underground nuclear astrophysics.

Astrophysics (LENA) located at the Triangle Universities Nuclear Laboratory. A Van de Graaf accelerator supplies proton beams of several hundred keV. Targets are produced with techniques that allow removal of surface impurity (reducing beam-induced background) and are monitored during the experiment. Prompt gamma rays are detected and analyzed. Examples are the ^{14}N(p, γ)^{15}O [10] and ^{17}O(p, γ)^{18}F [11]. Slightly different are reactions such as (p, α), often appearing in novae scenarios (subsection 1.5.2). These have challenges of their own which will not be addressed here (see, for example, [12] or [13] for details).

One effective solution to the cosmic ray background is to go underground. In Fig. 13.9 we show the interior of the Laboratory for Underground Nuclear Astrophysics (LUNA), located in the Gran Sasso mountain, as an example of such a facility. The accelerator produces α particles of several keV and, due to its location, the experimental hall is well shielded against cosmic rays. The ^{3}He(α, γ)^{7}Be was previously the most uncertain rate in the pp-cycle ($>$10%). With the new LUNA measurement [14] the uncertainty was reduced to 4%. This reaction was discussed in Chapter 1 (see Fig. 1.4).

In explosive scenarios one typically needs to know capture rates on radioactive nuclei. Focused experimental programs, coupled to radioactive beam facilities, have been developed to address these issues.

Radioactive beam facilities such as Louvain-la-Neuve are often used for direct measurements in inverse kinematics. An example is ^{18}Ne(α, p)^{21}Na [15]. For this

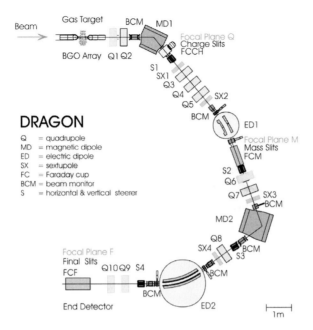

Fig. 13.10. DRAGON mass separator enabling measurements for astrophysics, at ISAC-TRIUMF. Reprinted from S. Engl *et al.*, *Nucl. Instr. Methods A* **553** (2005) 491, Copyright (2005), with permission from Elsevier.

experiment, an ISOL beam (see Section 13.1) of several tens of MeV of ^{18}Ne is produced, and then slowed by degraders, to interact with a helium gas target. For long-lived isotopes, radioactive targets are an option, but the production and handling of a radioactive target is difficult and not many facilities are willing to use them, to avoid contamination (in many countries regulations prevent this).

Coupled to the ISAC-TRIUMF facility (see Section 13.1), the DRAGON mass spectrometer (see Fig. 13.10) was built to measure the rates of some particular capture reactions. A radioactive nuclear beam from ISAC goes through a gas target with either hydrogen or helium. In order that the beam may pass unobstructed, an opening is located on either side of the target, which poses some technical problems because the beam-line should be kept close to vacuum. Gas is pumped away from the beam-line, but target thickness needs to be carefully controlled so that accuracy is not lost. The products of the capture reaction go through the mass spectrometer, which separates the products from the original beam, and are guided into the detector. An example of an experiment using this setup is the ^{26}Al$(p, \gamma)^{27}$Si [16], an important reaction to understand the observed galactic abundances of ^{26}Al.

Last but not the least, fusion reactions involving heavier nuclei can have a critical role in several stellar burning scenarios. Stable-beam accelerators can be used, and an experiment using this type of facility measured the $^{12}C + ^{12}C$ fusion by γ-rays originating from the α,p and n evaporation [17].

13.3.2 Neutron beams

For the s-process and the r-process, it is important to know (n, γ) reaction rates, as discussed in Chapter 1. These have traditionally been measured at small van der Graaf accelerators which, through a charge-exchange reaction such as $^{7}Li(p,n)^{7}Be$, produce a small flux of neutrons with well-defined energy. An example of this application can be found at Karlsruhe [18]. By contrast, a large astrophysics program at CERN, called n_TOF (neutron time-of-flight), is devoted to a systematic study of (n, γ) reactions. There, a high-intensity pulsed proton beam of 20 GeV is used to produce neutrons through spallation on a lead target. This produces a very intense white neutron spectrum, which is monitored by measuring time of flight relative to the production pulse. As sketched in Fig. 13.11, the first experimental vault is nearly 200 m downstream from the production target, which enables a very good energy resolution for the neutrons. Beam energies ranging from extremely low (1 eV) to very high (250 MeV) are all possible, and prompt γ-ray decay cascades are used to register the (n, γ) neutron capture events. Systematic studies are being performed

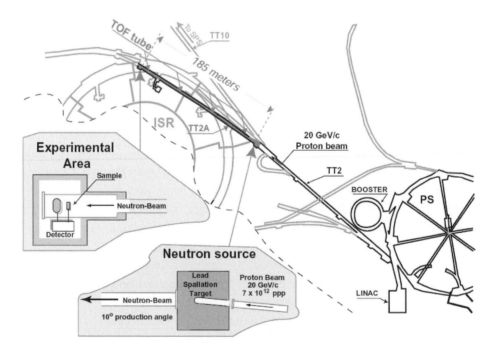

Fig. 13.11. Neutron time-of-flight facility at CERN n_TOF.

on a variety of targets. The recent neutron capture on ^{204}Pb [19] is a good example, where data was collected from 1 eV to 440 keV.

13.3.3 Photon beams

Sometimes, accurate measurements of the capture rate is more difficult than measuring the photo-nuclear cross section. Facilities coupled to electron beams have been developing photon beam capabilities to face this challenge. One possibility is to use an aluminum radiator to convert the electron beam into bremsstrahlung radiation. This is the method used at MAX-lab, in Sweden, a tagged-photon facility. The recoil electrons are detected and used to tag the energy of individual photons, which are collimated before hitting the target. An application of this method was the measurement of ^{4}He(γ, n) [20], which used a liquid target. The outgoing neutrons were detected in liquid-scintillator arrays.

Polarized photons have the added value of providing information on the analyzing power and consequently more details about the reaction mechanisms. The High Intensity Gamma-ray Source (HIGS) at the Duke Free-Electron Laser (FEL) laboratory produces pulses of polarized photons. Electrons from a linear accelerator are injected into a storage ring. These go through an undulator, which causes the bunch to produce a pulse of linearly polarized photons. These photons interact with a second electron bunch and undergo Compton scattering, which allows a clear separation of photon energy by scattering angle. The resulting Compton-scattered photon beam has a very small energy spread at a given angle [21]. The first measurement of the d(γ, n)p analyzing power was performed using this method [22].

Exercises

13.1 Consider the elastic scattering of 50 MeV protons on a ^{7}Be radioactive target. If the charged-particle detection array covers from $10° - 60°$ in the laboratory, what is the angular coverage in the center-of-mass frame?

13.2 Consider ^{7}Be as now the beam with the detection setup the same as before. What would be the angular coverage in the center-of-mass frame?

References

[1] Ian G. Brown, *The Physics and Technology of Ion Sources*, New York: Wiley-VCH (2nd edn.).
[2] T. J. M. Symons *et al.*, *Phys. Rev. Lett.* **42** (1979) 40.
[3] B. Jonson, H. L. Ravn and G. Walter, *Nucl. Phys. News* **3** (1993) and references therein.
[4] J. Aystö, *Nucl. Phys. A* 693 (2001) 477.

[5] D. Habs *et al.*, *Nucl. Instr. Meth.* **B204** (2003) 739.

[6] Thomas P. Wangler, *RF Linear Accelerators*, New York: John Wiley and Sons.

[7] D. J. Morrissey *et al.*, *Nucl. Instr. Meth.* **B204** (2003) 90.

[8] P. Santi *et al.*, *Phys. Rev. C* **67** (2003) 024606.

[9] C. Nociforo *et al.*, *Phys. Lett. B* **605** (2005) 79.

[10] R. C. Runkle *et al.*, *Phys. Rev. Lett.* **94** (2005) 082503.

[11] C. Fox *et al.*, *Phys. Rev. C* **71** (2005) 055801.

[12] K. Y. Chae *et al.*, *Phys. Rev. C* **74** (2006) 012801R.

[13] D. W. Bardayan *et al.*, *Phys. Rev. C* **74** (2006) 045804.

[14] D. Bemmerer *et al.*, *Phys. Rev. Lett.* **97** (2006) 122503.

[15] D. Groombridge *et al.*, *Phys. Rev. C* **66** (2002) 055802.

[16] C. Ruiz *et al.*, *Phys. Rev. Lett.* **96** (2006) 252501.

[17] E. F. Aguilera *et al.*, *Phys. Rev. C* **73** (2006) 064601.

[18] K. Wisshak *et al.*, *Phys. Rev. C* **73** (2006) 015802.

[19] C. Domingo-Pardo *et al.*, *Phys. Rev. C* **75** (2007) 015806.

[20] B. Nilsson *et al.*, *Phys. Rev. C* **75** (2007) 014007.

[21] V. N. Litvinenko *et al.*, *Phys. Rev. Lett.* **78** (1997) 4569.

[22] E. C. Schreiber *et al.*, *Phys. Rev. C* **61** (2000) 061604R.

14

Spectroscopy

> There are two possible outcomes: if the result confirms the hypothesis,
> then you've made a measurement. If the result is contrary to the
> hypothesis, then you've made a discovery.
>
> *Enrico Fermi*

In order to learn about the internal properties of nuclei, some kind of reaction
is necessary. The most interesting questions concern the arrangements of nucleons
inside the nucleus, and, for this purpose, **transfer reactions** have been the standard
procedure for examining the **single-particle structure** of nuclei and extracting
spectroscopic factors. Transfer experiments in the sixties and seventies were
abundant, but they suffered a decrease of popularity following the shutdown of
some low-energy laboratories. They are now becoming popular again to study
exotic nuclei, as more intense beams are produced in ISOL facilities. In this
chapter we primarily discuss the various features of transfer reactions. We present
the standard theory used to analyze these reactions, namely the DWBA, look at
its advantages and limitations, and then consider other approaches that handle
higher-order effects. We compare transfer probes with **electron knockout** and
nuclear knockout reactions. At the end of the chapter we briefly discuss **charge-
exchange reactions**, which are used to extract Fermi or Gamow-Teller transition
strengths.

14.1 Transfer spectroscopy

We begin by discussing *standard DWBA theory* for describing transfer reactions.
This theory is most useful for reactions that probe the surface regions of the
nuclei, and try to measure the spectroscopic factors of the single-particle states. We
later consider *more peripheral reactions*, and also *more central reactions*. More
peripheral reactions (at low energies or at forward angles) tend to probe just the
asymptotic tail of the transfer wave functions, whereas the more central reactions

typically involve more than one transfer or inelastic step, and require theory beyond the DWBA.

14.1.1 DWBA transfer theory

Transfer reactions have been used as a standard means to extract information on the valence orbitals of a nucleus, at least for the surface-dominated reactions. We first discuss the traditional method to analyze transfer experiments which makes use of DWBA theory. Supposing the angular distributions are similar, the spectroscopic factors are extracted from the data by a ratio of cross sections:

$$\frac{d\sigma^{\text{exp}}}{d\Omega} = S^{\text{exp}}\frac{d\sigma^{\text{DWBA}}}{d\Omega}. \tag{14.1.1}$$

Whereas the normalization of the cross section determines the spectroscopic factor, the shape of the angular distribution can often be used to determine the orbital angular momentum of the single-particle state. Spectroscopic factors can then be compared with structure calculations, such as *ab-initio* calculations or the standard shell model (where an internal closed shell is assumed).

Next, we present the details of a DWBA calculation, collecting elements introduced in earlier chapters. Of relevance are the concepts of a vector-form T matrix introduced in subsection 3.3.2, transfer couplings covered in subsection 4.5.1 and overlap functions in Section 5.3.

Let us consider the transfer process $A(a,b)B$, where $a = b + v$ transfers a cluster v onto the target A, forming a bound state $B = A + v$. The relevant coordinates are shown in Fig. 14.1. We first consider the exact transfer matrix element, which can be equally written in the prior form

$$\mathbf{T}_{\text{prior}}^{\text{exact}} = \langle \Psi^{(-)\text{exact}} | V_{vA} + V_{bA} - U_i | \Phi_{I_b:I_a}(\mathbf{r}_{vb})\chi_{\mathbf{k}_i}(\mathbf{R}_i)\rangle, \tag{14.1.2}$$

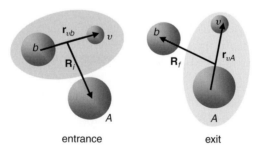

entrance   exit

Fig. 14.1. Coordinates in the entrance and exit partitions for the transfer $A(a,b)B$ where $a = b + v$ and $B = A + v$.

or the post form

$$\mathbf{T}_{\text{post}}^{\text{exact}} = \langle \Phi_{I_A:I_B}(\mathbf{r}_{vA}) \chi_{\mathbf{k_f}}^{(-)}(\mathbf{R}_f) | V_{vb} + V_{bA} - U_f | \Psi^{\text{exact}} \rangle. \qquad (14.1.3)$$

The DWBA theory consists in replacing one of the exact solutions $\Psi^{(\pm)\text{exact}}$ of a three-body problem by a distorted wave multiplying a corresponding bound state: $\Phi(\mathbf{r})\chi(\mathbf{R})$. The T matrix for the transfer process is thus

$$\mathbf{T}_{fi}^{\text{DWBA}} = \langle \chi_f^{(-)}(\mathbf{R}_f) \Phi_{I_A:I_B}(\mathbf{r}_f) | \mathcal{V} | \Phi_{I_b:I_a}(\mathbf{r}_i) \chi_i(\mathbf{R}_i) \rangle, \qquad (14.1.4)$$

where $\mathcal{V} = V_{vA} + V_{bA} - U_i$ in the prior form, or $\mathcal{V} = V_{vb} + V_{bA} - U_f$ in the post form. The interaction V_{bA} is the core-core interaction between b and A, and V_{vb} and V_{vA} are the binding potentials in the entrance and exit channels, respectively. The U_i and U_f are the entrance and exit optical potentials that are used to generate the corresponding distorted waves χ_i and χ_f. Within the transfer operator $\mathcal{V}$, the differences $V_{bA} - U_i$ and $V_{bA} - U_f$ are called the *remnant terms* in prior and post couplings, respectively.

The physical inputs are thus: (i) the optical potentials for the incoming and outgoing distorted waves, U_i and U_f; (ii) a specification of the overlap functions $\Phi_{I_A:I_B}(\mathbf{r}_{Av})$ and $\Phi_{I_b:J_p}(\mathbf{r}_{bv})$, which typically are described by single-particle states in a Woods-Saxon potential with the depth fitted to reproduce a state with the correct binding energy and the appropriate nlj quantum numbers; (iii) the transfer operator, which, if using the same binding potentials and optical potentials, will require as an additional ingredient only the core-core interaction.

The standard procedure for surface-dominated reactions thus contains three steps:

(i) Extract the optical potentials from elastic scattering data, for the correct nucleus and at the right energy (or for a nearby nucleus and the closest energy available). Alternatively, sometimes it is preferable to use potentials fitted to several data sets of elastic scattering on the relevant nucleus but covering a range of energies. This approach improves the confidence level as it avoids resonances or other anomalous effects at isolated energies.

(ii) Calculate the single-particle orbitals using the standard values for the radius and diffuseness ($r \approx 1.2$ fm and $a \approx 0.65$) and normally adding a spin-orbit force, as we did in Section 4.2. Adjust the Woods-Saxon parameters to reproduce the empirical binding energies.

(iii) Compute the differential cross section within the DWBA. If done correctly, post and prior couplings should give the same result (see page 104). Compare with experimental data, and, all going well, extract S^{exp} by Eq. (14.1.1).

The most commonly used transfer reactions is (d,p). A useful compilation of (d,p) reactions, accompanied by a systematic analysis, can be found in [1], who also provide a web database containing a large number of sets of transfer data. They show [1] that predictions of shell model are in good agreement with the spectroscopic factors extracted by this standard transfer-reaction analysis method.

In the following subsection we present particular features of transfer reactions, and discuss how this analysis changes for the more peripheral and the more central collisions. In each case, we discuss what sort of information can be extracted from the analysis of such experiments.

14.1.2 Q-value sensitivity

Although not explicit in Eq. (14.1.4), there is a strong dependence of the magnitude of the transfer cross section on the Q-value for the reaction. For the transfer of a neutral particle, the cross section is largest when $Q = 0$, so Q-value matching enhances the population of particular excited states. For instance, (p,d) and (^{3}He,α) transfers populate very different states of the product nucleus because their Q-values are 2.2 MeV and 19.8 MeV respectively. The (^{3}He,α) reaction will extract much more deeply bound neutrons from the target nucleus. The larger the energy mismatch, the smaller the cross section, as seen in Fig. 14.2.

14.1.3 Angular momentum sensitivity

To see the angular momentum selectivity in transfer reactions most clearly, it is useful to take a specific case and make several approximations. We consider the A(d,p)B reaction in post form for transfers to a specific bound orbital angular momentum l_B. For this argument, we neglect the remnant term, so that the transfer operator V is simply $V_{np}(r)$, the binding potential for the deuteron. This potential is

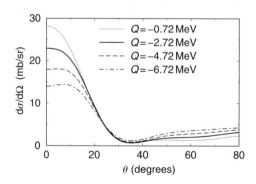

Fig. 14.2. Transfer cross sections for different Q-values for ^{12}C(d,p)^{13}C at 20 MeV. The Q-value is varied arbitrarily.

of short range, so we may often use the zero-range approximation for the deuteron: $V_{np}\phi_d(\mathbf{r}_i) = D_0\,\delta(\mathbf{R}_f - \mathbf{R}_i)$, where $D_0 \simeq -122.5\,\mathrm{MeV\,fm^{3/2}}$.

For a simple demonstration, we now use a plane-wave approximation for the entrance and exit channels ($U_i = U_f = 0$), for in this case the transfer T matrix reduces to simply

$$\mathbf{T}_{fi}^{\mathrm{PWBA}} = D_0 \int e^{i\mathbf{q}\cdot\mathbf{R}}\Phi_{I_A:I_B}(\mathbf{R})\mathrm{d}\mathbf{R}$$

$$= \sum_{l=0}^{\infty} i^l(2l+1)\int F_l(0,qR)/(qR)P_l(\cos\theta)\Phi_{I_A:I_B}(\mathbf{R})\mathrm{d}\mathbf{R}. \qquad (14.1.5)$$

where $l = l_B$ is the angular momentum transferred. Here, $\mathbf{q}$ is the momentum transfer that increases with scattering angle according to $q^2 = p_i^2 + p_f^2 - 2p_i p_f\cos\theta$, with p_i and p_f the momenta of the incoming and outgoing particles, and θ the scattering angle. The angular part of the second integral selects only those l values that appear in the $\Phi_{I_A:I_B}(\mathbf{R})$ bound state: just one value for the simplest overlap functions. As a function of momentum transfer q (directly related to θ), the cross section is

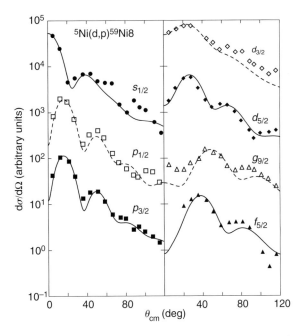

Fig. 14.3. Dependence of the transfer angular distribution on the transferred angular momentum for $^{58}\mathrm{Ni(d,p)}^{59}\mathrm{Ni}$ at 8 MeV, with data from [2]. Reprinted from [3], with permission.

essentially proportional to $|F_l(0, qR)|^2$ for a single l value. Even when distorting optical potentials are included, there will typically be dominant components whose shapes allow one to determine the main angular momenta of the overlap functions. In Fig. 14.3 we show the more realistic variation of the DWBA angular distribution depending on the angular-momentum transfer: the larger the angular-momentum transfer, the smaller the cross section, and the more its peak shifts away to larger angles. For $l = 0$ (s-waves) there is a peak at $\theta = 0$.

14.1.4 Extraction of asymptotic normalization coefficients

Consider again the transfer reaction $A(a, b)B$, where the nucleus of interest is $B = A + v$. From subsection 4.5.1, outside the nuclear potentials of radius R_n the asymptotic behavior of the overlap function $\Phi_{I_A:I_B}(\mathbf{r})$ is proportional to a Whittaker function as

$$\Phi_{I_A:I_B}(\mathbf{r}) \overset{r > R_n}{=} C_{lj}^{A:B} W_{-\eta_v, l+1/2}(2k_v r)/r, \tag{14.1.6}$$

where $k_v = \sqrt{2 \mu_{Av} \varepsilon_{Av}/\hbar^2}$, with ε_{Av} the binding energy for $B \to A + v$. The constant $C_{lj}^{A:B}$ is the asymptotic normalization coefficient (ANC) defined in Eq. (4.5.24).

Although the overlap function $\Phi_{I_A:I_B}(\mathbf{r})$ is a many-body quantity, we generally try to write it as proportional to some single-particle wave function, and introduce many-body effects just through the constant of proportionality, which in this case is the normalization of the state

$$\Phi_{I_A:I_B}(\mathbf{r}) = A_{\ell s j}^{j I_A I_B} u_{lj}(\mathbf{r}), \tag{14.1.7}$$

where the single-particle wave function $u_{lj}(\mathbf{r})$ is normalized to unity. The asymptotic form of $u_{lj}(\mathbf{r})$ is identical to Eq. (14.1.6) with an asymptotic normalization b_{lj} such that $C_{lj}^{A:B} = A_{lsj}^{j I_A I_B} b_{lj}$.

The main contribution to the norm of an overlap function comes from the nuclear interior, whereas the ANC is a peripheral quantity. If the reaction were *completely* peripheral, the DWBA transition matrix would be

$$\mathbf{T}_{fi}^{\text{DWBA}} = C_{lj}^{A:B} \langle \chi_\alpha^{(-)}(\mathbf{R}_f) W_{-\eta_v, l+1/2}(2k_v r)/r |\mathcal{V}| \Phi_{I_b:J_p}(\mathbf{r}_{bv}) \chi_{\alpha_i}(\mathbf{R}_i) \rangle. \tag{14.1.8}$$

Peripheral reactions are therefore ideal for extracting ANC values [4], as the differential cross section becomes proportional to $(C_{lj}^{A:B})^2 = S_{lj}^{A:B} b_{lj}^2$ and not to the spectroscopic factor $S_{lj}^{A:B} = (A_{lsj}^{j I_A I_B})^2$ alone. If spectroscopic factors are extracted from peripheral reactions, the uncertainty from the single-particle potential

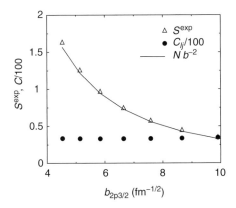

Fig. 14.4. Dependence of the spectroscopic factor and the asymptotic normalization coefficient on the single-particle parameters as a function of b for fits to the transfer cross section ^{48}Ca(d,p)^{49}Ca(g.s.) at 13 MeV. The line is an inverse quadratic to guide the eye.

parameters (which determine b_{lj}) will be extremely large. This phenomenon is illustrated in Fig. 14.4: while the fitted spectroscopic factor (triangles) varies with b_{lj}, the fitted ANC (black circles) remains constant. This is a good example of an indirect method for measuring neutron capture rates. The ^{48}Ca(d,p)^{49}Ca can be used to determine ^{48}Ca(n, γ)^{49}Ca as mentioned in the context of the *s*-process in Chapter 1.

ANCs have a unique astrophysical interest: they determine the direct capture rate $A(v, \gamma)B$ at the limit of zero relative energy. Experimental programs for extracting ANCs from peripheral transfer reactions have been implemented (for example, at Texas A&M, Oak Ridge and Prague). To ensure peripherality, targets heavier than deuterons are typically used, such as ^{14}N. Another method proposed for extracting ANCs consists of measuring the transfer in heavy-ion collisions at sub-Coulomb energies, with the advantage that the kinematics would be direct [5, 6]. This method has not yet been explored experimentally using radioactive beams.

14.1.5 Extraction of spectroscopic factors

The ability to extract spectroscopic factors in transfer reactions depends on whether the reaction mostly takes place in the surface, in the periphery, or more in the interior of the nucleus. Figure 14.5 sketches the sensitivity to the different radial regions of the overlap function of ^{19}C for the reaction ^{18}C(d,p)^{19}C at beam energies of 5, 15 and 30 MeV. The sensitivity also depends on the mapping between scattering angles and impact parameters.

For surface-dominated reactions, if the optical potentials have been determined through elastic scattering, the Q-value is extracted from the energy spectrum, the

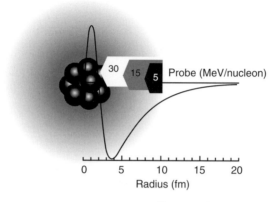

Fig. 14.5. Radial sensitivity of the reaction ^{18}C(d,p)^{19}C to different parts of the overlap function ^{19}C relative to the core ^{18}C: at 5 MeV/u only ^{18}C-n distances up to 8 fm are probed (completely peripheral); at 15 MeV/u the reaction becomes sensitive to the overlap function down to 5 fm; at 30 MeV the reaction is sensitive to the overlap function in the deep interior. Reprinted from [3], with permission.

l-values are known by shape matching, and there is a good guess as to the structure of the overlap function, then it is possible to calculate the transition amplitude in DWBA. It is common to extract S^{exp} by direct comparison with the data, fitting up to the first minimum (corresponding to an impact parameter grazing the surface).

As one moves to larger angles (scattering from impact parameters smaller than the nuclear radius, so more interior), the DWBA is no longer expected to provide reasonable results, even for the angular shape of the cross section (this is true for transfer but also inelastic studies, charge exchange, etc.).

If only the very forward angles are used, by contrast, it may be that the transfer is completely peripheral and thus no longer sensitive to the interior. Under these circumstances, one should rather extract the asymptotic normalization coefficient.

Whatever the impact parameter, if the state being investigated is a pure single-particle state, one configuration should suffice to describe the data. In many cases, however, the state being investigated will not be a single-particle state, and then all contributing orbitals need to be taken into account. These contributions should be added coherently, as interference may affect the results, and so for a multi-configuration state the extraction of the spectroscopic factor is not so clear. A possible procedure is thus to take predictions for the spectroscopic amplitudes from the shell model (or some other microscopic model) and form the transition amplitude by the coherent sum of the components using those model amplitudes. If the various components have different shapes, the experimental angular distribution may help to impose limits on the relative strengths of their specific contributions.

14.1.6 Dependence on optical potentials

DWBA transfer cross sections are typically very sensitive to the optical potential parameters. The angular distributions from different optical potentials can vary significantly, as shown in Fig. 14.6. Moreover, different optical potential parametrizations can provide spectroscopic factors differing by factors of 2 or 3. Consequently, it is very important to pin these down as much as possible. When studying stable nuclei, transfer experiments were often accompanied by elastic scattering measurements for the required channels at the relevant energies. It has been argued [7] that the use of elastic data for a single nucleus and at a single energy may produce an unreasonable optical parameter set, and that a preferable approach is to fit simultaneously a set of elastic data for a range of energies. There is often more than one minima in the χ^2 function, and a multiple-data fit reduces the chances of spurious variations of the potential parameters with energy.

To better illustrate the situation we consider a systematic compilation and re-analysis of the ^{12}C(d,p)^{13}C data [8], for beam energies ranging from 2 MeV to 70 MeV. The published spectroscopic factors differed by up to a factor of three. The analysis of [8] was performed within zero-range DWBA, using global optical potentials such as CH89 [9]. Global potentials reduce significantly the fluctuations in the results, although individual fits may not be so close. When general trends are intended rather than an accurate cross section for a given energy, it is best to use consistently a global parametrization. A further improvement can be obtained using a Jeukenne-Lejeune-Mahaux (JLM) folding procedure for calculating the nucleonic optical potentials ([10], see subsection 5.2.3) which reduces the number of parameters, although some fine tuning of the imaginary part is still necessary.

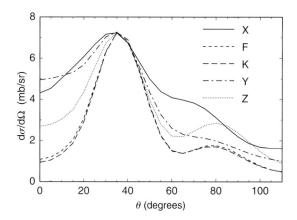

Fig. 14.6. Dependence of the spectroscopic factor on optical parameters for ^{40}Ca(d,p)^{41}Ca(g.s.) at 11 MeV. Optical potentials are taken from [12]. All cross sections are renormalized to emphasize the different angular dependence.

In [8], the deuteron potentials were calculated from the nucleon-target potentials following the Johnson and Soper adiabatic prescription ([11], see subsection 7.1.2) to take deuteron breakup into account. Deuteron breakup effects are expected to increase with beam energy, so the adiabatic approximation (subsection 7.1.5) is very useful to obtain a simple description of the transfer reaction that goes well beyond DWBA.

Whether there is the adequate elastic data to determine the optical parameters or not, it is important to evaluate the uncertainty in the extracted spectroscopic factor caused by U_i and U_f. This dependence tends to increase if the remnant contribution is large. If the potential depths are approximately proportional to the number of nucleons in the projectile (≈ 50 MeV for nucleons, ≈ 100 MeV for deuterons, etc.) then there will typically be some cancellations in the remnant terms. If some shallow potentials are used, by contrast the effects of including the remnant will be large, even if they reproduce elastic scattering.

14.1.7 Dependence on single-particle parameters

The uncertainties of the transfer cross section from their dependence on the details of the radial behavior of the overlap function are usually neglected. This may be appropriate for bound states in stable nuclei, where electron scattering provides detailed density distributions from which the radii and surface thickness can be fairly well determined, but uncertainties in the single-particle parameters are expected to increase nearer the neutron and proton driplines. *Ab-initio* calculations of overlap functions of light loosely bound nuclei are consistent with a Woods-Saxon-shape effective interaction, but often the standard geometrical parameters $r = 1.2$ fm and $a = 0.65$ fm need large adjustments. One problem of using many-body calculations for the overlap functions is connected to precision: *ab-initio* calculations so far are only just beginning [13] to provide accurate results for radii outside the nuclear size, and in addition they are restricted to light nuclei.

The uncertainties induced by ignorance of the geometrical parameters (r, a) are correlated, and are therefore partly described in terms of the single-particle ANC b_{lj}. Generally in transfers, more than 50% of the transfer cross section comes from the tail of the overlap function, and this translates into the cross sections depending more on b_{lj}^2. In the past, no attempt to constrain the asymptotic part was made: single-particle parameters where tuned to produce unit spectroscopic factors for closed-shell nuclei, at the cost of arbitrary ANCs of the overlap functions. Simultaneous analysis at high and low energies can help to pin down b_{lj} and consequently reduce the uncertainty in the single-particle parameters [14, 15].

In the limit of a very peripheral reaction, the dependence of the DWBA cross section on b_{lj} is quadratic, since the relevant part of the square of the wave function, contributing to the cross section, is proportional to $(C_{lj}^{A:B})^2 = S_{lj}^{A:B} b_{lj}^2$. As the spectroscopic factor is extracted as a ratio of the data to the DWBA prediction, the extracted spectroscopic factor will depend on b_{lj}^{-2}, as illustrated by the line in Fig. 14.4. By varying the potential geometry, a range of b_{lj} can be generated, and if we use these to predict the DWBA cross section, we obtain a spectroscopic factor which depends strongly on b_{lj}. Also plotted in Fig. 14.4 is the corresponding ANC (divided by 100) which is constant, regardless of the potential geometry chosen. From Fig. 14.4 we would predict that spectroscopic factors extracted from ^{48}Ca(d,p)^{49}Ca(g.s.) at 13 MeV will be very dependent on the single-particle parameters.

14.1.8 Higher-order corrections

There are several possible types of excitations that give rise to higher-order corrections. On one hand there are distortions and/or breakup of bound states, and on the other the excitations of collective states (vibrational or rotational).

When channel couplings are strong, we need to go beyond one-step approximations and allow for projectile and/or target excitations. It may be necessary to describe the entrance or exit channels of the transfer process in a coupled-channels formalism. If the strongly coupled states are excited bound states of a collective nucleus, a suitable framework is that of the coupled-channels Born approximation (CCBA, see subsection 6.2.2). If, on the other hand, the projectile system is loosely bound, it may have an (even more) loosely bound excited state, and it will surely couple strongly to the continuum. The bound-state excitations can be treated as excitations within the standard CCBA formalism. Methods for including single-particle excitations into the continuum (breakup) need additional attention in determining the excitation couplings and were discussed in Chapter 8. A particular method for including breakup is the adiabatic method presented in subsection 7.1.2, which includes deuteron breakup while keeping the calculations simple.

In single-particle excitations, we must know the spectroscopic *amplitudes* within the overlap functions (rather than spectroscopic factors) because the contributions from various components of the overlap function of $A = B + v$ need to be added coherently as

$$\Phi_J^B(\xi_A, \mathbf{r}) = \sum_{ljI} A_{lsj}^{jIJ} u_{lj}(\mathbf{r}) \, \Phi_I^A(\xi_A). \tag{14.1.9}$$

This mixes different states I of the core A. In Fig. 14.7, we illustrate possible couplings contributing to the ^{12}C(d,p)^{13}C reaction [16]. Including the first excited state of ^{12}C produces a three-component $1/2^-$ ground state for ^{13}C. If many such

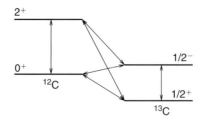

Fig. 14.7. Example of couplings included in the calculation of the transfer reaction
$^{12}C(d,p)^{13}C$ in [16].

components were important, it would be impossible to extract all amplitudes A_{lsj}^{jlJ} from the transfer angular distribution. Flux may also feed through the different states of the composite system B, for example, through the excited states of ^{13}C shown in Fig. 14.7. The situation becomes then even more intricate, and it is rather difficult to disentangle reaction from structure. Good practice is thus to take shell model predictions for all the spectroscopic amplitudes, with the appropriate phases, and compare the overall predictions with the data.

Added complexity arises when the transfer coupling is so strong that coupled reaction channels (CRC) methods are required. Then, because there are contributions from transfer 1-step, 2-step, etc., the dependence of the differential cross section on the spectroscopic factor is no longer linear.

The best prescription is again to take the most reliable structure prediction (usually shell model for single-nucleon transfer and a microscopic cluster model for multi-nucleon transfer) and use it with the best reaction model, compute the differential cross section and compare cross sections directly to the data. One should then vary some of the less-certain parameters in the structure model to reproduce the experimental results, and thereby infer on the accuracy of the structure ingredients.

14.2 Knockout spectroscopy

In the study of stable nuclei, electrons were used to study the configuration of the ground state, as an energetic electron can probe deep inside the nucleus without being much absorbed. Nuclear transfer reactions were also used for the same purpose, but typically the reactions are then more surface peaked. In Fig. 14.8, the probability for $(e,e'p)$ knockout on ^{49}Ca in the second row is compared to the probability for $(d,^{3}He)$ transfer in the last row, as functions of the ^{48}Ca-n distance. Results for the extracted spectroscopic factors from $(e,e'p)$ and transfer do not always agree, but efforts [17] have been made to understand the source of disagreement.

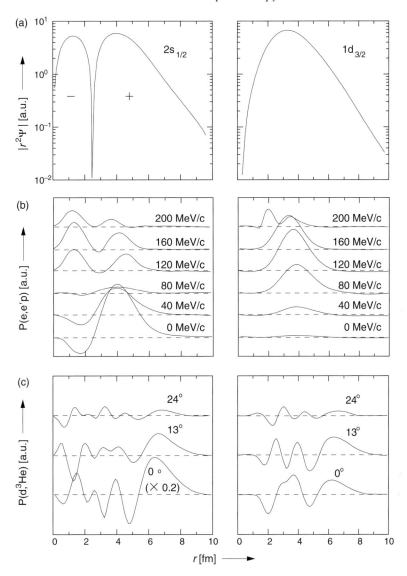

Fig. 14.8. (a) Bound-state wave function for $2s_{1/2}$ (left) and $1d_{5/2}$ (right) in ^{48}Ca. (b) The sensitivity of (e,e′p) to these bound-state wave functions. (c) The sensitivity of (d,^{3}He) to these bound-state wave functions, from [17]. Reprinted from G. J. Kramer, *et al.*, *Nucl. Phys. A* **679** (2001) 267. Copyright (2001), with permission from Elsevier.

More recently, nuclear knockout with γ-coincidences ($A(a, b\gamma)X$) has become an alternative method [18]. One-particle nuclear knockout reactions are those where a projectile A is incident on a light target, and the state of the $A-1$ system and its momentum distribution are measured by the detection system. In Fig. 14.9 we show

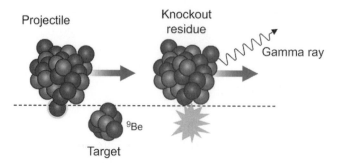

Fig. 14.9. Schematic of a nuclear knockout reaction. Reprinted from [3] with permission.

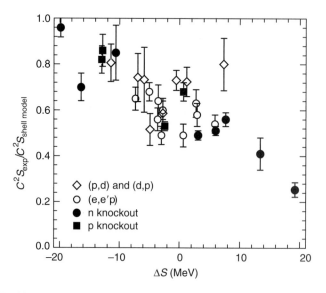

Fig. 14.10. Single-particle spectroscopic strengths as a function of mass number: comparison of electron knockout, nuclear knockout and (p,d) transfers as a function of the difference between the neutron and the proton separation energies, for nuclei with masses $12 \leq A \leq 49$. ΔS is the difference in neutron and proton separation energies, for neutron spectroscopic factors and vice versa for proton spectroscopic factors. Reprinted from [3], with permission.

a diagram of a nuclear knockout reaction, and many examples for both neutron and proton knockout can be found in the survey [18]. Spectroscopic factors extracted from nuclear knockout experiments with stable beams are in agreement with the (e,e′p) values.

A detailed comparison of spectroscopic factors extracted from transfer, electron knockout and nuclear knockout is shown in Fig. 14.10. Direct comparisons are limited because electrons cannot be used to probe nuclei far from stability in fixed targets; because those exotic nuclei have short lifetimes, they can only be used in colliding-beam experiments. Such experiments may become feasible in the next generation of machines (for example, GSI has plans for an electron-scattering facility with rare-isotope beams).

It is simpler to compare the two standard nuclear methods for extracting the spectroscopic factors, transfer reactions versus knockout measurements at higher energy. Whilst knockout is limited to ground-state spectroscopy, with transfer one can study both ground and excited states. Also, transfer experiments probe both particle and hole states (through pickup and stripping), while knockout can only study hole states. The agreement between the spectroscopic factors extracted through these two methods is shown in Fig. 14.10.

While transfer data is traditionally analyzed within DWBA, the analysis of knockout is usually performed with the eikonal approximation (see Section 7.2). In nuclear knockout reactions, a nucleus A is incident on a target T and loses one or more nucleons, and the momentum distribution of the $A-1$ system is measured (see subsection 8.3.1), with an accompanying gamma ray whenever the knockout leads to an excited state of the core. In this way the residual population of each individual core state I can be determined. The measured momentum distributions are inclusive, however, as they include stripping (transfer) as well as diffraction (elastic breakup), but both components of the single-particle cross sections can be calculated within the eikonal model (see subsection 7.2.4ff). The essential ingredients in the calculations are the S matrices as a function of impact parameter b, which are often built by folding methods, or from empirical potentials that should reproduce the corresponding elastic scattering [19]. The shape of the momentum distributions relates to the angular momentum transfer. As in transfer, the normalization is connected to the spectroscopic factor.

In order to obtain the angle-integrated theoretical cross section to be compared with experiment from an initial state i of nucleus A to a final state f of the $A-1$ system, each single-particle cross section $\sigma^{\mathrm{sp}}(\ell j)$ is multiplied [19] by the corresponding spectroscopic factor $S_{\ell j}^{fi}$ from the shell model:[1]

$$\sigma_{fi}^{\mathrm{th}} = \sum_j S_{\ell j}^{fi} \sigma^{\mathrm{sp}}(\ell j). \tag{14.2.1}$$

One main drawback of the method is that Coulomb contributions are estimated independently and added incoherently. Also, it is assumed that stripping and diffraction processes do not interfere, which may not always be the case.

---

[1] Note that shell-model spectroscopic factors using harmonic oscillator basis states should be multiplied by a center-of-mass correction $(A/(A-1))^N$ where $N = 2n + l$ is the harmonic-oscillator quantum number, if the extraction method of Eq. (5.1.14) is not used.

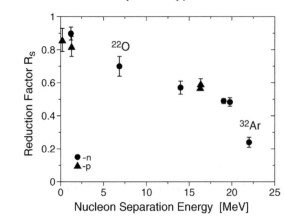

Fig. 14.11. Reduction factors for a selected group of nuclei as a function of the binding energy of the knockout projectile [20]. Reprinted with permission from A. Gade *et al.*, *Phys. Rev. Lett.* **93** (2004) 042501. Copyright (2004) by the American Physical Society.

Nevertheless, the eikonal method has provided a simple and consistent platform for analyzing the systematic knockout studies performed on dripline nuclei [18].

By comparing σ_{fi}^{th} to experiment, a consistent reduction factor has been observed $R_s = S_{\ell j}^{exp}/S_{\ell j}^{th} < 1$ (see Fig. 14.11). This may be explained theoretically as the effect of short-range correlations, or other types of correlations not included in the standard shell model, based on effective interactions. For stable nuclei this reduction factor is approximately 0.6, and it approaches unity for very loosely bound systems. Research on this topic continues (e.g. [20]).

In summary, knockout reactions, as transfer reactions, probe single-particle properties of the nucleus, and provide a good testing ground for shell model predictions (for example the occurrence of magic numbers). Since the shell model provides much input to astrophysics, these reactions are a natural and important component of astrophysical research programs.

14.3 Inelastic spectroscopy

The cross section for exciting a nucleus from its ground state into a bound excited state relates to the intrinsic structure of the nucleus, usually expressed as a reduced transition strength $B(i \rightarrow f)$ (see Section 4.4). If the process of excitation is one step and can be treated within DWBA, then, as with the transfer case, the normalization of the predicted DWBA cross section to the experimental cross section provides a transition strength. Just as in transfer reactions, the shape of the angular distribution again provides information on the angular momentum of the state.

Many cases exist, though, where strong couplings between states are present. Then a coupled-channel formulation of the problem is necessary. In that case, the extraction of the transition strengths between the various relevant states is no longer straightforward, and relies on input from a structure model, such as the shell model or mean-field models.

If the probe used to excite the nucleus is light, the reaction is nuclear dominated, and one can extract information of the nuclear transition strength. Proton inelastic scattering is often studied with this purpose in mind (see, for example, the extraction of quadrupole strength of ^{12}Be [21]). Often one is interested in the electromagnetic strengths connecting states. One might assume that the nuclear matter and the charged matter are equally distributed, which means that the nuclear deformation is the same as the Coulomb deformation, but this is not a good approximation if the nucleus is far from stability. Then it is best to also use a heavy probe, because the excitation then occurs through the Coulomb field, and the analysis should probe the required electromagnetic properties.

Nuclear collectivity is an intriguing phenomenon that continues to attract attention. Quantitative indicators of collectivity are the electromagnetic transition probabilities. Collective couplings, introduced in subsection 4.4.1, can be determined experimentally by measuring either the decay lifetime, or the Coulomb excitation cross section [22, 23]. The Coulomb excitation ('Coulex') technique was originally devised to measure transition probabilities of the target at sub-Coulomb energies chosen to ensure a nuclear-free measurement. More recently, a growing interest in understanding the development of collectivity on the nuclear driplines has led to a Coulomb excitation program at higher energies, where the beam particle, now the object of study, is excited through the virtual photon field with a well-known heavy target. The energies are high enough to inhibit multi-step effects and data is taken only at very forward angles, where one expects to be free from nuclear interference. The analysis of such experiments is often performed within the semiclassical theory of Alder and Winther introduced in Section 7.3, although there are cases where quantum effects and couplings need to be considered [24]. An application of the method can be found in [25] where the excitation of the unstable nucleus ^{30}S was studied to extract B(E2) strength.

As opposed to collective properties, one may also be interested in single-particle excitations (see subsection 4.4.2). Those couplings connect single-particle states rather than states which belong to a collective band. As before, heavy probes will allow the excitation to happen mostly through the Coulomb field, and here provide a direct measure of the single-particle electromagnetic properties of the nucleus. One example of a very strong single-particle transition probability is the B(E1) between the ground state and the only bound excited state of ^{11}Be. An early use of Coulomb excitation to extract this strength can be found in [26].

14.4 Breakup spectroscopy

Many capture rates involving charged particles that are relevant for astrophysics are needed at low energies where the Coulomb repulsion makes the cross section forbiddingly small. It is therefore convenient to devise indirect methods from which to extract the required information, so here we discuss two of these indirect methods.

14.4.1 Coulomb dissociation method

Through detailed balance, the photo-disintegration reaction $p + \gamma \rightarrow c + v$ is directly proportional to the capture reaction $c + v \rightarrow p + \gamma$ according to

$$\frac{\mathrm{d}\sigma_{\mathrm{capt}}}{\mathrm{d}E_\gamma} = \frac{2(2I_p + 1)}{(2J_c + 1)(2J_v + 1)} \frac{k_\gamma^2}{k^2} \frac{\mathrm{d}\sigma_{\mathrm{dis}}}{\mathrm{d}E} \tag{14.4.1}$$

where k derives from the relative energy ε of $v + c$, and k_γ is the wave number of the photon, defined by $k_\gamma = E_\gamma/\hbar c = (\varepsilon + Q)/\hbar c$. Generally, $k_\gamma \ll k$ so that $\sigma_{\mathrm{dis}} \gg \sigma_{\mathrm{cap}}$, which makes it easier to measure photo-dissociation in comparison with direct capture. The Coulomb dissociation method [27] consists of inducing the dissociation in the field of virtual photons generated by a heavy target $p + t \rightarrow v + c + t$. If the Coulomb breakup process is one-step, and dominated by a single multipole transition λ, it is possible to give separate factors for the kinematics of the reaction and the photo-dissociation rates. Using, for example, the semiclassical theory of Alder and Winther introduced in Chapter 7, we have

$$\frac{\mathrm{d}^2\sigma_{\mathrm{cd}}}{\mathrm{d}\Omega \mathrm{d}E_\gamma} = \frac{1}{E_\gamma} \frac{\mathrm{d}n_{E\lambda}}{\mathrm{d}\Omega} \frac{\mathrm{d}\sigma_{\mathrm{dis}}(E\lambda)}{\mathrm{d}E}. \tag{14.4.2}$$

The quantity $\mathrm{d}n_{E\lambda}/\mathrm{d}\Omega$ is called the *virtual photon number* [28], and has analytical expressions in semiclassical perturbation theory (Section 7.3) either in relativistic or non-relativistic kinematics.

We can write the photo-dissociation cross section explicitly [30] in terms of the reduced transition probability (introduced in Chapter 4) as

$$\frac{\mathrm{d}\sigma_{\mathrm{dis}}(E\lambda)}{\mathrm{d}E} = \frac{(2\pi)^3(\lambda + 1)}{\lambda((2\lambda + 1)!!)^2} \left(\frac{E_\gamma}{\hbar c}\right)^{2\lambda - 1} \frac{\mathrm{d}B(E\lambda)}{\mathrm{d}E}. \tag{14.4.3}$$

These Eqs. (14.4.1, 14.4.2, 14.4.3) would suggest that a measurement of Coulomb dissociation should directly determine the capture rate. An example of the application of this method is shown in Fig. 14.12 for determining the capture rate

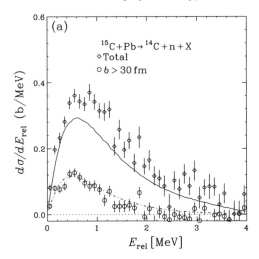

Fig. 14.12. Relative energy distribution following the breakup of ^{15}C on ^{208}Pb [29]. Comparison of the data with the virtual photon method. Reprinted from T. Nakamura *et al.*, *Nucl. Phys. A* **722** (2003) 301c. Copyright (2003), with permission from Elsevier.

of ^{14}C(n,γ)^{15}C. The capture cross section extracted indirectly through Coulomb dissociation and the direct measurements are compared in Chapter 1 in the context of the *r*-process. However, there are some complications that need to be considered.

In most cases, the direct-capture cross section is dominated by $E1$ transitions, whereas the Coulomb dissociation contains non-negligible $E2$ contributions, and different multipolarities of the strong Coulomb field interfere with each other. More importantly, many capture reactions happen at very large impact parameters and are thus Coulomb dominated, whereas the Coulomb dissociation usually contains nuclear contributions. Experimentalists restrict their Coulomb-dissociation measurements to forward-angle regions and small relative energies to try to minimize these problems. In some cases, the nuclear contributions are scaled from reactions on a light target and subtracted from the measured dissociation cross sections in order to extract a 'pure Coulomb' contribution. It has been shown [31], however, that, in particular for loosely bound systems, scaling the nuclear contribution is not reliable, and also that nuclear interference effects are large, even for restricted angular cuts, because of nuclear diffraction from small impact parameters to small scattering angles. The best way to obtain reliable results, therefore, consists of using the best reaction-theory model, including nuclear and Coulomb consistently and multipole interference (such as CDCC introduced in Chapter 8), and tuning the projectile structure potential such that the measured cross sections are reproduced. The resulting structure model for the projectile will then determine the multipole response $dB(E1)/dE$ required to calculate the capture rates at the necessary astrophysical energies, as for example in [32].

14.4.2 Extracting an asymptotic normalization coefficient

As mentioned above, charged-particle capture rates for astrophysics are very strongly hindered by Coulomb repulsion, so astrophysical capture rates must often be extrapolated to the low thermal energies needed in stellar environments. Due to the uncertainties in the extrapolation to low energies, a method to determine the astrophysical rate at zero energy would be very useful. It turns out that the S-factor at zero energy is dependent on the ANC but not on the continuum properties. An experimental measurement of the ANC will therefore determine $S(E=0)$ for a given reaction of interest [33, 34].

We have seen that the ANC values can be extracted by suitable transfer experiments. In addition, recent work shows that breakup predictions are sensitive mainly to the asymptotic properties of the projectile [35, 36, 37, 38]. Both the ANC of the initial bound state and the phase shifts in the continuum play significant roles in the dissociation process. If the continuum structure is sufficiently well known to pin down the phase shifts, and the breakup mechanism is well modeled, normalization of the breakup cross section should provide an accurate way to determine the ANC.

14.5 Charge-exchange spectroscopy

In Chapters 1 and 12, we saw that some weak interaction rates are needed for determining β-decay rates back to stability in element nucleosynthesis. These are related to the spin and isospin responses of nuclei, which is usually measured through charge-exchange reactions. The simplest charge-exchange reaction is the (p,n) reaction with $\Delta T_z = -1$. Alternatively, one can also use (^{3}He,t) or heavy-ion charge exchange. For populating states with $\Delta T_z = +1$, one can use (n,p) reaction rates, but (d,2p) and (t,^{3}He) have also been used on stable nuclei, and more recently (^{7}Li,^{7}Be) has been attempted on unstable systems. For an overview of experiments on charge exchange see [39]. The (p,n) and (n,p) reactions probe the response in the nuclear interior, whereas reactions with the composite probes ($A \geq 3$) are more surface peaked, with (d,2p) being in between [40]. Obviously (d,2p) has the additional complication of a three-body final continuum state. The use of composite systems as probes adds complexity to the reaction model, so here we start with the (p,n) reaction. We use the Fermi and the Gamow-Teller transition operators introduced in subsection 5.4.2 ($\tau_1 \cdot \tau_2$ and ($\sigma_1 \cdot \sigma_2$) ($\tau_1 \cdot \tau_2$) respectively), and their reduced transition probabilities defined by Eqs. (5.4.2) and (5.4.3).

For (p,n) reactions at energies above $\sim100\,\mathrm{MeV}$, and for the cross sections at forward angles, the reaction can be assumed to take place in one step. Thus the transitions between target states $|I_i M_i\rangle$ and $|I_f M_f\rangle$ are typically evaluated in first-order distorted wave approximation using an effective interaction between the proton and the nucleons in the target [42, 43]. At backward angles the two-step exchange transfer mechanism becomes dominant, so in first approximation, for extracting Fermi (F) or Gamow-Teller (GT) strengths, only forward angles would be useful. When nucleons are identical, one needs to include exchange effects associated with antisymmetrization. The general transition amplitude is

$$T_{fi}^{\mathrm{DW}} = \int \chi_f^{(-)}(\mathbf{k}_f,\mathbf{r}_p)^* \, U_{fi} \, \chi_i(\mathbf{k}_i,\mathbf{r}_p) \, \mathrm{d}\mathbf{r}_p, \qquad (14.5.1)$$

with $U_{fi} = \langle I_f M_f, \frac{1}{2}m_f | \sum_j V_{pj}(1-P_{pj})|I_i M_i, \frac{1}{2}m_i\rangle$, where P_{pj} represents the permutation $p \leftrightarrow j$ of spatial, spin and isospin coordinates. It can be shown that the transition amplitude of Eq. (14.5.1) can be factorized into a structure part and a reaction part, as explained in [44]. Here again, elastic scattering data should be used to fix the optical parameters, as the dependence on optical potentials can be as large as 20%. An example of DWBA predictions for a charge-exchange reaction is shown in Fig. 14.13 [41], which connects back to the r-process as discussed in Chapter 1.

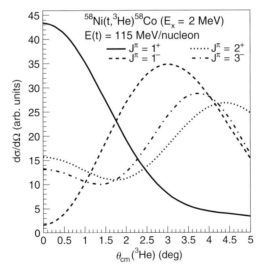

Fig. 14.13. Differential cross section for the charge-exchange reaction $^{58}\mathrm{Ni}(t,^3\mathrm{He})$ as a function of the angle of $^3\mathrm{He}$ in the center-of-mass frame [41]. Different lines correspond to the contributions of different partial waves. Reprinted with permission from A. L. Cole *et al.*, *Phys. Rev. C* **74** (2006) 034333. Copyright (1997) by the American Physical Society.

One can further simplify Eq. (14.5.1) assuming plane waves or the eikonal approximation. If the momentum transfer could be zero, there would be a direct proportionality between the cross section and the F/GT strength [45]:

$$\frac{d\sigma}{d\Omega}(q=0) = N\frac{E_i E_f}{(\pi \hbar c^2)^2}|\mathcal{J}_{\sigma\tau}|^2 B(F/GT), \qquad (14.5.2)$$

with E_i, E_f the relative energy in the entrance or exit channel. The $\mathcal{J}_{\tau/\sigma\tau}$ is the volume integral of either the τ component (for Fermi transitions) or the $\sigma\tau$ component (for Gamow-Teller transitions) of the effective nucleon-nucleon interaction and N is a factor that corrects for the plane wave approximation, and accounts for the nuclear distortion. Generally, to fix N, Gamow-Teller results need to be calibrated using strengths known directly from β-decay [45]. The cross section at $\frac{d\sigma}{d\Omega}(q=0)$ can never be measured for reactions with non-zero Q-value, so a common practice is to use cross sections measured at $0°$ (where q is smallest) and combine them with DWBA ratios to extrapolate to $q=0$ (this is important for composite particles). Another approximation is to neglect the exchange component (or simplify it).

We make some additional comments pertaining particularly to the Gamow-Teller transitions. When using composite probes, there is a larger number of spin and isospin combinations allowed, and further complications arise when several combinations contribute to the same level. If J is the change in total spin of the target in the charge-exchange reaction, $\mathbf{J} = \mathbf{L} + \mathbf{S}$ (the transferred total angular momentum is the sum of the orbital angular momentum and the transferred spin). This is the case in the example shown in Fig. 14.13. For GT transitions we are most interested in $S = 1$, but there may often be several L components contributing to the same state. If there is no interference, components can often be disentangled as done in [40]. However, when tensor terms are included in the operator, a large source of uncertainty comes from the interference between the various L components which cannot be determined experimentally.

In order to circumvent these uncertainties, the best approach is (as before) to introduce the relative amplitudes from a shell-model calculation, and to use the overall comparison of the theory predictions with experiment to conclude about the accuracy (or otherwise) of those shell-model wave functions.

Exercises

14.1 Consider the transfer reaction ^{82}Ge(d,p)^{83}Ge measured in inverse kinematics at 4 MeV/u [46] with the aim of extracting a spectroscopic factor. The deuteron wave function can be calculated using a simple Gaussian interaction $V(r) = -72.15\exp(-(r/1.484)^2)$ that binds the proton and the neutron in an s-state at 2.2 MeV.

(a) Analyze the shell structure of ^{83}Ge and determine the (n, l, j) orbital for the transferred neutron. Use a standard radius $R = 1.2A^{1/3}$ fm and diffuseness $a = 0.65$ fm, fix the spin-orbit strength to 7 MeV with the same geometry and adjust the Woods-Saxon depth V_{ws} to reproduce the correct binding energy for this single-particle state.

(b) Use the Koning and Delaroche global parametrization and the Johnson and Soper approximation to determine the optical potentials in the entrance and exit channels.

(c) Calculate the DWBA transfer cross section in zero range. How large is the finite-range correction within the local energy approximation?

(d) Calculate the cross section in finite range without remnant and compare with the estimates obtained above.

(e) Calculate the DWBA finite range with full complex remnant and compare post form with prior form.

(f) Compare your calculations of 1-step DWBA with the data and extract a spectroscopic factor from the normalization of the cross section $S^{\mathrm{exp}} = \frac{\sigma^{\mathrm{exp}}}{\sigma^{\mathrm{DWBA}}}$. How do your results compare with those in [46]?

(g) Iterate the transfer coupling in a coupled reaction channel procedure. How many iterations do you need for convergence?

14.2 Proton inelastic scattering of ^{32}Si at 42 MeV/u was studied to determine the B(E2) [47]. In that work a deformation of $\beta_2 = 0.28$ was extracted for ^{32}Si.

(a) Obtain the optical potential for the problem from the global parametrization of Koning and Delaroche.

(b) Calculate the inelastic scattering in DWBA and compare the differential cross sections when you include only nuclear deformation with those when both the nuclear and the Coulomb potentials are deformed.

(c) How sensitive are the results to the choice of optical potential?

(d) Perform a full coupled-channels calculation and estimate the error in the β_2 extracted in the analysis of [47].

(e) Compare your results with the data in Fig. 2 of [47] and adjust β_2 to fit the data.

References

[1] M. B. Tsang, J. Lee and W. G. Lynch, *Phys. Rev. Lett.* **95** (2005) 222501; URL:groups.nscl.msu.edu/nscl_library/pddp/database.html.

[2] M. S. Chowdhury and H. M. Sen Gupta, *Nucl. Phys. A* **205** (1973) 454.

[3] *Isotope Science Facility*, NSCL white paper 2007.

[4] H. M Xu, *et al.*, *Phys. Rev. Lett.* **73** (1994) 2027.

[5] M. A. Franey *et al.*, *Nucl. Phys.* **A324** (1979) 193.

[6] C. R. Brune *et al.*, *Phys. Rev. Lett.* **83** (1999) 4025.

[7] J. P. Schiffer *et al.*, *Phys. Rev.* **164** (1967) 1274.

[8] X. D. Liu *et al.*, *Phys. Rev. C* **69** (2004) 064313.

[9] R. L. Varner *et al.*, *Phys. Rep.* **201** (1991) 57.

[10] J.-P. Jeukenne, A. Lejeune and C. Mahaux, *Phys. Rev. C* **15** 10 (1977).

[11] R. C. Johnson and P. J. R. Soper, *Phys. Rev. C* **1** (1970) 976.

[12] L. L. Lee *et al.*, *Phys. Rev.* **136** (1964) B971.

[13] S. Quaglioni and P. Navratil, *Phys. Rev. Lett.* **101** (2008) 092501.

[14] A. M. Mukhamedzhanov and F. M. Nunes, *Phys. Rev. C* **72** (2005) 017602; *Phys. Rev. C* **72** (2005) 014610.

[15] D. Y. Pang, F. M. Nunes and A. M. Mukhamedzhanov, *Phys. Rev. C* **75** (2007) 024601.

[16] F. Delaunay, F. M. Nunes, W. G. Lynch and M. B. Tsang, *Phys. Rev. C* **72** (2005) 014610.

[17] G. J. Kramer, H. P. Blok and L. Lapikas, *Nucl. Phys. A* **679** (2001) 267.

[18] P. G. Hansen and J. A. Tostevin, *Annu. Rev. Nucl. Part. Sci.* **53** (2003) 219.

[19] J. A. Tostevin, *J. Phys. G* **25** (1999) 735.

[20] A. Gade *et al.*, *Phys. Rev. Lett.* **93** (2004) 042501.

[21] H. Iwasaki *et al.*, *Phys. Lett. B* **481** (2000) 7.

[22] T. Glasmacher, *Annu. Rev. Nucl. Part. Sci.* **48** (1998) 1.

[23] J. M. Cook, T. Glasmacher and A. Gade, *Phys. Rev. C* **73** (2006) 024315.

[24] F. Delaunay and F. M. Nunes, *J. Phys. G* **34** (2007) 2207.

[25] P. D. Cottle *et al.*, *Phys. Rev. Lett.* **88** (2002) 172502.

[26] T. Nakamura *et al.*, *Phys. Lett. B* **394** (1997) 11.

[27] G. Baur, C. Bertulani and H. Rebel, *Nucl. Phys.* **A458** (1986) 188.

[28] C. A. Bertulani and G. Baur, *Nucl. Phys.* **A442** (1985) 739.

[29] T. Nakamura *et al.*, *Nucl. Phys. A* **722** (2003) 301c.

[30] H. Esbensen and G. Bertsch, *Nucl. Phys.* **A600** (1996) 37.

[31] M. Hussein, R. Lichtenthäler, F. M. Nunes and I. J. Thompson, *Phys. Lett. B* **640** (2006) 91.

[32] N. C. Summers and F. M. Nunes, *Phys. Rev. C* **78** (2008) 011601R; *Phys. Rev. C* **78** (2008) 069908.

[33] L. Trache, F. Carstoiu, C. A. Gagliardi and R. E. Tribble, *Phys. Rev. Lett.* **87** (2001) 271102.

[34] L. Trache, F. Carstoiu, C. A. Gagliardi and R. E. Tribble, *Phys. Rev. C* **69** (2004) 032802.

[35] S. Typel and G. Baur, *Phys. Rev. Lett.* **93** (2004) 142502.

[36] S. Typel and G. Baur, *Nucl. Phys. A* **759** (2005) 247.

[37] P. Capel and F. M. Nunes, *Phys. Rev. C* **73** (2006) 014615.

[38] D. Baye and E. Brainis, *Phys. Rev. C* **61** (2000) 025801.

[39] M. N. Harakeh and A. van der Woude 2001, *Giant Resonances: Fundamental High-Frequency Modes of Nuclear Excitations*, New York: Oxford University Press.

[40] R. G. T. Zegers *et al.*, *Phys. Rev. Lett.* **99** (2007) 202501.

[41] A. L. Cole *et al.*, *Phys. Rev. C* **74** (2006) 034333.

[42] W. G. Love and M. A. Franey, *Phys. Rev. C* **24** (1981) 1073.

[43] M. A. Franey and W. G. Love, *Phys. Rev. C* **31** (1985) 488.

[44] F. Osterfeld, *Rev. Mod. Phys.* **64** (1992) 491.

[45] T. N. Taddeucci *et al.*, *Nucl. Phys.* **A469** (1987) 125.

[46] J. S. Thomas *et al.*, *Phys. Rev. C* **76** (2007) 044302.

[47] P. D. Cottle *et al.*, *Phys. Rev. Lett.* **88** (2002) 172502.

15

Fitting data

An approximate answer to the right problem is worth a good deal more than an exact answer to an approximate problem.

John Tukey

Given the R-matrix, DWBA or coupled-channels theories of the previous chapters, we now discuss how to vary any unknown parameters to fit experimental measurements. The fitting parameters of a reaction model could be those which specify Woods-Saxon potentials, or they could be structural parameters such as deformations or spectroscopic factors. Sometimes the parameters linearly determine the model predictions (such as spectroscopic factors in one-step DWBA), but more often the predictions depend non-linearly on the model parameters (especially those defining the potentials). When there are multiple non-linear parameters, more systematic methods are needed.

Data fitting is thus the systematic variation of the parameters p_j ($j = 1, \ldots, P$) of a theory in order to find a parameter combination which minimizes the discrepancies with experiment. This is most commonly done by minimizing the χ^2 measure of these differences, as defined below, and in Section 15.5 we discuss strategies for this minimization.

15.1 χ^2 measures

15.1.1 Specifications

We want to compare theoretical with experimental cross sections, and use this comparison to improve the theory input parameters $\{p_j\}$ so that observed and predicted cross sections agree better. Let the experimental cross sections be $\sigma^{\exp}(i)$ from observations at angles θ_i and (possibly identical) energies E_i, $i = 1, 2, \ldots, N$. Suppose that these experimental results $\sigma^{\exp}(i)$ are randomly distributed, with standard deviations $\Delta\sigma(i)$ as absolute errors. If a theoretical model predicts

403

values $\sigma^{th}(i)$ at the same angles and energies, then the standard measure of their discrepancy is the $\mathcal{X}^2$ sum

$$\mathcal{X}^2 = \sum_{i=1}^{N} \frac{(\sigma^{th}(i) - \sigma^{exp}(i))^2}{\Delta\sigma(i)^2}. \tag{15.1.1}$$

If the theory and experiment agreed exactly, then $\mathcal{X}^2 = 0$, but that is statistically unlikely because of the random errors $\Delta\sigma(i)$ in each data point. Even if the theory is completely accurate, the most likely outcome is that the discrepancy between theory and experiment will be of the order of $\Delta\sigma(i)$, one standard deviation error in the experimental value. This implies that the best achievable $\mathcal{X}^2$ value is obtained when each fraction $(\sigma^{exp}(i) - \sigma^{th}(i))^2/\Delta\sigma(i)^2$ is of the order of unity, so that $\mathcal{X}^2 \sim N$. For this reason, we usually describe a fit in terms of the value of $\mathcal{X}^2/N$: the $\mathcal{X}^2$ per degree of freedom.

The best to be expected in practice is that $\mathcal{X}^2/N \sim 1$. If $\mathcal{X}^2/N \gg 1$, then the data is not yet described fully by theory. If $\mathcal{X}^2/N \ll 1$, then the statistical improbability of this 'perfect fit' implies that the errors $\Delta\sigma(i)$ have been overestimated. This would occur if systematic or scaling errors had been included by mistake in the value of $\Delta\sigma(i)$ for each individual error. The values of $\Delta\sigma(i)$ are supposed to be the one-standard-deviation errors and all statistically independent, which means that even the best curve should miss every third error bar.

If there are uncertainties in the overall scaling of the experimental cross section, then the actual scaling factor s should be regarded as another parameter to be fitted. If the expected value of s is $E[s]$ (usually unity) but the 1σ uncertainty around $E[s]$ is Δs, then the $\mathcal{X}^2$ sum should have an extra term $(s - E[s])^2/\Delta s^2$, and appear as

$$\mathcal{X}^2 = \frac{(s - E[s])^2}{\Delta s^2} + \sum_{i=1}^{N} \frac{(\sigma^{th}(i) - s\,\sigma^{exp}(i))^2}{s^2\Delta\sigma(i)^2}. \tag{15.1.2}$$

The $\mathcal{X}^2$ per degree of freedom is now $\mathcal{X}^2/(N+1)$ because of the additional 'data point' $E[s]$.

15.1.2 Multivariate theory

Multivariate normal distribution

In order to understand the principles underlying χ^2 measures, we need first to know the general probability distribution for a set of random variables ('variates') x_i.

We use the notation $E[X]$ as the *expectation value* for some expression X depending on x, the value averaged over the probability density function $f(x)$ for X:

$$E[X] \equiv \int X f(x)\mathrm{d}x, \tag{15.1.3}$$

so the variance is

$$\Delta^2 = E[(x - \mu)^2] = E[x^2] - 2\mu E[x] + \mu^2 = E[x^2] - \mu^2, \tag{15.1.4}$$

and the standard deviation $\Delta = \sqrt{E[(x - \mu)^2]}$ is the root mean square deviation from the mean.

If we have no other knowledge, we would generally assume that the variates x_i follow *normal* distributions, for which the standard probability density function for a single variate is

$$f(x) = \frac{1}{\sqrt{2\pi}\,\Delta} \exp\left[-\frac{(x - \mu)^2}{2\Delta^2}\right], \tag{15.1.5}$$

when the mean is $\mu = E[x]$ and the standard deviation Δ where $\Delta^2 = E[(x - \mu)^2]$ is the variance.

For many correlated variables $\mathbf{x} = \{x_1, \ldots, x_N\}$, their joint probability distribution is the more general formula [1, p. 62]

$$f(\mathbf{x}) = (2\pi)^{-\frac{N}{2}} |\mathbf{V}|^{-\frac{1}{2}} \exp\left[-\frac{1}{2}(\mathbf{x} - \boldsymbol{\mu})^T \mathbf{V}^{-1}(\mathbf{x} - \boldsymbol{\mu})\right], \tag{15.1.6}$$

where $\boldsymbol{\mu}$ is the vector of means, $\mathbf{V}$ is the symmetric *covariance matrix*

$$\mathbf{V}_{ij} = E[(x_i - \mu_i)(x_j - \mu_j)], \tag{15.1.7}$$

and $|\mathbf{V}|$ is its determinant. The diagonal elements $\mathbf{V}_{ii}$ of the covariance matrix are the individual variances Δ_i^2, and its off-diagonal elements are

$$\mathbf{V}_{ij} = \rho_{ij}\Delta_i\Delta_j, \tag{15.1.8}$$

where ρ_{ij} are the correlation coefficients between the fluctuations of the data points x_i and x_j about their respective means.

The $\mathcal{X}^2$ sum

The probability that a single data point x_i with variance Δ_i^2 is correctly fitted by a prediction y_i may be found using Eq. (15.1.5):

$$f_i(y_i) = \frac{1}{\sqrt{2\pi}\,\Delta_i} \exp\left[-\frac{(x_i - y_i)^2}{2\Delta_i^2}\right]. \tag{15.1.9}$$

For many statistically independent data points x_i, the joint probability of being correctly fitted by y_i is the product of such single probabilities, giving

$$P_{\text{tot}} = (2\pi)^{-\frac{N}{2}} \Delta^{-1} \exp\left[-\frac{1}{2} \sum_{i=1}^{N} \frac{(x_i - y_i)^2}{\Delta_i^2} \right] \tag{15.1.10}$$

$$= (2\pi)^{-\frac{N}{2}} \Delta^{-1} \exp\left[-\frac{1}{2}\mathcal{X}^2 \right], \tag{15.1.11}$$

once we define $\Delta = \prod_i^N \Delta_i$ and

$$\mathcal{X}^2 = \sum_i^N \frac{(x_i - y_i)^2}{\Delta_i^2}, \tag{15.1.12}$$

using the definition of Eq. (15.1.1). Here, we write x_i as the data value for which y_i is the theoretical prediction.

Comparison of Eq. (15.1.11) with Eq. (15.1.6) shows that a generalized definition of the $\mathcal{X}^2$ sum, suitable for correlated data, is

$$\mathcal{X}^2 = (\mathbf{x} - \mathbf{y})^T \mathbf{V}^{-1} (\mathbf{x} - \mathbf{y}) \tag{15.1.13}$$

$$= \sum_{ij=1}^{N} (x_i - y_i)[\mathbf{V}^{-1}]_{ij}(x_j - y_j). \tag{15.1.14}$$

where $\mathbf{V}$ is the covariance matrix of the data points. It is thus called the *data covariance matrix*.

In both cases, *minimizing* $\mathcal{X}^2$ is equivalent to maximizing P_{tot}, the probability that the statistical means of the data points $\mathbf{x}$ are described by the predictions $\mathbf{y}$.

The χ^2 distribution

If we add together N squares of independent normal distributions z_i, with zero mean and unit variance, then the sum

$$\mathcal{X}^2 = \sum_{i=1}^{N} z_i^2 \tag{15.1.15}$$

has what is called a $\chi^2_{(N)}$ *distribution*. This is the well-known probability distribution already defined in subsection 11.1.1:

$$f(\mathcal{X}^2) = \frac{1}{2\Gamma(\frac{N}{2})} \left(\frac{\mathcal{X}^2}{2} \right)^{\frac{N}{2}-1} e^{-\mathcal{X}^2/2}, \tag{15.1.16}$$

which has mean, variance and standard deviation

$$E[\mathcal{X}^2] = N; \quad V(\mathcal{X}^2) = 2N; \quad \sigma(\mathcal{X}^2) = \sqrt{2N}. \tag{15.1.17}$$

The $\Gamma(z)$ is the gamma function, $\Gamma(z) = (z-1)!$

For large $N \gtrsim 20$, the $\chi^2_{(N)}$ distribution is approximately normal with the means and variances of Eq. (15.1.17). Hence $\mathcal{X}^2/N$ is also approximately normal with mean 1, and standard deviation $\sqrt{2/N}$.

Perfect and non-perfect fits

The optimal fit of a set of data occurs when the theory predicts exactly the statistical means of all the experimental values. This is not to give the actual values on any particular occasion, but the average if the same experiment were repeated many times at the same angle and the same energy. In this 'perfect' case we have $y_i = E[x_i] = \mu_i$, and the value of $\mathcal{X}^2$ is a sum of terms

$$z_i^2 = \frac{(x_i - y_i)^2}{\Delta_i^2} = \frac{(x_i - \mu_i)^2}{\Delta_i^2}, \tag{15.1.18}$$

so the z_i would have zero means and unit variances. If these were normally distributed, then the sum $\mathcal{X}^2 = \sum_i^N z_i^2$ would follow the $\chi^2_{(N)}$ distribution, which we saw above has mean N and variance $2N$. This is the reason we stated earlier that the best to be expected in practice is that $\mathcal{X}^2/N \sim 1$, and that (for a good fit) deviations both above and below unity are statistically improbable and must thus have other reasons.

Non-perfect fits will therefore yield larger $\mathcal{X}^2/N \gg 1$ values, and will have been constructed from z_i^2 with different statistical properties. If the z_i means are non-zero but their variances are still unity, then $\mathcal{X}^2$ will follow what is called [1, p. 66] a 'non-central χ^2 distribution with N degrees of freedom.'

More generally again, when the parameters giving the predictions y_i have been estimated by those values which minimize $\mathcal{X}^2$, the distribution $\mathcal{X}^2$ is no longer $\chi^2_{(N)}$. The minimum value of $\mathcal{X}^2$ is then the sum of squares of correlated, non-central random variables of non-zero mean, which are not, in general, normally distributed, and hence difficult to calculate. However, we can still numerically search for a minimum of $\mathcal{X}^2$ as a function of the theory parameters, to find $\mathbf{p}^0 = \{p_1^0, \ldots, p_P^0\}$.

Properties of the $\mathcal{X}^2$ minimum

Once we have found a smooth local $\mathcal{X}^2$ minimum for parameter values $\{p_m^0\}$, the $\mathcal{X}^2$ function will have zero first derivatives. It will hence be well described by a multivariate Taylor series, which up to the first-order terms is

$$\mathcal{X}^2(p_1,\ldots,p_P) \approx \mathcal{X}^2(p_1^0,\ldots,p_P^0) + \tfrac{1}{2}\sum_{m,n=1}^{P}\mathbf{H}_{mn}(p_m - p_m^0)(p_n - p_n^0)$$

$$\equiv \mathcal{X}^2(\mathbf{p}^0) + \tfrac{1}{2}(\mathbf{p} - \mathbf{p}^0)^T\,\mathbf{H}\,(\mathbf{p} - \mathbf{p}^0), \tag{15.1.19}$$

where the matrix of second derivatives, the *Hesse matrix* $\mathbf{H}$, is

$$\mathbf{H}_{mn} = \frac{\partial^2}{\partial p_m \partial p_n}\mathcal{X}^2(p_1,\ldots,p_P). \tag{15.1.20}$$

If we now substitute Eq. (15.1.19) into Eq. (15.1.11), we find that, as the parameters are varied around the minimum point, the fitting probability P_{tot} changes to

$$P_{\text{tot}} = \frac{1}{(2\pi)^{\frac{N}{2}}\Delta}e^{-\frac{\mathcal{X}^2(\mathbf{p}^0)}{2}}\exp\left[-\tfrac{1}{4}\sum_{mn}^{P}(p_m - p_m^0)\mathbf{H}_{mn}(p_n - p_n^0)\right]. \tag{15.1.21}$$

This probability of the data being fit by the theory can also be used for *fixed* data points, to show how the fit varies when the *theory* parameters are slightly changed.

The fitted parameters p_m^0 will in general have correlated errors, even if the original data had uncorrelated errors. We may therefore define a *parameter* covariance matrix by

$$\mathbf{V}_{mn}^{\text{p}} = E[(p_m - p_m^0)(p_n - p_n^0)] \tag{15.1.22}$$

for $m, n = 1,\ldots,P$, the number of parameters. If we compare Eq. (15.1.21) with Eq. (15.1.6), we see that it describes a multivariate normal distribution for the parameters p if we identify

$$(\mathbf{V}^{\text{p}})^{-1} = \tfrac{1}{2}\mathbf{H} \quad\text{or}\quad \mathbf{V}^{\text{p}} = 2\,\mathbf{H}^{-1}. \tag{15.1.23}$$

This important equation allows us to extract the correlated variances of the extracted parameters from the properties of the $\mathcal{X}^2$ surface around its minimum point.

Moreover, from Eq. (15.1.5), we find that the 1σ deviations away from a mean are when the exponential in the normal formula is $\tfrac{1}{2}$. This implies that the set of parameter combinations that are 1σ away from the minimum may be determined from

$$\tfrac{1}{2}\sum_{mn}^{P}(p_m - p_m^0)\mathbf{H}_{mn}(p_n - p_n^0) = 1, \tag{15.1.24}$$

and hence, even more simply by using Eq. (15.1.19), when

$$\mathcal{X}^2(p_1,\ldots,p_P) = \mathcal{X}^2(p_1^0,\ldots,p_P^0) + 1. \qquad (15.1.25)$$

The sets of parameters satisfying this condition will in general be multi-dimensional ellipses around the minimum point. These results for parameter correlations will be used in subsection 15.5.3.

15.2 Fitting cross-section harmonic multipoles

The simplest description of an angular cross section is as a linear combination of Legendre polynomials

$$\sigma(\theta) = \sum_{\Lambda \geq 0} a_\Lambda P_\Lambda(\cos\theta), \qquad (15.2.1)$$

for some real coefficients a_Λ which could be determined from experimental data by e.g. χ^2 minimization. The comparison of theory with experiment is most often done by comparing cross sections directly, but is sometimes performed by comparing the fitted and predicted values for these coefficients, a procedure which is most practical at low energies where not too many Λ values are needed.

To find the theoretical predictions for these coefficients, we note that in Chapter 3 the formulae (3.1.46) and (3.2.21) give the scattering amplitude $f(\theta)$ as linear combinations of Legendre polynomials: of $P_L(\cos\theta)$ in the single-channel case of Eq. (3.1.46), and of associated Legendre polynomials $P_L^M(\cos\theta)$ in the multi-channel case of Eq. (3.2.21). If the square modulus $\sigma = |f(\theta)|^2$ for the cross section is expanded into a sum over two partial wave indices, then each term will contain a product of two Legendre polynomials. We will see below how the sum of all these products can be rewritten as a sum exactly of the form (15.2.1).

Legendre expansion does not work with the point-Coulomb scattering amplitude, since the partial-wave series of Eq. (3.1.76) with Coulomb phase shifts does not converge. We therefore restrict ourselves to neutron elastic scattering, or else a non-elastic cross section. From the general form of Eq. (3.1.46), we have

$$\sigma(\theta) = \left| \frac{1}{k} \sum_{L=0}^{\infty} (2L+1) P_L(\cos\theta) \mathbf{T}_L \right|^2$$

$$= \frac{1}{k^2} \sum_{LL'} (2L+1)(2L'+1) P_L(\cos\theta) P_{L'}(\cos\theta) \mathbf{T}_L^* \mathbf{T}_{L'}. \qquad (15.2.2)$$

The product of two Legendre polynomials for the same argument is

$$P_L(z)P_{L'}(z) = \sum_{\Lambda=|L-L'|}^{L+L'} \langle L0, L'0|\Lambda 0\rangle^2 P_\Lambda(z), \qquad (15.2.3)$$

where $\langle L0, L'0|\Lambda 0\rangle$ is a Clebsch-Gordan coefficient. Non-zero contributions require even values of $L + L' + \Lambda$.

We thus derive the general form of Eq. (15.2.1) with coefficients

$$a_\Lambda = \frac{1}{k^2} \sum_{LL'} (2L+1)(2L'+1)\langle L0, L'0|\Lambda 0\rangle^2 \mathbf{T}_L^* \mathbf{T}_{L'}. \qquad (15.2.4)$$

Note that a_0 is proportional to the angle-integrated cross section of Eq. (3.1.50) by $\sigma_{\text{int}} = 4\pi a_0$, so often it is convenient to write

$$\sigma(\theta) = \frac{\sigma_{\text{int}}}{4\pi}\left[1 + \sum_{\Lambda>0} \bar{a}_\Lambda P_\Lambda(\cos\theta)\right] \qquad (15.2.5)$$

where $\bar{a}_\Lambda = a_\Lambda/a_0 = 4\pi a_\Lambda/\sigma_{\text{int}}$. These coefficients are a convenient way of describing and fitting the angular variations in cross sections obtained at low energies, when there are not too many partial waves involved.

When some of the initial and final participating nuclei have non-zero spin, the above procedure may be repeated starting from Eq. (3.2.21). Because the cross section (3.2.17) in this case is a sum over all the magnetic substates of the nucleus, the Legendre expansion of the cross section is still in terms of simply the P_L, as in Eq. (15.2.1). [1]

15.3 Fitting optical potentials

15.3.1 Sensitivities

Fitting the parameters of optical potentials to angular distributions is not always as simple as expected, because the sensitivities of the angular distributions to the optical potential vary a lot with the angles and radii under consideration.

The most obvious example is the diminishing effects of optical parameters on the forward-angle elastic cross sections, due to the fact that the potentials go to zero as $-V_{ws}\exp(-(R - R_{ws})/a)$ outside the sum of the radii of the nuclei. The Coulomb repulsion at radius R scatters to forward angles $\theta \approx 2\eta/Rk$ by Eq. (3.1.78), so cross sections at small angles should become equal to Rutherford $\sigma \sim \sigma_{\text{Ruth}}$ of Eq. (3.1.82). This fact can often be used to check the normalization of the

---

[1] The Legendre expansions for tensor analyzing powers $T_{kq}(\theta)$ of Eq. (3.2.28), by contrast, require associated Legendre polynomials $P_L^M(z)$ for $0 \leq M \leq q$.

experimental cross sections, by comparison to the Rutherford cross section at the smallest angles.

The spin-orbit potentials described in subsection 4.3.2 have their biggest effects on the vector analyzing powers $iT_{11}(\theta)$, but still have a small influence on the elastic cross section $\sigma(\theta)$. If a polarized projectile or target enters into the reaction, then it is essential to include the spin-orbit forces for that nucleus.

15.3.2 Ambiguities in optical potentials

There are three principal kinds of ambiguities in fitting optical potentials, one kind associated with phase shifts at low energies, another with potential volume integrals at low energies, and the third concerning surface properties in heavy-ion scattering.

The scattering at the lowest energies is determined by the phase shifts, and the pattern of phase shifts is largely unchanged if the potential is changed in depth so that the nearest bound or resonant state in a given partial wave remains the same distance from threshold. If the potential is made significantly deeper, then a bound state would have an extra node, and guidance from shell-model filling orders may then be used to resolve these ambiguities.

At medium energies, it is often possible to make small changes in V_{ws} and R_{ws}, while keeping $V_{ws}R_{ws}^2$ approximately constant, and not change the quality of the scattering fit. In these cases, the *volume integral* of the optical potential

$$\mathcal{J} = \int V(\mathbf{r})d\mathbf{r} = 4\pi \int_0^\infty V(r)r^2 dr \qquad (15.3.1)$$

is determined more precisely than the detailed parameters. Nucleon scattering potentials for targets of size A are generally expected to scale as $\mathcal{J} \sim A\mathcal{J}_0$ for $\mathcal{J}_0 \approx 450$ MeV at zero energy.

For collisions of heavier nuclei, 'grazing' collisions are those in which the semiclassical impact parameter $b = L/k$ is comparable to the sum of the radii of the interacting nuclei. Any collisions closer than grazing will almost certainly lead to strong imaginary parts, because of the great probability of flux leaving the elastic channel. In such reactions, the diffraction that occurs for grazing impact parameters will be governed by just the tail of the optical potentials at $R \gtrsim R_{ws} + a$. In this region, the real part is

$$V(R) \approx -V_{ws}e^{-(R-R_{ws})/a_{ws}} = -V_{ws}e^{R_{ws}/a_{ws}}e^{-R/a_{ws}} \qquad (15.3.2)$$

and hence R_{ws} and V_{ws} may be varied together as long as $V_{ws}e^{R_{ws}/a_{ws}}$ is constant, because only the exponential tail is important. In this case, it is good practice to plot $V(R)$ in the surface region for a set of fits, to see in which radial region its value is 'well determined.' See the examples for ^{16}O [2] and ^{6}He scattering [3] on lead around the Coulomb barrier.

15.4 Multi-channel fitting

15.4.1 Elastic fits

If a fitted optical potential is used as a diagonal potential in a coupled-channels system, then the predicted elastic cross sections will be different from the one-channel case, because of the extra coupling terms. Consider again the two-channel scenario on page 330. In this case, the dynamic polarization potential will have an imaginary part that reflects the loss of flux from the elastic channel 1 into the non-elastic channel 2, and will also have a real part that may modify potential barriers and hence any low-energy tunneling probabilities.

This means that we must distinguish *optical potentials*, which fit some experimental data, from the *bare potentials* U_1, U_2 that are the diagonal potentials to be used in a coupled set of equations such as Eq. (11.5.1). Because of the dynamic polarization effects arising from the couplings, the bare potentials will be different from any optical potentials. The bare potentials will typically have weaker imaginary parts than an optical potential, since the imaginary part of a bare potential is used to describe the loss of flux out of the *model space*, which is less than the flux leaving the elastic channel since the non-elastic channel 2 is still part of the coupled-channels model space.

Different strategies may be used to deal with this difference, in order to solve coupled-channels equations while fitting elastic data. *One* is to initially ignore the DPP, start with a bare potential equal to an optical one, and then later make small corrections to the parameters of the bare potential to improve the fit. Typically, the strength of its imaginary part will be reduced by small fractions in order to restore the elastic fit. A *second strategy* is redo the entire fitting of the optical potential, now including all of the equations (11.5.1) within the search process. This is probably the most accurate strategy overall. A *third* strategy, however, may still prove useful: this is to artificially set to zero the reverse coupling V_{12} that feeds back to the elastic channel. This means that the elastic channel still reproduces the result of the optical potential, so this strategy is another form of the distorted-wave Born approximation introduced on page 103. This step does not require the couplings to be small, as we only argue that the DPP is already included within the optical potential, since it fits the experimental scattering, and therefore the elastic fit should not be disturbed by couplings because this would count the coupling effects twice.

15.4.2 Fitting inelastic scattering

To fit the inelastic scattering of a nucleus to states in its rotational band by the collective mechanisms discussed in subsection 4.4.1, we normally proceed in two steps. *First* we use a simple first-order rotational model with quadrupole couplings to construct transition potentials between states of the rotational band, the only unknown parameter being the deformation β_2 (or deformation length δ_2, or the Coulomb reduced matrix element) which linearly scales the potential. This transition potential will have both nuclear and Coulomb contributions, but these usually scale in the same way. Then, comparison with the overall magnitudes of the observed inelastic cross sections will give a good estimate of the deformation. *Afterwards*, the fit can be improved by detailed adjustments of the optical potential, inclusion of the deformations non-linearly as in Eq. (4.4.27), possible β_4 effects, along with the coupled-channels effects from multiple couplings. If necessary, the Coulomb and nuclear deformations may be made different: the Coulomb matrix elements will particularly affect the forward-angle cross sections in a way that strongly depends also on the excitation energies of the levels in the rotational band.

15.4.3 Transfer fits

Transfer reactions are individually rather weak processes, and so most commonly are treated in first order in some kind of Born approximation. As mentioned in Chapters 3 and 14, that approximation is equivalent to neglecting some set of reverse couplings.

The *simplest* transfer model, therefore, is the one-step DWBA, in which optical potentials are used for both the entrance and exit channels and the transfer coupling is taken to first order, to couple just the entrance to the exit channel (and not also the reverse). This then becomes exactly the DWBA considered on pages 103 and 380, where the entrance optical potential reproduces some elastic data of these nuclei at the needed or a nearby energy. In first-order DWBA, the cross section is exactly proportional to the spectroscopic factor S according to Eq. (14.1.1). We might therefore attempt to find the empirical spectroscopic factor as that which scales the theoretical ($S=1$) curve to best reproduce the observed transfer angular distribution, especially its forward angle or largest peak, which might contribute most to the total cross section.

The *second-simplest* model is one in which other inelastic couplings are permitted to act to higher or all orders, but the transfer coupling still only to first order. This scheme is the CCBA, or *coupled-channels Born approximation*. One of the participating nuclei, for example, might have a rotational band that can be easily excited. Note first that if all the couplings were only from and to the ground state, then these rotational effects would have already been included in the entrance optical potential. A more detailed treatment is necessary, therefore, if there are transfers *to or from excited states of the collective band*. In this case, at least the CCBA scheme is essential, as multiple routes to a transfer state are possible. There could be transfers from ground state to ground state, or first an excitation and then a transfer, or an excitation followed by transfer followed by a de-excitation. These higher-order routes with increasing numbers of steps become individually less significant, but may be essential if for some reason the direct route is suppressed on structural grounds, or if there are *many* such intermediate steps energetically allowed. One complication to note when multiple routes may contribute, is that then the *relative phases* become important, as the amplitudes for the different processes may interfere constructively or destructively.

The detailed shape of a transfer cross section is affected by these higher-order effects, and also by the features of the bound-state wave function that comes, for example, from the geometry of its binding potential. Transfers at sub-barrier energies are expected to probe only the remote tail of the bound-state wave function, where its amplitude is governed mostly by the asymptotic normalization coefficient (ANC) defined in Eq. (4.5.24). As shown in subsection 14.1.4, we may expect low-energy reactions to determine these ANC values to good accuracy. At higher energies, above the Coulomb barrier, there will be more penetration to the surface and perhaps into the interior, so more global properties of the transfer wave function govern the cross section. We saw in Section 14.2 how a wider selection of experiments at different energies is needed to probe the entire wave function.

15.4.4 A Progressive Improvement Policy

We may summarize the above strategies for the detailed fitting of experimental data as arising from a 'Progressive Improvement Policy.' Fitting data, by this policy, should start with the simplest data and the simplest reaction model. In most cases this should be elastic data and some optical potential, preferably over a range of energies so that local peculiarities or resonances will not upset the modeling. Then for inelastic excitation, transfer or breakup, one-step DWBA should be used to fit specific final states that can be reached from the initial bound state. Only *after* that has been done and the results studied, should more complicated models be employed, such as those with inelastic excitations in the entrance or exit channels, or those with two-step or coupled-channels transfers.

Only in special cases should coupled channels be started at the beginning, such as those where the entrance optical potential is unknown and has to be calculated dynamically, including the effects of many inelastic or breakup couplings.

The progressive improvement policy is also particularly useful during χ^2 searches, as we see next.

15.5 Searching

The task of searching is to take a defined set of theoretical parameters which may be modified (perhaps with experimental absolute normalization factors), and numerically vary these in a systematic way while monitoring the total $\mathcal{X}^2$ value, until $\mathcal{X}^2$ becomes as small as possible. For linear variations this can be achieved by solving simultaneous equations, but for non-linear problems as here, iterative searches are necessary. A widely used tool to achieve this is the program MINUIT [4]. Within this suite the most efficient minimizing technique is called MIGRAD, and this is a gradient search method using Fletcher's switching variation [5] to the Davidson-Fletcher-Powell variable metric algorithm [6]. It approaches a local minimum closely and generates parabolic error estimates, which will be true errors when the $\mathcal{X}^2$ function around the minimum is accurately quadratic with respect to each parameter according to Eq. (15.1.19).

15.5.1 Strategies

It is useful to remember the following strategies for χ^2 searching:

(i) The process can be restarted from any intermediate stage, if subsequently either strange or no results were found. Intermediate sets of parameter values should therefore be saved during the search.

(ii) Ambiguities in searching can be dealt with by means of *grid searches*. That is, if for example there is a mutual ambiguity between R_{ws} and V_{ws}, then searches should be conducted for finite increments in R_{ws}. In this case, the potentials for each R_{ws} value will, when plotted, often cross at a 'sensitive region' for the scattering. Without grid searches, searching ambiguities will give highly correlated errors between the related variables. These correlations will be discussed below.

(iii) Discrete ambiguities, such as those which change the number of nodes in the wave functions for low-energy scattering, require specific restarting of a gradient search with increased or decreased potentials. The required potential depths may be estimated by finding what is needed to change the number of nodes in a one-channel bound state.

(iv) If some features of the data are never fitted properly, then the error bars for these data points can be temporarily reduced by a factor of say 10, to increase their importance in the χ^2 sum. You can then learn what in the model is sufficient to fit those features, and perhaps reconsider the model to make this easier when the error bars are restored to their realistic values.

(v) If a parameter is given a limited range and then ends up near one end of this extent, then the MIGRAD procedure often gives a spurious convergence with respect to its variation because of the way it internally maps the limited range. Apparently converged results with one variable near its limit should always be repeated with a wider range allowed.

(vi) Sometimes, instead of searching for two correlated variables, they should be combined into one variable. For example, in one-step DWBA the two projectile and target spectroscopic factors must be combined (or else one of them fixed). Similarly, if both a bound state and scattering data are to be fitted with the same potential, it may be possible to find the bound state first and reset the potential depth to that necessary for this exact eigenstate.

During the search process also, we may apply the Progressive Improvement Policy outlined above. The initial stages of a χ^2 search should use the simplest reaction theory that gives plausible results for the measurements of interest. There is no harm in making approximations to the scattering theory at early stages (such as zero-range rather than finite-range transfers, or omitting Coulomb excitations at large radii, or using the DWBA rather than finding coupled-channels solutions), and a lot of time can be saved by making these approximations initially. Then, after an apparent convergence, more details of the reaction theory can be inserted, and/or the existing theory calculated more accurately. After these model improvements have been made, the search procedure can be restarted for progressively finer readjustments.

15.5.2 Theoretical expectations and the Bayesian method

Theoretical expectations may arise if we want to include the effects of prior experimental results. Some of those prior results may be built into strong theoretical expectations – such as the radii of nuclei as measured by completely independent and verified electron-scattering experiments. Others may be simply the current experiment at another energy. Lifetime measurements may have given deformations, and other transfer reactions may have suggested spectroscopic factors. In all cases, this information should be taken into account when performing the current search.

In the $\mathcal{X}^2$ method, the results from other experiments may be used to give values and error bars for some of the current search parameters. As long as the parameter 1σ errors from those experiments are known, they may be used to constrain a present search by including an extra contribution to the $\mathcal{X}^2$ sum, as in the form shown in Eq. (15.1.2).

A more systematic method for including prior statistical information about the theoretical fitting parameters is to use the *Bayesian method* for the fitting. Such a procedure is used, for example, in the SAMMY code [7] for R-matrix fitting. This yields the same results as $\mathcal{X}^2$-fitting if there are no theoretical expectations, or at the beginning of a search when the dependences are non-linear and the parameters are unknown. It requires input uncertainties for all its parameters (whether small, or effectively infinite), and the action of the Bayesian method is to reduce these parameter uncertainties. The method has the advantage of not being sensitive to the grouping of experimental data, unlike the $\mathcal{X}^2$ method.

15.5.3 Error estimates from $\mathcal{X}^2$ fitting

Using the $\mathcal{X}^2$ expression of Eq. (15.1.1), the Eq. (15.1.25) tells us that the simplest estimate of the error in each of the fitted parameters is that variation which increases

$\mathcal{X}^2$ from its converged minimum by unity (so $\mathcal{X}^2/N$ is increased by $1/N$ for N data points). These parameter errors give a very useful estimate how the parameters are fixed by the given data. Note that if the errors $\Delta\sigma(i)$ are all overestimated by a factor A, then the resulting parameter errors from the fit will also be overestimated by the same factor A.

To examine correlations between the fitted parameters, we need to look at the *error matrix*: the parameter covariance matrix $\mathbf{V}^p$, such as that calculated by Eq. (15.1.23). This is twice the inverse of the Hesse matrix $\mathbf{H}_{mn}$ of partial second derivatives of $\mathcal{X}^2$ defined in Eq. (15.1.20). The errors based on the error matrix take account of all the parameter correlations, but not their non-linearities (the 2σ error will be exactly twice the 1σ error estimate). When the covariance error matrix has been calculated, the parameter errors are the square roots of the diagonal elements of this matrix, which include the effects of correlations with the other parameters.

These correlation effects are included because the inverse of the error matrix, the second derivative Hesse matrix $\mathbf{H}_{mn}$ of Eq. (15.1.20), has as diagonal elements the second partial derivatives $\partial^2(\mathcal{X}^2)/\partial p_m^2$ with respect to one parameter at a time. These diagonal elements are not therefore coupled to any other parameters, but when the matrix is inverted, the diagonal elements of the inverse contain contributions from all the elements of the second derivative matrix, and thus include the correlation effects on the individual errors.

It is also instructive to look at the matrix of *parameter correlation coefficients*

$$\rho_{mn}^p = \frac{\mathbf{V}_{mn}^p}{\sqrt{\mathbf{V}_{mm}^p \mathbf{V}_{nn}^p}} \tag{15.5.1}$$

found from Eq. (15.1.8). Correlation coefficients very close to one (greater than 0.99) indicate a difficult or false fit, and occur especially if there are severe ambiguities between two or more search parameters. These would occur in an ill-posed problem with more free parameters than can be determined by the model and the data.

Exercises

15.1 Elastic scattering of ^{8}Li on ^{208}Pb has been measured near the Coulomb barrier [8] at five separate beam energies.

(a) Fit the elastic within the optical model at each single energy taking as initial parameters for the Woods-Saxon interactions: $V_R = 40\,\text{MeV}$; $r_R = 1.2\,\text{fm}$; $a_R = 0.6\,\text{fm}$ and a volume imaginary part as $V_I = 40\,\text{MeV}$; $r_I = 1.2\,\text{fm}$; $a_I = 0.6\,\text{fm}$. To begin with, neglect the spin-orbit interaction. Do not start with a six-parameters fit but rather gradually allow for parameter variations.

(b) Compare the quality of the optical potentials obtained in this way. Are they energy dependent? Do they scale with the radius of the target?

(c) Try also an optical-potential fit to the three energies simultaneously and comment on your results.

(d) Study the sensitivity of your final optical parameters to the initialization. In particular, analyze the correlation between parameters.

(e) Repeat the procedure assuming a surface imaginary component instead. Determine which of the parameter sets (volume versus surface) is best pinned down.

(f) Include spin-orbit term and perform a new fit. Can you improve the description of the data with this term?

15.2 Fit δ_2 and δ_4 deformations to the $\alpha + {}^{20}$Ne inelastic cross sections at 104 MeV of [9], in the (a) one-step DWBA, (b) two-step DWBA, and (c) full coupled-channels cases. Can you fit Coulomb and nuclear deformations separately? How are the results correlated?

15.3 Using a minimization search to fit a spectroscopic factor to the transfer cross section of Ex. 14.1 (or one of your choice), using first DWBA and later CRC reaction models. If there is elastic data at that energy, fit the elastic and transfer cross sections simultaneously.

15.4 Fit an R-matrix pole to the p-wave resonant phase shifts for n + α scattering of [10]. Repeat with the s-wave non-resonant phase shifts. Compare the quality of the agreement with that from fitting a local potential similar to that used in [11].

References

[1] W. T. Eadie, D. Drijard, F. James, M. Roos and B. Sadoulet 1971, *Statistical Methods in Experimental Physics*, Amsterdam: North-Holland.

[2] I. J. Thompson, M. A. Nagarajan, J. S. Lilley and M. J. Smithson, *Nucl. Phys.* **A505** (1989) 84.

[3] O. R. Kakuee *et al.*, *Nucl. Phys. A* **765** (2006) 294.

[4] F. James, *MINUIT, Function Minimization and Error Analysis: Reference Manual*, CERN Program Library Long Writeup D506, Computing and Networks Division, CERN, Geneva, Switzerland. URL:http://wwwasdoc.web.cern.ch/wwwasdoc/minuit/minmain.html.

[5] R. Fletcher, *Comput. J.* **13** (1970) 317.

[6] R. Fletcher and M. J. D. Powell, *Comput. J.* **6** (1963) 163.

[7] N. M. Larson, *Updated Users Guide for Sammy: Multilevel R-Matrix Fits to Neutron Data using Bayes Equations*, Oak Ridge National Laboratory, ORNL/TM-9179/ R7, 2007.

[8] J. J. Kolata *et al.*, *Phys. Rev. C* **65** (2002) 0546016.

[9] H. Rebel, G. W. Schweimer, G. Schatz, J. Specht, R. Lohken, G. Hauser, D. Habs and H. Klewe-Nebenius, *Nucl. Phys.* **A182** (1972) 145.

[10] J. E. Bond and F. W. K. Firk, *Nucl. Phys.* **A287** (1977) 317.

[11] J. M. Bang and C. Gignoux, *Nucl. Phys.* **A313** (1979) 119.

Appendix A

Symbols

Symbol	Meaning	Eq., §	Page
a	matching radius	(3.1.28)	55
a_0	scattering length	(3.1.95)	70
a_i	imag. potential diffuseness	(3.1.108)	74
a_r	real potential diffuseness	(3.1.107)	74
a_{fi}	transition amplitude	(7.3.2)	245
a_Λ	Legendre coefficient	(15.2.1)	409
$\mathbf{b}, b$	impact parameter	Box 3.3	64
c	speed of light	(1.1.1)	1
c_L	lower cutoff scale	(6.3.6)	207
d_c	closest approach	(7.3.5)	245
d_i	distances to c.m.	(9.2.13)	280
e	unit electric charge	(3.1.70)	62
e_α	coupled eigenenergy	(6.5.18)	222
$e(\Omega_c)$	detector efficiency	(8.2.37)	270
e_p	R-matrix pole energy	(10.2.11)	299
f	scattering amplitude	(2.4.9)	45
$f(\mathbf{k}'; \mathbf{k})$	from $\mathbf{k}$ to $\mathbf{k}'$	(3.1.56)	60
$\tilde{f}$	with flux factors	(2.4.15)	46
f_c	point Coulomb	(3.1.76)	63
f_n	nuclear	(3.1.87)	65
f_{nc}	nuclear + Coulomb	(3.1.87)	65
f^{PWBA}	plane wave	(3.3.50)	101
f_ε	with ε-exchange	(3.4.15)	110
f_S	spin-S exchange	(3.4.17)	111
f^{WKB}	WKB	(7.4.13)	249
$\tilde{f}(\mathbf{k})$	Fourier transform of $f(\mathbf{r})$	(5.2.3)	188
f	β-decay factor	§5.4.2	196

(*Cont.*)

Symbol	Meaning	Eq., §	Page
$f(R)$	$2\mu/\hbar^2$-scaled $[V-E]$	(6.1.5)	201
$f(z)$	partial sum polynomial	(6.3.33)	214
g_{J_T}	spin weighting factor	(3.2.31)	84
g_V	coupling constant vector	(5.4.3)	197
g_A	axial vector	(5.4.6)	197
g_α	coupled eigensolution	(6.5.18)	222
h	radial step size	§6.1	201
$\mathbf{j}$	current	§2.4	42
$\hat{\mathbf{j}}$	per solid angle	(2.4.13)	46
$\mathbf{j}_{\text{free}}$	free field	(3.1.98)	71
$\mathbf{j}_q$	electric	(3.1.101)	71
$\mathbf{j}_2$	two-body	(3.1.103)	72
j_γ	photon	(3.5.30)	119
j	nucleon $\ell \pm s$	§4.2	132
k	momenta in c.m. frame	(2.4.4)	44
k_γ	photon momentum	(3.5.17)	116
k_B	Boltzmann constant	§1.2	6
$k(R)$	local wave number	§6.1	202
ℓ	nucleon partial wave	§4.2	132
m	mass	§1.1.1	1
m_u	unit atomic mass	(4.4.7)	143
n	number density	(1.2.2)	7
n^*	with excited states	(12.3.15)	358
n_γ	photons	(12.1.33)	349
n_{occ}	occupation number	(5.1.16)	180
n_Q	quantum concentration	(12.3.9)	356
n	number of radial nodes	§4.2	132
n_A^B	number of identicals	(5.3.13)	194
o	'post' or 'prior'	(4.5.6)	151
$\hat{\mathbf{p}}$	momentum operator	(3.5.18)	117
p_i	initial momentum	§14.1.3	382
p_f	final momentum	§14.1.3	382
$\mathbf{p}, p_i$	fitting parameters	(15.1.19)	408
q	electric charge	(3.1.101)	71
$\mathbf{q}, q$	momentum transfer	(7.2.9)	239
$\mathbf{q}_{\ell',\ell}^{T,v}$	transfer form factor	(4.5.13)	153

Symbol	Meaning	Eq., §	Page
q	WKB phase	(7.4.4)	247
$\mathbf{r}$	vector within a nucleus	(4.2.4)	133
$\mathbf{r}_{vt}$	valence-target vector	(4.4.29)	149
$\mathbf{r}_{ct}$	core-target vector	(4.4.29)	149
r_i	reduced imaginary radius	(3.1.108)	74
r_r	reduced real radius	(3.1.107)	74
r_{rms}	root mean square radius	(5.1.24)	183
r_m	matching radius	(6.4.6)	216
r	reaction rate (per sec)	(12.1.2)	341
s	nucleon spin	§4.2	132
$\mathbf{s}$	core-(nn) radius	Fig. 9.3	277
s	scaling factor	(15.1.2)	404
t	time	§1.2	6
$\hat{t}$	isospin operators $\hat{\boldsymbol{\tau}}/2$	(3.4.4)	107
$\hat{\mathbf{t}}$	projectile total isospin	(4.6.4)	159
$t_{\frac{1}{2}}$	half-life	(12.1.7)	342
t_α	symmetrizing mass factor	(6.5.22)	223
u	atomic mass unit	§1.1.1	1
$u_{\ell sj;b}$	radial wave function	(4.2.4)	133
$\hat{u}_{\ell sj;b}$	continuum	(4.2.12)	136
$u_{\ell sj}^{jI_AI_B}$	overlap	(5.3.4)	191
$\tilde{u}_p$	bin	(8.2.1)	259
$\mathbf{v}, v$	velocity	§2.3	34
v_p	projectile	(7.2.4)	238
v	valence particle	§5.3.1	190
$v_{\ell sj}^{jI_AI_B}$	normalized overlap	(5.3.7)	192
$w(R)$	channel eigensolutions	(6.5.2)	218
$w(r)$	rescaled wave function	(7.4.2)	247
x	partition index	§3.2.1	75
x	rescaled radius	(7.4.2)	247
$\mathbf{x}$	scaled nn distance	(9.2.8)	278
$y(R)$	trial solution	(6.1.6)	201
$\mathbf{y}$	scaled c(nn) distance	(9.2.8)	278
z_n	auxiliary radial function	(6.1.10)	202
A	mass number of nucleus	§1.1.1	1
A	beam amplitude	(2.4.8)	45

(*Cont.*)

Symbol	Meaning	Eq., §	Page
$A^{J_{\text{tot}}M_{\text{tot}}}_{\mu_{p_i}\mu_{t_i}}$	incoming coefficient	(3.2.9)	79
$\mathbf{A}$	vector potential	(3.5.2)	114
$\mathbf{A}$	time independent	(3.5.11)	115
$A^{jI_AI_B}_{\ell sj}$	spectroscopic amplitude	(5.3.6)	192
$\tilde{A}^{jI_AI_B}_{\ell sj}$	without isospin	(5.3.11)	193
A_n	scattering expansion	(6.5.6)	219
$\hat{\mathbf{A}}_{nn}$	antisymmetrization operator	(9.3.7)	286
$\mathbf{A}$	level matrix	(10.3.28)	310
B	binding energy	(1.1.1)	1
$B(E\lambda)$	reduced transition probability	(4.4.3)	142
$\mathcal{B}_{\alpha'}$	branching ratios	(11.3.11)	321
C_L	Coulomb constant	(3.1.59)	61
C_ℓ	asymptotic normalization coefficient	(4.5.24)	155
C	isospin amplitude	(5.3.11)	193
C_{jk}	2-body identical fraction	(12.2.2)	351
C_{jkl}	3-body identical fraction	(12.2.3)	351
D	sub-Coulomb vertex	(4.5.20)	155
D	mean level spacing	(11.3.1)	319
D_0	zero-range vertex	(4.5.16)	154
D_c	head-on closest approach	Box 3.3	64
E	relative energy	(2.3.6)	35
E_b	eigenstate	§2.1	28
E_p	complex pole	(3.1.94)	68
E_r	real resonance	(3.1.92)	66
E_r^f	formal	(10.2.15)	299
E_r^{obs}	observed	(10.2.21)	300
E_{tot}	total	(2.3.2)	35
E_γ	photon	(3.5.17)	116
E_{SM}	shell model	(5.1.12)	178
E_α	within channel α	(6.2.6)	203
E_G	Gamow	(12.1.18)	344
E_0	effective burning	(12.1.19)	344
F_L	regular Coulomb function	Box 3.1	52
$_1F_1$	confluent hypergeometric	(3.1.61)	61

Symbol	Meaning	Eq., §	Page
$\mathcal{F}_\lambda$	multipole-λ form factor	(4.3.1)	138
$\mathcal{F}^{\ell'\ell\Lambda v}_{\lambda,K'KT}$	transfer	(4.5.15)	153
$\mathsf{F}(R)$	$2\mu/\hbar^2[V-E]$ matrix	(6.1.5)	201
$\widehat{\mathcal{F}}_{M'M}$	bin breakup amplitude	(8.2.14)	263
$\mathcal{F}$	spin cutoff factor	(11.4.2)	327
G^+	Green's function	(3.3.7)	93
G_1^+	Green's function with potential	(3.3.40)	99
G_L	irregular Coulomb function	Box 3.1	52
$\mathcal{G}_\ell$	Gamow factor	(7.4.22)	251
$\mathcal{G}_\gamma^{LM}$	angular function	(9.3.3)	285
G_P	partition function	(12.3.13)	357
$H_L^\pm$	Hankel function	Box 3.1	53
$\mathsf{H}^\pm$	diagonal matrix of $H_L^\pm$	(6.3.5)	206
$\mathbf{H}$	magnetic vector field	(3.5.1)	114
$\mathbf{H}_{mn}$	Hesse matrix	(15.1.20)	408
H	Hamiltonian operator	(3.2.37)	85
$H_{\gamma p}$	$\gamma \leftarrow$ particle	(3.5.17)	116
$H_{p\gamma}$	particle $\leftarrow \gamma$	(3.5.23)	118
H_{intr}^λ	λ-multipole part	(4.3.1)	138
H_F	Fermi part	(4.6.3)	159
H_{GT}	Gamow-Teller part	(4.6.6)	160
H_{LSJT}	general multipole	(4.6.8)	161
H_A	of nucleus A	(5.1.2)	174
H_{eff}	effective	(5.1.5)	175
H_α	within channel α	(6.2.6)	203
H_{3b}	three-body	(8.1.2)	255
H_{core}	core	(9.2.5)	277
I	nucleus total spin	(3.2.1)	28
$\mathcal{I}$	energy averaging interval	(11.3.1)	319
$\mathbf{J}_{\text{tot}}$	total angular momentum	§3.2.1	75
$\mathbf{J}_p$	projectile $\mathbf{L}+\mathbf{I}_p$	§3.2.1	75
$\mathcal{J}$	potential volume integral	(15.3.1)	411
$\mathcal{J}_0$	per nucleon	(15.3.1)	411
$\mathbf{K}$	c.m. momentum	(2.3.20)	39
$\mathbf{K}_L$	partial-wave K matrix	(3.1.47)	58
K_0	incident wave number	(7.1.9)	231
K	hyperangular momentum	(9.2.19)	231

(Cont.)

Symbol	Meaning	Eq., §	Page
$\hat{L}$	angular momentum operator	(3.1.3)	49
L	partial wave integer	(3.1.7)	51
$\bar{L}$	typical L value	(6.3.6)	207
L	logarithmic matrix	(10.3.2)	306
M	z-projection state	(3.1.54)	60
M_{tot}	total z-projection	(3.2.1)	77
$\mathcal{M}^q_{JM}$	electric multipole operator	(4.7.21)	166
$\mathcal{M}_{\text{rigid}}$	moment of inertia	(4.4.7)	143
N	neutron number	§1.1.1	1
$\mathcal{N}$	oscillator quanta	Table 5.1	178
$\hat{N}_{\alpha'\alpha}$	norm overlap operator	(3.2.50)	88
$N_i(j)$	production of i from j	§12.2.1	351
$N_i(jk)$	production of i from jk	§12.2.1	351
N_A	Avogadro's number	§12.2.1	352
N_u	$1/m_u$	§12.2.1	352
$\hat{O}$	polarization observable	(3.2.23)	83
P	four-momentum	(2.3.21)	40
$P_{[n,m]}$	Padé matrix	(6.3.34)	214
P_{fi}	transition probability	(7.3.2)	245
P_ℓ	penetration probability	(7.4.17)	250
P	penetrability	(10.2.5)	298
$\mathcal{P}$	probability distribution	§11.1.1	315
$\mathcal{P}^{\text{PT}}_\alpha$	Porter-Thomas	(11.1.1)	315
$\mathcal{P}_n$	χ^2 with n d.o.f.	(11.1.3)	316
P_{pj}	permutation operator	(14.5.1)	399
P_{tot}	total joint probability	(15.1.11)	406
Q	energy released	(1.1.3)	3
Q	quadrupole moment	(5.1.36)	186
Q_{sp}	single-particle moment	(5.1.40)	186
$\mathbf{Q}_{\lambda\mu}$	electric multipole op.	(5.1.35)	185
$Q^2(r)$	rescaled potential	(7.4.3)	247
$\mathbf{R}, R$	separation of nuclei	(2.3.1)	35
R_n	outside potential	§3.1.1	49
$\mathbf{R}_x$	in partition x	§3.2.1	75
$\tilde{R}$	in body fixed frame	§4.4.8	144

Symbol	Meaning	Eq., §	Page
$\mathbf{R}_c$	core-core	(4.5.9)	152
$R_{\min}$	lower cutoff	§6.1	202
R_0	surface radius	§4.4.8	144
$\mathbf{R}_L$	partial-wave R matrix	(3.1.28)	55
$\mathbf{R}_{\alpha\alpha'}$	R matrix	(6.5.23)	223
$\mathbf{R}_\alpha^u$	uncoupled diagonal	(6.5.36)	225
$\mathbf{R}^{\mathrm{hyb}}$	hybrid	(10.4.1)	312
R_r	real potential radius	(3.1.107)	74
R_{nl}	oscillator radial form	(5.1.9)	176
R	WKB phase derivatives	(7.4.6)	248
$R(E)$	S-factor for neutrons	(12.1.30)	348
R_s	reduction factor	(14.2.1)	393
$S(E)$	astrophysical S-factor	(1.1.4)	5
S_{eff}	effective	(12.1.27)	346
$\mathbf{S}$	center-of-mass vector	(2.3.1)	35
$\mathbf{S}$	channel spin	§3.2.1	75
S	shift function	(10.2.4)	298
S^0	$S - a\beta$	(10.2.14)	299
$\mathbf{S}_P$	Poynting vector	(3.5.27)	118
$\mathbf{S}_L$	partial-wave S matrix	(3.1.30)	55
$\mathbf{S}^n$	nuclear	(3.1.84)	65
$\mathbf{S}_{\alpha\alpha_i}^{J_{\mathrm{tot}}\pi}$	S matrix	(3.2.10)	80
$\tilde{\mathbf{S}}_{\alpha\alpha_i}^{J_{\mathrm{tot}}\pi}$	with flux factors	(3.2.12)	81
S^{eik}	eikonal	(7.2.13)	240
S_{fi}^{OL}	optical limit	(7.2.18)	242
$S_{\ell sj}^{jI_AI_B}$	spectroscopic factor	(5.3.9)	192
$\tilde{S}_{\ell sj}^{jI_AI_B}$	without isospin	(5.3.12)	193
S^{exp}	experimental	(14.1.1)	380
$\hat{T}$	kinetic energy operator	(3.1.2)	49
T	temperature	§1.2	6
T	transfer multipole	(4.5.12)	152
$\mathbf{T}$	a Jacobi coordinate set	Fig. 9.4	279
T_9	temperature (GK)	§1.2	6
T_{Qq}^{xpt}	tensor analyzing power	(3.2.28)	84
T_r	tensor force: radial	(4.3.12)	141
T_p	momentum	(4.3.12)	141

(*Cont.*)

Symbol	Meaning	Eq., §	Page
T_L	angular-momentum	(4.3.12)	141
$\hat{\mathbf{T}}$	total isospin operator	(3.4.5)	107
$\hat{\mathbf{T}}$	target total isospin	(4.6.4)	159
$\mathbf{T}_L$	partial-wave T matrix	(3.1.43)	58
$\mathbf{T}_{\alpha\alpha_i}^{J_{tot}\pi}$	T matrix	(3.2.14)	81
$\tilde{\mathbf{T}}$	with flux factors	(3.2.14)	81
$\mathbf{T}^{PWBA}$	plane wave	(3.3.49)	100
$\mathbf{T}^{DWBA}$	distorted wave	(3.3.60)	103
$\mathbf{T}_{\alpha\alpha_i}^{dw\text{-}prior}$	distorted wave prior	(3.3.63)	104
$\mathbf{T}_{\alpha\alpha_i}^{dw\text{-}post}$	distorted wave post	(3.3.64)	104
$\mathbf{T}$	vector form T matrix	(3.3.31)	97
$\mathbf{T}^{(1)}$	from potential 1	§3.3.3	97
$\mathbf{T}^{(1+2)}$	from potentials 1+2	(3.3.33)	97
$\mathbf{T}^{2(1)}$	from potl. 2 with 1	(3.3.36)	98
$\mathbf{T}^{exact}$	exact	(7.1.19)	234
$\mathbf{T}_{prior}^{exact}$	exact prior	(14.1.2)	380
$\mathbf{T}_{post}^{exact}$	exact post	(14.1.3)	381
$\mathbf{T}^{sp}$	special model	(7.1.20)	234
$\mathbf{T}_{\alpha0}^{prior\ DW}$	prior DWBA	(8.2.18)	267
$\mathbf{T}_{\alpha0}^{prior\ BP\text{-}DW}$	prior DWBA (bare)	(8.2.20)	267
$\mathbf{T}_{\alpha0}^{1step}$	1-step	(8.2.22)	267
$\mathbf{T}_{\alpha0}^{cdcc}$	CDCC	(8.2.24)	267
$\widehat{\mathcal{T}}_{M'M}^{p}$	bin	(8.2.29)	268
$\mathbf{T}_{fi}^{PWBA}$	PWBA	(14.1.5)	383
$T(E\lambda)$	transition rate	(5.4.1)	196
$\mathcal{T}_\alpha$	transmission coefficient	(11.3.9)	320
$\check{U}$	$2\mu/\hbar^2$-scaled potential	(3.3.5)	93
U_{pt}	folded potential	(5.2.1)	187
U_{ct}	core-target potential	(7.1.1)	230
U_{vt}	valence-target potential	(7.1.1)	230
U_{opt}^α	optical potential	(8.2.19)	267
U_{pol}	polarization potential	(8.2.19)	267
V	potential	§3.1.1	49
V_c	point-Coulomb	(3.1.70)	62
V_C	actual Coulomb	(4.4.19)	146

Symbol	Meaning	Eq., §	Page
V_i	imaginary depth	(3.1.108)	74
V_m	mean field	(5.1.17)	181
V_n	neutron depth	(4.1.5)	131
V_p	proton depth	(4.1.5)	131
V_r	real depth	(3.1.107)	74
V_{so}	spin-orbit depth	(4.3.7)	140
V_{ss}	spin-spin depth	(4.3.11)	141
V_0	isoscalar part	(4.1.5)	131
V_T	isovector part	(4.1.5)	131
$\mathcal{V}_x$	interaction part in x	(3.2.38)	86
V_{fi}^{λ}	coupling multipole λ	(4.3.5)	139
$\hat{V}^o_{\alpha'\alpha}$	channel o coupling	(3.2.48)	88
$V^{(2)}$	two-body	§5.1	173
$V^{(3)}$	three-body	§5.1	173
V_{M3Y}	M3Y effective	(5.2.15)	190
V_{DPP}	polarization potential	(11.5.25)	336
V_{telp}	trivial equivalent	(11.5.5)	331
V_{welp}	weighted equivalent	(11.5.6)	331
$V_{\ell'L':\ell L}^{\Lambda v}$	transfer kernel	(4.5.14)	153
$\mathbf{V}_{ij}$	covariance matrix	(15.1.7)	405
W	imaginary potential part	(3.1.108)	74
$W_{\ell+\frac{1}{2},\eta}$	Whittaker function	(4.5.24)	155
$W_{\alpha\alpha'}$	width fluctuation factor	(11.3.4)	319
X_L	branching ratio	(4.5.28)	158
$\mathbf{Y}_{\Lambda J}^M$	vector spherical harmonic	(3.5.38)	121
$\mathbf{Y}$	matrix of trial solutions	(6.3.1)	205
$\mathbf{Y}$	a Jacobi coordinate set	Fig. 9.4	279
Y_i	abundance ratio	(12.2.7)	352
Z	proton number	§1.1.1	1
$\mathbf{Z}$	photon wave function	(3.5.31)	119
α	channel index in J basis	(3.2.1)	77
β	v/c	(2.3.31)	41
β	channel index in S basis	(3.2.2)	77
β	logarithmic derivative	(6.5.1)	218
β	nuclear state	(11.3.17)	324
β_q	fractional deformation	(4.4.9)	144
γ	photon	(1.2.1)	7

(*Cont.*)

Symbol	Meaning	Eq., §	Page
γ	$1/\sqrt{1 - \beta^2}$	(2.3.30)	41
γ_n	reduced width amplitude	(6.5.14)	221
γ_n^2	reduced width	(6.5.15)	221
γ_W^2	Wigner limit	(10.2.25)	303
γ_{sp}^2	single particle	(10.2.26)	303
γ	three-body channel index	(9.2.23)	282
δ_L	partial-wave phase shift	(3.1.38)	57
δ_{res}	resonant	(3.1.91)	66
δ_{bg}	background	(3.1.91)	66
δ^{WKB}	WKB	(7.4.12)	249
$\bar{\delta}_p$	nuclear + Coulomb	(8.2.28)	268
δ_R	R-matrix	(10.2.10)	298
δ_q	deformation length	(4.4.9)	144
δ_{bs}	back shift	(11.4.1)	327
χ_L	projectile-target scattering radial wave function	(3.1.9)	51
$\tilde{\chi}$	adiabatic wave function	(7.1.13)	233
χ	eikonal phase	(7.2.6)	238
$\chi_{\gamma\gamma'}$	3-body scattering waves	(9.4.20)	293
$\tilde{\chi}_{\gamma\gamma'}$	bins	(9.3.4)	286
ϵ_b	excitation energy	§2.1	28
ε	exchange index	§3.4.2	108
$\varepsilon_k^{(j)}$	epsilon algorithm	(6.3.35)	214
ε_n	channel eigenenergy	(6.5.2)	218
η	Sommerfeld parameter	(3.1.71)	63
κ	hypermomentum	(9.3.2)	285
λ	transition multipole	(4.3.1)	138
λ	decay constant	(12.1.6)	342
μ	reduced mass	(2.3.4)	35
μ	z-projection states	(3.2.1)	77
μ	statistical mean	(15.1.5)	405
ν	$2\eta k$	(6.4.2)	216
π	parity	§2.1	28
ϕ	azimuthal angle	§2.4	42
$\phi_{I_p\mu_p}^{xp}$	p'th state of projectile	§3.2.1	75
$\phi_{I_t\mu_t}^{xt}$	t'th state of target	§3.2.1	75

Symbol	Meaning	Eq., §	Page
ϕ	scalar e/m potential	(3.5.2)	114
$\phi^m_{\ell sj;b}$	nucleon wave function	(4.2.4)	133
ϕ_K	intrinsic deformed state	(4.4.15)	145
$\phi_{I_A:I_B}$	overlap function $\langle A|B\rangle$	(5.3.1)	191
ϕ	eikonal modulation	(7.2.1)	237
$\tilde{\phi}$	bin wave functions	(8.2.10)	262
ϕ	transformation angle	(9.2.27)	284
ϕ	hard-sphere phase shift	(10.2.7)	298
ϕ	relative velocity dist.	(12.1.16)	343
ψ	relative wave function	(2.3.20)	39
ψ_c	pure Coulomb	(3.1.72)	63
$\psi_{\alpha\alpha_i}$	coupled channels	(3.2.8)	79
ρ	scaled radius kR	(3.1.21)	53
ρ_{tp}	turning point	(3.1.67)	62
ρ	hyper-radius	(9.2.11)	279
ρ	density		
$\hat{\rho}$	matrix	(3.2.25)	83
ρ_q	charge	(3.5.1)	114
ρ_n	neutron	(5.1.25)	183
ρ_0	profile	(5.1.33)	185
ρ_{int}	internal	(5.1.25)	183
ρ_{mat}	matter	(5.1.25)	183
ρ_{ld}	of levels	(11.4.1)	327
ρ_{eff}	effective range	(4.5.20)	155
ρ_{ps}	phase space	(8.2.34)	269
ρ^p_{mn}	correlation matrix	(15.5.1)	417
$\boldsymbol{\rho}_i$	vectors w.r.t. center mass	§5.1	173
σ_L	Coulomb phase shift	(3.1.64)	61
σ_{spc}	spin cutoff parameter	(11.4.3)	327
σ	cross section	(1.1.4)	5
σ_A	absorption	(3.1.111)	75
σ_{el}	integrated elastic	(3.1.50)	59
σ_R	reaction	(3.1.113)	75
σ_{Ruth}	Rutherford	(3.1.82)	64
σ_{nc}	nuclear + Coulomb	(3.1.82)	64
σ_{xpt}	in channel *xpt*	(3.1.82)	64
σ^{pol}_{xpt}	for polarized beam	(3.2.27)	83

(Cont.)

Symbol	Meaning	Eq., §	Page
σ_{tot}	total $\sigma_R + \sigma_{\text{el}}$	(3.2.34)	85
σ_X	outcome X	(4.5.28)	158
σ_{bu}	breakup	(7.2.24)	244
σ_{str}	stripping	(7.2.25)	244
σ^*	to all final states	(12.3.16)	358
σ^{DWBA}	DWBA	(14.1.1)	380
σ^{exp}	experimental	(14.1.1)	380
σ^{th}	theoretical	(14.2.1)	393
σ_{sp}	single-particle	(14.2.1)	393
σ_{cap}	capture	(14.4.1)	396
σ_{photo}	γ disintegration	(14.4.2)	396
σ_{cd}	Coulomb dissociation	(14.4.1)	396
τ	resonance lifetime	§2.1	28
$\hat{\tau}_{Qq}$	spherical tensor	(3.2.26)	83
$\hat{\boldsymbol{\tau}}$	isospin operators	(3.4.4)	107
θ	scattering angle from $\hat{\mathbf{z}}$	§2.4	42
$\theta_{\alpha\beta}$	fraction of 'post'	(6.3.20)	210
θ	hyperangle	(9.2.11)	279
θ_W^2	Wigner ratio	(10.2.27)	303
θ_u	unit concentration	(12.3.17)	359
$\boldsymbol{\xi}$	internal coordinates	§3.2.1	75
ξ	Glauber thickness	(9.4.16)	292
$\boldsymbol{\xi}_\mu$	complex unit vectors	(3.5.36)	120
ζ_γ	photon radial wfn.	(3.5.37)	120
ω_{fi}	$(E_f - E_i)/\hbar$	(7.3.3)	245
$\mathcal{X}^2$	χ^2 measure	(15.1.1)	404
$\hat{\Delta}_{pq}$	shift level matrix	(10.3.24)	309
Δ	width of burning peak	(12.1.22)	345
Δ_{prior}	remnant term in prior	§14.1.1	380
$\Delta\sigma$	cross section uncertainty	(15.1.1)	404
Γ	resonance width	§2.1	28
Γ^f	formal	(10.2.15)	299
Γ^{obs}	'observed'	(10.2.21)	300
Γ_{tot}	total formal	(10.3.14)	307
$\hat{\Gamma}_{pq}$	width level matrix	(10.3.25)	309
Θ	Coulomb phase	Box 3.2	61

Symbol	Meaning	Eq., §	Page
Θ_S	Slater determinant	(5.1.7)	176
Φ	system bound state	(3.5.16)	116
Φ_A^{HF}	in Hartree Fock	(5.1.18)	181
Φ_A^{SM}	in shell model	(5.1.11)	177
Φ_i	Maxwell-Boltzmann dist.	(12.1.8)	342
Ψ	system wave function	(2.3.18)	38
$\Psi_{xJ_{\text{tot}}}^{M_{\text{tot}}}$	in partition x	(3.2.1)	77
$\overline{\Psi}_{J_{\text{tot}}}^{M_{\text{tot}}}$	over all partitions	(3.2.5)	78
$\Psi_{x_ip_it_i}^{\mu_{p_i}\mu_{t_i}}$	with m-state labels	(3.2.7)	79
Ψ_ε	with ε-exchange	(3.4.10)	110
Ψ^{ad}	adiabatic	(7.1.5)	231
Ψ^{sp}	special model	(7.1.17)	234
$\Psi^{(n)}$	Faddeev components	(8.1.1)	255
Ψ^{CDCC}	CDCC	(8.2.6)	261
Ψ_i^{ad}	adiabatic channel	(9.4.8)	289
Ψ^{ad}	total adiabatic	(9.4.9)	290
Ω	solid angle	§2.4	42
Ω_α	source term	(3.3.2)	92
Ω_5	$\{\theta, \hat{\mathbf{x}}, \hat{\mathbf{y}}\}$	(9.2.15)	281
Ω	hard-sphere phase matrix	(10.3.5)	306
Σ	detector area	§2.4.1	42

Notation

Symbol	Meaning	Eq., §	Page
$\hat{T}$	operator	§3	48
$\hat{\mathbf{z}}$	unit vector	§2.4.4	45
$\hat{x}$	$\sqrt{2x+1}$	§4.3.1	138
D_{MK}^I	rotation matrix	(4.4.15)	145
$L_n^a(z)$	Laguerre polynomial	(5.1.10)	176
$P_L(z)$	Legendre polynomial	(3.1.7)	51
Y_L^M	spherical harmonic	(3.1.54)	60
$\langle L_1M_1, L_2M_2|LM \rangle$	Clebsch-Gordon coef.	(3.2.1)	77

(*Cont.*)

Symbol	Meaning	Eq., §	Page
$W(abcd;ef)$	Racah coefficient	(3.2.4)	78
$\langle L_f \parallel Y_\lambda \parallel L_i \rangle$	reduced matrix element	(4.3.3)	138
$\Delta(\lambda, I_i, I_f)$	triangular inequality	(4.4.1)	142
$J_\nu(z)$	cylindrical Bessel	(7.2.12)	239
$\overline{\mathbf{S}}(E)$	energy average over $\mathcal{I}$	(11.5.8)	332
$\langle \sigma v \rangle$	thermal average	(12.1.3)	341
$E[X]$	statistical average	(15.1.2)	404
$\Gamma(z)$	gamma function	(15.1.17)	407

Appendix B

Getting started with FRESCO

B.1 General structure

FRESCO is a general-purpose reaction code, created and frequently updated by Ian Thompson. The code calculates virtually any nuclear reaction which can be expressed in a coupled-channel form. There is a public version of the code which can be downloaded from the website **www.fresco.org.uk**. FRESCO is accompanied by SFRESCO, a wrapper code that calls FRESCO for data fitting and SUMBINS and SUMXEN, two auxiliary codes for integrated cross sections. Although we do not include it here, in the same site you can also find XFRESCO the front-end program to FRESCO for X-window displays.

Its original version was written in Fortran 77 but some important sections were ported to Fortran 90. An important part concerns the input, which now uses *namelist* format, making it much easier to view the relevant variables. In this section we will discuss the general namelist format of the FRESCO input.

There are several different layers of output produced by FRESCO. The default output contains the most important information concerning the calculation, repeating the input information, and the resulting observables but most detailed information is contained in the generated fort files, including files ready for plotting purposes. At the end of this section we present the list of output produced by FRESCO.

B.1.1 Input file

Input files contain five major namelists regarding different aspects of the calculation: *fresco, partition, pot, overlap, coupling*. The first is for general parameters, the second for defining the properties intrinsic to the projectile and the target, the third for potentials, the fourth for the radial overlap functions and the last for the couplings to be included. Keep in mind that in some inputs, you may not find all these namelists. As FRESCO can calculate rather intricate processes, input files can sometimes look daunting. However, with the namelist format, you do not need to define all variables but only those that are relevant for your example. Below we introduce the contents of each namelist and their purpose. Detailed instructions are given in the FRESCO input manual on the website.

heading

Every input file starts with a heading (80 characters) that should describe and identify the reaction to be calculated, with perhaps some detail of the method and states included. The following line begins with NAMELIST to indicate the subsequent style of input.

&fresco

This section introduces the parameters involved in the numerical calculations. It contains the radial information: the step with which the coupled-channel equations are integrated (*hcm*) and the radius at which the integrated wave function gets matched to the asymptotic form (*rmatch*). Whenever non-local kernels are involved, a few more parameters are needed (*rintp, hnl, rnl, centre*), but these will be discussed in subsection B.2.4.[1]

In this namelist you also find general options for the calculations and the desired observables. Cross sections are calculated from the angular range *thmin–thmax* (in degrees) in steps of *thinc*. This is also where you define the number of partial waves in the calculation, by providing the initial and final total angular momentum: *jtmin, jtmax*.[2]

absend controls the convergence. If in the interval $\max(0, jtmin) < J < jtmax$ the absorption in the elastic channel is smaller than *absend* mb for three consecutive $J\pi$ sets, the calculation stops. When *absend* < 0, it takes the full J-interval.[3]

There are many control variables which trace intermediate steps in the calculation (starting from zero, increasing values will give more in depth information). Here we shall mention just a few of the most frequently used. For printing the coupled partial

---

[1] It is often useful to introduce lower radial cutoffs in the calculations, especially for scattering below the Coulomb barrier. With *cutr* (fm) or *cutl* you introduce a lower radial cutoff in the coupled-channel equations. Whereas *cutr* is the same for all partial waves, *cutl* allows you to define an *L*-dependent cutoff (*L* is the total angular momentum of the set). The code will use $\max(cutl*L*hcm, cutr)$. If *cutr* is negative, the lower cutoff is put at that distance inside the Coulomb turning point. Finally, *cutc* (fm) removes off-diagonal couplings inside the given radius.

[2] To enable greater speed and flexibility, you can define a number of angular momentum intervals *jump(i)*, *i*=2,5 and the steps with which you want to perform the calculation *jbord(i)*, *i*=2,5. Note that *jump(1)*=1 and *jbord(1)=jtmin*, so that the first interval is calculated fully. The omitted J values are provided by interpolation on the scattering amplitudes $A(m'M' : mM; L)$ prior to calculating cross sections. *jsets* is a variable that enables the calculation of positive parity (*jsets*='P') or negative parity only (*jsets*='M' or 'N'), for each energy. If *jsets* $= 0$, ' ', *or F*, no restriction is made.

[3] Sometimes, for accurate elastic scattering cross sections it is only necessary to include the elastic channel. This can be done with the option *jtmin*<0. Then, in the range $J < \mathrm{abs}(jtmin)$, transfers and excited states are ignored in the calculation.

waves for each J^π (total angular momentum and parity) use *chans*; for details of the coupling coefficients *listcc*, and for S matrices use *smats* (absorption & reaction cross sections for successive partitions and excitations are output when *smats*$\geq$ 1, and elastic S-matrix elements are ouput when *smats*$\geq$ 2).

The variable *xstabl*>0 prints the cross sections and tensor analyzing powers up to rank k=*xstabl* for all excitation levels in all partitions in Fortran file 16 (usually called fort.16).

Finally, and most importantly, this is the place where the beam energy is specified through *elab*. If you want a calculation at several energies, you can use the array *elab(i) i=1,4* and *nlab(i), i=1,3* to define the boundary points and the number of intermediate energy steps between *elab(i)* and *elab(i+1)*.

By default, the code assumes the elastic channel, the channel with the incoming plane wave, is the first excitation of the first partition. You can change this by using *pel>1* for the partition number and *exl>1* for the excitation within that partition. Also, *elab* refers by default to the energy of the projectile *lin=1*, but you can easily change the calculation to inverse kinematics by setting *lin=2*.

& partition

In *partition* you introduce all the mass partitions and the corresponding channels to be considered in the reaction. In the simplest case, elastic scattering within the optical model, you introduce just one partition, including the details of the projectile (*namep, massp, zp*) and the details of the target (*namet, masst, zt*). The Q-value for the reaction is given with *qval* (MeV) and the number of states that you want to include in this partition is *nex*. Below defining each partition, you have to introduce the *nex* associated pairs of states (at least one). This is done through another namelist &*states*.

&states

Each pair of states is a specific combination of one state of the projectile and one state of the target. So this is the place where you introduce the spin, parity and excitation energy of these states: (*jp, ptyp, ep*) for projectile and (*jt, ptyt, et*) for target. The variables *bandp* and *bandt* are synonyms for *ptyp* and *ptyt* respectively. The optical potential for the distorted wave for $p + t$ relative motion is given by the index *cpot*, also defined here. This namelist is repeated as many times as necessary, to introduce all the pairs of states you wish to include in the calculation. When repeating the &*states* namelist, if one of the bodies stays in the same state, you should not introduce spin, parity and excitation energy again, but just set *copyp* or *copyt* to refer to the &*states* namelist in which the original state was first introduced.

&pot

This namelist contains the parameters for the potentials to be used in the reaction calculation, either for bound single-particle states or optical potentials. The namelist is repeated for each term in the potential. To identify the potential, there is an index *kp*, and all the components with a given *kp* value are added together to produce the potential used. So, when calculating the distorted waves, *cpot* will refer to one of the *kp*. The same will be used when calculating the bound single-particle states with *kbpot* (in *&overlap*). Each term in the potential is characterized by a type and a shape, followed by parameters $p(i), i = 1, \ldots, 6$. Traditionally, we define the Coulomb term first (*type=0*, *shape=0* for a charged sphere). Then $p(1)=ap$ and $p(2)=at$ correspond to mass number of the projectile and the target needed for the conversion of the reduced radii into physical radii $R = r(ap^{1/3}+at^{1/3})$. The $p(3)=r_c$ is the reduced Coulomb radius. Note that the same mass factor $(ap^{1/3} + at^{1/3})$ is used in all terms of a given potential. *type=1* corresponds to the volume nuclear interaction, with *shape=0* for Woods-Saxon shape.[4] The parameters for the real part are $p(1)=V_0$, $p(2)=r_0$, $p(3)=a_0$ (for depth in MeV, the reduced radius in fm and the diffuseness in fm), while the parameters for the imaginary part are $p(4)=W_i$, $p(5)=r_i$, $p(6)=a_i$. A surface nuclear interaction is introduced with *type=2* and the spin-orbit with *type=3*, for the projectile, and *type=4* for the target.[5]

When the potential does not have an analytic form, it is useful to read it in numerically. This can be achieved setting *shape=7,8,9* to read from file fort.4 the real or the imaginary part of the potential, or the full complex potential, respectively.

& overlap

Overlap functions are needed in single-particle excitation calculations or in transfer calculations. The overlaps can refer to bound states and scattering states, but we will leave the latter for Section B.2.3. Every *&overlap* begins with an index *kn1*. The overlap function introduced in Section 5.3 tells us how the composite nucleus B looks relative to its core A. The composite nucleus and the core are in partition *ic1* and *ic2* respectively, and refer to the projectile (*in=1*) or target (*in=2*). In the simpler case *kind=0*, we ignore the spin of the core and take $|(l, sn)j\rangle$ coupling.[6] The overlap has *nn* number of nodes (including the origin), *l* relative angular momentum, *sn* for the spin of the additional fragment (typically a neutron or proton *sn=1/2*), and total angular momentum *j*. The potential used in the calculation of the state is that indexed *kbpot*. You can also introduce the binding energy *be* if you want

---

[4] A large number of standard shapes are predefined of which we mention *shape=2* for Gaussian $\exp[-(r-R_0)^2/b^2]$ (with $p(2) = r_0$ and $p(3) = b$) as the most common alternative.

[5] Tensor interactions and projectile/target deformation can be introduced with *type=5-11*. We will return to this in subsection B.2.2.

[6] Multi-channel spin couplings are also available with *kind=3*.

the potential to be adjusted to reproduce the binding energy (*isc*=1 for adjusting the depth of the central part). If no rescaling is needed, set *isc*=0. A spectroscopic amplitude for the overlap can be set to the value $\sqrt{nam}*ampl$ if both of these are non-zero. This amplitude can also be introduced after &*coupling* in the namelist &*cfp*.

For printing more detailed information into the standard output, there is a trace variable *ipc*. Its default value is zero, and, as it increases, it provides more detailed information on the overlap function.[7]

&*coupling*

Couplings are calculated with the information given in this namelist and include general spin transfer (*kind*=1); electromagnetic couplings (*kind*=2); single particle excitations (*kind*=3, 4 for projectile and target respectively); transfer couplings (*kind*=5, 6, 7, 8 for zero-range, local energy approximation, finite-range and non-orthogonality corrections respectively). The coupling is from all states in partition *icfrom* to all states in partition *icto*. Couplings are included in the reverse direction unless *icto* < 0. For specific options of the coupling we use the parameters *ip1, ip2, ip3* and for choices of the potentials in the operator there are *p1, p2* parameters. More detail on this namelist and others that follow will be given with specific examples.

B.1.2 Output files

The main output file (fort.6 or stdout) contains first of all a representation of all the parameters read from the input file. It will provide a summary of the calculation of the overlap functions (including binding energy, depth of the adjusted potential, rms radius and asymptotic normalization coefficient) and the coupling matrix elements. For each beam energy, it provides some information relative to the kinematic variables in the reaction followed by the contribution to the cross section of each partial wave. Integrated cross sections and angular distributions are printed at the end of the file.

Also at the end of the standard output file, as a reminder to the user, is a list of other files that were created during the run with additional information. Here we mention a few: fort.16 contains all angular distributions in a graphic format (to be read by XMGR or XMGRACE); fort.13 contains total cross sections for each channel; fort.56 contains the total absorptive, reaction and non-elastic cross section for each angular momentum. Separate cross sections are included in files 201, 202, ... in the order they were specified. A full list of file allocations is given in Table B.1.

---

[7] Radial wave functions of bound states can be obtained by setting *ipc* odd, intermediate iterations with *ipc*≥3, and the final iteration with *ipc*>0.

Table B.1. *File allocation for the inputs and outputs for* FRESCO.

File	Routines	Use
2	SFRESCO	search specification file
3	FREADF, FR	temporary namelist file
4	INTER	input external KIND=1,2 form factors
	POTENT	input external potentials
5		standard input
6		standard output
7	DISPX	S-matrix elements
13	FR	total cross sections for each Elab
16	CRISS	tables of cross sections
17	FR	output scattering waves
20–33	For users	(eg bound states, amplitudes)
34	POTENT	output potentials
35	FR	astrophysical S-factors for E_{cm}
36	CRISS	scattering Legendre coefficients
37	CRISS	scattering amplitudes
38	DISPX	cross sections for each $J\pi$
39	FR	2 cross sections for each E_{cm}
40	FR	all cross sections for each E_{lab}
41	SOURCE	source terms at each iteration
42	SOURCE	bin wave functions for each E
43	INFORM	bin phase shifts as k functions
44	INFORM	bin phase shifts as E functions
45	ERWIN	scattering phase shift as E functions
46	INFORM	ANC ratios and bound wave functions
47		reduced matrix elements
48	FR	misc log file
55	INFORM	single-particle wave functions
56	FR	J fusion, reaction and nonelastic
57	FR	output CDCC amplitudes
58	INFORM	bound-state wave functions
59	INFORM	bound-state vertex functions
60–62	RMATRIX	trace of R-matrix calculations
66	INTER	KIND=1 non-local formfactor
71	FR	phase shifts as E_{lab} functions
75	FR	astrophysical S-factors for Elab
89	MULTIP	all coupling potentials
105	FCN	χ^2 progress during fitting
106	FCN	parameter snapshots during fitting
200	CRISS	elastic cross section if not fort.201
201–210	CRISS	cross sections (cf 16) of up to 10 states
301	CDCIN	new FRESCO input
303	SFRESCO	input search file

Table B.1. *(Cont.)*

File	Routines	Use
304	SFRESCO	output plot file
305	CDCIN	new input from CDCC, col format
306	SFRESCO/FRXX0	input FRESCO file
307	SFRESCO/FRXX0	initial output FRESCO file
308	SFRESCO/FRXX0	main output FRESCO file

B.2 Learning through examples

B.2.1 Elastic scattering

As an elastic scattering example, we chose the proton scattering on ^{78}Ni within the optical model. This exotic nucleus is an important waiting point in the *r*-process (Chapter 1). The input for our example is shown Box B.1. The calculations are performed up to a radius of *rmatch*=60 fm and partial waves up to *jtmax*=50 are included. Three beam energies are calculated. For this case, only one partition is needed with the appropriate ground states specified (the proton is spin 1/2 and positive parity, and the ^{78}Ni, being an even-even nucleus, has $J^{\pi} = 0^{+}$). The only remaining ingredient is the potential between the proton and ^{78}Ni (indexed *cpot*=*kp*=1) which contains a Coulomb part and a nuclear real and imaginary part. The results can be found in the standard output file, but it is easier to plot the fort.16 file to obtain Fig. B.1.

```
p+Ni78 Coulomb and Nuclear;
NAMELIST
 &FRESCO hcm=0.1 rmatch=60
          jtmin=0.0 jtmax=50 absend= 0.0010
          thmin=0.00 thmax=180.00 thinc=1.00
          chans=1 smats=2 xstabl=1
          elab(1:3)=6.9 11.00 49.350 nlab(1:3)=1 1 /

&PARTITION namep='p' massp=1.00 zp=1
          namet='Ni78' masst=78.0000 zt=28 qval=-0.000 nex=1  /
&STATES jp=0.5 bandp=1 ep=0.0000 cpot=1 jt=0.0 bandt=1 et=0.0000  /
&partition /

&POT kp=1 ap=1.000 at=78.000 rc=1.2  /
&POT kp=1 type=1 p1=40.00 p2=1.2 p3=0.65 p4=10.0 p5=1.2 p6=0.500  /
&pot /
&overlap /
&coupling /
```

Box B.1 FRESCO **input for the elastic scattering of protons on** ^{78}**Ni at several beam energies**

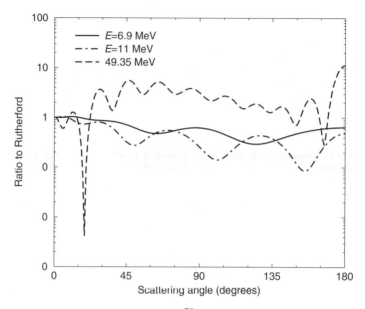

Fig. B.1. Elastic scattering of protons on ^{78}Ni at several beam energies, calculated with input from Box B.1.

B.2.2 Inelastic scattering

Inelastic scattering exciting collective states can be illustrated with the example $^{12}C(\alpha, \alpha)^{12}C(2^+)$, where the carbon nucleus gets excited into its first excited state. This reaction can provide complementary information to one of the most important reactions in astrophysics (subsection 1.3.2), the α-capture reaction on ^{12}C. In this type of inelastic reaction, only one partition is needed, but it contains two states. The projectile state is not changed (*copyp*=1) but the appropriate spin, parity and excitation energy need to be introduced for the target. The input is shown in Box B.2.

In order for the reaction to happen, the potential needs to contain a tensor Y_{20} part to enable the target transition $0^+ \rightarrow 2^+$. This is done assuming a rotor model for the target and, in the input, only a deformation length needs to be introduced. For deforming a projectile *type*=10, while for deforming a target, *type*=11. Here the deformation length δ_2 is *p(2)*=1.3 fm, as twice highlighted in Box B.2. The optical potential introduced includes a Coulomb term, the nuclear real term, and a nuclear imaginary with a volume (*type*=1) and a surface part (*type*=2). Each part needs to be deformed, and only the two nuclear parts are deformed. If Coulomb deformation were needed, an additional line after the Coulomb potential would have to be introduced with the same format, except that, instead of the deformation length, the reduced matrix element of Eq. (4.4.25) should be given.[8] As the proton

---

[8] If you do not want to assume a rotational model, you can introduce these couplings (either deformation length or matrix element) for each initial-to-final state through *type*=12, 13 for projectile and target respectively. In

```
alpha+c12 -> alpha+c12* @ 100 MeV; nuc def
NAMELIST
&FRESCO hcm=0.050 rmatch=20.000 rintp=0.20
        jtmin=0.0 jtmax=40 absend= 0.01
        thmin=0.00 thmax=180.00 thinc=1.00
        iter=1 ips=0.0 iblock=0
        chans=1 smats=2  xstabl=1
        elab(1)=100.0 /

&PARTITION namep='alpha' massp=4.0000 zp=2
           namet='12C'   masst=12.000 zt=6 qval=0.000 nex=2   /
&STATES jp=0. bandp=1 ep=0.0000 cpot=1 jt=0.0 bandt=1 et=0.0000 /
&STATES copyp=1                 cpot=1 jt=2.0 bandt=1 et=4.4390 /
&partition /

&POT kp=1 ap=4.000 at=12.000 rc=1.2  /
&POT kp=1 type=1  p1=40.0 p2=1.2 p3=0.65 p4=10.0 p5=1.2 p6=0.500  /
&POT kp=1 type=11 p1=0.0  p2=1.3 /
&POT kp=1 type=2  p1=0.00 p2=1.2 p3=0.65 p4=6.0 p5=1.2 p6=0.500  /
&POT kp=1 type=11 p1=0.0  p2=1.3 /
&pot /
&overlap /
&coupling /
```

Box B.2 FRESCO **input for the inelastic excitation of** ^{12}C **by** α **particles at 100 MeV**

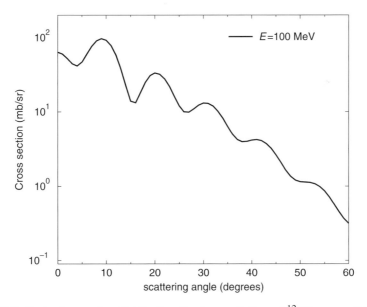

Fig. B.2. Inelastic angular distribution for the excitation of ^{12}C by α particles at 100 MeV obtained with the input of Box B.2.

this case, give a &*step* namelist specifying *ib, ia, k, str* for a coupling from state *ib* to state *ia*, multipolarity *k* and strength *str* (*str* is the reduced matrix element for Coulomb transitions and the reduced deformation length for nuclear transitions).

and neutrons do not necessarily have the same spatial distribution, the deformation parameters will, in general, not be the same.

The example shows a DWBA calculation as *iter*=1. You could check the validity of the DWBA by including higher-order terms in your Born expansion (increasing *iter*) or performing a full coupled-channels calculation (*iter*=0, *iblock*=2). Results for the inelastic excitation of ^{12}C are shown in Fig. B.2.

B.2.3 Breakup

Breakup calculations can be modeled as single-particle excitation into the continuum. In this example we show a typical CDCC calculation as described in Section 8.2. It calculates the breakup of ^{8}B into p+^{7}Be, under the field of ^{208}Pb at intermediate energies. The input is shown in Box B.3. The breakup of ^{8}B has been measured many times with the aim of extracting the proton capture rate on ^{7}Be.

Several new ingredients need to be explained. First of all, due to the long range of the Coulomb interaction it is very important to include the effect of couplings out to large distances. Instead of integrating the CDCC equations up to very large radii, we introduce *rasym*. Setting *rmatch<0* tells the code that the integration of the equations should be done up to *rmatch*, numerically, but these should then be matched with coupled-channel Coulomb functions up to *rasym*. Also important are the partial waves. For these intermediate energies, many partial waves need to be included and it is useful, instead of calculating each single one, to interpolate between them. This can be done with *jump* and *jbord*. In this example, we start with *jtmin*=0 until $j = 200$ in steps of 1, for $jt = 200$–300 use steps of 10, for $jt = 300$–1000 use steps of 50, and for $jt = 1000$–9000 use steps of 200. With the inclusion of so many partial waves, the strong repulsion at short distances can introduce numerical problems. This is avoided with a radial cutoff *cutr* $= -20$ fm, where the minus sign puts the cutoff 20 fm inside the Coulomb turning point.

This example contains only *s*-waves in the continuum, sliced into 20 energy bins. Other partial waves (p, d, f are needed for convergence) are left out of this example to make it less time consuming (beware, it will still take a few minutes in a desktop computer!). Since in general there will be many channels involved, it is convenient to drop off channels/couplings whenever they are weak. This is done through *smallchan* and *smallcoup*. To perform a full CDCC calculation, *iter*=0 and *iblock*=21.

The continuum of ^{8}B is binned into discrete excited states of positive energy, so under the first partition the namelist *states* needs to be repeated for each bin, with appropriate excitation energy and quantum numbers. Since in this example

```
CDCC 8B+208Pb ; nuclear and coulomb s-wave breakup
NAMELIST
 &FRESCO hcm= 0.01 rmatch= -60.000 rintp=0.15 rsp=  0.0 rasym= 1000.00 accrcy= 0.001
    jtmin=  0.0 jtmax=  9000.0 absend= -50.0000
    jump =    1    10    50    200
    jbord=   0.0  200.0  300.0 1000.0 9000.0
    thmin=  0.00 thmax= 20.00 thinc=  0.05 cutr=-20.00
    ips= 0.0000 it0= 0 iter= 0 iblock= 21 nnu= 24 smallchan= 1.00E-12  smallcoup= 1.00E-12
    elab=   656.0000    pel=1 exl=1 lab=1 lin=1 lex=1 chans= 1 smats= 2 xstabl= 1 cdcc= 1/
 &Partition namep='8B' massp=  8. zp=  5 nex= 21  pwf=T namet='208Pb' masst=208. zt= 82 qval=  0.1370/
 &States jp= 1.5 ptyp=-1 ep=  0.0000 cpot=  1 jt= 0.0 ptyt= 1 et=  0.0000/
 &States jp= 0.5 ptyp= 1 ep=  0.1583 cpot=  1 copyt= 1/
 &States jp= 0.5 ptyp= 1 ep=  0.2180 cpot=  1 copyt= 1/
 &States jp= 0.5 ptyp= 1 ep=  0.3260 cpot=  1 copyt= 1/
 &States jp= 0.5 ptyp= 1 ep=  0.4830 cpot=  1 copyt= 1/
 &States jp= 0.5 ptyp= 1 ep=  0.6889 cpot=  1 copyt= 1/
 &States jp= 0.5 ptyp= 1 ep=  0.9438 cpot=  1 copyt= 1/
 &States jp= 0.5 ptyp= 1 ep=  1.2478 cpot=  1 copyt= 1/
 &States jp= 0.5 ptyp= 1 ep=  1.6007 cpot=  1 copyt= 1/
 &States jp= 0.5 ptyp= 1 ep=  2.0027 cpot=  1 copyt= 1/
 &States jp= 0.5 ptyp= 1 ep=  2.4536 cpot=  1 copyt= 1/
 &States jp= 0.5 ptyp= 1 ep=  2.9536 cpot=  1 copyt= 1/
 &States jp= 0.5 ptyp= 1 ep=  3.5025 cpot=  1 copyt= 1/
 &States jp= 0.5 ptyp= 1 ep=  4.1005 cpot=  1 copyt= 1/
 &States jp= 0.5 ptyp= 1 ep=  4.7474 cpot=  1 copyt= 1/
 &States jp= 0.5 ptyp= 1 ep=  5.4434 cpot=  1 copyt= 1/
 &States jp= 0.5 ptyp= 1 ep=  6.1884 cpot=  1 copyt= 1/
 &States jp= 0.5 ptyp= 1 ep=  6.9824 cpot=  1 copyt= 1/
 &States jp= 0.5 ptyp= 1 ep=  7.8253 cpot=  1 copyt= 1/
 &States jp= 0.5 ptyp= 1 ep=  8.7173 cpot=  1 copyt= 1/
 &States jp= 0.5 ptyp= 1 ep=  9.6583 cpot=  1 copyt= 1/
 &Partition namep='7Be' massp=  7. zp=  4 nex=-1 pwf=T namet='209Pb' masst=209.0000 zt=83 qval=  0./
 &States jp= 0.0 ptyp= 1 ep=  0.0000 cpot=  2 jt= 0.0 ptyt= 1 et=  0.0000/
 &Partition /
 &Pot kp= 1 type= 0 shape= 0 p(1:3)=   1.0000  0.0000  2.6500 /
 &Pot kp= 2 type= 0 shape= 0 p(1:3)= 208.0000  0.0000  1.3000 /
 &Pot kp= 2 type= 1 shape= 0 p(1:6)= 114.2000  1.2860  0.8530  9.4400  1.7390  0.8090 /
 &Pot kp= 3 type= 0 shape= 0 p(1:3)= 208.0000  0.0000  1.3000 /
 &Pot kp= 3 type= 1 shape= 0 p(1:6)=  34.8190  1.1700  0.7500 15.3400  1.3200  0.6010  /
 &Pot kp= 4 type= 0 shape= 0 p(1:3)=   1.0000  0.0000  2.3910 /
 &Pot kp= 4 type= 1 shape= 0 p(1:3)=  44.6750  2.3910  0.4800 /
 &Pot kp= 4 type= 3 shape= 0 p(1:3)=   4.8980  2.3910  0.4800 /
 &Pot /
 &Overlap kn1=  1 ic1=1 ic2=2 in= 1 kind=0 nn= 1 l=1 sn=0.5 j= 1.5 nam=1 ampl=  1.00 kbpot= 4 be=  0.1370 isc= 1 ipc=0 /
 &Overlap kn1=  2 ic1=1 ic2=2 in= 1 kind=0 l=0 sn=0.5 j= 0.5 nam=1 ampl=  1.00 kbpot= 4 be= -0.0182 isc=12 ipc=2 nk=  20 er= -0.0344 /
 &Overlap kn1=  3 ic1=1 ic2=2 in= 1 kind=0 l=0 sn=0.5 j= 0.5 nam=1 ampl=  1.00 kbpot= 4 be= -0.0771 isc=12 ipc=2 nk=  20 er= -0.0834 /
 &Overlap kn1=  4 ic1=1 ic2=2 in= 1 kind=0 l=0 sn=0.5 j= 0.5 nam=1 ampl=  1.00 kbpot= 4 be= -0.1850 isc=12 ipc=2 nk=  20 er= -0.1324 /
 &Overlap kn1=  5 ic1=1 ic2=2 in= 1 kind=0 l=0 sn=0.5 j= 0.5 nam=1 ampl=  1.00 kbpot= 4 be= -0.3419 isc=12 ipc=2 nk=  20 er= -0.1814 /
 &Overlap kn1=  6 ic1=1 ic2=2 in= 1 kind=0 l=0 sn=0.5 j= 0.5 nam=1 ampl=  1.00 kbpot= 4 be= -0.5479 isc=12 ipc=2 nk=  20 er= -0.2304 /
 &Overlap kn1=  7 ic1=1 ic2=2 in= 1 kind=0 l=0 sn=0.5 j= 0.5 nam=1 ampl=  1.00 kbpot= 4 be= -0.8028 isc=12 ipc=2 nk=  20 er= -0.2794 /
 &Overlap kn1=  8 ic1=1 ic2=2 in= 1 kind=0 l=0 sn=0.5 j= 0.5 nam=1 ampl=  1.00 kbpot= 4 be= -1.1067 isc=12 ipc=2 nk=  20 er= -0.3284 /
 &Overlap kn1=  9 ic1=1 ic2=2 in= 1 kind=0 l=0 sn=0.5 j= 0.5 nam=1 ampl=  1.00 kbpot= 4 be= -1.4596 isc=12 ipc=2 nk=  20 er= -0.3774 /
 &Overlap kn1= 10 ic1=1 ic2=2 in= 1 kind=0 l=0 sn=0.5 j= 0.5 nam=1 ampl=  1.00 kbpot= 4 be= -1.8616 isc=12 ipc=2 nk=  20 er= -0.4264 /
 &Overlap kn1= 11 ic1=1 ic2=2 in= 1 kind=0 l=0 sn=0.5 j= 0.5 nam=1 ampl=  1.00 kbpot= 4 be= -2.3125 isc=12 ipc=2 nk=  20 er= -0.4754 /
 &Overlap kn1= 12 ic1=1 ic2=2 in= 1 kind=0 l=0 sn=0.5 j= 0.5 nam=1 ampl=  1.00 kbpot= 4 be= -2.8125 isc=12 ipc=2 nk=  20 er= -0.5245 /
 &Overlap kn1= 13 ic1=1 ic2=2 in= 1 kind=0 l=0 sn=0.5 j= 0.5 nam=1 ampl=  1.00 kbpot= 4 be= -3.3614 isc=12 ipc=2 nk=  20 er= -0.5735 /
 &Overlap kn1= 14 ic1=1 ic2=2 in= 1 kind=0 l=0 sn=0.5 j= 0.5 nam=1 ampl=  1.00 kbpot= 4 be= -3.9594 isc=12 ipc=2 nk=  20 er= -0.6225 /
 &Overlap kn1= 15 ic1=1 ic2=2 in= 1 kind=0 l=0 sn=0.5 j= 0.5 nam=1 ampl=  1.00 kbpot= 4 be= -4.6064 isc=12 ipc=2 nk=  20 er= -0.6715 /
 &Overlap kn1= 16 ic1=1 ic2=2 in= 1 kind=0 l=0 sn=0.5 j= 0.5 nam=1 ampl=  1.00 kbpot= 4 be= -5.3023 isc=12 ipc=2 nk=  20 er= -0.7205 /
 &Overlap kn1= 17 ic1=1 ic2=2 in= 1 kind=0 l=0 sn=0.5 j= 0.5 nam=1 ampl=  1.00 kbpot= 4 be= -6.0473 isc=12 ipc=2 nk=  20 er= -0.7695 /
 &Overlap kn1= 18 ic1=1 ic2=2 in= 1 kind=0 l=0 sn=0.5 j= 0.5 nam=1 ampl=  1.00 kbpot= 4 be= -6.8413 isc=12 ipc=2 nk=  20 er= -0.8185 /
 &Overlap kn1= 19 ic1=1 ic2=2 in= 1 kind=0 l=0 sn=0.5 j= 0.5 nam=1 ampl=  1.00 kbpot= 4 be= -7.6843 isc=12 ipc=2 nk=  20 er= -0.8675 /
 &Overlap kn1= 20 ic1=1 ic2=2 in= 1 kind=0 l=0 sn=0.5 j= 0.5 nam=1 ampl=  1.00 kbpot= 4 be= -8.5763 isc=12 ipc=2 nk=  20 er= -0.9165 /
 &Overlap kn1= 21 ic1=1 ic2=2 in= 1 kind=0 l=0 sn=0.5 j= 0.5 nam=1 ampl=  1.00 kbpot= 4 be= -9.5173 isc=12 ipc=2 nk=  20 er= -0.9655 /
 &Overlap /  !
 &Coupling icto= 1 icfrom= 2 kind=3 ip1= 2 ip2= 0 ip3= 0 p1=   3.0000 p2=  2.0000 /
 &Coupling /
```

Box B.3 FRESCO **input for the breakup of** ^{8}B **on** ^{208}Pb **at 82 MeV/u**

we are not interested in the second partition, it does not get printed with the option of negative *nex*. Several new variables are needed when defining the bins (as in subsection 8.2.1): negative *be* provides bins with energy relative to threshold $|be|$, with a width *er*, and an amplitude $\sqrt{nam}*ampl$. To characterize the weight function of the bin we use *isc* (*isc*=2 for non-resonant bins, and *isc*=4 for resonant bins). Note that here, the same potential is used for the ^{8}B bound and continuum states. This need not be the case.

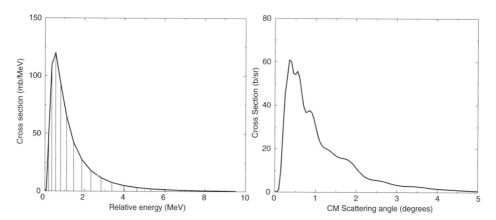

Fig. B.3. Breakup of ^{8}B on ^{208}Pb at 82 MeV/u. Left: p–^{7}Be relative energy distribution. Right: center-of-mass angular distribution. Both are obtained with the input of Box B.3.

```
8B+208Pb ; N+C breakup with q=0,1,2
CDCC
 &CDCC
  hcm=0.01 rmatch=-60 rasym=1000 accrcy=0.001 absend=-50
  elab=656
  jbord= 0 200 300 1000 9000
  jump = 1  10 50 200
  thmax=20 thinc=0.05 cutr=-20 smats=2 xstabl=1
  ncoul=0 reor=0 q=2
  /
 &NUCLEUS part='Proj' name='8B' charge=5 mass=8
            spin=1.5 parity=-1 be = 0.137 n=1 l=1 j=1.5 /
 &NUCLEUS part='Core' name='7Be' charge=4 mass=7 /
 &NUCLEUS part='Valence' name='proton' charge=1 mass=1 spin=0.5/
 &NUCLEUS part='Target' name='208Pb' charge=82 mass=208 spin=0 /

 &BIN spin=0.5 parity=+1 start=0.001 step=0.50 end=10. energy=F l=0 j=0.5/
 &BIN /

 &POTENTIAL part='Proj' a1=1 rc=2.65 /
 &POTENTIAL part='Core'    a1=208 rc=1.3 v=114.2 vr0=1.286 a=0.853 w=9.44 wr0=1.739 aw=0.809 /
 &POTENTIAL part='Valence' a1=208 rc=1.3 v=34.819 vr0=1.17 a=0.75 w=15.340 wr0=1.32 aw=0.601/
 &POTENTIAL part='Gs' a1=1 rc=2.391 v=44.675 vr0=2.391 a=.48 vso=4.898 rso0=2.391 aso=0.48 /
```

Box B.4 FRESCO **input for the breakup of** ^{8}B **on** ^{208}Pb **at 82 MeV per nucleon
(short version)**

After defining the overlaps, coupling parameters are introduced: *kind=3* stands for single-particle excitations of the projectile (*kind=4* would be for the target), *ip1* is the maximum multipole order in the expansion of the couplings (Eq. (8.2.13)) included, *ip2=0,1,2* for Coulomb and nuclear, nuclear only and Coulomb only, respectively, and *ip3* makes specific selections of couplings with default *ip3=0* when all couplings are included.[9] For the interactions in the coupling matrix, the core-target is potential index *p1=3* and the valence-target is potential index *p2=2*.

Angular distributions of the cross sections for each energy bin can be found in fort.16. To obtain a total angular distribution one needs to sum over all bins (use sumbins < fort.16 > xxx.xsum). For the breakup example shown here, the resulting total angular distribution is plotted in Fig. B.3(left). If you are interested in the energy distribution, fort.13 contains all angular integrated cross sections for each bin. In general, for each energy, a sum over all ℓ partial waves within the projectile is necessary (use sumxen < fort.13 > xxx.xen). In Fig. B.3(right) we show the energy distribution for the ^{8}B breakup here considered. In addition, it is useful to look at fort.56 (cross section per partial wave L) to ensure that enough partial waves are included in the calculation.

Defining a long list of bin states and overlaps can be easily automated. The revised CDCC style of input has been developed specifically for large CDCC calculations, and transforms a simpler input into the standard input we have just gone through. The simpler input would then look like Box B.4.

B.2.4 Transfer

Transfer reactions are often used to extract structure information to input in astrophysical simulations. Here we consider the ^{14}N(^{17}F,^{18}Ne)^{13}C transfer reaction at 10 MeV per nucleon. This reaction was measured with the aim of extracting the asymptotic normalization coefficient of specific states in ^{18}Ne, which in turn provides a significant part of the rate for ^{17}F(p, γ). The proton capture reaction on ^{17}F appears in the *rp*-process in novae environments (subsection 1.5.2). The ratio of the proton capture rate and the decay rate of ^{17}F is also very important for the understanding of galactic ^{17}O, ^{18}O and ^{15}N. The input for the transfer example is given in Box B.5.

A few important new parameters need to be defined when performing the transfer calculation. Because the process involves a non-local kernel $V_{fi}^o(R', R)$ of Eq. (4.5.7), in addition to the radial grids already understood, we need to introduce *rintp, hnl, rnl,centre*. The *rintp* is the step in R, *hnl, rnl* are the non-local step and the

---

[9] If *ip3=1*, there are no reorientation couplings for all but the monopole, if *ip3=2*, only reorientation couplings are included, and if *ip3=3* it includes only couplings to and from the ground state. More options exist but are not presented here.

```
n14(f17,ne18)c13 @ 170 MeV;
NAMELIST
 &FRESCO  hcm=0.03 rmatch=40. rintp=0.20 hnl=0.1 rnl=5.00 centre=0.0
          jtmin=0.0  jtmax=120 absend=-1.0
          thmin=0.00 thmax=40.00 thinc=1.00
          iter=1 nnu=36
          chans=1 xstabl=1
          elab=170.0  /

 &PARTITION namep='f17'  massp=17. zp=9 namet='n14'  masst=14. zt=7 nex=1  /
 &STATES jp=2.5 bandp=1 ep=0.0 jt=1.0 bandt=1 et=0.0000  /

 &PARTITION namep='ne18' massp=18. zp=10 namet='c13'  masst=13. zt=6 qval=3.6286 nex=1  /
 &STATES jp=0. bandp=1 ep=0.0  cpot=2 jt=0.5 bandt=-1 et=0.0000  /
 &partition  /

 &POT kp=1 ap=17.000 at=14.000 rc=1.3  /
 &POT kp=1 type=1 p1=37.2 p2=1.2 p3=0.6  p4=21.6 p5=1.2 p6=0.69  /
 &POT kp=2 ap=18.000 at=13.000 rc=1.3  /
 &POT kp=2 type=1 p1=37.2 p2=1.2 p3=0.6  p4=21.6 p5=1.2 p6=0.69  /
 &POT kp=3 at=17 rc=1.2  /
 &POT kp=3 type=1 p1=50.00 p2=1.2 p3=0.65   /
 &POT kp=3 type=3 p1=6.00  p2=1.2 p3=0.65   /
 &POT kp=4 at=13 rc=1.2  /
 &POT kp=4 type=1 p1=50.00 p2=1.2 p3=0.65   /
 &POT kp=4 type=3 p1=6.00  p2=1.2 p3=0.65   /
 &POT kp=5 ap=17.000 at=13.000 rc=1.3  /
 &POT kp=5 type=1 p1=37.2 p2=1.2 p3=0.6  p4=21.6 p5=1.2 p6=0.69  /
 &pot  /

 &Overlap kn1=1 ic1=1 ic2=2 in=1 kind=0 nn=1 l=2 sn=0.5 j=2.5 kbpot=3 be=3.922  isc=1 ipc=0 /
 &Overlap kn1=2 ic1=2 ic2=1 in=2 kind=3 nn=1 l=1 sn=0.5 ia=1 ib=1 j=1.0 kbpot=4 be=7.5506 isc=1  /
 &overlap  /

 &coupling icto=-2 icfrom=1 kind=7 ip1=0 ip2=-1 ip3=5  /
 &CFP in=1 ib=1 ia=1 kn=1 a=1.00  /
 &CFP in=2 ib=1 ia=1 kn=2 a=1.00  /
 &CFP  /
 &coupling  /
```

Box B.5 FRESCO **input for the transfer reaction** $^{14}N(^{17}F,^{18}Ne)^{13}C$ **at 10 MeV/u**

non-local range in $R' - R$, respectively, and centered at *centre*. Gaussian quadrature is used for the angular integrations in constructing the non-local kernels, and *nnu* is the number of the Gaussian integration points to be included.

In this example the core has non-zero spin, and in order to generate the appropriate overlap of the composite nucleus ^{14}N, it is necessary to take into account, not only the angular momentum of the neutron but also the spin of ^{13}C. This can be done with *kind*=3 in the overlap definition where the spin of the core *ia* and of the composite *ib* need to be specified. The coupling scheme is $|(l_n, s_n)j, I_A; I_B|$.

The only other new part of the input concerns the transfer coupling itself, as all other parts (partitions, potentials and overlaps) have already been previously presented. Transfer couplings are defined in the namelist &*coupling* by *kind*=5,6,7 for zero-range, low-energy approximation and finite range, respectively. For finite-range transfers, *ip1*=0,1 stands for post or prior, *ip2*=0,1,−1 for no remnant, full real remnant and full complex remnant respectively and *ip3* denotes the index of the core-core optical potential. If *ip3*=0 then it uses the optical potential for the first pair of excited states in the partition of the projectile core.

Following the &*coupling* namelist, we need to define the amplitudes (coefficients of fractional parentage) of all the overlaps to be included in the calculation. Here, this is done with &*cfp* where *in*=1,2 for projectile or target, *ib/ia* corresponds to the state index of the composite/core and *kn* is the index of the corresponding overlap function. So the first &*cfp* refers to the $\langle {}^{17}\text{F}|{}^{18}\text{Ne}\rangle$ overlap and the second &*cfp* refers to the $\langle {}^{13}\text{C}|{}^{14}\text{N}\rangle$ overlap.

The angular distribution obtained from our example is presented in Fig. B.4.

B.2.5 Capture

Capture reactions are of direct interest in astrophysics as repeatedly pointed out in Chapter 1. Although the electromagnetic operator is well understood, coupling effects may be non-trivial and require focused work. Here we pick a neutron capture reaction that is completely dominated by $E1$: ${}^{14}\text{C}(\text{n},\gamma){}^{15}\text{C}$. This reaction was first introduced in Chapter 1 in the context of the *r*-process. In our example, the capture is calculated at 50 different scattering energies, from 5 keV up to 4 MeV. The input is presented in Box B.6.

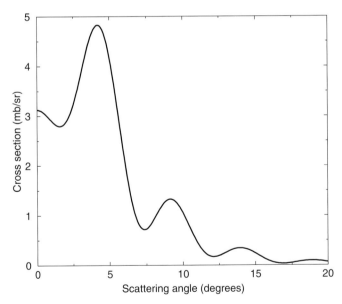

Fig. B.4. Transfer cross section for ${}^{14}\text{N}({}^{17}\text{F},{}^{18}\text{Ne}){}^{13}\text{C}$ at 10 MeV/u calculated with the input of Box B.5.

```
14C(n,g)15C E1 only
NAMELIST
&Fresco hcm= 0.100 rmatch=100
     jtmin=0 jtmax=4.5 absend=-1
     thmin=0 thmax=0 iter=1
     elab(1)= 0.005 4.005 nlab=50/

&PARTITION namep='neutron' massp=1.0087 zp=0 nex=1 namet='14C' masst=14.0032 zt=6 /
&STATES jp=0.5 ptyp=1 ep=0 cpot=1 jt=0.0 ptyt=1 et=0/

&PARTITION namep='Gamma' massp=0 zp=0 nex=1 namet='15C' masst=15.0106 zt=6 qval=1.218 /
&STATES jp=1.0 ptyp=1 ep= 0 cpot=3 jt=0.5 ptyt=1 et=0/
&partition /

&Pot kp=1 type= 0 shape= 0 p(1:3)=  14.0000  0.0000  1.3000 /
&Pot kp=1 type= 1 shape= 0 p(1:3)=  57.0000  1.7  0.7000 /
&Pot kp=1 type= 3 shape= 0 p(1:3)=   0.0000  1.7  0.5000 /
&Pot kp=2 type= 0 shape= 0 p(1:3)=  14.0000  0.0000  1.2000 /
&Pot kp=2 type= 1 shape= 0 p(1:3)=  55.7700  1.2230  0.5000 /
&Pot kp=2 type= 3 shape= 0 p(1:3)=   5.0000  1.2230  0.5000 /
&pot /

&OVERLAP kn1=1 ic1=1 ic2=2 in=-2 kind=0 nn=2 l=0 sn=0.5 j=0.5 kbpot=2 be=1.218 isc=1 /
&overlap /

&COUPLING icto=2 icfrom=1 kind=2 ip1=-1 ip2= 1/
&cfp in=2 jb=1 ia=1 kn=1 a=1.000  /
```

Box B.6 Fresco **input for neutron capture by** ^{14}C

For capture reactions, the first partition is defined in the usual way, but in the second partition, the projectile should be *Gamma* (with spin *jp*=1 and positive parity) and *cpot* should refer to a non-existing potential in order that there be no photon potential. The $2s_{1/2}$ ^{15}C overlap is defined in *&overlap*. Electromagnetic one-photon couplings are defined through *kind*=2. Therein, *ip1* refers to the multipolarity of the transition and *ip2*=0,1,2 for including both electric and magnetic transitions, electric only and magnetic only, respectively. If *ip1* > 0, all multipolarities up to *ip1* are included, otherwise only |*ip1*| is calculated.

There are several outputs available specifically for astrophysics. In Fig. B.5 we plot the cross section for the ^{14}C(n, γ)^{15}C capture reaction as a function of center-of-mass energy (found in fort.39). For charged-particle reactions, astrophysical S-factors are also available (see Table B.1).

B.3 Runtime errors

In a complicated modeling computer program like Fresco, accurate results cannot be obtained if there are obvious numerical errors either in the input, or produced during the calculation. Problems may occur at energies very much below the Coulomb barrier, or at relativistic energies, since Fresco should not be used in

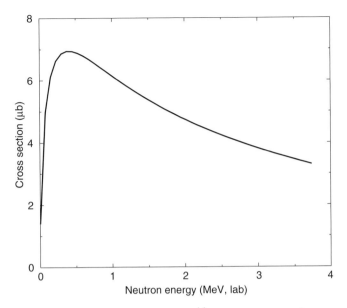

Fig. B.5. Neutron capture cross section for ^{14}C as a function of neutron energy and calculated with the input of Box B.6.

these cases. The program is written to stop when large numerical inaccuracies occur, but no results can be trusted until they are examined to see that they are not sensitive to further increases to the maximum radius *rmatch*, maximum partial waves *jtmax*, maximum non-local range *rnl*, and further decreases in the radial step size *hcm* and lower radial cutoff parameter *cutl*. We give some guidance on these parameters below.

When the program reads the input file, typing errors in variables or their values will cause the program to stop after printing out the complete namelist so you can see which variables have been successfully read in. Alternatively, if the words NAMELIST and CDCC in the second line of the input are written in lower case ('namelist' or 'cdcc'), the compiler's own error handling will be used instead. You can also look at the input echo in fort.3, to see up to which line the input has been read successfully.

During the running of FRESCO, a number of induced or cancellation errors can be detected. These are:

Step size too large: If k is the asymptotic wave number for a channel, then its product with the step size should be sufficiently small for the Numerov integration method to avoid large errors: $k*hcm < 0.2$.

Bound-state search failure: Bound states are found in subroutine eigcc by a Newton-Raphson method with at most 40 iterations. Such states can be found at a specific energy by varying some part of the potential: this part cannot be zero.

Insufficient non-local width: If the non-local coupling form factors are too large when $|R - R'| \geq rnl$, then *rnl* should be increased as recommended.

L-transfer accuracy loss: In the transfer Legendre expansion Eq. (4.5.12), if $\ell + \ell'$ is too large, there can be cancellation errors between different multipoles T. This can be remedied by increasing the input parameter *mtmin* to use the slower *m*-dependent method of [1].

Matching deficiency: The program will print an error message if the nuclear potentials are not smaller than 0.02 MeV at the outer matching radius, and stop if they are larger than 0.1 MeV. Increase *rmatch*, or correct some potential that does not decay sufficiently fast.

Iteration failure in solving coupled equations: If in some partial wave set, more than *iter* iterations still do not appear to converge, then all cross sections will be affected. Either use Padé acceleration (see page 214), increase *iter* or make it negative so that the 'best' intermediate value is used, or increase *ips* slightly. The detailed progress of the iterations can be seen by setting *smats* ≥ 5.

Accuracy loss in solving coupled equations: If any of the channels are propagating in a classically forbidden region, there will be loss of linear independence of the separate solutions $\{Y_{\alpha\beta}(R)\}$ of Eq. (6.3.2), as all solutions will tend to become exponentially increasing. This is monitored in the subroutine **erwin** during the summation Eq. (6.3.3), and will lead to a halt if the cancellation errors are expected to be more than 3%. In this case, increase the lower cutoff parameters *cutl* or *cutr*, decrease the matching radius *rmatch* if possible, or else use the R-matrix expansion method of page 217 to solve the equations. An extreme error occurs if the simultaneous equations from the matching conditions are singular.

Internal parameter error: Sometimes, the precalculation of array sizes is inadequate. For advanced users these can be improved by selective specification of the maxcoup(1:3) and expand(1:11) input arrays.

At the end of a Fresco calculation, a final 'accuracy analysis' is presented. This rechecks that the step size *hcm* is small enough, and that *rnl* is large enough. Then, using the Coulomb trajectories described in Box 3.3 on 64, it calculates the minimum scattering angles expected to be accurate because of the finite values of *rmatch* and *jtmax*.

B.4 Fitting data: Sfresco

A first calculation of cross sections using Fresco will rarely be near the experimental data. Perhaps the reaction model is too simplified, or perhaps the input parameters are not accurate enough. The optical potential, binding potentials, spectroscopic amplitudes or R-matrix reduced widths could well be adjusted to see if the agreement between theory and experiment can be improved. The program Sfresco searches for a χ^2 minimum when comparing the outputs of Fresco with sets of data, using the Minuit search routines as discussed on page 414.

The inputs for Sfresco specify the Fresco input and output files, the number and types of search variables, and the experimental data sets to be compared with.

These experimental data can be (*type*=0) an angular distribution for fixed energy, (1) an excitation and angular cross-section double distributions, or (2) an excitation cross section for fixed angle. They could also be (3) an excitation function for the total, reaction, fusion or inelastic cross section, (4) an excitation phase shift for fixed partial wave, (5) a desired factor for bound-state search, or (6) even a specific experimental constraint on some search parameter.

The simplest and most common fitting requirement is to determine an optical potential to fit the observed elastic scattering angular distribution. For example, to find a proton optical potential to fit cross sections for scattering on ^{112}Cd at

```
p + 112Cd elastic
NAMELIST
&FRESCO hcm= 0.100 rmatch= 20.000 jtmin= 0.0 jtmax= 200.0
    thmin= 0.00 thmax=180.00 thinc= 2.00 xstabl= 1
    elab(1)=  27.9 /
&PARTITION namep='Proton ' massp= 1.0000 zp= 1 nex= 1
    namet='112Cd  ' masst=112.0000 zt= 48 qval= 0/
&STATES jp= 0.5 ptyp= 1 ep= 0.0000 cpot= 1 jt= 0.0 ptyt= 1 et= 0.0000/
&partition /
&pot kp= 1 type= 0 p(1:3)=  112.000 0.0000 1.2000 /
&pot kp= 1 type= 1 p(1:6)=  52.500 1.1700 0.7500 3.5000 1.3200 0.6100 /
&pot kp= 1 type= 2 p(1:6)=  0.000 0.0000 0.0000 8.5000 1.3200 0.6100 /
&pot kp= 1 type= 3 p(1:3)=  6.200 1.0100 0.7500 /
&pot /
&overlap /
&coupling /
```

Box B.7 FRESCO **input file** p-cd.frin **for proton scattering on** ^{112}Cd

```
'p-cd.frin' 'p-cd.frout'
4 1
&variable kind=1 name='r0' kp=1 pline=2 col=2 potential=1. step=0.01/
&variable kind=1 name='V' kp=1 pline=2 col=1 potential=50.0 step=0.1/
&variable kind=1 name='W' kp=1 pline=2 col=4 potential=5.0 step=0.1/
&variable kind=1 name='WD' kp=1 pline=3 col=4 potential=10. step=0.1/
&data iscale=0 idir=1 lab=F abserr=T/
22.    0.548    0.044
26.    0.475    0.024
30.    0.481    0.014
38.    0.447    0.009
50.    0.144    0.004
66.    0.499    0.010
70.    0.248    0.005
86.    0.463    0.014
90.    0.485    0.015
106.   0.087    0.003
110.   0.135    0.004
130.   0.161    0.005
&
```

Box B.8 SFRESCO **input** p-cd.search **for proton scattering on** ^{112}Cd

27.90 MeV, we start with the normal FRESCO input of Box B.7. The cross sections will be calculated at the experimental angles, not those specified here. In order to vary the real and imaginary potential strengths in this input, as well as the real radius, we have the **search file** for SFRESCO of Box B.8. This search file begins by identifying the previous FRESCO input and naming the temporary output file, then giving the number of search variables and the number of experimental data sets.

There are 3 &*pot* namelist lines in Box B.7 for the nuclear parts of potential $kp=1$, so the variables of the interaction potentials (*kind*=1) in the search are identified by the specification of *kp* and *pline* in the &*variable* namelist in Box B.8, along with *col* for the index to the *p* array. The *potential* value gives the initial value for the search, and *step* the initial magnitude for trial changes. Deformations can as well be searched upon. There are also parameters *valmin* and *valmax*,[10] to limit the range of that variable. Spectroscopic amplitudes to be varied (*kind*=2) are identified by the order *nfrac* in which they appear in the FRESCO input in a &*cfp* namelist, and then by their initial value *afrac*. Other variable *kind*s are described in the FRESCO input manual.

Experimental data sets are identified by their *type* specifications as listed above, and then by *data_file* for name of data file with data, which can be '=' for this search file, '<' for stdin (the default is '='). Then *points* gives the number of data points (default: keep reading as many as possible), *lab* is T or F for laboratory angles and cross sections (default false), and *energy* is lab energy for a *type*=0 dataset (default: use *elab(1)* from &*fresco* namelist). The *abserr* is true for absolute errors (default: false). Next, *idir* is -1 for cross-section data given as astrophysical S-factors, 0 for data given in absolute units (the default), and 1 as ratio to Rutherford. Finally, *iscale* is -1 for dimensionless data, and 0 absolute data in units of fm^2/sr, 1 for b/sr, 2 for mb/sr (the default) and 3 for μb/sr.

With these two files (Boxes B.7 and B.8), SFRESCO is invoked interactively. First the name of the search file in Box B.8 is given, then commands as in Table B.2. A minimum set of commands will be **p-cd.search / min / migrad / end / plot**. The output plot file (default name **search.plot**) also contains the final values of the searched variables. The fitted parameters in this example are $V = 52.53$ MeV, $r0 = 1.179$ fm, $W = 3.46$ MeV and $WD = 7.43$ MeV with $\chi^2/N = 2.19$. The initial and final fits to the data are shown in Fig. B.6.

B.5 System requirements, compilations and installation

The website accompanying this book contains a complete distribution set of files for FRESCO and SFRESCO, SUMBINS and SUMXEN. It contains the input and output files for all the examples in this Appendix B, along with a version of Appendix

---

[10] Both or none of these must be present.

Table B.2. SFRESCO *input commands. Words in* `typewriter font` *are to be replaced with user values.*

Command	Operation
Q	query status of search variables.
SET `var val`	set variable number `var` to value `val`.
FIX `var`	fix variable number `var` (set `step=0`).
STEP `var step`	unfix variable `var` with step `step`.
SCAN `var val1 val2 step`	scan variable `var` from value `val1` to `val2` in steps of `step`.
SHOW	list all datasets with current predictions and χ values.
LINE `plotfile`	write file (default: **search.plot**) with theoretical curves only.
READ `file`	read plot output `file` for further searches, if not:
READ `snapfile`	if the name of `snapfile` contains the string 'snap',
	read last set of snap output `snapfile` from a previous **fort.105**.
ESCAN `emin emax estep`	scan lab. energy in incident channel.
MIN	call MINUIT interactively.
MIGRAD	in MINUIT, perform MIGRAD search.
END	return to SFRESCO from MINUIT.
PLOT `plotfile`	write file (default: **search.plot**) with data and theory curves.
EX	exit (also at end of input file).

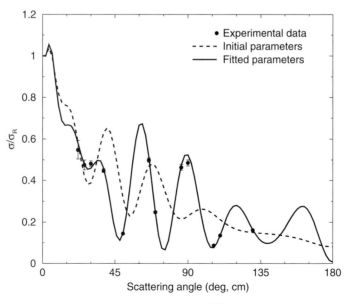

Fig. B.6. Initial and fitted proton scattering on ^{112}Cd at 27.9 MeV.

B itself. The distribution files also contain input and output files for a range of test cases for your installation, as well as copies of the detailed input manuals for FRESCO and MINUIT.

The distribution set contains a set of precompiled binaries for FRESCO and SFRESCO, as well as the source code. To compile the source, you will need a FORTRAN compiler for at least Fortran 90 or 95. The code is in the directory fres/source/, where the script mk attempts to find the best compiler for your system, and then compile in a subdirectory named by your system architecture *arch* and the compiler chosen using the makefile. You may have to edit the makefile to set FFLAGS for your compiler, and set TIME and FLUSH according to your system libraries. After compilation, mk install copies the binaries to a standard bin/*arch*/ directory for execution in other places.

The website www.fresco.org.uk will be regularly updated to include descriptions of any further changes needed or advisable for the programs. All of the information on the website is published under the conditions of the GNU Public License described at www.gnu.org/copyleft/gpl.html.

References

[1] T. Tamura, T. Udagawa, K. E. Wood and H. Amakawa, *Comput. Phys. Commun.* **18** (1979) 63.

Select bibliography

Theory of Reactions

N. Austern 1970, *Direct Nuclear Reaction Theories*, New York: Wiley-Interscience.

J. R. Taylor 1972, *Scattering Theory*, New York: Wiley.

A. Bohr and B. R. Mottelson 1975, *Nuclear Structure, Vol. 1*, Reading, MA: Benjamin.

H. Goldstein 1980, *Classical Mechanics*, 2nd edn. Reading: Addison-Wesley.

G. R. Satchler 1983, *Direct Nuclear Reactions*, Oxford: Clarendon Press.

J. M. Eisenberg and W. Greiner 1987, *Nuclear Theory: Nuclear Models, Vol. 1*, Amsterdam: North-Holland.

M. L. Goldberger and K. M. Watson 1992, *Collision Theory*, New York: Dover.

P. Frobrich and R. Lipperheide 1996, *Theory of Nuclear Reactions*, Oxford: Clarendon Press.

C. A. Bertulani and P. Danielewicz 2004, *Introduction to Nuclear Reactions*, Bristol: Institute of Physics.

P. Descouvemont 2004, *Theoretical Models for Nuclear Astrophysics*, Hauppauge, NY: Nova Science Publishers.

Nuclear Astrophysics

D. D. Clayton 1968, *Principles of Stellar Evolution and Nucleosynthesis*, New York: McGraw-Hill.

C. E. Rolfs and W. S. Rodney 1988, *Cauldrons in the Cosmos*, Chicago: University of Chicago.

J. J. Cowan, F.-K. Thielemann, J. W. Truran, *Physics Reports* **208** (1991) 267.

K. Langanke and M. Weischer, *Reports on Progress in Physics* **64** (2001) 1657.

C. Iliadis 2007, *Nuclear Physics of Stars*, Weinheim: Wiley.

Mathematics

M. Abramowitz and I. A. Stegun 1964, *Handbook of Mathematical Functions*, Washington: National Bureau of Standards.

W. T. Eadie, D. Drijard, F. James, M. Roos and B. Sadoulet 1971, *Statistical Methods in Experimental Physics*, Amsterdam: North-Holland.

W. H. Press, S. A. Teulosky, W. T. Vetterling and B. P. Flannery 1992, *Numerical Recipes in Fortran 77*, 2nd edn., Cambridge: Cambridge University Press (reprinted with corrections in 1999).

Computer Programs

F. James, *MINUIT Reference Manual, Function Minimization and Error Analysis*, CERN
   Program Library Long Writeup D506, Version 94.1 Edition August 1998,
   URL:wwwinfo.cern.ch/asdoc/minuit.

M. Herman, in Workshop on Nuclear Reaction Data and Nuclear Reactors: "Physics,
   Design and Safety," N. Paver, M. Herman and A. Gandini (eds.), Trieste, Italy, (2001)
   137. URL: www.nndc.bnl.gov/empire.

A. J. Koning, S. Hilaire and M. C. Duijvestijn, *TALYS-1.0*, Proceedings of the
   International Conference on Nuclear Data for Science and Technology – ND2007,
   April 22–27 2007, Nice, France (2008). URL:www.talys.eu.

Index